Case Studies of Breeding Strategies in Major Plant Species

Edited by Haiping Wang

Published in London, United Kingdom

Case Studies of Breeding Strategies in Major Plant Species
http://dx.doi.org/10.5772/intechopen.100882
Edited by Haiping Wang

Contributors
Arif Atak, Gizachew Haile Gidamo, Ahmed M. Hassanein, Palanisamy Dhamotharan, Sampathkumar Sruthi, Sakthivel Surya Krishna, Subburaj Monica, Bratislav Stankovic, Xianzhong Feng, Muhammad Hafeez Ullah Khan, Priyanka Shanmugavel, Gowtham Ramasamy, Geethalakshmi Vellingiri, Rajavel Marimuthu, Kalaimagal Thiyagarajan, Devender Sharma, Anroop Kaur, Sukhpreet Kaur, Ajinder Kaur, Navraj Kaur Sarao, Ashok Kumar, Rahul Dev Gautam, Ravi Kumar, Ujala Kashyap, Pawan Kumar, Sanatsujat Singh, Satbeer Singh, Chiebuka Uzoebo, Nallathambi Premalatha, Debadatta Panda, Latha Ananda Lekshmi, Rachel Lissy Vargheese, Mahadevan Kumar, Lakshmanan Mahalingam, Suman Dutta, Vignesh Muthusamy, Rajkumar U. Zunjare, Firoz Hossain, Raman Selvakumar, Monika Punia, Lalit Kumar Rolaniya, Ram Lal Jat, Satya Prakash, Suhel Mehandi, Harmeet S. Janeja, Imene Rajhi, Bechir Baccouri, Fatma Rajhi, Moez Amri, Guido Flamini, Haythem Mhadhbi, John Charles Aru, Scovia Adikini, Ronald Kakeeto, Williams Esuma, Mosses Kasule Faiso, Sam Omaria, Francis Okiasi, Michaeal Adrogu Ugen, Eric Manyaza, Abdul Fattah, Elisa Winanda, Idaryani, Salim, Sri Wahyuni Manwan, Wanti Dewayani, Ayyub Arrahman, Herniwati Herniwati, Nurjanani, M. Yasin, Sahardi Mulia, Warda, M. Basir Nappu, Abdi Negara, Asni Ardjanhar, M. Sudjak Saenong, Amiruddin Syam, Muh. Taufik, Andi Yulyani Fadwiwati, Muslimin Muslimin, Sitti Nuraeni, Anup Kumar Sarkar

First published in London, United Kingdom, 2023 by IntechOpen
IntechOpen is the global imprint of INTECHOPEN LIMITED, registered in England and Wales, registration number: 11086078, 5 Princes Gate Court, London, SW7 2QJ, United Kingdom

British Library Cataloguing-in-Publication Data
A catalogue record for this book is available from the British Library

Additional hard and PDF copies can be obtained from orders@intechopen.com

Case Studies of Breeding Strategies in Major Plant Species
Edited by Haiping Wang
p. cm.
Print ISBN 978-1-80356-104-2
Online ISBN 978-1-80356-105-9
eBook (PDF) ISBN 978-1-80356-106-6

For EU product safety concerns:
IN TECH d.o.o., Prolaz Marije Krucifikse Kozulić 3, 51000 Rijeka, Croatia,
info@intechopen.com or visit our website at intechopen.com.

Meet the editor

Haiping Wang is a full-time research scientist and professor in the Department of Germplasm Resources, Institute of Vegetables and Flowers, Chinese Academy of Agricultural Sciences (IVFCAAS). He is also the chief of the innovation team of the Department of Germplasm Resources. He conducts research on vegetable genetics to collect germplasm to preserve the diversity of the National Mid-term Genebank for Vegetable Germplasm Resources, China. His areas of interest include genetic diversity and phylogenetic development of vegetable germplasm resources, construction of core vegetable germplasm, and innovative utilization of vegetable germplasm based on modern biotechnology. He also collaborates with the garlic and ginger production industries as well as with consumers. Prof. Wang is the author/co-author of more than 100 publications in scientific journals and 11 book chapters. He is also a reviewer for several journals.

Contents

Preface

Plant breeding, as one of the most popular agricultural technologies, has played a significant role in the development of human civilizations. Conventional plant breeding has significantly improved crop yield, disease resistance, and adaptability to the environment by selecting new varieties, thereby contributing to agricultural production and social development. In modern agriculture, plant breeding has been widely applied. As the genetic diversity of the varieties used gradually decreases, the bottleneck effect of traditional breeding is becoming increasingly apparent, and it is difficult to cultivate breakthrough new varieties using conventional breeding techniques. The innovation of biotechnology has greatly promoted the development of modern breeding. With the development of molecular biology, genomics, systems biology, synthetic biology, and biotechnology, design breeding technology has emerged and developed rapidly.

This book presents the latest findings and research on the theories, strategies, and technologies of plant breeding. It includes nineteen chapters organized into four sections. The first section, "Genetic Resources for Plant Breeding", includes one chapter covering the basic information and are view of plant genetic resources for plant breeding. The second section, "Breeding Theory and Strategy", discusses basic, traditional, and modern strategies for plant breeding. The third section, "Breeding Practice and Cases", presents some reviews and case studies. The final section, "The Perspective for Plant Breeding", discusses the latest technologies and development trends in plant breeding.

I would like to extend thanks to all authors who contributed to this book. My sincere thanks also to Author Service Manager Ms. Ana Cink at IntechOpen for her assistance throughout the publication process.

Haiping Wang
Institute of Vegetables and Flowers,
Chinese Academy of Agricultural Sciences,
Beijing, China

Section 1

Genetic Resources for Plant Breeding

Chapter 1

Revolution in Plant Genetic Resources

Chiebuka Uzoebo

Abstract

Plant genetic resources are the use of plant materials, such as seed, pollen, and other plant organs, which have potential value for food and agriculture. The future of crop improvement programs relies on the quality of plant materials. Globally, agriculture and food production are confronted with diverse issues, such as climate change, genetic erosion, land degradation, loss of biodiversity, and human encroachment. A wider genetic diversity research will mitigate these challenges and improve crop production. In today's science innovative approaches, such as the use of molecular markers, cryopreservation, genebanks, and relevant molecular assays are changing the face of collating data, preparation, processing, and sorting of genetic resources. The objective of this review article is to discuss the revolutionary approaches to plant genetic resources and how they will help in the improvement of agricultural production.

Keywords: plant genetic resource, biodiversity, research, technology, conservation

1. Introduction

Plant genetic resources date to about 10,000 years ago when man invented agriculture. People started differentiating variations in plants and later domesticated them through natural means of selection [1]. These plants became our foremost crops. Our planet houses around 310,000 described species of plants and possibly an overall estimate of 400,000 species. About 5000 plants have been harnessed by modern man for food, clothes, shelter, and other needs. And as our population increases, we become increasingly dependent on plants for survival. Today about 150 plant species are what humans dearly need for food and essential needs with only about 12 of those plants proving 80% of the world's food. Some of them are wheat, rice, barley, oats, millet, cotton seed, potato, cassava, yam, soybean, common beans, tomatoes, onion, sugarcane, melon, banana, and others [2].

This diversity of species is concentrated into areas of unusual richness and exhibits variation at both a global and a regional scale. The taxonomic diversity of plants is usually highest in the tropic with high amount of rainfall. The species found in one habitat can be used to identify the characteristics of the conservation priorities of that habitat which also mirrors the uniqueness of the ecosystem. For us to achieve measurable progress, a range of actors will need to generate, access, integrate, and synthesize data that is widely dispersed across organizational and international boundaries, and work through international partnerships that bring together complex portfolios of skills, sources of information, and perspectives [3, 4].

Over the years, plant conservation has often been reduced to an activity for government agencies involved in forestry management. Halewood et al. [3], outlined the benefits of establishing a conservation center for plants. They include promoting an integrated approach to plant conservation; utilizing and promoting professional skills; developing collaborative relationships with protected area networks, government agencies, parastatals; and the omics revolution in the biological sciences has considerable potential for changing the flows of information, the nature of partnerships, and the range of products that can be generated through plant genetic resource conservation.

In germplasm conservation, the method of collection initially captures maximum variation of plant materials. This procedure can be carried out either in the wild or controlled environment [5]. Ex situ conservation is reliable usually in seed banks, where they are cryopreserved. Additionally, technologies for generating and analyzing large quantities of genotypic, phenotypic, and environmental data are evolving at accelerating rates, so too are technologies and methods for synthesizing genetic materials [6].

2. Utilization of plant genetic resources in food sustainability

The sustainable use of genetic resources is critical to food security and sustainability. Globally, the improvement of food production has been successful. However, biodiversity seems to have been neglected. Biodiversity influences food production, as it ensures adequate and quality soil for optimum productivity and supplies invaluable genetic resources for all crops [7].

The world has been struggling to provide quality nutrition and access to safe water and eradicating all forms of malnutrition according to the sustainable millennium goals (SDGs). In 2020, an estimated 811 million people faced hunger attributed majorly to COVID-19 pandemic. Another report predicted that if global food security is not treated as a matter of urgency an additional 660 million people may suffer from hunger by 2030 [8].

Humans' inalienable rights would be realized when there is enough food for sustenance. From a broader aspect, it is pertinent for countries to provide access or means to sufficient food and potable water for their citizens. Great strides need to be shown in boosting food production, providing genetic resources, and widening the biodiversity of food crops. Government needs to enact favorable laws and policies and create solid institutional framework to ensure the access to genetic resources and agricultural materials. As no country can adequately sustain food production without a robust and sustainable genetic resource [8].

Molecular tools have proven to overcome some of the bottlenecks experienced in agriculture. Molecular techniques have continued to answer previously unanswered questions in taxonomy, breeding, etc. Techniques, such as the use of microsatellites and single nucleotide polymorphism (SNP), amplified fragment length polymorphism (AFLP), and random amplified polymorphic DNA (RAPD) are efficiently used in diversity study and in pest and disease resistance, high yield and salt and drought tolerance breeding programs [9, 10].

3. Biodiversity of plant genetic resources

Biodiversity is the bedrock of food security. The purpose of sustaining a functional biodiversity is to downgrade the threat of genetic erosion of important genetic resources. The protection of our biodiversity is critical as it directly affects food security. Plant genetic resources are at perennial risk of genetic erosion, which leads to loss of valuable genes, concomitantly loss of biodiversity. Some of the elements responsible for loss of biodiversity include climate change, floods, droughts, fire hazards, and urbanization to mention a few [11, 12].

The necessary practices for protecting targeted plant genetic resources, some of which may be harvested for food, in nature are locating the species, describing the status of their conservation, and actively managing and monitoring the populations where they occur in nature [13]. This is particularly critical as the genetic diversity of PGRFA in nature is being eroded by various factors, that is, loss and fragmentation of habitats and extreme weather events that may be linked to climate change [13].

Genetic erosion of plant genetic resources has been problematic both for primordial and modern agriculture and has been broadly documented. Several issues highlighted above, such as natural and man-made disasters are largely responsible for this immense loss. China in 1949 recorded a loss of about 10,000 wheat varieties; the USA in the 1970s lost about 95% of cabbage, 91% of maize, and 81% of tomatoes all to genetic erosion. It is noteworthy to mention that the cultivation of high-yielding variety causes genetic uniformity and is a pointer to the spread of diseases. This has been reported in wine grape and corn blight by the Information Bulletin, ICSC II, 1996.

4. Conventional approach to conservation

Before technological approaches to conservation were birthed, humans were conservation inclined. There has always been a need to preserve food materials for commercial purposes and to ensure their continual survival. Msuya and Kideghesho [14], outlined nine traditional conservation practices by the locals for medicinal plants. They include domestication, beliefs in sacred forests, beliefs in sacredness of trees, respect for cultural forests, protection of plants at burial rites, selective harvesting, secrecy, use of energy-saving traditional stoves, and collection of deadwood for firewood. The setbacks in these crude practices were climatic factors, pests and diseases, poverty, development activities, and changing agricultural policies. Traditional approaches are not sustainable and cannot meet our ever-growing population. Genetic erosion and poor management are factors delimitating traditional crops [15].

5. Advances in plant genetic resources

The conservation and use of plant genetic resources are important to the continued maintenance and improvement of agricultural and forestry production and, thus, to sustainable development and poverty alleviation. The objective of plant genetic resources conservation is to preserve as broad a sample of the extant genetic diversity of target species as is scientifically and economically feasible, including currently recognized genes, traits, and genotypes [16].

Germplasm banks are storage repositories equipped with facilities for long time storage. Some facilities have capacity to store genetic materials for 25 years and more. The brain behind conservation of plant genetic resources is to have a lot of variation. Variation in genetic resources affords farmers and researchers with options for breeding and other programs. Diversity of both landraces and introduced varieties ensure that the food global basket is never empty [12, 17].

Molecular and in vitro culture techniques are great tools. In vitro culture such as tissue culture provides multiplication of plantlets or clones of endangered plants. Tissue culture generates plant free from viruses, bacteria, or fungi. Molecular markers have been deployed in diversity study and for generating data for plant fingerprint. They are used to identify cultivars or landraces; used to discover important genes of interest and characterization of species [5]

The integration of big data into breeding programs is revolutionary. Generating sequence information is no longer a bottleneck to crop improvement. Phenotypic characterization has historically been more problematic, but increasingly, molecular phenotypes can be used as indicators of physiological or performance phenotypes, while quantitative imaging techniques using remote sensing can directly measure plant architectural and stress response characteristics in a variety of experimental set-ups [18].

6. Biotechnological approach

In modern agriculture, improved crops are farmers' preferred choice. Crops with improved yield ability, resistance to pests and diseases, and reduced environmental impacts are usually desirable. These are achievable with traditional methods but can be improved and enhanced using biotechnological approaches. The approach is genetically fingerprinting varieties from the wild and landraces. These plants are sequenced, which enables researchers to know which genes are important in conferring the trait that is needed. Utilization of these genes for agricultural purpose is a huge task because it involves a lot of data that the bioinformaticians need to deal with. This computational approach will be used to collect data from plants in gene banks and analyze promising plants for further analysis [19].

In an ever-growing population, to meet food demand, a sustainable system for food production is necessary. A very exquisite technique is the use of in vitro culture and cryopreservation, which is making it easy to conserve genetic resources especially seeds since most are difficult to conserve. Also, techniques such as enzyme-linked immunosorbent assay (ELISA) is used in testing healthy seeds against pathogens. Another method is tissue culture, used for eliminating systemic diseases like viruses for germplasm conservation. Polymerase chain reaction (PCR) and other molecular approaches are proving useful improvements in collection, accessioning, and resolving taxonomic discrepancies in relationship [19].

Molecular techniques are continuously evolving and their application in determining variation has been successfully applied in plant breeding. Molecular techniques for detecting variation include restriction fragment length polymorphism (RFLP), use for cutting short sequences of interest, and the use of PCR-based techniques such as amplified fragment length polymorphism; random amplified polymorphic DNA and simple sequence repeats have also proved very effective on genetic diversity study. These are all used to improve the state of plant genetic resources [19].

7. Cryopreservation

Cryopreservation is the storage of plant materials at ultra-low temperatures in liquid nitrogen (-196°C). The plant cells kept at this low temperature are devoid of metabolic activities and cell division. Therefore, the materials can be kept or stored for long time without any changes in their cellular structure [20].

Cryopreservation can be a primary or secondary storage technique. For certain plant materials such as embryonic culture that lose their capacity to move to the next stage of embryo formation, it could be a form of primary storage. When used for the conservation of plant genetic resources, it is a secondary storage form and usually as a form of backup or reserve for plant species [21]. The prospect of cryogenic technology is promising as it is important in conservation of genetic uniformity, preservation of rare genomes, sustaining disease-free plant materials, maintenance of morphogenetic potentials, and delaying aging of plant materials. The above techniques are all forms of secondary storage, which have tremendous commercial benefits [19, 21].

8. The evolution of genebanks

The erosion of genetic diversity of plant species is a global concern and a threat to food security. This has continued to stretch the stability of agriculture globally and negatively impacted market demand. The creation of genebanks in the 20th century was strategic and a recovery move to conserve local varieties (landraces). Ever since its creation, there have been some bottlenecks limiting the progress [22], outlined the outcome of the genebank workshop held in Spain in 2014. At the workshop, stakeholders summarized some of the shortcomings of genebanks since their creation. They include inefficient coordination of species across genebanks; insufficient phenotyping, genotyping, and epiphenotyping; and noticing unnoticed duplicates and lack of enough funding among other challenges outlined.

Today, there are about 1750 genebanks in the world housing millions of plant accessions and their wild relatives. Globally about 7.4 million plant accessions are banked ex situ in over 1750 genebank facilities. In conservation and utilization of crop diversity, genebanks are invaluable. It supports germplasm exchange, international ex situ collections, mining of genetic resources, and safeguarding of distinct species [23, 24].

9. Integration of farmers

The variation of crops has multifaceted impact on farmers as it influences their choice. Farmers' choices are influenced by certain traits, such as high yield, pest and disease resistance, nutritional values, and processing and taste qualities. However, due to poor and scanty research, farmers are usually left with their local varieties, which they continue to cycle for years. Consequently, militating their production in general [25, 26].

Primarily, the benefits of preserving plant genetic resources in genebanks are to assist farmers, especially rural farmers. Farmers need to work with plant genetic experts in choosing crop reproduction systems, cycle time of landraces, and genotype and phenotype characterization [25].

Many farmers are presently left with little option to practice modern agriculture both as a way of life and a form of social interaction. Farmers have been indirectly forced onto the global mono-cultural system of industrial agriculture, as evidentin their neglect of the traditional landraces, because of radically shrinking space for their (farmers') relevance and operation in the global food system. It has become very necessary moving forward for farmers to be schooled on genetic diversity of crops. The loss of potential of invaluable resources is hampering development of agriculture, especially in third-world countries. If plant genetic resources are well managed by all stakeholders, it will help reduce the high cost of food, reduce the cost of production, and improve both farmers and society at large [26–29].

10. Conservation strategy

The initiation of programs to study modern germplasm conservation strategy has been instituted in many international bioscience centers. This has been prompted by the loss of plant genetic resources [20]. Currently, biodiversity is currently being lost at up to 1,000 times the natural rate. Some scientists and researchers are now referring to the crisis as the "Earth's sixth mass extinction," comparable to the last great extinction crisis 65 million years ago. These extinctions are irreversible and present a serious threat to our health. Identification and management of protected areas is the pivot of biodiversity conservation. We must ensure that collection methods are able to capture most variation and also techniques that reduce genetic erosion [20, 30].

The use of in situ and ex situ methods in conservation of plant genetic resources have been widely used. In situ techniques have been successfully used in collecting small zygotic embryos and taking them back in sterile state to the research laboratories. Samples are preserved and remain in good condition afterward. Ex situ methods like storage of seeds, the use of botanical gardens, and genebanks have recorded huge success in the conservation of plant genetic resources, especially in tree crops. Crops like banana, cassava, potato, and yam that do not easily produce seeds are better to conserve in field genebanks. For the sake of loss of information or sample identity, it is better to conserve duplicate samples to avoid total loss in event of calamity or destruction (Withers and [20, 31, 32]).

Human activities have continued to threaten the survival of our biodiversity. This continued pressure has resulted in rise in the number of species under threat. Factors, such as weed infestation and introduction of new species have been implicated in narrowing plant diversity. Urbanization and globalization are other technical factors threatening conservation of plant genetic resources with orphan crops being the hardest hit. We must be proactive in our approach to preserve our biodiversity, because biodiversity is an important factor for food security [5, 33].

11. Discussion and conclusion

We live in a technologically advancing world that is having significant impact on the conservation of plant genetic resources. Concerns in agriculture about the loss of genetic resources and loss of genetic diversity propelled the response from scientists globally. This necessitated the use of advance technology. The use of advance biotechnological techniques, such as molecular marker technology, enzyme assays, cryopreservation, and modern genebanks have recorded huge success.

No concept is universally correct, and more than one may be appropriate in any context [34], underscoring the need for plant geneticists and crop scientists to work together. As there is need to study, understand, and enhance the value of plant genetic resources through research. For a successful collaboration, there is need to understand the full extent of plant diversity and analysis of the best technological approach to adopt in conserving plant genetic resources [11]. This collaboration should also be extended between Africa and the international community as much of Africa's biodiversity is still understudied.

The use of modern techniques in plant genetic resources especially biotechnological techniques, genebank, and cryopreservation methods have been highly beneficial in improving conservation and management of plant genetic resources. Areas, such as diversity gap data, gene pool coverage,

and molecular markers technology are vital in advancing the science of how plant genetic resources can be properly managed [20]. This review article recommends a sustained and proactive strategy in sharing genetic diversity data among scientists and the integration of farmers into global network of food security. It is important to add vital plant biological information, such as genotypic, phenotypic, and epigenetics data into the database for easy access and traceability. Farmers should be able to access this information with ease too. Additionally, empowering local farmers with requisite biotechnological tools and knowledge and other advanced methods of conserving plant genetic resources will go a long way in sustaining global efforts on food security [16, 26, 28, 32, 35–47].

Author details

Chiebuka Uzoebo
Botany Department, University of Ibadan, Nigeria

*Address all correspondence to: chiebuka2020@gmail.com

References

[1] Ford-Lloyd, Jackson. Plant genetic resources: An introduction to their conservation and use. 1986. Available from: www.semanticschool.org/paper/plant-genectic-resources

[2] Wilkes G. Germplasm conservation towards the year 2000 potential for new crops and enhancement of present crops. In: Yeatman C, Kefton D, Wilkes G, editors. Plant Genetic Resources: A Conservation Imperative. Washington D.C., USA: American Assoc. for the Advancement of Sciences; 1894

[3] Halewood M, Chiurugwi T, Sackville R, Kurtz B, Marden E, Patron N, et al. Plant genetic resources for food and agriculture: Opportunities and challenges emerging from the science and information technology revolution. New Phytologist. 2018;**127**(4):1409-1419

[4] Hammond E. Digital Genebankers Plan to Ignore UN Request on the Impact of Genomics and Synthetic Biology on Access and Benefit Sharing: A Preliminary Report. Penang, Malaysia: Third World Network; 2016

[5] Singh C, Poornima R, Rajkumar B. Pin conservation of plant genetic resources. Agricultural Research & Technology: Open Access Journal. 2017;**555718**. DOI: 10.19080/ARTOAJ.2017.07.555718

[6] Policy Science for Environment. Synthetic Biology and Biodiversity: Future Brief 15. Bristol, UK: For the European Commission DG Environment by the Science Communication Unit, UWE; 2016

[7] IAEA. Mutant variants database of joint FAO/IAEA division in nuclear techniques in food and agriculture. 2020. Available from: www.mvd.iaea.org/#!Home

[8] FAO. Coherent food security responses: Incorporating right to food into global and regional food security initiatives. 2011. Available from: www.fao.org/in-action/right-to-food-global

[9] Ibrahim A, Mohammed A, Haseeb K, Homaidan A. A brief review of molecular techniques to assess plant diversity. International Journal of Molecular Sciences. 2010;**11**(5):2079-2096

[10] Puchta H. Applying CRISPR/Cas for genome engineering in plants: The best is yet to come. Current Opinion in Plant Biology. 2017;**36**:1-8

[11] Ramanatha V, Hodgkin T. Genetic Diversity and Conservation and Utilization of Plant Genetic Resources. Serdang, Malaysia and Principal Scientist, GRST, IPGRI, Rome, Italy: IPGRI-APO; 2001

[12] Singh C, Poornima R, Rajkumar B. Recent advances in conservation of plant genetic resources. Agricultural Research & Technology: Open Access Journal. 2017;7:555718. DOI: 10.19080/ARTOAJ.2017.07.555718

[13] FAO. The Future of Food and Agriculture: Trends and Challenges. Rome, Italy: Food and Agriculture Organization of the United Nations; 2017. p. 163

[14] Msuya TS, Kideghesho JR. The role of traditional management practices in enhancing sustainable use and conservation of medicinal plants in west usambara mountains, Tanzania. Tropical Conservation Science. 2009:**2**(1):88-105. DOI: 10.1177/194008290900200109

[15] Ogwu MC, Osawaru ME, Ahana CM. Challenges in conserving and utilizing plant genetic resources (PGR). International Journal of Genetic and Molecular Biology. 2014;**6**(2):16-22

[16] Karp A, Kresovich S, Bhat K, Ayad G, Hodgkin T. Molecular tools in plant genetic resources conservation: A guide to the technologies. IPGRI Bulletin. 2012;**2**:23

[17] Godwin S, Gurtowski J, Ethe-Sayers S, Deshpande P. Oxford nanopore sequence, hybrid error correction and def nove assembly of a Eucharyotic genome. Cold Spring Habour Labs Journals. 2015;**2015**:013490-013490

[18] Dedeurwaerdere T. Institutionalizing global genetic resource commons for food and agriculture. In: Halewood M, Lopez Noriega I, Louafi S, editors. Crop Genetic Resources as a Global Common: Challenges in International Law and Governance. Oxford, UK: Routledge; 2013. pp. 369-391

[19] Nukari A, Uosukainen M, Rokka M, Antoniu K. Cryopreservation techniques and their application in vegetatively propagated crop plants in Finland. Agriculture and Food Science. 2009;**18**:117-128

[20] Rao K. Plant genetic resources: Advancing conservation and use through biotechnology. African Journal of Biotechnology. 2004;**3**(2):136-145

[21] Reed B. Plant Cryopreservation: A Practical Guide. Corvallis, OR: U.S. Department of Agriculture; 2008

[22] Diez M, Lucia R, Isaura M, Elena M, Mallor C, Casana F. Plant genebanks: Present situation and proposal for improvement: The case of Spanish network. Frontiers in Plant Science. 2018;**9**:1794

[23] Elbert A, Engels J. Plant biodiversity and genetic resources matter. Plants (Basel). 2020;**9**(12):1706

[24] Parvaze A, Zargar M, Salgotra K. Roles of Gene Banks in Maintaining Crop Genetic Resources. Germane-British London: Springer Nature; 2020. pp. 211-224

[25] Evenson R, Gollin D. Assessing the impact of the green revolution, 1960 to 2000. Science. 2003;**300**:758-762

[26] Schiml S, Puchta H. Revolutionizing plant biology: Multiple ways of genome engineering by CRISPR/Cas. Plant Methods. 2016;**12**:8

[27] Bello W. The Food Wars. London: Verso; 2009

[28] Oguamanam C. Plant genetic resources interdependence: Re-integrating farmers into the global food system. In: Liljeblad J, Kennedy A, editors. London: Routledge; 2016. pp. 143-162

[29] Saad N, Sperling L, Ashby J. Farmers and plant genetic resources. Biotechnology. 2012;**8**:291

[30] Astley D. Preservation of genetic diversity and accession integrity. Field Crops Research. 1992;**29**:205-224

[31] Englemann F. In vitro conservation methods. In: Ford-Lloyd BV, Newbury HJ, Callow JA, editors. Biotechnology and Plant Genetic Resources: Conservation and Use. UK: CABI; 1997. pp. 119-126

[32] Roberts E. Predicting the viability of seeds. Seed Science and Technology. 1973;**1**:499-514

[33] UNESCO. Conservation and sustainable use. 2021. Available from: https://en.unesco.org/themes/biodiversity/conservation

[34] Meng H, Smale, Bellon M, Grimanelli D. Definition and measurement of crop diversity for economic analysis. In: Smale M, editor. Farmers, Gene Banks, and Crop Breeding. The Netherlands: Kluwer Academic Publishers; 1998

[35] Allard RW. Genetic changes associated with the evolution of adaptedness in cultivated plants and their wild progenitors. Journal of Heredity. 1988;**79**:225-238

[36] Anon. World population worldometer. 2020. Available at: https://www.worldometers.info/world-population

[37] Frankel OH, Hawkes JG. Crop Genetic Resources for Today and Tomorrow. Cambridge, UK: Cambridge University Press; 1975

[38] Halewood M. What kind of goods are plant genetic resources for food and agriculture? Towards the identification and development of a new global commons. International Journal of the Commons. 2013;**7**:278-312

[39] Hamrick L, Godt M. Allozyme diversity in plant species. In: Brown AHD, Clegg MT, Kahler AL, Weir BS, editors. Plant Population Genetics, Breeding and Genetic Resources. Sunderland: Sinauer Associates Inc.; 1990. pp. 43-63

[40] Hamrick L, Godt M, Sherman-Broyles S. Factors influencing levels of genetic diversity in woody plant species. New Forests. 1992;**6**:95-124

[41] Holness S, Hamer M, Magos Brehm J, Raimondo D. Priority areas for the in-situ conservation of crop wild relatives in South Africa. Plant Genetic Resources: Characterization and Utilization. 2019;**17**(2):115-127. DOI: 10.1017/S1479262118000503

[42] International Atomic Energy Agency. Plant biodiversity and genetic resources. 2021. Available from: https://www.iaea.org/topics/plant-biodiversity-and-genetic-resources

[43] Mba C, Dullo E, Nnadozie K. Plant genetic resources for food and agriculture for sustainable development: Plant genetic resources: A review of current research and future needs. 2021:3-34

[44] Nguyen KL, Grondin A, Courtois B, Gantet P. Next-generation sequencing accelerates crop gene discovery. Trends in Plant Science. 2019;**24**(3):263-274

[45] Nukari A, Uosukainen M, Rokka M, Antoniu K. Cryopreservation techniques and their application in vegetatively propagated crop plants in Finnland. Agriculture and Food Science. 2009;**18**:117-128

[46] Saad N, Sperling L, Ashby J. Farmers and plant genetic resources. Biotechnology. 2012:VIII

[47] Varshney RK, Pandey MK, Bohra A, Singh VK, Thudi M, Saxena RK. Toward the sequence-based breeding in legumes in the post-genome sequencing era. Theoretical and Applied Genetics. 2019;**132**:797-816

Section 2

Breeding Theory and Strategy

Chapter 2

Genomic Selection: A Faster Strategy for Plant Breeding

Gizachew Haile Gidamo

Abstract

Many agronomic traits, such as grain yield, are controlled by polygenes with minor effects and epistatic interaction. Genomic selection (GS) uses genome-wide markers to predict a genomic estimate of breeding value (GEBV) that is used to select favorable individuals. GS involves three essential steps: prediction model training, prediction of breeding value, and selection of favorable individual based on the predicted GEBV. Prediction accuracies were evaluated using either correlation between GEBV (predicted) and empirically estimated (observed) value or cross-validation technique. Factors such as marker diversity and density, size and composition of training population, number of QTL, and heritability affect GS accuracies. GS has got potential applications in hybrid breeding, germplasm enhancement, and yield-related breeding programs. Therefore, GS is promising strategy for rapid improvement of genetic gain per unit time for quantitative traits with low heritability in breeding programs.

Keywords: genomic selection, training population, breeding population, prediction accuracies, plant breeding

1. Introduction

Since the 1990s, when promising analysis results for tagging genes or mapping QTL led to the development of marker-assisted selection (MAS). MAS has become a popular strategy in plant breeding. In the identification of underlying key genes in gene pools and their transfer to desirable traits in many plant breeding programs, marker-assisted selection and molecular breeding, have been applied. The use of MAS has shown some flaws, such as extensive selection schemes and the inability to catch "minor" gene effects when looking for crucial marker-QTL relationships. As a result, improving traits with complex inheritance, such as grain yield and abiotic stress tolerance, using MAS is difficult.

Genomic selection, also known as genome-wide selection, is a strategy that employs genotypic data from throughout the entire genome to accurately predict any trait, allowing for the selection of a favorable individual [1]. The most suitable individual is chosen based on a genomic estimate of breeding values (GEBVs). Breeding values are a popular and widely used measure in the animal breeding industry. Breeding values are defined as the "sum" of the estimated genetic deviation and the weighted total of the estimated breed effect [2], which are predicted using phenotypic data from family pedigrees based on the additive infinitesimal model. The success of selection in animal breeding, particularly in cattle and pigs, was aided by the infinitesimal genetic model and quantitative genetics. In the estimation of breeding value in animal breeding, the best linear unbiased prediction (BLUP) and Bayesian framework are often utilized. Following the introduction of genome sequencing in several model animals, a novel method for selection dubbed GEBV was developed [1]. In this chapter, principles of genomic selection and their application as a faster strategy for plant breeding is presented.

1.1 Phases of genomic selection

For large effect alleles, molecular marker technology has aided QTL identification, marker-assisted introgression, and selection, but not for low heritability quantitative traits, which still require considerable field testing. In the case of low heritability quantitative traits, locus identification and effect estimation are difficult to predict. New statistical methods that account for such uncertainty in genomic selection were used to make the best predictions. When classical marker-assisted selection and genomic selection are compared, the core framework is similar, including both breeding and training phases [3, 4]. In genomic selection, there are three crucial processes [3]:

1. Prediction model training and validation. Some lines in a population under selection are referred to as training sets. The training population is made up of germplasm that has both phenotypic and genome-wide marker data and is used to estimate marker effect and cross-validate results (**Figure 1**). These data were used to develop a statistical model that links variance in detected genotypes marker loci to variation in individual phenotypes. The training set's statistical

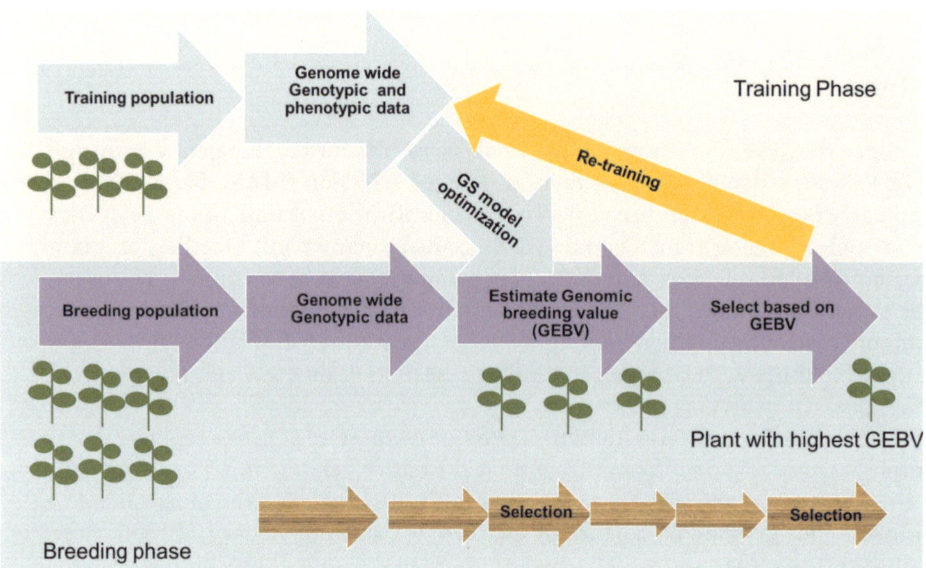

Figure 1.
Phases of genomic selection. Genome wide genotypic and phenotypic information from training population allows GS model optimization and breeding value estimation.

analysis evaluates allele effect in all loci at the same time. The second group of lines, known as selection candidates (breeding materials), are genotyped in order to calculate their GEBV.

2. Prediction of breeding value. The genomic estimate of breeding value (GEBV) is calculated using a mixture of marker effects estimations and marker data from a single cross. For an individual, GEBV is the sum of all marker effects incorporated in the model. These GEBVs were used to make selection. The line is the unit of evaluation as long as phenotype is a selection factor. When allele effect prediction is used as a selection criterion, the allele becomes the evaluation unit. The maximum GS accuracies were achieved in the presence of a large training population. The training population consists of numerous generations of training and comprises of parents or very recent ancestors of the population under selection. The use of markers for effect measurement is one of the key differences between traditional MAS and genomic selection. Only important markers are used in traditional MAS for effect estimation during QTL identification and selection. The non-significant markers were ruled out of consideration. However, all of the markers were utilized in genomic selection to capture the whole additive genetic variance. As a result, a more precise and small-effect QTL can be identified for use in breeding programs.

3. Selection based on predicted GEBV. The single cross is subjected to GEBV-based selection. GS made it possible to calculate breeding value directly from genotype rather than phenotype. When limited seed production prevents the application of selection and recombination in the F1 generation, a modified recurrent selection method using GS among F2 individuals is proposed in crop plants.

2. Statistical methods used in genomic selection

Finding a causal association between a genetic element and the characteristic of interest is a common selection framework used in all pedigree-based phenotypic selections, classic marker-assisted selection, and genomic selection. Typical QTL mapping and MASs either overstate or ignore the marker effect. The desire to use high-density genotyping technology for complicated trait prediction led to the creation of genomic selection. Avoiding marker selection minimizes bias during effect estimation and genetic value computations, according to Meuwissen et al. [1]. Because marker selection produces a bigger predictor effect (P) than the number of data (smaller n). When there is not enough degree of freedom, the ordinary least square estimator fails to estimate all of the predictors' effects, resulting in an over-fitted model. As a result, the ordinary least square's prediction performance suffers. To address this issue, various genomic selection models have been developed. Models like the shrinkage methods, variable selection models, kernel approaches, and dimension reduction methods can be mentioned.

2.1 The basic genetic model and variance decomposition

The basic genetic model that relates the phenotype (P) of an individual with summation of the genetic values (G) by assuming that only effects of the genetic factors were inherited to the next generations. The genetic values include genetic, dominance and epistatic effects and the residual environmental effect (E). It is mathematically denoted as:

$$P = G + E \tag{1}$$

In absence of G and E interaction, the covariance between G and E becomes zero. Therefore, the phenotypic variance V (P) can be expressed as [2]:

$$V(P) = V(G) + V(E) + 2COV(G, E) \tag{2}$$

$$V(P) = V(G) + V(E) + 0 = V(G) + V(E) \tag{3}$$

The GEBV is generally equal to G.

2.2 Heritability

The fraction of phenotypic variation (V(P)) owing to variation in genetic value (V (G)) is known as heritability. It assesses how well a population's phenotypic characteristics are passed on to the following generation. There are two ways to explain heritability: broad and narrow sense approaches. The fraction of phenotypic variation owing to genetic value is captured by broad-sense heritability (H^2). It concentrates on all genetic influences, including additive, dominance, and epistatic effects. Therefore, it can be mathematically represented as

$$H^2 = \frac{V(G)}{V(P)} \tag{4}$$

While the narrow-sense heritability (h^2) captures only the proportion of genetic variation that is due to additive genetic effect (V (A)) and the residual effect variance denoted as V(ε). It is represented by $h^2 = \frac{V(G)}{V(P)}$. Therefore, for h^2, the genetic model can be rewritten as:

$$V(P) = V(A) + V(\varepsilon) \tag{5}$$

where V (ε) represents the residual effects that are not included in the additive genetic effect (A) such as the dominant and epistatic effects.

The narrow-sense heritability is the most important in plant selection because it accounts for nearly all of the genetic variance that affect response to selection (close to 100%). Meuwissen et al. [1] suggested that V(A) might be broken down into various DNA markers effect like V(A1), V(A2), V(A3), and so on. This made it easier to calculate the breeding value of a plant using markers that covered the complete genome.

2.3 Breeding value

In animal breeding, the word "breeding value" refers to how many beneficial genes one animal passes on to its progeny. The genotype value and the breeding value can be equivalent. However, owing to dominance or epistatic situations, this is not always the case. Alleles at the loci that affect phenotype are heritable. Knowing the effect of an allele in a population can assist in predicting the progeny's phenotype. The deciding variables of a given trait in a population are allele frequencies and the effect of each genotype that includes the allele. It is also referred to as the allele's average effect.

An individual's breeding value is the total of the average effects of all the alleles the individual bears [3]. An AB heterozygote, for example, has a breeding value of 3 if an A allele is worth +5 and a B allele is worth −2. It is an individual's genetic value added together. Breeding value (BV) can alternatively be described as the departure of offspring's phenotypic mean value from the population phenotypic mean value using the narrow sense heritability concept (h^2). This can be expressed numerically as:

$$BV = \overline{m}_0 + h^2(y_i - m_0) = m_0 + (y_i - m_0)\frac{V(A)}{V(P)} \tag{6}$$

where y_i is the phenotypic value of individual i (i = 1, 2, ... n) and mo denotes the population's mean phenotypic value. Estimate of breeding value (EBV) is a term used to describe a breeding value that is estimated based on heredity. Genomic selection, on the other hand, employs genome-wide markers to evaluate genotype effect and breeding value, resulting in GEBVs (genomic estimate of breeding values) [2].

3. Models

3.1 The linear model

A linear model or its extension can be used to describe the causal link between phenotype and genotype. For the pair of observed phenotype and genotype of the marker of ith individual (y_i, $x1_i$), i.e. (y1, x11), (y2, x12), .., (yN, x1N) in the training population, which assumes N individuals and M biallelic markers. N individuals' phenotypes are normally distributed, and based on their marker genotype, they get an additional normally distributed phenotypic value of β1, depending on their marker genotype. The phenotype (y_i) can be modeled using genetic value $g_i = x1_i\beta1$ as a parametric regression on marker covariate $x1_i$ as follows: $y_i = \beta0 + x1_i\beta1 + \varepsilon_i$, where, β0 is the intercept (overall mean) and β1 is the marker effect (regression coefficient), $x1_i$ is the genotype value of marker 1 for individual i. The values of β0 and β1 are the parameters that need to be determined, and εi is an error term that is usually assumed to have a normal distribution with a mean of zero. To determine the unknown parameters, least-squares estimation, such that the summation of εi^2, that is an error function $E = \sum_i(yi - \beta o - x1i\beta1)^2$, is minimized and the line is fitted to the phenotype. However, applying the model for P and N number of markers and individuals, respectively, result over fitting. To avoid overfitting, a penalty term is introduced in the error function, i.e.,

$$E = \sum_{i=1}^{N}\left(yi - \sum_{j=0}^{M} xij\beta j^2\right) + \lambda\sum_{j=0}^{M} |\beta j|^q) \tag{7}$$

where, effect of the penalty term is controlled by λ.

To incorporate genome-wide markers in the model, the above formula can be extended into a multiple linear regression model, which gives the following formula [2]:

$$y_i = \beta0 + x1_i\beta1 + x2_i\beta2\cdots xm_i\beta m_i + \varepsilon_i \tag{8}$$

$$y_i = \beta_0 + \sum_{j=0}^{M} xji\beta j + \varepsilon i \tag{9}$$

where y_i = the phenotypic value of the individual i and xji is the genotype value of the jth marker in i^{th} individual. The coefficient βj is the effect of marker j on the phenotype or regression of y_i on the jth marker covariate x_{ij} and ε_i is the random error assumed [2]. $X0_i = 1$ is a dummy variable. Similarly, the coefficients were determined by minimizing the error function,

$$E = \sum_{i=1}^{N} \left(yi - \sum_{j=0}^{M} xji\beta j^2 \right) \tag{10}$$

In genomic selection, the focus is given to calculations of the genome enhanced breeding value rather than the exact location of the QTL; therefore, using the link function of linear model assumption, which provides relationship between linear predictor and the mean of the distribution function and error variance of regression, it can be rewritten as [2]:

$$y_i = \sum_{j=0}^{M} xji\beta j + \varepsilon i, \tag{11}$$

A number of models, including random regression best linear unbiased prediction (RR-BLUP), least absolute shrinkage and selection operator (LASSO), reproducing kernel Hilbert spaces (RKHS) and support vector machine regression, Bayesian methods, and collaborative filtering recommender system [5] have been developed using the above fundamental concepts. The majority of GS models aim to reduce the cost function [6].

3.2 Evaluating genomic prediction accuracy

Candidates for selection have no phenotypic information. As a result, their GEBV predictive performance may be evaluated using either a group of validation individuals with highly accurate EBVs and many progenies or cross validation. Both methods necessitate a reference population that contains both marker genotypes and phenotypic information.

3.2.1 Correlation studies between GEBV and observed EBV value

The r(GEBV: EBV) correlation between the GEBVs and empirically determined breeding values (observed) is used to assess the GEBVs' prediction accuracy (predicted). The EBV can be produced in a number of ways, the most basic of which is as a phenotypic mean. This relationship establishes a direct link between GEBV prediction accuracy and selection response, as well as a rough estimate of selection accuracy. Other statistics are occasionally used, such as mean-square error (MSE). The correlation between GEBV and true breeding value (TBV), that is, r(GEBV:TBV) is used to quantify genomic selection accuracy. Due to the fact that we can only measure r(GEBV: EBV), we must transform this value to an estimate of r(GEBV:TBV). To do so,

$$r(GEBV : EBV) = r(GEBV : TBV) * (EBV : TBV) \tag{12}$$

This assumption is accurate if the TBV is the only component that the GEBV and the EBV have in common. In other words,

$$GEBV = TBV + e1 \tag{13}$$

and

$$EBV = TBV + e2 \tag{14}$$

where e1 and e2 are uncorrelated error residuals, the assumption holds. If the training and validation data were obtained in the same setting, the assumption may be broken. In that instance, a common component of error in both GEBV and EBV would be generated by genotype by environment (GxE) interaction, biasing their correlation higher. To obtain accurate estimations of GEBV prediction accuracy, training and validation data should be collected in various environments. The r(EBV:TBV) correction accommodates for the fact that the EBV in the validation set is not error-free. Within the validation set, r(EBV:TBV) equals the square root of heritability (h) when the EBVs are phenotypes [7].

3.2.2 Evaluating GEBV accuracy through cross validation (CV)

Cross validation is used in GS research to evaluate GEBV accuracy on empirical data (CV). The reference population is divided into subsets in cross validation, such as a training set and a validation/testing set. Similar genetic backgrounds and relationships of validation and selection individuals to the reference population are required for cross validation, so that the accuracies achieved for selection candidates resemble those estimated using the reference population. The size of the subset determines accuracy; higher sizes usually result in lower sampling variance of anticipated and observed correlations [8].

The number of observations in each set varies, but a fivefold CV is frequently employed, in which the data set is divided into five sets at random, four of which are combined to form the training set, and the remaining set is designated as the validation set. Each subset of the data is used as a validation set once, and the model's correctness should be evaluated before it is applied to the breeding population. To do so, the majority of the training population is utilized to build a prediction model, which is then used to estimate the genomic estimated breeding values of the remaining individuals in the training population based solely on genotypic data. This allows researchers to "test" and develop the prediction model to ensure that it has high enough prediction accuracy that future predictions can be trusted. Once validated, the model is frequently used to calculate GEBVs of lines for which genotypical but not phenotypical information is available [8, 9].

3.3 Factors affecting genomic selection accuracy

The response of genomic selection is the result of numerous elements that contribute to the accuracy of GEBV estimation. These components are intricately linked in a comprehensive and complex way. The extent and distribution of linkage disequilibrium between individuals, as well as model performance, sample size and relatedness, marker density, gene effect, heritability, and genetic design are all factors to consider.

a. Marker density

Marker density and TP sizes required for satisfactory accuracy are heavily influenced by factors such as effective population size and QTL number. Minimum number of markers that cover the complete genome were used based on LD decay, with at least one marker in LD with each gene area. When there were a lot of LD and dense markers, the prediction was better [10]. However, unless the marker density is extremely low, marker density has minimal effect on prediction accuracy within families. Furthermore, some GS models, such as Bayes B, do not require a particularly dense marker for good breeding value prediction. The required marker density is also determined by the type of marker. For example, bi-allelic markers like SNP required two to three times the density of multi-allelic markers like SSR [3, 4, 11].

b. Size and composition of training population

GS accuracy is affected by the size of the training population. Up to the highest size possible, Vanraden et al. [12] found that the connection between accuracy and training population (TP) size was nearly linear. In other words, when the training population size was big, the maximum GS accuracies were achieved. Furthermore, population structure, training population age, and numerous generations of training all have an impact on accuracy. A near ancestor or parents, older lines, related lines, and multiple generations of training have good accuracy [4, 10]. Additionally, using a pooled training set of heterotic groups could improve accuracy [13]. As a result, the under-selection population's parents or recent ancestors can be used as the training population in a repeated generation of training to achieve high accuracy.

To maintain accuracy when using landrace or exotic germplasm in GS, very high marker density and a large training population size are required [1]. In addition, the training population's unrelatedness and single crosses cause marker effects become inconsistent. Due to the presence of various alleles, allelic frequencies, genetic background, or epistatic interaction, erroneous assessment of marker effect and GEBVs may occurs [14].

c. Number of QTL

The number of QTL and trait heritability determines the appropriate marker density and training population size. Even traits with low heritability can be accurately predicted in the context of a large training population [15]. For this prediction, a model like BLUP can be employed, which captures a lot of modest effect QTL that may not be in LD with the marker.

d. Heritability

Lower GEBV accuracies are associated with low heritability of a trait. High accuracy can only be maintained in the case of low heritability traits (particularly for h^2 0.4) by utilizing a large training population with many phenotypic data [10, 15]. Consider a population with an effective size of 1000 individuals and an accuracy of 0.70. If the heritability, h^2, is 0.2, it is expected that the training population (TP) size will need to be 9000, however, if h^2 is 0.50, a TP size of fewer than 3000 will be required. Responses to genomic selection were 18–43 percent higher than MARS across varied population sizes, QTL numbers, and heritability [16].

e. Linkage disequilibrium (LD)

LD refers to the nonrandom linkage of alleles at different loci. Marker density and GS accuracy can be estimated using the rate of LD decay across the genome. It has been found that for high heritability traits, an average nearby marker LD value (r^2) of 0.15 is sufficient, but increasing the r^2 value to 0.2 enhances GEBV prediction accuracy for low heritability traits.

f. Model used

Several published GS studies compared the accuracy of various statistical models. The disparity in prediction accuracies is negligible. However, some studies have found that the prediction accuracies of various models vary greatly, as seen in rice hybrid breeding [17–20].

3.4 Approaches to improve genetic gain and GS accuracies

3.4.1 Using biparental populations

With no group structure, biparental populations have a high level of LD between markers and trait alleles. Three shorter GS cycles can be completed in one year by using full-sib families. Genetic gain per unit time was improved in biparental populations from rapid selection cycles (C), according to studies made in maize [21]. Under drought conditions, maize hybrids generated from C3 cycles yielded 7.3 percent more than C0, according to Beyene et al. [21]. In winter wheat biparental populations, Lozada et al. [4] found a 10% increase in responsiveness to selection using genomic section relative to phenotypic selection. Similar studies have been conducted in oat by Asoro et al. [22] and wheat by Rutkoski et al. [23].

3.4.2 Using multi-parental populations and multi-environment models

GS rapid cycles of multi-parental crosses were performed in diallelic fashion to form cycle 0 in the CIMMYT maize breeding program. With two selection cycles each year and two location experiments, it suggested an improvement in genetic gain in which the study predicted a 0.1 t/h per year yield gain over a period of 4.5 years [24]. However, when comparing the C4 cycle to the C0 cycle, a decrease in genetic diversity was seen [24].

3.4.3 Combining GS with high-throughput phenotyping

In addition to genotyping data, accurate phenotypic data is required for genomic prediction model training to achieve the desired accuracy. For large-scale field-based accurate phenotypic data collection, a number of high-throughput phenotyping technologies have been built. These platforms are based on image and distant or proximal sensor technologies. Infrared thermometry and thermal imaging; visible/near-infrared spectroradiometry; and red, green, and blue light color digital photography are the three types of technologies in use for high throughput phenotyping. Their deployment is determined by the trait of interest and experimental design in the field. These technologies' data can be used as the primary input for model training. It is feasible to quantify high-density phenotypes over time and space using distant or proximal sensing by applying high throughput phenotyping such as canopy hyperspectral reflectance in a large number of breeding lines. This can improve the precision and intensity of selection, as well as the selection response, while lowering the phenotyping expenses. Lozada et al. [4] have proved in wheat that combining GS with high throughput phenotyping results in the highest accuracy for grain yield. The advantage of this imaging method is that large numbers of phenotypes can be screened for complex phenotypic expression and secondary traits that are genetically connected with grain yield at a low cost during early-generation testing. Juliana et al. [25] claim that by utilizing high throughput phenotyping, they were able to achieve a 60 percent increase in genetic gain for wheat yield and secondary characteristics.

3.4.4 Using historical data

Predicting the performance of new lines can be done by using phenotypic data from relatives and ancestors for model training that accounts for GxE interaction in multi-location research [26]. Historical data from breeding programs can be used effectively to increase genomic selection accuracy, particularly when the training set is adjusted to include only the most informative individuals from the target testing set [27].

3.4.5 Genotype imputation

For genotyping, genomic selection uses a high sample size and a dense marker set. In such data sets, missing data is a problem. Missing data were dealt with in one of four ways: (1) repeating genotyping in missing regions, (2) adapting analysis methods to accommodate missing data, (3) eliminating SNPs and/or samples with missing data, or (4) inferring the missing data (imputation).

Imputation of genotype is useful in a variety of situations. First, genotyping by sequencing, which is regarded as a low-cost genotyping method, typically yields a large number of markers at a low cost, but with a high proportion of missing data due to the poor genome sequencing depth. As a result of the imputation, the data set is full and ready for further study [28]. Second, utilizing low density genotyping and a closely similar reference panel genotyped at high density, imputation can enable GEBV prediction without a significant loss of accuracy. Using this in silico genotyping technique, low density genotyping in GS can be done without sacrificing accuracy. Imputation, on the other hand, may pose the danger of biases and inaccuracies [29].

Haplotype tagging is the simplest technique for genotype imputation [30]. In this strategy, a tag from the reference panel was chosen so that the majority of known (common SNPs) have a r^2 of less than 0.8 with the tag SNP. To identify shared haplotypes, the sequences of the reference panel haplotypes were compared to the genotyped markers. The missing genotypes were then filled in by copying alleles found in a matching reference haplotype (called FILLIN method) [29].

For imputation, a number of statistical approaches have been developed. These include the expectation maximization, Bayesian, LinkImpute, LD k-nearest neighbor imputation (LD KNNi) and entropy methods. These methods integrate models of recombination by partitioning markers into haplotype blocks. The tree-based imputation infers on the basis of perfect phylogeny and pairwise haplotype dissimilarity rather than haplotype structure [31].

3.5 Applications of genomic selection

3.5.1 GS for breeding of quality traits and yield

Grain yield is a crucial economic feature that has been studied in most GS studies of crops such as wheat. Grain yield is a complicated quantitative trait that is impacted by interactions between genes and surroundings and is regulated by a number of genes with little effects. GS has been shown to be important in grain yield studies in cereals. Prediction accuracies have been improved by include GxE effects in models [32]. Furthermore, GS aided in the cost reduction of phenotyping for malt quality in barley breeding [33].

3.5.2 GS and breeding for disease resistance

In terms of boosting intricate quantitative disease resistance, the GS method has a lot of potential for crop breeders. Pathogens find it difficult to overcome quantitative disease resistance because it is governed by a large number of genes with minor effects. Wheat rust, fusarium head blight, and rice blast resistance are three of the most well-studied diseases using the GS approach [18, 34].

3.5.3 GS for germplasm enhancement

Alleles for cultivar development can be found in abundance in gene bank accessions. Identification of these alleles is costly and time-consuming, and it necessitates extensive pre-breeding operations. Germplasm augmentation initiatives can begin with landraces by crossing them with elite testers. High genome-enabled prediction accuracy may be attained with GS, which may aid breedings in introducing valuable genetic variants. This supports the use of GS to introduce landrace accessions into elite germplasm and create gene pools and populations suited for pre-breeding and germplasm improvement [35].

3.5.4 GS for hybrid breeding

In hybrid breeding, parental selection is a critical issue. The performance of prospective crosses of a given parent set with genotyped parents and a small number of crosses examined in the field can be improved by employing whole genome markers in GS. This lowers the expense of hybridization and field testing of all possible hybrids. GS can also be used to predict hybrid performance and assist in hybrid selection. The predicted hybrids can be tested in the field and released as superior hybrids if they pass the test. There are just a few papers indicating the use of genomic selection for hybrid breeding in maize and rice [17, 20].

4. Conclusion

Before creating phenotypes, traditional cultivar development in plant breeding necessitates understanding of biological function. GS enables breeding without mapping and characterization of genes/QTLs at a low cost in order to gain functional data. GS can result in high genetic gain per unit time in crop breeding programs by enhancing GEBV accuracies through employing dense markers, increasing training population size, trait heritability, adopting a good GEBV prediction model, and using imputation techniques.

Acknowledgements

I acknowledge the support from the Department of Biotechnology, Addis Ababa Science and Technology University, Addis Ababa, Ethiopia.

Nomenclature

MAS	Marker-assisted selection
GS	Genomic selection
BV	Breeding value
GEBV	Genomic estimated breeding value
TBV	True breeding value
CV	Cross-validation
TP	Training population
BP	Breeding population
LD	Linkage disequilibrium
C	Selection cycle

Author details

Gizachew Haile Gidamo
Department of Biotechnology, Biotechnology and Bioprocess Centre of Excellence, College of Biological and Chemical Engineering, Addis Ababa Science and Technology University, Addis Ababa, Ethiopia

*Address all correspondence to: gizachew.haile@aastu.edu.et

References

[1] Meuwissen THE, Hayes BJ, Goddard ME. Prediction of total genetic values using genome-wide dense marker maps. Genetics. 2001;**157**:1819-1829. DOI: 10.1093/genetics/157.4.1819

[2] Nakaya A, Isobe S. Will genomic selection be a practical method for plant breeding? Annals of Botany. 2012; **110**:1303-1316. DOI: 10.1093/aob/mcs109

[3] Jannink J, Lorenz AJ, Iwata H. Genomic selection in plant breeding: From theory to practice. Briefings in Functional Genomics. 2010;**9**:166-177. DOI: 10.1093/bfgp/elq001

[4] Lozada DN, Mason RE, Sarinelli JM, Brown-Guedira G. Accuracy of genomic selection for grain yield and agronomic traits in soft red winter wheat. BMC Genetics. 2019;**20**:82. DOI: 10.1186/s12863-019-0785-1

[5] Lozada DN, Carter AH. Genomic selection in winter wheat breeding using a recommender approach. Genes. 2020; **11**:779. DOI: 10.3390/genes11070779

[6] Heslot N, Yang H-S, Sorells ME, Jannick JL. Genomic selection in plant breeding: A comparison of models. Crop Science. 2012;**52**:146-160. DOI: 10.2135/cropsci2011.06.0297

[7] Storlie E, Charmet G. Genomic selection accuracy using historical data generated in a wheat breeding program. The Plant Genome. 2013;**6**:1-9. DOI: 10.3835/plantgenome 2013. 01.0001

[8] Zhao Y, Gowda M, Liu W, Würschum T, Maurer HP, Longin FH, et al. Accuracy of genomic selection in European maize elite breeding populations. Theoretical and Applied Genetics. 2012;**124**:769-776. DOI: 10.1007/s00122-011-1745-y

[9] Heffner EL, Jannink JL, Sorrells ME. Genomic selection accuracy using multifamily prediction models in a wheat breeding program. The Plant Genome. 2011;**4**:65-74. DOI: 10.3835/plantgenome2010.12.0029

[10] Liu X, Wanga H, Wanga H, Guoa Z, Xua X, Liua J, et al. Factors affecting genomic selection revealed by empirical evidence in maize. The Crop Journal. 2018;**6**:341-352. DOI: 10.1016/j.cj.2018.03.005

[11] Solberg TR, Sonesson AK, Woolliams JA, Meuwissen TH. Genomic selection using different marker types and densities. Journal of Animal Science. 2008;**86**:2447-2454. DOI: 10.2527/jas.2007-0010

[12] VanRaden PM, Van Tassell CP, Wiggans GR, Sonstegard TS, Schnabel RD, Taylor JF, et al. Reliability of genomic predictions for north American Holstein bulls (invited review). Journal of Dairy Science. 2009; **92**:16-24. DOI: 10.3168/jds.2008-1514

[13] Technow F, Bürger A, Melchinger AE. Genomic prediction of northern corn leaf blight resistance in maize with combined or separated training sets for heterotic groups. Genes Genomes Genetics. 2013;**3**:197-203. DOI: 10.1534/g3.112.004630

[14] Bernardo R. Molecular markers and selection for complex traits in plants: Learning from the last 20 years. Crop Science. 2008;**48**:1649-1664. DOI: 10.2135/cropsci2008.03.0131

[15] Hayes B, Bowman P, Chamberlain A, Goddard M. Invited review: Genomic selection in dairy cattle: Progress and challenges. Journal of Dairy Science. 2009;**92**:433-443. DOI: 10.3168/jds.2008-1646

[16] Bernardo R, Yu JM. Prospects for genomewide selection for quantitative traits in maize. Crop Science. 2007;**47**: 1082-1090. DOI: 10.2135/cropsci2006.11.0690

[17] Cui Y, Li R, Li G, ZhanG F, Zhu T, Zhang Q, et al. Hybrid breeding of rice via genomic selection. Plant Biotechnology Journal. 2020;**18**:57-67. DOI: 10.1111/pbi. 13170

[18] Huang M, Balimponya EG, Mgonja EM, McHale LK, Luzi-Kihupi A, Wang GL, et al. Use of genomic selection in breeding rice (Oryza sativa L.) for resistance to rice blast (Magnaporthe oryzae). Molecular Breeding. 2019; **39**:114. DOI: 10.1007/s11032-019-1023-2

[19] Islam MS, Fang DD, Jenkins JN, Guo J, McCarty JC, Jones DC. Evaluation of genomic selection methods for predicting fiber quality traits in upland cotton. Molecular Genetics and Genomics. 2020;**295**:67-79. DOI: 10.1007/s00438-019-01599-z

[20] Xu Y, Wang X, Ding X, Zheng X, Yang Z, Xu C, et al. Genomic selection of agronomic traits in hybrid rice using an NCII population. Rice. 2018;**11**:32. DOI: 10.1186/s12284-018-0223-4

[21] Beyene Y, Semagn K, Mugo S, Tarekegne A, Babu R, Meisel B, et al. Genetic gains in grain yield through genomic selection in eight bi-parental maize populations under drought stress. Crop Science. 2015;**55**:154c163. DOI: 10.2135/cropsci2014.07.0460

[22] Asoro FG, Newell MA, Beavis WD, Scott MP, Tinker NA, Jannink JL. Genomic marker-assisted and pedigree-BLUP selection methods for beta-glucan concentration in elite oat. Crop Science. 2013;**53**:1894-1906. DOI: 10.2135/cropsci2012.09.0526

[23] Rutkoski J, Singh RP, Huerta-Espino J, Bhavani S, Poland J, Jannink JL, et al. Genetic gain from phenotypic and genomic selection for quantitative resistance to stem rust of wheat. Plant Genome. 2015;**8**:1-10. DOI: 10.3835/plantgenome2014.10.0074

[24] Zhang X, Pérez-Rodríguez P, Burgueño J, Olsen M, Buckler E, Atlin G, et al. Rapid cycling genomic selection in a multi-parental tropical maize population. Genes Genomes Genetics. 2017;**7**:1-12. DOI: 10.1534/g3.117.043141

[25] Juliana P, Montesinos-López OA, Crossa J, Mondal S, González Pérez L, Poland J, et al. Integrating genomic-enabled prediction and high-throughput phenotyping in breeding for climate-resilient bread wheat. Theoretical and Applied Genetics. 2019;**132**:177-194. DOI: 10.1007/s00122-018-3206-3

[26] Dawson JC, Endelmana JB, Heslot N, Crossad J, Poland J, Dreisigacker S, et al. The use of unbalanced historical data for genomic selection in aninternational wheat breeding program. Field Crops Research. 2013;**154**:12-22. DOI: 10.1016/j.fcr.2013.07.020

[27] Atanda SA, Olsen M, Burgueño J, Crossa J, Dzidzienyo D, Beyene Y, et al. Maximizing efficiency of genomic selection in CIMMYT's tropical maize breeding program. Theoretical and Applied Genetics. 2021;**134**:279-294. DOI: 10.1007/s00122-020-03696-9

[28] Rutkoski JE, Poland J, Jannink JL, Sorrells ME. Imputation of unordered markers and the impact on genomic selection accuracy. Genes Genomes Genetics. 2013;**3**:427-439. DOI: 10.1534/g3.112.005363

[29] Li Y, Willer C, Sanna S, Abecasis G. Genotype imputation. Annual Review of

Genomics and Human Genetics. 2009; **10**:387-406. DOI: 10.1146/annurev.genom.9.081307.164242

[30] Johnson GCL, Esposito L, Barratt BJ, Smith AN, Heward J, Genova GD, et al. Haplotype tagging for the identification of common disease genes. Nature Genetics. 2001;**29**:233-237. DOI: 10.1038/ng1001-233

[31] Roberts A, McMillan L, Wang W, Parker J, Rusyn I, Threadgill D. Inferring missing genotypes in large SNP panels using fast nearest-neighbour searches over sliding windows. Bioinformatics. 2007;**23**:401-407. DOI: 10.1093/bioinformatics/btm220

[32] Juliana P, Singh RP, Braun HJ, Huerta-Espino J, Crespo-Herrera L, Govindan V, et al. Genomic selection for grain yield in the CIMMYT wheat breeding program status and perspectives. Frontiers in Plant Science. 2020;**11**:e 564183. DOI: 10.6084/m9.figshare.12350000.v1

[33] Schmidt M, Kollers S, Maasberg-Prelle A, Großer J, Schinkel B, Tomerius A, et al. Prediction of malting quality traits in barley based on genome-wide marker data to assess the potential of genomic selection. Theoretical and Applied Genetics. 2015. DOI: 10.1007/s00122-015-2639-1

[34] Ornella L, Singh S, Perez P, Burgueño J, Singh R, Tapia E, et al. Genomic prediction of genetic values for resistance to wheat rusts. The Plant Genome. 2012;**5**:136e148. DOI: 10.3835/plantgenome2012.07.0017

[35] Gorjanc G, Jenko J, Hearne SJ, Hickey JM. Initiating maize pre-breeding programs using genomic selection to harness polygenic variation from landrace populations. BMC Genomics. 2016;**17**:30. DOI: 10.1186/s12864-015-2345-z

Chapter 3

Molecular Techniques for Essentially Derived Varieties

Bratislav Stankovic

Abstract

The term "essentially derived varieties" (EDVs) was introduced by the International Union for the Protection of New Varieties of Plants (UPOV) to prevent the exploitation of minor changes in relation to the initial plant variety (IV), without the holder of the IV right being able to share in the revenues. A plant variety is deemed an EDV when it is predominantly derived from the IV, or from a variety that is itself predominantly derived from the IV, while retaining the expression of the essential characteristics that result from the genotype of the IV. Molecular markers can be used to characterize plant genetic resources and to provide measures of genetic (dis)similarities between plant varieties. Genetic distance estimates based on molecular markers are a preferred approach to estimate genetic conformity between putative EDVs and their IVs. Numerous publications have shown the advantages of molecular markers, their high throughput, high map resolution, and high repeatability for determination of EDVs. They help reduce the legal uncertainty surrounding the EDV concept, thus providing a more predictable business environment that allows genetic diversity to be surveyed with ever-increasing effectiveness, improving the selection of new genotypes that are optimally able to perform in target agricultural environments.

Keywords: essentially derived variety (EDV), UPOV, molecular markers, intellectual property, plant variety protection, plant breeders' rights

1. Introduction

The developments of precise genetic engineering tools, such as sequencing platforms, *in planta* gene editing, and molecular marker-assisted breeding, have made possible the recent rapid advances in agricultural biotechnology. These are all exciting accomplishments as plant breeders are racing worldwide to meet the needs of the increasing human population against the backdrop of industrialization, climate change, and water scarcity. Improved understanding of the genetic make-up underpinning the phenotype, variability, and adaptability of desirable crops and ornamentals offers unprecedented opportunities for highly targeted genetic engineering approaches.

Traditional plant breeding can be an arduous and long process. However, highly related, similar, and/or identical plant varieties can subsequently easily be produced. Accordingly, plant breeders need protection to recover their investments. Plant breeding has conventionally offered challenges for intellectual property (IP) protection due to a number of technical and legal factors, which include the definition and the verification of whether the breeding of a new(er) plant variety constitutes a new innovation—or not. Just a few countries (e.g., USA and Japan) issue utility patents for plant varieties, thus this type of IP rights protection is not a feasible option for developing countries.

The International Union for the Protection of New Varieties of Plants (UPOV) was established as an international organization by the 1961 International Convention for the Protection of New Varieties of Plants [1]. UPOV strives to recognize the rights of plant breeders on an international basis and to provide and promote an effective system of plant variety protection, with the aim of encouraging the development of new varieties of plants. Through periodic revisions, the UPOV system has been gradually strengthening breeders' rights by adding crop species, restricting farm-saving of seed, and extending the scope of protection. Seventy-eight countries are UPOV member states as of November 2021. UPOV defines a "plant variety" as a more precisely defined group of plants, selected from within a species, with a common set of characteristics [1].

In a globalized economy, the problem of plant variety derivation and the need for an appropriate protection system was identified decades ago. A plant variety should be declared distinct only if it differs in important characteristics from all other known varieties. The concept of "essentially derived varieties" (EDVs) [of plants] was introduced by UPOV to refine the scope of plant breeders' rights. The goal was to prevent fraudulent practices and the exploitation of minor changes in relation to the initial plant variety (IV), for example by genetic engineering or mutagenesis, without the holder of the initial variety right being able to share in the revenues. According to the UPOV's definition in Article 14(5)(b), which describes the scope of the breeders' rights, "a variety shall be deemed to be essentially derived from another variety ("the initial variety") when (i) it is predominantly derived from the initial variety, or from a variety that is itself predominantly derived from the initial variety, while retaining the expression of the essential characteristics that result from the genotype or combination of genotypes of the initial variety, (ii) it is clearly distinguishable from the initial variety and (iii) except for the differences which result from the act of derivation, it conforms to the initial variety in the expression of the essential characteristics that result from the genotype or combination of genotypes of the initial variety" [1]. This definition was in 2017 further clarified with the issuance of explanatory notes on essentially-derived varieties under the 1991 Act of the UPOV Convention [1]. To honor the IV breeder's rights, the breeder of the EDV must reach an agreement with the holder of the IV before exploitation. Where derivation occurs, it is important to be able to distinguish between subsequent (second, third, etc.) plant varieties that have been "derived" from an existing variety, which are allowed; and those that have been "essentially derived," which are not allowed [2]. The EDV concept aims to ensure a fair return on investment to the breeder of the IV.

Molecular genetic markers are increasingly becoming a powerful scientific tool that can be used to characterize plant genetic resources and to provide comparison(s) of genetic similarities between various plant varieties. In particular, genetic distance estimates based on molecular markers are gradually becoming a preferred approach to estimating genetic conformity between putative EDVs and their IVs.

This chapter reviews the use of molecular markers for the determination of "EDV" as used by UPOV. It summarizes recent publications that demonstrate the advantages afforded by the use of molecular markers, their high throughput, high map resolution, and high repeatability for the determination of EDVs. The overarching goal of these efforts is to help provide a predictable business environment that will allow a greater breadth of genetic diversity to be surveyed with ever-increasing effectiveness to continually improve abilities to select new genotypes that are optimally able to perform in target agricultural environments. It also highlights criticisms that address the use of molecular markers for determination of EDVs.

2. Fundamentals

Determining whether a plant variety is "essentially derived" has largely been left to scientists and plant breeders who favor a quantitative approach that assesses genetic similarity using "genetic thresholds," "genetic distances" and "genetic coefficients." Unfortunately, science is not settled on the most appropriate method of assessing whether a given plant variety is essentially derived [2]. As such, a key challenge facing plant IP is to traverse the gap between science and law as it relates to the essential derivation of plant varieties. Clarity is necessary to ensure the effectiveness of essential derivation as a legal concept and to instill confidence in the plant breeders' rights scheme. The importation of notions of quality into the assessment of essential derivation by the judiciary has a number of advantages, including the ability to meet the needs of developments in plant breeding techniques and providing plant breeders with greater certainty in relation to whether new varieties are essentially derived [2].

Who decides whether a plant variety is essentially derived from another? In case of dispute, the holder of the initial variety has the initial burden of demonstrating that another variety is derived from their variety. In an application procedure, there authorities granting plant variety rights are not in the best position to determine whether a given variety is an EDV. The IP rights protection system is generally non-interventionalist; an EDV need not be protected *per se*, and may, therefore, entirely elude the attention of the authority granting plant variety rights. Since EDVs concern the scope of the protection, it is the responsibility of the holder of the IV plant breeder's rights to enforce the right, and/or settle with the breeder of the variety which is essentially derived from the initial one. In case of disputes related to EDVs, if the parties cannot agree, then the courts will have the final word. However, there is little by way of case law available, which might indicate that either in most cases the parties settle the matters amongst themselves, and/or the anticipated costs of legal action do not justify a possible profit that such an action might provide.

In Article 14(5)(c), the UPOV convention provides some guidance on how EDVs might be obtained. Exemplary modes of obtaining EDVs are the selection of a natural or induced mutant, or of a somaclonal variant, the selection of a variant individual from plants of the initial variety, backcrossing, or transformation by genetic engineering [1]. The EDV need not be directly derived from the initial variety; it may also be bred by using one or more other varieties predominantly derived from the initial variety. Two principal criteria need to be met by an EDV: (i) there must have been an act of derivation, and (ii) there must be a significant extent of similarity at the phenotypical level. The latter is relatively easier to establish; the former—not so much.

The phenotype-based assessments of eventual essential derivation are time-consuming, laborious, and costly processes that involve field and greenhouse comparisons. As well, various factors limit the expression and use of phenotypic traits. The measurement of morphological traits may be influenced by environmental factors, resulting in reduced precision and discriminatory power. This is especially disadvantageous for the evaluation of disease resistance or complex quantitative traits, such as flowering time.

One of the key advantages of molecular marker-assisted breeding is that it increases the breeding efficiency compared to conventional breeding based on phenotypic selection. Molecular markers thus allow for the identification and selection of individual plants with high reliability—in any context. The use of molecular marker data to assist in the determination of EDVs has already received significant attention, as molecular markers are an important means to help determine EDV status [3]. In a few instances, crop-specific experimental guidelines have even been developed; see *infra*. Because the comparison of two varieties is by definition a relative endeavor, some of the relevant questions include: how deep should the marker analysis be? What should be the reference framework? What should be a representative sample of markers for a given phenotypic trait? Is the choice of marker technology important? What about the number of markers used? How general vs. species-specific should the marker analysis be? Fortunately, some of the above points have been addressed in the literature.

The choice of marker technology does not appear to be crucial. In several cases, the analysis of the same dataset with different technologies led to same conclusions. Any DNA marker technology can do the job, as long as the technology and performance have proven to be accurate. Furthermore, an increased number of markers will give a more accurate representation of the genome. Due to the ever-increasing abundance of reference genomes and advances in molecular technologies, functional markers that target specific alleles responsible for phenotype variation can and will be developed.

In disputes relating to plant breeders' rights, the initial burden is upon the owner of the IV who should provide *prima facie* proof of essential derivation and consequently claim the dependence of the new variety. This is precisely where molecular marker data can provide an indication that can be interpreted in relation to a trigger point to reverse the burden of proof [4].

Genetic distance estimates based on molecular markers are a preferred approach to estimating genetic conformity between putative EDVs and their corresponding initial varieties (IVs). For example, genetic similarity between two varieties can be suitably measured using both amplified fragment length polymorphisms (AFLPs) and simple sequence repeats (SSRs) [3]. To counterbalance the advantages and disadvantages of AFLP and SSRs, a complementary application of the two marker systems is suggested for discrimination between EDVs and independently derived varieties (IDVs) [5].

Critics of this approach and of these rights (the use of molecular markers for determination of EDVs in particular, and EDVs in general) point to the relatively dominant position of a very small number of corporations in the global seed market, the impact of this domination on farmers, and the role of IP in promoting these perceived corporate interests, often at the expense of environmental and social justice concerns [6]. Critics argue that the extension of plant breeders' rights to EDVs is controversial as it creates tension with the breeders' exemption [7]. This chapter submits that the determination of whether a given plant variety is an EDV is best approached as a holistic endeavor: both genetic comparisons and morphological differences between the varieties in question should be considered.

The extension of the protection for initial varieties to derived varieties was intended as an exceptional provision, which should be interpreted restrictively by its nature. In a time when plant gene editing is becoming routine, it is possible to use a single locus conversion and to introduce a single transgene into a variety. Such a modification could even be precisely introduced into the IV via new breeding techniques (NBTS, e.g., CRISPR/Cas-mediated genome editing, or mutagenesis). The newly obtained variety would thus have only a single genetic difference (single locus) vs. the initial variety, but a cleverly introduced single transgene could result in multiple (i.e., more than a few) downstream gene and protein expression modifications, for example, changes to a metabolic pathway, regulation of a chosen key transcription factor, etc. It could thus result in a clearly distinct phenotype (e.g., flower color, fruit size, etc.). Therefore, the determination of whether a derived variety differs significantly from the initial should be conducted on a case basis, on a crop basis, and on a species basis. Conversely, in instances where there are major morphological differences established between the varieties in question, there may not be a need to conduct a genetic comparison; a phenotypic comparison might suffice.

3. Particulars

Possible starting points for the assessment of EDVs include the establishment of predominant derivation (e.g., evidence of genetic conformity with the IV), and conformity of the essential characteristics (evidence of conformity in the expression of the essential characteristics of the IV). Essential characteristics means heritable traits that contribute to the principal features, performance, or value of the variety, and/or characteristics that are important from the perspective of the producer, seller, supplier, buyer, recipient, or user. These essential characteristics may be different in different crops and species. They also may or may not be phenotypic characteristics used for the examination of distinctness, uniformity, and stability (DUS) [1]. The assessment of conformity is essentially a judgment call that is based on essential characteristics. The number of differences that results from the act of derivation should be minimal (one or very few). The derived variety must retain almost the totality of the genotype of the initial variety and be different from the IV by a very limited number of characteristics.

Numerous studies have demonstrated the utility of molecular genetic markers for the determination of EDVs. For example, DNA fingerprinting has been used to illustrate the possibilities and limitations in disputes on alleged fraud and essential derivation of ornamental plants, including roses and *Phalaenopsis* orchids. Following AFLP analysis of putative parents, the parties (breeders) settled the dispute(s) [8]. AFLP, as a multilocus marker technology, enables quick random screening of the plant genome. The effectiveness of a primer combination to reveal good AFLP patterns for a certain plant species depends on three factors: (i) the occurrence of the restriction sites recognized by the used enzymes; (ii) the number of selective bases added to the primer in the function of the genome size (five to six selective bases); and (iii) the GC content of the DNA [8].

To invest in research and product development, companies value the certainty afforded by IP protection. Property rights with respect to plant varieties are not an exception. Accordingly, a thorough review of the issues important to the private sector with respect to genomic approaches and IP protection for EDVs was recently provided [9]. Notably, the American Seed Trade Association (ASTA) and the French Seed Association (UFS, Union Française des Semenciers) jointly embarked on a project using an Illumina® 56,000 SNP (single nucleotide polymorphism) chip to profile a set of maize inbred lines including those of historic and current importance. The results of this study provided a list of thousands of publicly available SNP loci, as a repository for the selection of desired sets of markers to measure genetic similarities (or distances) between pairs of inbred lines. These loci can find utility, particularly for the purpose of resolving questions in regard to the possible status as an EDV [9]. These molecular tools can be very robust, and their utility has been demonstrated in numerous case studies. For example, by providing markers (e.g., SNP loci) that are judiciously selected, it is possible to identify a relatively small set of SNP loci that collectively have a very high power of discrimination among maize inbred lines, and which, therefore, could be used to measure uniformity and stability (genetic purity) for example, as few as 16 SNPs can discriminate among more than 400 pioneer proprietary inbred lines. Indeed, marker-based distances are more appropriate than pedigree records to assess genetic relationships among maize inbreds [9].

The international seed federation (ISF) works to represent the interests of the seed industry at a global level, including farmers, growers, industry, and consumers. ISF promotes the establishment and protection of IP rights for seeds, plant varieties, and associated technologies. ISF has adopted arbitration procedure rules tailored to the technical and legal aspects of essential derivation in plant breeding, using plant species-specific protocols and markers. Because the determination of genetic similarity thresholds varies between species depending on the genetic variability, the breeding procedures, and the used molecular tools, distinct guidelines and experimental protocols for handling disputes on essential derivation in various crops (e.g., cotton, lettuce, maize, oilseed rape, and perennial ryegrass) have been developed by the ISF [10]. For example, the ISF has supplied a list of 3072 SNP markers used in the guidelines provided for essential derivation in maize [11]. Furthermore, the key issues that should be addressed by technical experts to define molecular marker sets for establishing thresholds for EDV arbitration by the ISF have also been published [12].

Perhaps due to its economic significance, much of the studies on the characterization of the essential derivation of plant varieties have so far been conducted in maize [13–15]. These include the study of inbreds developed without and with a selection from F_2 populations [13], the selection and evaluation of an optimal panel of SSR loci in maize [14], and a comparison of SSR and SNP data in different maize genotypes, using European and North American germplasm, with the help of a set of 163 SSR loci previously identified to help determine EDV status [15].

The increased availability and cost-efficiency of DNA-based markers in recent years make them attractive options to explore their use to supplement or even ultimately replace, existing morphological-, phenotype-, and protein-based approaches. Indeed, exploring such options is the focus of UPOV's working group on "biochemical and molecular techniques and DNA profiling in particular" (BMT). A recent review argues for the adoption of a DNA-based system which is endorsed by the UPOV, and highlights that the efficiency of phenotype-based assessments of plant variety protection and registration could be improved by the integration of DNA-based testing [16]. So far, two competing models, namely "characteristic-specific molecular markers" and "combining phenotypic and molecular distances in the management of variety collections" have been favored [16].

Plant variety protection is also of high relevance for the horticultural community. In cases of plant varieties with a relatively narrow gene pool, it is possible to use molecular markers as a technique for the identification of EDVs in highly identical genotypes. For example, in a case involving *Calluna vulgaris* (common heather), a successful marker comparison was performed by means of random amplified polymorphic DNA (RAPD) and intersimple sequence repeats (ISSR) fingerprinting while using 168 mono- and polymorphisms. Accordingly, this methodology has been recommended for the future establishment of proof-of-essential-derivation, not only for *C. vulgaris*, but also generally for other vegetatively propagated crops [17].

The concept of protecting the rights of breeders through EDVs has itself been the subject of criticism, as being directed to narrowing the breeder's exemption while expanding the exclusive rights of first-generation breeders [2]. Furthermore, the methodology of using molecular markers for the determination of EDVs has also specifically been attacked. It has been argued that utilizing molecular markers to assess essential derivation could actually provide a mechanism to undermine the intention of essential derivation—which is to discourage copying and free riding. Ironically, a breeder may use marker-assisted breeding to evade a declaration of essential derivation, through a clever selection of a molecular marker profile that is 'sufficiently different' from the initial variety. For instance, if the genetic threshold for a given variety is 90%, it may be possible to ensure that subsequent varieties will show genetic thresholds of 85% or less, despite there only being minor changes made to the initial variety. While the new variety may be quantitatively outside the boundary established for essential derivation, it may still draw on the qualitative features of the existing variety for its commercial appeal to the industry [2].

4. Discussion and prospects

The UPOV system introduced in 1991 the concept of essential derivation and EDV. Some non-UPOV member countries (e.g., India, Malaysia, and Thailand) have also independently introduced the concept of essential derivation. China, a UPOV member operating under the 1978 Convention, is introducing EDVs via seed laws [18]. The EDV concept is alive and well. Due to the relative complexity associated with its definition, the implementation of the EDV concept has sometimes been challenging.

The use of molecular genetic markers can bolster the science-preferred quantitative approach to the determination of EDVs. Threshold values for genetic conformity can be determined by using similarities that are calculated from molecular marker characterizations of varieties [19]. The use of marker-based (dis)similarities in essential derivation cases raises statistical questions, such as how to define a reference population of varieties within which potential essential derivation disputes could occur, how many marker loci to use for a required precision, and how to define a threshold value on the basis of the observed distribution of similarity values. Initial results from special studies undertaken to answer these questions in lettuce and barley have been published [19]. It has already been shown that, at least for a subset of DUS phenotypic traits, molecular markers can be robustly used as a tool for the determination of all these three components of the DUS testing in the diploid crop species barley [20]. This is important since crop-specific DUS test guidelines have been developed and followed for plant variety assessment, to register a new variety. Plant varieties can differ significantly, and the phenotypic differences depend on the genetic constitution but also on the sample size, levels of confidence, reject numbers, etc. [21].

The study and applications of genomics continue to drive our understanding and characterization of germplasm collections, gene function, phenotypic expression and adaptability, the identification of agronomically favorable alleles, and the creation of improved cultivars. Expanding our knowledge of the genetic basis of agronomic traits will help breeders to more efficiently explore and harness plant genetic resources. It is becoming clear that criteria based solely on plant morphology and phenotype are insufficient or even not applicable for the determination of EDVs. Molecular markers should help quantify the differences between similar and/or related plant varieties, increase the certainty of essential plant derivation, and assuage the concerns with respect to the scope of IP protection afforded by the concept of EDV. Accordingly, it would be reasonable to expect that the use of molecular markers in EDV determination and EDV-related disputes will increase in the future.

Crop production plays a key role in our society's needed and expanding sustainability goals. The development of new plant varieties is not only a scientific endeavor, but very much a social, political, and economic one as well. As well, IP rights regimes are established to achieve societal goals with a broad impact. The ability to obtain adequate protection of newly developed plant varieties as IP can encourage investments into agricultural biotechnology and specifically plant breeding. It is thus of paramount importance that innovations in plant breeding receive the appropriate innovation incentives. The advent of new molecular breeding techniques has reignited discussions about the scope of afforded plant breeders' rights. This might require modernization and redefining of the concept and definition of EDVs [22]. Technological advances that facilitate the simultaneous introduction of several transgenes might push the boundaries and fundamentally challenge the concept of essential derivation; perhaps a breeder exception should be formulated on a crop-by-crop basis [18]. Decisions about plant breeders' rights should be considered in conjunction with broader policy support, to ensure to the coexistence of a responsive public agricultural research system with a diverse and competitive private seed sector. They also depend on a well-functioning legal system with enforceable contracts between plant growers and merchants, the promotion of responsible business practices, and the encouragement of engagement of professional associations in agribusiness [23].

Refinements of the used molecular techniques should be expected in the future. To increase sensitivity and accuracy, simultaneous use of various and/or multiple types of markers/assays could be performed. It will be beneficial for the community of plant breeders to reach a consensus on the use of (sub)sets of plant species-specific markers that would be deemed necessary and sufficient for determining genetic resources [4]. It will also be important to establish a minimum level of relative marker profile similarity as a threshold for distinctness. Such a threshold must be based upon the use of a set of markers (e.g., SNPs) that are identified as meeting the criteria of (i) collectively providing for fairly even genome coverage, and (ii) having a proven ability to discriminate among varieties, even among those that are very closely related by pedigree [8].

Molecular genetic markers are ever-improving tools that are making an impact in the ongoing debate on the scope of IP rights protection for EDVs. They introduce a higher degree of scientific certainty. This is good because critics argue that protection for EDVs brings uncertainty in assessing what is an EDV, which then becomes problematic for the plant-breeding businesses globally. Breeders need simple rules, which are easy to understand and enforce. Some have even argued that small farmers in the developing world are in an inferior position, as they are not capable of making such comparisons and assessments [22]. Furthermore, the technological and economic environment in which plant breeders operate will likely continue to change and will become more complex. In that context, these changes may well cause plant breeders and policymakers to seek to further adjust the *sui generis* IP system for plants (and particularly for EDVs), which system can optimally contribute to—and complement—other forms of IP protection for plants [3].

Exciting days lie ahead for plant breeders.

Author's contribution

Bratislav Stankovic contributed to this article in his personal capacity. The views and comments expressed herein are solely the opinion of the author, do not reflect the performance of duties in the author's official capacity, and are not endorsed by, nor should be construed as, any viewpoint official or unofficial of the United States Patent and Trademark Office. The author confirms to the best of his knowledge that no information contained herein is privileged, confidential, or classified.

Author details

Bratislav Stankovic
United States Patent and Trademark Office, Alexandria, VA, USA

*Address all correspondence to: bratislav.stankovic@fulbrightmail.org

References

[1] UPOV. 1991. Available from: https://www.upov.int/portal/index.html. [Accessed: 02 May 2022]

[2] Sanderson J. Essential derivation, law, and the limits of science. Law in Context: A Socio-Legal Journal. 2006;**24**:34-53

[3] Smith JSC, Jones ES, Nelson BK. The use of molecular marker data to assist in the determination of essentially derived varieties. In: Tuberosa R, Graner A, Frison E, editors. Genomics of Plant Genetic Resources. Dordrecht: Springer; 2014;**1**:49-65

[4] Noli E, Soccorsa Teriaca M, Conti S. Criteria for the definition of similarity thresholds for identifying essentially derived varieties. Plant Breeding. 2013;**132**:525-531. DOI: 10.1007/s11032-011-9582-x

[5] Heckenberger M, Muminović J, van der Voort JR, Peleman J, Bohn M, Melchinger AE. Identification of essentially derived varieties obtained from biparental crosses of homozygous lines. III. AFLP data from maize inbreds and comparison with SSR data. Molecular Breeding. 2006;**17**:111-125. DOI: 10.1007/s11032-005-3851-5

[6] Sell SK. Corporations, seeds, and intellectual property rights governance. In: Clapp J, Fuchs D, editors. Corporate Power in Global Agrifood Governance. Cambridge, Massachusetts-London: MIT Press; 2009. pp. 189-191

[7] Kock MA. Essentially derived varieties in view of new breeding technologies—Plant breeders' rights at a crossroads. GRUR International. 2021;**70**:11-27. DOI: 10.1093/grurint/ikaa156

[8] De Riek J, Dendauw J, Leus L, De Loose M, Van Bockstaele E. Variety protection by use of molecular markers: Some case studies on ornamentals. Plant Biosystems. 2001;**135**:107-113. DOI: 10.1080/11263500112331350720

[9] Smith JSC, Jones ES, Nelson BK, Phillips DS, Wineland RA. Genomic approaches and intellectual property protection for variety release: A perspective from the private sector. In: Tuberosa R, Graner A, Frison E, editors. Genomics of Plant Genetic Resources. Dordrecht: Springer; 2014;**1**:27-47

[10] International Seed Federation. Technical protocol for assessment of genetic distance for ISF guidelines for the handling of a dispute on EDV in diploid perennial ryegrass, ISF Secretariat. pp. 1-5. Available from: https://www.worldseed.org. [Accessed: 02 May 2022]

[11] International Seed Federation. Annex 2—List of SNPs Markers, ISF Secretariat. pp 1-13. Available from: https://www.worldseed.org. [Accessed: 02 May 2022]

[12] International Seed Federation. Issues to be Addressed by Technical Experts to Define Molecular Marker Sets for Establishing Thresholds for ISF EDV Arbitration, ISF Secretariat. pp 1-2. Available from: https://www.worldseed.org. [Accessed: 02 May 2022]

[13] Bernardo R, Kahler AL. North American Study on essential derivation in maize: Inbreds developed without and with selection from F_2 populations. Theoretical and Applied Genetics. 2001;**102**:986-992. DOI: 10.1007/s001220000479

[14] Kahler AL, Kahler JL, Thompson SA, Ferriss RS, Jones ES, et al. North American Study on essential

derivation in maize: II. Selection and evaluation of a panel of simple sequence repeat loci. Crop Science. 2010;**50**:486-503. DOI: 10.2135/CROPSCI2009.03.0121

[15] Rousselle Y, Jones E, Charcosset A, Moreau P, Robbins K, et al. Study on essential derivation in maize: III. Selection and evaluation of a panel of single nucleotide polymorphism loci for use in European and North American Germplasm. Crop Science. 2015;**55**:1170-1180. DOI: 10.2135/CROPSCI2014.09.0627

[16] Jamali SH, Cockram J, Hickey LT. Insights into deployment of DNA markers in plant variety protection and registration. Theoretical and Applied Genetics. 2019;**132**:1911-1929. DOI: 10.1007/s00122-019-03348-7

[17] Borchert T, Krueger J, Hohe A. Implementation of a model for identifying essentially derived varieties in vegetatively propagated *Calluna vulgaris* varieties. BMC Genetics. 2008;**9**(56):1-10. DOI: 10.1186/1471-2156-9-56

[18] Smith JSC. The future of essentially derived variety (EDV) status: Predominantly more explanations or essential change. Agronomy. 2021;**11**:1261. DOI: 10.3390/agronomy11061261

[19] Van Eeuwijk FA, Law JR. Statistical aspects of essential derivation, with illustrations based on lettuce and barley. Euphytica. 2004;**137**:129-137. DOI: 10.1023/B:EUPH.0000040510.31827.ae

[20] Saccomanno B, Wallace M, O'Sullivan DM, Cockram J. Use of genetic markers for the detection of off-types for DUS phenotypic traits in the inbreeding crop, barley. Molecular Breeding. 2020;**40**(13):1-10. DOI: 10.1007/s11032-019-1088-y

[21] Chakrabarty SK, Choudhury DR. DUS testing for plant variety protection: Some researchable issues. Indian Journal of Genetics. 2019;**79**(Suppl. 1):320-325

[22] Bostyn SJ. Plant variety rights protection and essentially derived varieties: A fresh proposal to Untie the Gordian Knot. GRUR International. 2020;**69**:785-802. DOI: 10.1093/grurint/ikaa094

[23] Wani SH, Devi TR, Sinhg NB, Harinhushan A, Khan H. Intellectual property rights system in plant breeding. Journal of Plant Science and Research. 2013;**29**:112-122

Chapter 4

Speed Breeding: A Propitious Technique for Accelerated Crop Improvement

Priyanka Shanmugavel, Gowtham Ramasamy, Geethalakshmi Vellingiri, Rajavel Marimuthu and Kalaimagal Thiyagarajan

Abstract

Development of climate-resilient genotypes with high agronomic value through conventional breeding consumes longer time duration. Speed breeding strategy involves rapid generation advancement that results in faster release of superior varieties. In this approach, the experimental crop is grown in a controlled environment (growth chambers) with manipulation provisions for temperature, photoperiod, light intensity, and moisture. The generation of the crop cycle can be hastened by inducing changes in the physiological process such as photosynthesis rate, flowering initiation, and duration. Speed breeding eases multiple trait improvement in a shorter span by integration of high-throughput phenotyping techniques with genotype platforms. The crop breeding cycle is also shortened by the implementation of selection methods such as single-seed descent, single plant selection, and marker-assisted selection.

Keywords: accelerated breeding, controlled environment, crop Improvement, rapid generation advancement, speed breeding

1. Introduction

The increase in world population coupled with climatic fluctuations such as drought, flood, and high temperature poses a serious threat to food security [1]. Many researchers quoted the importance of enhancing the genetic gain of primary crops at a faster rate to meet the global food demands [2]. It remains a challenging task for plant breeders to evolve resilient varieties in a shorter period by employing conventional approaches. The slow progress in crop improvement is mainly attributed to long breeding cycles/generation [3]. To overcome the drawbacks involved in traditional methods and to safeguard food security, speed breeding concepts are now being adopted at large/small units for realizing a rapid genetic gain in many crop species.

The speed breeding techniques include the use of controlled environments with manipulation provisions for the light duration, intensity, and temperature. This serves as more advantageous for the plant breeder to hasten the crop development

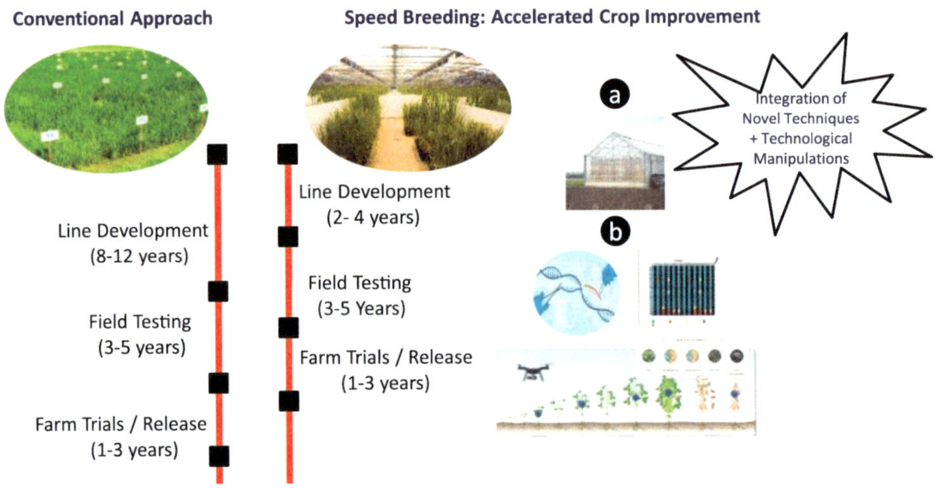

Figure 1.
Rapid generation advancement through speed breeding. a. Experimental crop grown under controlled environments. b. Use of high-throughput genotyping platforms; advanced phenotyping tools and other modern breeding techniques in speed breeding protocol.

in several major photosensitive crops [4]. The concept of stimulating an artificial environment for plant growth was first initiated by a team of botanists several years ago. Around 1980, similar protocols were again adopted by scientists of National Aeronautics and Space Administration (NASA) in collaboration with Utah State University to understand the accelerated crop growth cycle under constant light in the space station [5]. As an outcome, a new dwarf variety USU-Apogee was released by NASA in wheat [6]. In earlier crop improvement programs, the breeders employed few manipulations in conventional approaches such as the single-seed descent method [7], shuttle breeding [8], and haploid technique for rapid delivery of superior varieties. These were upgraded and combined with the use of other innovative technologies under the term speed breeding. Scientists achieved rapid generation advancement through the adoption of novel techniques such as marker-assisted selection, *in vitro* culture, high-throughput phenotyping, next-generation sequencing, genomic selection, and gene editing in the speed breeding protocols [9]. The speed breeding concept was first employed in *Triticum aestivum* (wheat) to investigate the seed dormancy trait under controlled conditions [10]. At present, speed breeding protocols are widely employed in several crops, including underutilized species [11]. Around six generations per year have been achieved in crops such as oat [12], barley [13], wheat [14], chickpea [15], faba bean, and lentil [16] through the implementation of speed breeding techniques. Speed breeding protocols allow for the integration of new techniques along with several manipulations in influencing factors (**Figure 1**), which have been briefly discussed in this chapter.

2. Speed breeding techniques

2.1 Crops under controlled environment

Speed breeding techniques involve deliberate manipulation of environmental conditions for the rapid advancement of crop cycle. The use of controlled growth chambers equipped with manipulation provisions for light intensity, temperature regime, photoperiod, soil moisture, carbon dioxide level and nutrition supply will influence/alter the plant physiological growth process [17]. Researchers employ these modifications in a crop improvement program to achieve increased generation per year. The early flowering was induced in IR 64 rice variety by altering the light exposure in the growth chamber [18]. Similarly, a photoperiod of 22 hours of light exposure reduced flowering duration in wheat genotypes [19]. The breeding strategy can be efficiently planned with photosensitive crops through the adoption of light-based speed breeding protocols. The quality of light delivered per day highly influences the photosynthesis rate, gas exchange, transpiration rate, stomatal activity, and other plant developmental processes [20]. Adoption of 360–650 μmol/m^2 light intensity with 400–700 nm of PAR (photosynthetic active radiation) was found successful in barley, wheat, chickpea, canola, and other major crops for early flowering and seed set [15]. The induction of early flowering was observed in legumes such as chickpea, faba beans, and pea with the use of blue and far-red light spectrums [21]. Early flowering was induced in groundnut by continuous exposure (24 hours) of 450 W lamps 25 days after germination [22].

The temperature variation plays a crucial role in the transition from vegetative to flowering stage in crop plants [23]. It influences the seed germination rate, plant growth, flowering period, seed set per cent, and maturity [24]. A temperature range of 12–30°C for germination and 25–30°C for other developmental processes (growth, flowering, and seed formation) is found suitable for most of the species [25]. Rapid plant development is observed on introducing the crop to altered temperature regime (17°C/32°C) and photoperiod in groundnut [22].

A shift from vegetative to reproductive phase is reported in crop plants at increased CO_2 levels [25]. Plants' response to CO_2 levels highly varies with the genotype of a species. The experimental genotype has to be evaluated with a critical range of CO_2 levels in growth chambers to determine the optimum value for induction of earliness in flowering. The breeding cycle was enhanced up to five generations per year in soybean by manipulating CO_2 supply (> 400 ppm) coupled with light exposure of 14 hours cycle in a growth chamber [26].

Most crop species exhibit early flowering and seed set on subjecting to moisture stress [27]. Modulation of soil moisture status in speed breeding protocol helps in rapid generation advancement of crop species. The high induction of grain filling and maturation is observed in barley, wheat, and chickpea on the gradual decrease of moisture status at the end of the flowering stage [15].

High-density planting is a low-cost strategy in speed breeding as it contributes to rapid generation turnover along with the maintenance of large population size. Crops raised at high density tend to compete with each other resulting in early induction of flowering and seed maturity [28]. The earliness in flowering at high density was reported in rice, sorghum, and cotton [25]. On contrary, many researchers found no deviations in flowering initiation at high-density planting [29]. Therefore, the genotypic responses need to be investigated in each species to validate the use of high-density planting as a component in speed breeding.

Application of plant growth hormones and essential nutrients tend to regulate flowering and seed set under *in vitro* conditions [30]. More breeding cycles per year can be generated through the use of growth regulators with other approaches. Around eight generations per year were obtained in lentil and faba bean with the use of plant growth regulators *viz*., auxin, cytokinin, and flurprimidol under modified temperature (22°C light/18°C dark) and photoperiod (18 hr. light/6 hr. dark) in growth chambers [31].

The immature seeds obtained from plants grown under speed breeding protocols with an extended duration of photoperiod (22 h of light) proved to be viable in wheat and barley [15]. A similar finding on early seed harvest was reported in wheat cultivars [32]. The advancement of subsequent generations can be hastened by the adoption of early harvest with other speed breeding techniques. The immature seeds (37 days after postanthesis) from plants grown under CO_2 supplementation exhibited a high germination rate similar to control in soybean [26]. Around 7–8 generations/year is achieved in lentils by integrating early harvest with the application of plant growth regulators [16].

2.2 Accelerated crop improvement through integration of novel approaches

Speed breeding is a feasible platform that allows the integration of modern approaches along with generation advancement techniques. The conventional breeding techniques (pedigree selection, mass selection, pure line selection, bulk selection, and recurrent selection) of line development require more number of inbreeding and selection processes. These methods were not found amenable for inclusion in speed breeding protocols [25]. The use of modern techniques coupled with high-throughput phenotyping platforms in speed breeding would highly augment the crop improvement program. The target-specific traits involved in biotic and abiotic stress can be improved at a faster rate by creating artificial environments with accurate phenotyping.

Few modifications in conventional selection methods proved efficient for inclusion in speed breeding protocol. The single plant selection method was employed in the handling of backcross progeny at earlier generations (F_2 and F_3). A rigid selection for the trait under transfer and characteristics of the recurrent parent was made in segregating generations (F_2 and F_3) after the first and third backcross. Each F_2 selected plant was harvested separately for the advancement of generation (F_3) following the progeny-row method. The inclusion of selection in the early generation reduced the number of backcrosses and thereby saves labor, time, and other resources. The modified backcross method was employed in barley for the rapid development of introgression lines [33]. The European barley cultivar (Scarlett) was crossed with other donor parents to evolve lines exhibiting resistance to blotch and leaf rust. The lines under evaluation were raised under growth chambers with continuous light exposure at 22°C. Similarly, the single plant selection in combination with the speed breeding protocol was followed in wheat for multiple trait improvement [34].

Single-seed descent serves as a promising selection approach for inclusion in speed breeding techniques in field and controlled environments. The attainment of homozygosity is accelerated through constant inbreeding of segregating population by forwarding a single seed of each individual to the next generation. It allows for the advancement of generations in growth chambers and small nursery fields [35]. The single-seed descent method provides the opportunity for high-density planting and proves to be a very effective strategy for resource-limited environments [36]. The popular rice cv. Nipponbare was developed by adopting a single-seed descent method with rapid generation techniques at growth chambers [37]. Around 450 inbred lines evolved rapidly under field conditions following the single-seed descent method in wheat [38]. No selection is imposed in any successive generation which may carry more inferior progenies in a population compared to other selection methods.

A slight deviation from the single-seed descent method was found successful in legume species. The selection of one pod per plant was followed from F_2 to F_4 generation instead of a single seed. Single-pod descent selection provides scope for maintaining each F_2 line in advanced generations compared to the single-seed descent method. It also possesses the advantage of early selection of pods, which is not feasible in the single-seed descent method. The mean yield of progenies developed from single-pod descent (7.96 g / plant) was higher compared to the single-seed descent (6.42 g/plant) selection method in soybean [39]. However, the conduct of preliminary test trials under controlled environments is required to validate the selection efficiency of the single-pod descent method in legume crops [25].

The precise identification of candidate genes has become feasible due to recent advancements in genotypic platforms and high-throughput phenotyping techniques. The development of mapping population (F_2, recombinant inbred line (RIL), and backcross) requires a longer generation time on conventional approaches. The inclusion of the speed breeding technique promotes rapid identification and validation of QTL (quantitative trait loci) [21]. It facilitates minimal backcross (1–2) to introgress the target gene in a superior genotype (over 99% of the recurrent genome). The use of marker-assisted selection (MAS) in speed breeding protocol facilitates gene discovery at a faster rate and thereby meets the challenges associated with food production. The SNP marker-assisted selection is combined with speed breeding protocols for rapid development of mapping population (BC_3F_3) associated with salinity tolerance in rice [40].

The marker-assisted selection is efficient only with a few QTLs exhibiting a major effect on the trait of interest. At present, researchers employ a genomic selection approach in the breeding strategies, which is effective for complex trait improvement. It paves way for the identification of several minor QTLs, which is involved in the governance of biotic and abiotic stress resistance. With the development of next-generation sequencing (NGS) technologies, the cost and time involved in genomic selection are drastically reduced [41]. The genomic-estimated breeding values (GEBVs) of individuals are estimated based on genotype and phenotype datasets of a training population. It results in high accuracy of measuring the genetic worth of an individual compared to other selection methods [42]. The rapid genetic gain was realized in wheat through the implementation of genomic selection with other speed breeding protocols [43]. Several haplotypes related to yield improvement have been identified in rice and many other species. Introgression of haplotype into superior cultivars requires more breeding cycles and is highly time-consuming. The haplotype breeding can be accelerated by the integration of speed breeding protocols with the genomic selection approach [9]. Speed breeding also serves as a promising strategy for the rapid advancement of generations in transgenic crops [44].

3. Challenges in adoption of speed breeding protocols

The use of speed breeding techniques for crop improvement demands high infrastructure equipped with control facilities for temperature, photoperiod, humidity, and other factors. It requires the need for expertise/skilled technicians for the maintenance of experimental crops in controlled conditions [45]. Lack of modern tools/techniques in underdeveloped countries, lack of continuous financial assistance, and unsupportive policies add up the concern toward adoption of speed breeding protocol in practice. Many experimental fields have reduced access toward a continuous supply of electricity. The use of energy-efficient LED bulbs and air conditioners under solar power with battery support may help to some extent for small infrastructures. The limited number of crosses and population is maintained under speed breeding due to high input and maintenance costs for infrastructure. Integrated research employing scientists from different organizations is needed to avoid duplications of work, minimize investments on resources, and help in support/sharing of specialized equipment.

4. Conclusion

The adoption of speed breeding protocols in crop improvement programs will hasten the breeding cycle to a great extent with improved selection efficiency. It promotes the rapid delivery of resilient varieties by integrating modern breeding techniques with generation advancement protocols. The superior genotype with improvement over multiple traits such as yield, quality, biotic, and abiotic stress resistance can be developed at a minimal period with the inclusion of high-throughput genotyping and phenotyping platforms in speed breeding. Many superior varieties have been rapidly developed in economically important species through the exploitation of speed breeding techniques. The inclusion of genomic selection approaches in speed breeding paved the way for the improvement of complex traits governing resistance. Few modified conventional approaches *viz.*, single plant selection, single-pod descent, and single-seed descent are included in speed breeding protocols which greatly reduced the limitations of long generation time, cost, and labor. The evolution of advanced genomic techniques coupled with rapid gene fixation approaches offers faster realization of genetic gain in crop breeding programs. In addition to accelerated progression toward the attainment of homozygosity, the speed breeding protocols also prove efficient in the rapid evaluation of genetically modified/transformed lines of a crop species. The standardized speed breeding protocols suitable for small environments are now available with modification provisions to meet the local needs. However, it still remains a less adopted choice in many developing countries due to cost-expensive infrastructure development, lack of trained professionals, unsupportive policies, no proper financial support from the public domain and lack of essential resources. With the coordination of multidisciplinary organization, speed breeding becomes an efficient tool to meet ever-challenging food demand under changing climatic conditions.

Acknowledgements

The authors are highly thankful to Dr. M. Raveendran, Director of Research, Tamil Nadu Agricultural University (TNAU), for his valuable suggestions toward this chapter. We also acknowledge Dr. R. Sudhagar, Associate Professor and Head, Sugarcane Research Station, TNAU, and Dr. K. Ganesamurthy, Professor and Head (Retd.), Department of Rice, TNAU, for rendering supportive documents.

Conflict of interest

The authors declare no conflict of interest in this chapter.

Thanks

The authors express their sincere gratitude to the Department of Plant Breeding and Genetics, Tamil Nadu Agricultural University, for providing scientific assistance on speed breeding techniques.

Author details

Priyanka Shanmugavel[1]*, Gowtham Ramasamy[2], Geethalakshmi Vellingiri[3], Rajavel Marimuthu[3] and Kalaimagal Thiyagarajan[4]

1 Department of Plant Breeding and Genetics, Tamil Nadu Agricultural University (TNAU), Coimbatore, India

2 Agro Climate Research Centre, Tamil Nadu Agricultural University, Coimbatore, India

3 Tamil Nadu Agricultural University, Coimbatore, India

4 Department of Millets, Tamil Nadu Agricultural University, Coimbatore, India

*Address all correspondence to: priya300593@gmail.com

References

[1] Ray DK, Mueller ND, West PC, Foley JA. Yield trends are insufficient to double global crop production by 2050. PlOS One. 2013;**1**:8

[2] Lin Z, Cogan NOI, Pembleton LW, Spangenberg GC, Forster JW, Hayes BJ, et al. Genetic gain and inbreeding from genomic selection in a simulated commercial breeding program for perennial ryegrass. Plant Genome. 2016;**9**:1-12

[3] Samantara K, Bohra A, Mohapatra SR, Prihatini R, Asibe F, Singh L, et al. Breeding more crops in less time: A perspective on speed breeding. Biology. 2022;**11**(2):275

[4] Singh H, Janeja HS. Speed breeding: A ray of hope for the future generation in terms of food security. European Journal of Molecular and Clinical Medicine. 2021;**8**(2):2653-2658

[5] Pfeiffer NE. Microchemical and morphological studies of effect of light on plants. Botanical Gazette. 1926;**81**(2):173-195

[6] Bugbee B, Koerner G. Yield comparisons and unique characteristics of the dwarf wheat cultivar 'USU-Apogee'. Advances in Space Research. 1997;**20**(10):1891-1894

[7] Brim CA. A modified pedigree method of selection in soybeans 1. Crop Science. 1966;**6**:220

[8] Borlaug N. Wheat Breeding and Its Impact on World Food Supply. Canberra, Australia: Australian Academy of Sciences; 1968. pp. 1-36

[9] Sharma S, Kumar A, Dhakte P, Raturi G, Vishwakarma G, Barbadikar KM, et al. Speed breeding opportunities and challenges for crop improvement. Journal of Plant Growth Regulation. 2022;**15**:1-4

[10] Hickey LT, Dieters MJ, DeLacy IH, Kravchuk OY, Mares DJ, Banks PM. Grain dormancy in fixed lines of white-grained wheat (*Triticum aestivum* L.) grown under controlled environmental conditions. Euphytica. 2009;**168**(3):303-310

[11] Chiurugwi T, Kemp S, Powell W, Hickey LT. Speed breeding orphan crops. Theoretical and Applied Genetics. 2019;**132**(3):607-616. DOI: 10.1007/s00122-018-3202-7

[12] Liu H, Zwer P, Wang H, Liu C, Lu Z, Wang Y, et al. A fast generation cycling system for oat and triticale breeding. Plant Breeding. 2016;**135**:574-579

[13] Hickey LT, Germán SE, Pereyra SA, Diaz JE, Ziems LA, Fowler RA, et al. Speed breeding for multiple disease resistance in barley. Euphytica. 2017;**213**(3):1-4. DOI: 10.1007/s10681-016-1803-2

[14] Mukade K. New procedures for accelerating generation advancement in wheat breeding. JARQ. 1974;**8**:1-5

[15] Watson A, Ghosh S, Williams MJ, Cuddy WS, Simmonds J, Rey MD, et al. Speed breeding is a powerful tool to accelerate crop research and breeding. Nature Plants. 2018;**4**(1):23-29. DOI: 10.1038/s41477-017-0083-8

[16] Mobini SH, Lulsdorf M, Warkentin T, Vandenberg A. Plant growth regulators improve *in vitro* flowering and rapid generation advancement in lentil and faba bean.

Invitro cellular and developmental biology. Plant. 2014;**51**:71-79

[17] Erskine W, Smartt J, Muehlbauer FJ. Mimicry of lentil and the domestication of common vetch and grass pea. Economic Botany. 1994;**48**(3):326-332

[18] Köhl K. Growing rice in controlled environments. Annals of Applied Biology. 2015;**167**(2):157-177

[19] Dubcovsky J, Loukoianov A, Fu D, Valarik M, Sanchez A, Yan L. Effect of photoperiod on the regulation of wheat vernalization genes VRN1 and VRN2. Plant Molecular Biology. 2006;**60**(4):469-480. DOI: 10.1007/s11103-005-4814-2

[20] Yang L, Wang D, Xu Y, Zhao H, Wang L, Cao X, et al. A new resistance gene against potato late blight originating from *Solanum pinnatisectum* located on potato chromosome 7. Frontiers in Plant Science. 2017;**8**:1729. DOI: 10.3389/fpls.2017.01729

[21] Ghosh S, Watson A, Gonzalez-Navarro OE, Ramirez-Gonzalez RH, Yanes L, Mendoza-Suárez M, et al. Speed breeding in growth chambers and glasshouses for crop breeding and model plant research. Nature Protocols. 2018;**13**(12):2944-2963. DOI: 10.1038/s41596-018-0072-z

[22] O'Connor DJ, Wright GC, Dieters MJ, George DL, Hunter MN, Tatnell JR, et al. Development and application of speed breeding technologies in a commercial peanut breeding program. Peanut Science. 2013;**40**(2):107-114. DOI: 10.3146/PS12-12.1

[23] Hatfield JL, Prueger JH. Temperature extremes: Effect on plant growth and development. Weather and Climate Extremes. 2015;**10**:4-10. DOI: 10.1016/j.wace.2015.08.001

[24] McClung CR, Lou P, Hermand V, Kim JA. The importance of ambient temperature to growth and the induction of flowering. Frontiers in Plant Science. 2016;**7**:1266. DOI: 10.3389/fpls.2016.01266

[25] Wanga MA, Shimelis H, Mashilo J, Laing MD. Opportunities and challenges of speed breeding: A review. Plant Breeding. 2021;**140**(2):185-194

[26] Nagatoshi Y, Fujita Y. Accelerating soybean breeding in a CO_2-supplemented growth chamber. Plant and Cell Physiology. 2019;**60**(1):77-84. DOI: 10.1093/pcp/pcy189

[27] Shavrukov Y, Kurishbayev A, Jatayev S, Shvidchenko V, Zotova L, Koekemoer F, et al. Early flowering as a drought escape mechanism in plants: How can it aid wheat production? Frontiers in Plant Science. 2017;**8**:1950

[28] Warnasooriya SN, Brutnell TP. Enhancing the productivity of grasses under high-density planting by engineering light responses: From model systems to feedstocks. Journal of Experimental Botany. 2014;**65**(11):2825-2834. DOI: 10.1093/jxb/eru221

[29] Fukushima A, Shiratsuchi H, Yamaguchi H, Fukuda A. Effects of nitrogen application and planting density on morphological traits, dry matter production and yield of large grain type rice variety bekoaoba and strategies for super high-yielding rice in the tohoku region of Japan. Plant Production Science. 2011;**14**(1):56-63. DOI: 10.1626/pps.14.56

[30] Bermejo C, Gatti I, Cointry E. In vitro embryo culture to shorten the breeding cycle in lentil (*Lens culinaris* Medik). PCTOC. 2016;**127**(3):585-590

[31] Mobini SH, Warkentin TD. A simple and efficient method of

in vivo rapid generation technology in pea (*Pisum sativum* L.) vitro cellular and developmental biology. Plant. 2016;**52**(5):530-536. DOI: 10.1007/s11627-016-9772-7

[32] Ferrie AM, Polowick PL. Acceleration of the breeding program for winter wheat. In: Accelerated Plant Breeding. Berlin: Springer; 2020. pp. 191-215

[33] Hickey LT, Dieters MJ, DeLacy IH, Christopher MJ, Kravchuk OY, Banks PM. Screening for grain dormancy in segregating generations of dormant × non-dormant crosses in white-grained wheat (*Triticum aestivum* L.). Euphytica. 2010;**172**(2):183-195. DOI: 10.1007/s10681-009-0028-z

[34] Alahmad S, Dinglasan E, Leung KM, Riaz A, Derbal N, Voss-Fels KP, et al. Speed breeding for multiple quantitative traits in durum wheat. Plant Methods. 2018;**14**(1):1-5. DOI: 10.1186/s13007-018-0302-y

[35] Arbelaez JD, Tandayu E, Reveche MY, Jarana A, van Rogen P, Sandager L, et al. Methodology: ssb-MASS: A single seed-based sampling strategy for marker-assisted selection in rice. Plant Methods. 2019;**15**(1):1-11. DOI: 10.1186/s13007-019-0464-2

[36] Funada M, Helms TC, Hammond JJ, Hossain K, Doetkott C. Single-seed descent, single-pod, and bulk sampling methods for soybean. Euphytica. 2013;**192**(2):217-226. DOI: 10.1007/s10681-012-0837-3

[37] Tanaka J, Hayashi T, Iwata H. A practical, rapid generation-advancement system for rice breeding using simplified biotron breeding system. Breeding Science. 2016;**2016**:15038. DOI: 10.1270/jsbbs.15038

[38] Pignone D, De Paola D, Rapanà N, Janni M. Single seed descent: A tool to exploit durum wheat (*Triticum durum* Desf.) genetic resources. Genetic Resources and Crop Evolution. 2015;**62**(7):1029-1035. DOI: 10.1007/s10722-014-0206-2

[39] Destro D, Bizeti HS, Garcia LA, Fonseca IC, Montalván R, Miglioranza É. Comparison between the SPD and the SPDS methods for segregating generation advancement in soybean. Brazilian Archives of Biology and Technology. 2003;**46**(4):545-551. DOI: 10.1590/S1516-89132003000400008

[40] Rana MM, Takamatsu T, Baslam M, Kaneko K, Itoh K, Harada N, et al. Salt tolerance improvement in rice through efficient SNP marker assisted selection coupled with speed-breeding. International Journal of Molecular Science. 2019;**20**(10):2585

[41] Poland JA, Endelman J, Dawson J, Rutkoski J, Wu S, Manes Y, et al. Genomic selection in wheat breeding using genotyping-by-sequencing. Plant Genome. 2012;**5**(3):103-113

[42] Crossa J, Pérez-Rodríguez P, Cuevas J, Montesinos-López O, Jarquín D, de los Campos G, et al. Genomic selection in plant breeding: Methods, models, and perspectives. Trends in Plant Science. 2017;**22**(11):961-975

[43] Voss-Fels KP, Herzog E, Dreisigacker S, Sukumaran S, Watson A, Frisch M, et al. "SpeedGS" to accelerate genetic gain in spring wheat. In: Miedaner T, Korzun V, editors. Applications of Genetic and Genomic Research in Cereals. Cambridge: Woodhead Publishing; 2019. pp. 303-327

[44] Flachowsky H, Le Roux PM, Peil A, Patocchi A, Richter K, Hanke MV. Application of a high-speed breeding technology to apple (Malus × domestica) based on transgenic early flowering

plants and marker-assisted selection. New Phytologist. 2011;**192**(2):364-377

[45] Shimelis H, Gwata ET, Laing MD. Crop improvement for agricultural transformation in Southern Africa. In: Sikora RA, Terry ER, Vlek PLG, Chitja J, editors. Transforming Agriculture in Southern Africa. New York: Routledge; 2019. pp. 97-103

Chapter 5

Non-Coding RNA and Its Prospective Utilization in Plant Breeding

Debadatta Panda, Latha Ananda Lekshmi, Rachel Lissy Vargheese, Nallathambi Premalatha, Mahadevan Kumar and Lakshmanan Mahalingam

Abstract

Non-coding RNA molecules are generally present in a dispersed manner throughout the genome. They may behave as long ncRNAs or convert into small RNAs of around 20–24 nts that are universally categorized using their size, function, or chromosomal position. ncRNAs are thought to play a vital role in regulating and modulating gene expression apart from their prospective role in several epigenetic mechanisms controlling specificity in biochemical pathways and phenotype development in clonal cells. They are also part of the natural defense system against viruses. ncRNA modulates genes by transcriptional and translational control of growth, development, and stress response alongside other RNA molecules. Some modes of action have unraveled in recent years. A lot more needs to be pondered upon for comprehending their involvement in the extremely intricate processes in a more wholesome manner. In this chapter, we will discuss the different ncRNA, their origin, classification, and their role in various physiological processes. Practical examples of the discovery of ncRNA in different crops and their functions have also been elucidated with the required details. The yield and quality enhancement, along with the better stress response being the aim of the crop improvement program, the prospective utilities of ncRNA are also explained in the subsequent part of the chapter.

Keywords: ncRNA, Epigentics, gene regulation, stress response, plant breeding

1. Introduction

RNA is the ribonucleic acid, one of the earliest formed molecules to shape life on earth. Being single-stranded, these are known to have a shorter life span and integrity compared to DNA. But, from other angles, these properties make RNA a magical molecule with a unique ability to work inside and outside the nucleus, leading to diverse sorts of roles in structural to regulatory aspects. In the case of the prokaryotes, most of the transcribed RNA is translated because of the smaller size of the genome and the variety and load of work endowed on it for the organism's survivability. But, in the case of the eukaryotes, as the genome size gets bigger, the actual coding part of the chromosomes becomes sparse and rather scattered, mostly. Secondly, the coding genes occupy only 1–2% of the genome. A very high percentage of eukaryotic genomes around up to 90% undergo transcription to produce RNA, nevertheless only a little portion of transcripts get actually translated into proteins [1].

Out of the total RNAs, non-coding RNA (ncRNA) pertains to active and functional RNA molecules that are not translated into proteins, although being transcribed from DNA. They encompass a wide range of RNA molecules having the potential to play in the regulation of gene expression [1]. They can act as long ncRNAs or be converted into s very smaller size RNA molecule proteins in eukaryotic cells. Mostly ncRNA is categorized worldwide based on its size, function, or genetic origin [1].

Based on the size of the molecule, their origin, functionality, and ncRNA are sorted into either the long non-coding RNAs (lncRNAs) having nucleotide lengths of more than 200 nucleotides or the small RNAs (sRNAs), which are comparatively quite smaller in size. Again the sRNAs are further classified into different types of RNA, such as miRNA, siRNA, piRNA, which will be disused further in detail with the various aspects of their origin, features, functions, and other aspects in the later section of the chapter.

ncRNAs are responsible for a wide range of biological functions. They control gene expression at the transcriptional, RNA processing, and translational levels by the particular structural aspects of RNA itself. Recent discoveries also show their role in various epigenetic phenomena affecting multiple physiological pathways and expression of particular phenotypes in a different situation. The fraction of the coding region of the protein genome varies considerably and is oppositely proportionate to the genome's size and complexity [2]. Many regulatory ncRNAs do have relatively high specificity of the target, whereas others serve as a major modulator of extensive regulatory signaling networks by acting across the genome [3]. They keep alien nucleic acids out of genomes and safeguard the genome against them. Non-coding RNAs target a single gene and pathways involving multiple genes at the genome level through distinct molecular mechanisms. Hence, these regulatory ncRNAs could be potential breeding targets for advanced breeding programs in plants [4]. They can regulate the synthesis of DNA and also take part in the reorganization of the genome. The biological activity of ribozymes and riboswitches is served by several ncRNAs that use the power of base pairing to interact with other nucleic acids preferentially [5].

Non-coding RNAs (ncRNAs), which act as a natural defense mechanism against—attacking viruses, have also been found as effectors in RNA-mediated gene silencing and hence now utilized in crop genetic modification [6]. The role of ncRNA has been observed in RNA interference and other regulatory mechanisms in plants; these provide a huge scope for the use of advanced molecular biology tools on these for enhancing the production potential of plants and modulation of growth and development of a plant to a certain extent. These have also been reported to influence the genes and biochemical pathways involving important traits like floral growth, maturation of seed, various biotic and abiotic stresses, along with pest and disease resistance processes. The ncRNA and their detailed structural to functional aspects are narrated meticulously in the upcoming sections of the chapter.

2. History of non-coding RNA

The beginning of the era of RNA dates back to the discovery of nucleic acids by Friedrich Miescher in the late 1860s. Later on, rigorous research was carried out about its structure, mode of action, and expression. Soon it became clear that majorly three kinds of RNA were involved in the protein production process, translation where mRNA was the carrier of genetic information ultimately translated into protein, and the process is assisted by tRNA and rRNA. Secondly, with the advent of newer technologies in molecular biology, several new RNA molecules appeared in the picture, out of which some had some role in the regulatory pathways or physiological pathways, for some other function is still unknown. Since, among all these only, mRNA codes for the protein, all other RNAs are termed non-coding RNA.

The first non-coding RNA to be discovered was the tRNA and its role in transferring amino acids was first observed by Paul C Zamecnik and Mahlon Hoagland in a cell-free system when RNA molecules were radioactively labeled [7]. Furthermore, it was the first non-coding RNA to be sequenced [8]. Later, in the early twenty-first century, many types of non-coding RNA, such as siRNA, miRNA, and piRNA, were discovered namely which had a role in gene regulation. During post-transcriptional gene silencing, a 25-nucleotide antisense RNA complementary to the target RNA was detected. This short interfering RNA, in virus-induced gene silencing, suppresses the production of viral proteins on binding with the target viral mRNA. This is a type of defense mechanism based on RNA against RNA and DNA viruses [9].

Caenorhabditis elegans is a completely sequenced nematode used as a model organism for many research programs. Out of its four larval stages, L1, L2, L3, and L4, it was found that the gene lin-4, the first miRNA discovered, was crucial for the transition of the larva from L1 to L2 [10].

H19 and Enod40 were the first eukaryotic lncRNA to be discovered [11]. The first plant lncRNA was discovered by Crespi in 1994. Long non-coding RNAs were first described during the whole genome sequencing and several types of lncRNA, such as Xist, Airn, MALATI, HOTAIR were discovered [12]. **Table 1** summarizes the different discovery events of the ncRNAs.

3. Classification of non-coding RNAs

The major category of grouping non-coding RNAs is based on their origin, nature of biogenesis, as well as based on its mechanism of action. The non-coding RNA transcripts can either perform housekeeping or regulatory functions. Those non-coding RNAs dynamically involved in cellular and ribosomal functions include tRNAs, snRNAs, rRNAs, and snoRNAs, whereas the regulatory ncRNAs are actively involved in most of the plant growth and development processes dealing with biotic and abiotic stress responses and plant immunity [16]. These regulatory ncRNAs, such as miRNAs, piRNAs, siRNAs, and lncRNAs, are transcribed from DNA and are primarily involved in transcriptional and post-transcriptional gene regulations [17]. The basic classification of ncRNA is presented in **Tables 2** and **3** described the brief details about the ncRNA. These regulatory ncRNAs cannot generally transcribe into proteins whereas housekeeping ncRNAs assist in protein translation [18].

S. No.	Type of ncRNA	Year of discovery	Scientists discovered	Remark
1.	tRNA	Before the 1960s	Paul C Zamecnik and Mahlon Hoagland	An adaptor molecule that mediates translation
2.	rRNA	Before 1965		Ribosomal RNA
3.	snRNA	1966		Small nuclear RNA
4.	snoRNA	1968		Small nucleolar RNA—U3 species—first snoRNA discovered
5.	siRNA	1999	Hamilton and Baulcombe	Small interfering RNA
	phasiRNA		Chen *et al.* [13]	
	tasiRNA	2004		Trans-acting siRNA
	easiRNA		Slotkin *et al.* [14]	Epigenetically-acting siRNA
	natsiRNA			
6.	miRNA	1993	Lee *et al.* [10]	Micro RNA
7.	piRNA	2006	Aravin *et al.* [15]	PIWI: P-element Induced WImpy testis in Drosophila
8.	lncRNA		Pachnis *et al.* [11]	H19, XIST, and HOTAIR

Table 1.
List of ncRNA and their brief history.

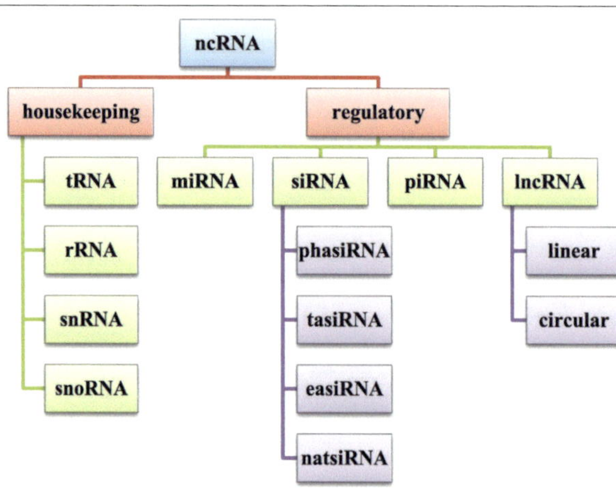

Table 2.
Classification of ncRNA.

4. Biogenesis of non-coding RNA

Basically, ncRNAs are consequence of various processes, such as the process of duplication, modification of transposons during the evolutionary process, pseudogenization of actual coding sequences, doubling of RNA viruses, part of some hairpin structures, double-stranded RNAs from heterochromatin regions and DNA repeats [19]. Despite having a similar structural build, the sRNA varies greatly in their method of biogenesis.

ncRNA		Remark
Housekeeping	tRNA	Actively involved in decoding genetic information from mRNA at the site of ribosome subunits during protein translation.
	rRNA	It is essentially a component of the ribosome, constituting small and large subunits of the ribosome. This ribozyme is RNA transcribed from ribosomal DNA (rDNA).
	snRNA	Small nuclear RNA (snRNA) is involved in the processing of pre-messenger RNA (hnRNA) in the nucleus, which is transcribed by either RNA polymerase II or RNA polymerase III, and the average length of snRNA is approximately 150 nucleotides.
	snoRNA	Small nucleolar RNAs have a role in tuning ribosomal and spliceosomal function by guiding ribose methylation and pseudouridylation at targeted nucleotide residues of ribosomal and small nuclear RNAs.
Regulatory	miRNA	microRNAs regulate gene expression post-transcriptionally. They generally bind to the 3'-UTR (untranslated region) of their target mRNAs and repress protein production by destabilizing the mRNA and translational silencing.
	siRNA	Small interfering RNA is sometimes known as short interfering RNA or silencing RNA is a class of double-stranded RNA, typically 20–s24 (normally 21) base pairs in length, and operating within the RNA interference (RNAi) pathway.
	i. hasiRNA	Trans-acting siRNAs are known to target complementary mRNAs for degradation and to function in development.
	ii.tasiRNA	Trans-acting siRNA, represses gene expression through post-transcriptional gene silencing in land plants.
	iii.easiRNA	Epigenetically activated small interfering RNAs (easiRNAs) from reactivated transposable elements triggered by miRNA.
	iii.natsiRNA	Natural antisense short interfering RNA. They are endogenous RNA regulators, which are between 21 and 24 nucleotides in length, and are generated from complementary mRNA transcripts, which are further processed into siRNA.
	piRNA	piwi-interacting RNA (piRNA) is the largest, expressed in animal cells. Mostly involved in the epigenetic and post-transcriptional silencing of transposable elements and other spurious or repeat-derived transcripts, but can also be involved in the regulation of other genetic elements in germ line cells.
	lncRNA	Long non-coding (lnc) RNAs are longer than 200 nt, which primarily interact with mRNA, DNA, protein, and miRNA and consequently regulate gene expression at the epigenetic, transcriptional, post-transcriptional, translational, and post-translational levels in a variety of ways.
	i. Linear	Maintaining nuclear structure integrity and positively or negatively regulating genes in *cis* or *trans* by recruiting transcription factors or chromatin-modifying complexes to DNA targets in the nucleus.
	ii.Circular	Function as a sponge to recruit miRNAs or transcriptional effectors to regulate target gene expression.

Table 3.
Brief details about the ncRNA classification.

4.1 miRNA

MicroRNAs are generated out of primary microRNAs (pri-miRNAs) by the action of RNA polymerase II. These precursor molecules have a double-stranded secondary structure which later on gets cleaved by DICER-LIKE 1 (DCL1) molecules to form single-stranded siRNA. These raw miRNAs is further processed in the nucleus before being exported to the cytoplasm [1]. Mature single miRNA that incorporates the RISC complex binds with other complementary mRNA sequences [20].

4.2 IsomiRNAs

IsomiRNAs (IsomiRNAs) are one of the variants of miRNAs that arise because of the inaccurate cleavage action performed through the RNase III enzyme. These may also be developed via the process of post-transcriptional RNA editing [21, 22].

4.3 siRNA

siRNAs are derived from long dsRNAs produced during varied mechanisms, such as folding in an inverted sequence, from a long non-coding RNA, hybridization of two fully or partially complementary sequences . This further gets processed by the action of DCL2, DCL3, and DCL 4 proteins, leading to the formation of 22 nt, 24 nt, and 21 nt siRNA, respectively [3]. RNA-dependent RNA polymerases 2 and 6 (RDR2, RDR6), and DNA-dependent RNA polymerases IV and V in plants also take part in the production of siRNAs [23].

4.4 phasiRNA and tasiRNA

In plants, phasiRNAs are the partially degraded product of mRNA as being passed through the RISC complex. DCL proteins are involved in this slicing process of dsRNAs to form a group of 21- or 24-nt siRNAs, termed phasiRNAs, *Trans*-acting siRNAs (tasiRNAs) are a class of DCL4-dependent 21-nt phasiRNAs generated from non-coding *TAS* transcripts [24, 25].

4.5 lncRNA

lncRNAs have appeared from intergenic, particularly regions giving rise to long intergenic ncRNAs (lincRNAs) and those developed from intronic regions giving intronic ncRNAs (incRNAs) [26]. lincRNAs and IncRNAs belong to conventional linear lncRNAs. Circular RNAs (circRNAs) generally develop from coding regions or intronic regions.

4.5.1 Linear-long non-coding RNAs

The lncRNAs, such as lincRNAs and IncRNAs are linear lncRNAs transcribed by Pol II. Because of having similar features as that of mRNA with a 5_ m7G cap and a 3_ poly (A) tail, they undergo similar modifications later on called mRNA mimics [27].

4.5.2 Circular long non-coding RNAs

Mostly these are circRNAs derived from back-splicing reactions of internal exons in pre-mRNAs and further move to the cytoplasm. Some other circRNA localized in the nucleus is produced from excised intron lariats that fail to be debranched [27].

5. Characteristic features of non-coding RNA

Non-coding RNAs are RNA molecules that are transcribed from DNA but cannot code for a protein. With the advances in transcriptomics and sequencing techniques, thousands of small and long non-coding RNAs were identified. They generally play a major role in gene regulation at the transcriptional or post-transcription level and some regulatory ncRNAs possess high target specificity and some ncRNAs are involved in epigenetic mechanisms too. A few major ones are discussed below.

5.1 tRNA (transfer RNA)

Next to mRNA which is the coding RNA, tRNA, and rRNA are considered more prominent. tRNA typically contains less than 100 nucleotides and, as their name indicates, their job is to carry an amino acid to the protein-synthesizing machinery. tRNA usually takes a clover leaf secondary structure, which forms a 3D L- shaped structure by stacking the helices. The structure of tRNA was found to be almost similar among different tRNA species. The secondary structure includes the acceptor arm, anticodon arm, T-arm and D-arm, and a variable arm. It was found that the size of the acceptor arm, anticodon arm, and T-arm were conserved whereas the D-arm and variable arm differ in their sizes [28]. This difference in the sizes of the variable arm led to the grouping of tRNAs into two classes. The major proportion of tRNA belongs to class I with less than 10 nucleotides in the variable loop; the class II tRNAs included $tRNA^{Ser}$, $tRNA^{Leu,}$ and $tRNA^{Tyr}$ with more than 10 nucleotides [8].

5.2 rRNA (ribosomal RNA)

It is one of the longest and most stable RNA molecules which form a major constituent of protein-synthesizing organelle comprising nearly 60% of ribosome's mass. Ribosomes of both prokaryotes and eukaryotes are made of smaller and larger subunits and they form a complex during translation. The smaller subunit of prokaryotes constitutes an RNA molecule with a Svedberg coefficient of 16S and its sedimentation rate is 30S after combining with other proteins. The larger subunit has two RNA molecules (5S and 23S) and they form a 50S subunit by binding with other proteins. On the contrary, the eukaryotic ribosome consists of 60S and 40S subunits with the larger subunit containing two long RNA molecules (18S and 28S) and the smaller subunit containing two short RNAs molecules (5S and 5.8S). A unique rRNA component, such as an undescribed helical structure, was found in the small subunit near the mRNA exit channel of 80S ribosome of *Trypanosoma cruzi*, a protozoan that causes Chagas disease. This structure was most likely involved in the binding of the ribosome to the 5'end of the mRNA facilitating translation [29].

5.3 snRNA (small nuclear RNA)

One of the most important post-transcriptional modifications is the splicing of pre-mRNA which is carried out by an RNA-protein complex known as spliceosome. Small nuclear RNAs form a part of this spliceosome and catalyze splicing [30]. On account of their sub-nuclear localization, snRNAs are grouped into spliceosomal uridylate snRNA, which is the most conserved among eukaryotes, small nucleolar RNA (snoRNA), and small Cajal-body-specific RNA (scaRNA) [31]. UsnRNA comprises U1–U6 being the most abundant and U7–U14 being the low abundant ones. The five major types U1, U2, U4, U5, and U6 are involved in the splicing of mRNA, whereas U3, U8, U13, and U14 are involved in the processing of mRNA [32]. Small nuclear ribonucleoprotein complexes (snRNPs) are formed by the association of each snRNA with one or more proteins.

5.4 snoRNA (small nucleolar RNA)

snoRNAs are functional non-coding RNAs with a length of 60–300 nucleotides, which are usually found near nucleoli and are prevalent in all eukaryotic organisms. Like snRNPs, they also form snoRNPs in association with a set of proteins [33]. snoRNA are majorly categorized into C/D box, snoRNAs which contain two conserved sequences box C (RUGAUGA and box D (CUGA) and direct 2′-*O*-ribose methylation, whereas H(ANANNA)/ACA box snoRNAs directs pseudouridylation. This classification is based on conserved sequence motifs [34]. The binding of fibrillarin, Nop56p, Nop58p, and 15.5 kDa/Snu13p snoRNP proteins are directed by the box C/D motif and form a kink turn, which is the most prevalent motif found in various RNAs. Proteins like dyskerin/Cbf5p, Gar1p, Nhp2p, and Nop10p are associated with the box H/ACA snoRNAs [35].

5.5 Other small non-coding RNA

Many types of small non-coding RNA have emerged in the last decade, but it is mainly classified into si (short interfering RNA), miRNA (microRNA), and piRNA (piwi-interacting RNAs). These are small non-coding RNAs with a length of about 20–30 nucleotides and form a protein complex with the Argonaute protein family and are present only in eukaryotes [36, 37].

5.5.1 miRNA and siRNA

Both miRNA and siRNA are initially part of a double-stranded RNA molecule with a guided strand and passenger strand. Their size is around 20–24 nucleotides only. A unique feature of siRNA is the occurrence of di-nucleotide overhang at the 3’OH. Phased siRNAs, trans-acting siRNAs, epigenetically activated siRNAs, and natsiRNAs are some of the types of siRNAs that play a role in regulating gene expression. A similar type of small non-coding RNA is the miRNA which is a small single-stranded RNA transcribed from DNA sequences into primary miRNA and processed into precursor miRNAs and finally becomes a mature miRNA [38]. Both siRNA and miRNA are almost similar in their biogenesis where an enzyme of the RNAse III family cleaves dsRNA into siRNA and miRNA. Respective RNA-induced silencing complexes are formed with the association of siRNA and miRNA termed siRSC and miRSC and are involved in gene regulation [39].

5.5.2 piRNA

piRNAs are also single-stranded with a length of 23–36 nucleotides and are more prevalent in animals. They bind to PIWI proteins that belong to the Argonaute protein family. The binding to PIWI and the independence from Dicer distinguished piRNA from siRNA and miRNA. piRNA is grouped into transposon-derived piRNA, miRNA-derived piRNA, lnc-derived piRNA, and *Caenorhabditis*-specific piRNA [40].

5.6 lncRNA (long non-coding RNA)

As previously said, non-coding RNA is of two types, small and long. lncRNA due to its length of more than 200 nucleotides develops complicated secondary and tertiary structures. They are prevalent either in the nucleus or in the cytoplasm of the cell. To sustain their function, structural conservation is more common than nucleotide sequence conservation. Furthermore, when compared to other non-coding RNAs the conservation is found to be less and low prevalence adds to the challenge of identifying and understanding the mode of action. Long non-coding RNA contains linear and circular lncRNA. They are synthesized from pre-mRNA by alternate splicing, which consists of a 5′ cap and 3′ tail. In circular ncRNA, the 5'end and 3′ end are linked forming a circle-like structure [41]. It was found that the level of expression varies between different types of lncRNA, some are organ and tissue-specific and the rest are expressed after encountering different external or internal stresses [42, 43].

6. Role of ncRNA in different physiological pathways in plants

Recent development in molecular biology tools has led to advanced research in the area of ncRNA, which in turn gave rise to newer insights about the various roles of ncRNA in plants. The finding and reports about these roles are summarized in **Table 4**.

6.1 ncRNAs determining plant yield and nutrition

The role of ncRNAs in various physiological traits and growth parameters is well studied, moreover also have an indirect influence on yield through these traits viz., tillering modulation and panicle branching related genes in rice through SPL transcription factors controlling, such as OsTB1 [66] and DEP1 [67] (regulatory non-coding article). Zhang *et al.* [68] reported overexpression of miR397 resulted in increased panicle branching with desired grain size suppressing the LACCASE gene. Apart from miRNA, several lncRNAs controlling photoperiod sensitive male sterility LDMAR (for long-day specific male-fertility-associated RNA) in rice are responsible for panicle development, floral organ development, sexual reproduction, and also control the premature programmed cell death of developing anthers.

It is interesting to understand the role of ncRNA that was earlier considered junk having a role in nutrient use efficiency as well as nutrient absorption efficiency. Both miRNAs and lncRNAs are well involved in phosphate metabolism and homeostasis. miR399 suppresses phosphate homeostasis genes PHO2 leading to increased uptake of phosphorous [69]. Similarly, lncRNAs in rice control the expression of the *OsPHO1* gene family during phosphate-deprived conditions [70]. Thus, modulating nutrient-related traits thereby contributes toward good yield performance. Not only

Type of ncRNA	Role	Example	Author
miRNA	The mutation of DCL1 in different plants leads to lethal embryos or to pleiotropic developmental effects, which are attributed mainly to the significant decrease in the miRNAs levels.	of Arabidopsis	Liu *et al.* [44]; Nodine and Bartel, [45]
	The transition of vegetative phase by regulating *SPL* genes in several angiosperm species.	Rice, maize, barley, soybean	
	Regulate the epigenomic machinery by regulating the expression of genes involved in the rearrangement of chromatin.	—	Choi *et al.* [46]
	Monocot-specific miR444 controls tillering.	Rice	Guo *et al.* [47]
	Monocot-specific miRNA is induced by nitrogen luxury conditions and regulates lodging resistance by targeting the lignin biosynthesis genes *ZmLACCASE 3* (*ZmLAC3*) and *ZmLAC5*.	Maize	
	miRNA represses auxin-responsive factors, promoting the development of lateral root growth and development.	Rice	Marin *et al.* [48]
	Overexpression results in developmental defects characterized by dwarf-ism, serrated leaves, and early flowering.	Arabidopsis Rice	Wu *et al.* [49]
siRNA	Involved in PTGS	—	Xie and Yu [50]
	DNA methylation	—	Nuthiattu *et al.* [51]
natsiRNA	Pathogen resistance	—	Katiyar-Agarwal *et al.* [52]
	Salt tolerance	—	Borsani *et al.* [53]
	Cell wall biosynthesis	—	Held *et al.* [54]
	Alter gene expression under environmental stress conditions.	—	Zhang *et al.* [55]
easiRNA	Heritable transcriptional gene silencing	—	

lncRNA	Response to environmental stimuli and stress	—	Wang *et al.* [56]
	Involved in protein and miRNAs hijack	—	Ariel *et al.* [57]
	Modulation of mRNA stability and translation	—	
	Modification of chromatin at diverse levels	—	
	Stress-responsiveness is induced by the infection with powdery mildew and by heat.	Wheat	Xin *et al.* [58]
	Tissue- or stress-specific expression	Foxtail millet Black cottonwood Chinese white poplar Soybean Peach *Brassica rapa*,	Qi *et al.* [59] Wang *et al.* [60] Yu *et al.* [61] Shuai *et al.* [62] Ye *et al.* [63] Chen *et al.* [64]
	Photoperiod-sensitive male sterility (PSMS)	Rice	Ding *et al.* [65]

Table 4.
Different ncRNA and their role in different processes in plant.

phosphorus but these ncRNAs are also reported to be involved in root nodulation thereby on nitrogen metabolic pathway [71].

6.2 ncRNAs in plant growth and development

Apart from their major role in stress response, these ncRNAs, such as miRNAs, are actively involved in the regulation of the transition of the vegetative phase by regulating *SPL* genes in several angiosperm species. For example, the vital role of miR172 in flowering control has been studied in rice, maize, barley, and soybean. In rice, the monocot-specific miR444 controls tillering [47] and participates in antiviral defense. miR528, a monocot-specific miRNA, is induced by nitrogen luxury conditions in maize and regulates lodging resistance by targeting the lignin biosynthesis genes *ZmLACCASE 3* (*ZmLAC3*) and *ZmLAC5* [72]. si RNAs, on the other hand, are involved in transcriptional gene silencing regulated by RNA-directed DNA methylation and also have a vital role in genome stability. siRNAs are also reported to monitor, genome compatibility and dosage during reproduction and hybridization [73].

7. Function of ncRNA concerning prospective areas of utilization of ncRNA in plant breeding

A major portion of eukaryotic transcriptomes comprised of ncRNAs, which were considered "junk" till the recent past, actually carry out significant roles in almost all the biological processes via regulation of gene expression at transcriptional and post-translational stages. Thus, these diverse ncRNA plays a vital role in plant development and environmental responses, which can be well employed in applied plant breeding and crop improvement. In drosophila, 30 UTR of mRNA is required for oogenesis [74]. Many studies also report the significant role of 30 UTR mRNAs in transmitting information required for cell proliferation as well as cell differentiation during cancers [75].

7.1 Abiotic and biotic stress response

These ncRNAs have considerable responsibility for plant stress response as well as plant immunity that can be better utilized in plant breeding for crop improvements. Jha *et al.* [76] have highlighted the differential expressions of ncRNAs in plants when they are ubjected to unfavorable conditions. Zhang *et al.* [68] identified that 17 microRNA were downregulated and 16 upregulated when subjected to drought stress at the seedling stage. A similar study was carried out on maize screened for drought conditions by Liu *et al.* [77]. When drought condition was induced, eight and seven miRNAs were enhanced in leaves and roots, respectively, whereas 13 and seven miRNAs were found to be suppressed in leaves and roots, respectively. In addition, a single miRNA can be upregulated and downregulated to express the same stress reaction. For example, in maize miR156, miR164, and miR171 undergo varying expressions when subjected to drought conditions. Similarly, differential expression was exhibited by miR156, and miR171 in rice was also reported [78, 79]. **Table 5** depicts the use of various miRNA in stress responses.

Sunar *et al.* [121] have found the role of siRNAs in response to abiotic stresses. Several studies of abiotic stress tolerance in wheat indicate that siRNA is upregulated when exposed to cold stress but are down-regulated when subjected to heat stress, NaCl and dehydration conditions [122]. Wang *et al.* [123] reported the contribution

Application	Crop	Gene	References
Viral diseases	Potato	HC-Pro (Potato virus-Y)	Waterhouse *et al.* [80]
		Silencing of viral coat protein	Missiou *et al.* [81]
	medicinally important papaya	Silencing of viral coat protein	Kertbundit *et al.* [82]
	Black gram	Hairpin construct of the non-coding intergenic region of mungbean yellow mosaic India virus	Pooggin *et al.* [83]
	Tobacco	Tobacco mosaic virus asRNA CP	Powell *et al.* [84]
	Pepper	PMMoV RNAi PMMoV replicase	Dalakouras *et al.* [85]
	Cassava	siRNAs designed to the replicase (Rep)-coding sequence of African cassava mosaic virus	Vanitharani *et al.* [86]
		Cassava brown streak disease (CBSD)	Patil *et al.* [87]
Bacterial diseases	*Arabidopsis thaliana*	Crown gall disease management strategy that targets the process of tumorigenesis (gall formation) by initiating RNAi of the iaaM and ipt oncogenes	Dunoyer *et al.* [88]
	Rice	Leaf blight RNAi OsSSI2	Younis *et al.* [89]
Fungal disease	*Nicotiana tabacum*	Downregulation of GST (glutathione S-transferases) enzyme against *Phytophthora parasitica var. nicotianae*	Hernández *et al.* [90]
	Apple	Apple scab fungus RNAi GFP & THN	Fitzgerald *et al.* [91]
	Wheat and Barley	Powdery mildew fungi *Blumeria graminis*	Nowara *et al.* [92]
	Rice	Sheath blight pathogen RNAi RPMK1-1/-2	Ila Mukul Tiwari *et al.* [93]
	Banana	*Fusarium oxysporum* f. sp. Cubense by RNAi-mediated knockdown of vital genes of fungus (velvet and Fusarium transcription factor 1)	Ghag *et al.* [94]
Insect resistance	Cotton	Cotton bollworm gut-specific-c cytochrome P450 gene CYP6AE14, which confers resistance to gossypol	Mao *et al.* [93]
	Corn	In plant expression of dsRNA against western corn rootworm larvae (*Diabrotica virgifera*)	Baum *et al.* [95] Mao *et al.* [93]
	Tobacco	dsRNA against EcR-USP (ecdysone receptor ultra-spiracle particle), AChE (acetylcholinesterase) and HR3 involved in the regulation of molting and development in *H. armigera*	Zhu *et al.* [96] Kumar [97] Xiong *et al.* [98]
Nematode attack		*Megalaima incognita* mitochondrial ATP synthase b subunit) silencing of root-knot nematodes	Huang *et al.* [99]
Male sterility	Tobacco	Downregulation of anther-specific gene TA29 by RNAi	Nawaz-ul-Rehman *et al.* [100]
	Tomato	Male sterility RNAi SmTAF10/13	Toppino *et al.* [101]
	Rice	hairpin RNA for OsGEN-L (OsGEN-like) gene	Moritoh *et al.* [102]
	Tomato	S-Adenosylmethionine decarboxylase (SAMDC), control of tapetal-specific A9 promoter using RNAi	Sinha and Rajam [103]

Application	Crop	Gene	References
Quality	Maize	Downregulation of lysine-poor zein gene via RNAi	Angaji *et al.* [104]
	Rice	Increased 2-Acety-1-pyrroline production by silencing OSBADH2 gene	Khandagale *et al.* [105]
	Cotton	hpRNA-mediated gene silencing of two fatty acid desaturase genes, stearoyl-acylcarrier protein D9-desaturase and oleoylphosphatidylcholine u6-desaturase for the fatty acid composition of cottonseed oil	Liu *et al.* [106]
	Potato	Silencing the β-carotene hydroxylase gene (BCH) to enhance β-carotene content	Eck *et al.* [107]
	Brassica napus	Silencing of DE-ETIOLATED1 (DET1) for increased levels of lutein, β-carotene, and zeaxanthin	Wei *et al.* [108]
	Tomato	Suppressing an endogenous photomorphogenesis regulatory gene, DET1, both carotenoid and flavonoid contents were increased	Davuluri *et al.* [109]
	Cassava	Removing linamarin, using RNAi silencing CYP79D1/D2	Meena *et al.* [110]
	Wheat	Enhancing amylose content using asRNA, targeted gene: Sbe2a	Sestili *et al.* [111]
	Rice	Reduce cadmium RNAi OsPCS1	Li *et al.* [112]
	Brassica	Reduce erucic acid RNAi BnFAE1	Shi *et al.* [113]
	Wheat	Reduce glutinin RNAi ɣ-gliadins	Gil-Humanes *et al.* [114]
Secondary metabolites	Rice	hpRNA from an inverted repeat for glutelin, leading to lower glutelin	Kusaba *et al.* [115]
	Papaver somniferum	Reduce the levels of the gene encoding the morphine biosynthetic enzyme salutaridinol 7-O-acetyltransferase (SalAT) led to the accumulation of the intermediate compounds, salutaridine and salutaridinol	Kempe *et al.* [116]
	Cotton	RNAi construct of the d-cadinene synthase gene of gossypol synthesis fused to a seed-specific promoter caused seed-specific reduction of gossypol	Sunilkumar *et al.* [117]
Keeping quality	Tomato	Chimeric RNAi-ACS construct designed to target ACS homologs effectively repressed the ethylene production	Gupta *et al.* [118]
	Tomato	SlSGR1 (encoding a STAYGREEN protein, retention of firmness and sustained cell membrane integrity and resulting in delayed fruit senescence	Luo *et al.* [119]
	Banana	Enhanced shelf-life RNAi MaMADS1/S2	Elitzur *et al.* [120]

Table 5.
Application of RNAi in crop quality and stress breeding.

of lncRNA in abiotic stress response by utilizing two distinct mechanisms, either they block the miRNA interactions with their target by mimicking as competitive endogenous RNAs. This mechanism of abiotic stress tolerance was reported in rice. The alternative mechanism followed by lncRNAs is antisense lncRNAs interact with sense mRNAs, forming double-stranded RNAs thereby preventing the expression of the gene. Zhang *et al.* [124] mentioned such interactions in drought stress studies of maize.

7.2 Plant immunity

ncRNAs also regulate plant disease resistance by switching on downstream R-genes, as well as the genes responsible for pathogenesis-related proteins or phenolic compounds or phytoalexins, and several other phytohormones signals in response to pathogen attack. miRNAs are engaged in Resistance gene (R-gene) regulation, whose activation is essential at the time of invasion by the pathogen. miRNAs, such as miR482, are downregulated during pathogen infection in potato, whereas its overexpression may lead to hypersensitive reactions [125]. **Tables 6** and **7** shows the crop-wise representation of different miRNA and other ncRNAs in maintaining plant immunity.

Other than stress breeding, the aspects of immunity, ncRNA, especially miRNA, have been explored in detail for their potential use in different attributes of crop breeding like enhancing the quality of the yield and its keeping quality, deposition of the secondary metabolites. It can also be used to induce male sterility in the plant, which is an important aspect of hybrid breeding. **Table 5** encapsulates different reports of usage of miRNA in various crops in the above-mentioned aspects of crop breeding.

8. Different repositories and databases of ncRNA research

In the recent past, non-coding RNAs have grabbed the attention of many researchers for their role in gene regulation. Different databases for housekeeping non-coding RNA and regulatory ncRNA have been developed which provided the scientists with a lot of information for their functional study. The sequence of tRNA was the one first compiled and published in 1989 with 455 tRNA sequences and 981 tRNA genes [156]. The primitive database for non-coding RNA was ncRNAdb which had 30,000 sequences with no specificities [157] and later the amount of information drastically increased and individual databases for each non-coding RNA are developed, such as silva for rRNA, miRNase for miRNA, snOPY for small nucleolar RNA.

Countless novel techniques had been emerged in the past decade, which led to the discovery of several new ncRNAs and ncRNA genes that are functionally characterized by modern biotechnological tools. Eukaryotic sRNAs, such as miRNA and siRNA, are short sequenced RNA molecules with a length of a maximum of 25 nucleotides. Detection algorithms of sRNA from RNA seq data involve mapping of single or paired-end reads to a reference genome later converted into genome-wide distribution. This is feasible when the size of sRNA is small and where the distribution remains uniform throughout the transcript. However, in bacteria with a lengthier sRNA ranging from 50 to 350 nucleotides, algorithms have to be designed in such a way to overcome the challenges imposed due to the extremely variable number of small transcripts. One such tool is APERO (analysis of paired-end RNA-seq output), which is used to detect bacterial sRNAs from the sequence data of RNA neglecting the need of converting the reads to genome-wide coverage which leads to the loss

Crop	Function and target	References
Arabidopsis	Trigger phasiRNA production target is RPS5	Boccara *et al.* [126]
	Regulate immune receptor targeting PPRL	Katiyar-Agarwal *et al.* [52]
	As transcription factor targeting GRFs	Soto-Suarez *et al.* [127]
	Regulate receptor-like kinase targeting ARLPK1/ ARLPK2	Niu *et al.* [57]
	ROS accumulation, targeting genes; PPR1/PPR2 and At5g38850/At3g04220	Park *et al.* [128] Nie *et al.* [129]
	PR gene expression; MEMB12	Zhang *et al.* [130]
	Callose deposition; MET2	Salvador-Guirao *et al.* [131]
	Hormone; TIR1/AFB2/AFB3	Navarro *et al.* [132]
	miRNA biosynthesis pathway; SERRATE	Niu *et al.* [57]
Rice	Act as transcription factor for the targeted genes NF-YAs, NAC60 and IPA1	Li *et al.* [133] Wang *et al.* [134] Liu *et al.* [135]
	ROS accumulation; ASCORBATE OXIDASE	Wu *et al.* [136] Yao *et al.* [137]
	PR gene expression; Nramp6	Campo *et al.* [138] Sanchez-Sanuy *et al.* [139]
	Callose deposition; ARF16	Li *et al.* [140]
Tobacco	Trigger phasiRNA production, target is N-gene	Li *et al.* [141] Deng *et al.* [142]
	Trigger phasiRNA production, target is EU713768	de Vries *et al.* [143]
Brassica	Regulate immune receptor, targets BraTNL1	He *et al.* [144]
Medicago	Trigger phasiRNA production, Medtr4g023400/ Medtr4g014580/ Medtr5g071220	Zhai *et al.* [145]
Soybean	Regulate immune receptor	Cui *et al.* [146]
Barley	Trigger phasiRNA production, MLA1	Liu *et al.* [147]

Table 6.
miRNA in plant immunity.

Crop	Function and target	References
Arabidopsis	Regulate immune receptor targeting TOE1/TOE2	Zou *et al.* [148]
Rice	Act as transcription factor for ST1	Zhang *et al.* [149]
Tomato	Regulate immune receptor	Jiang *et al.* [150] Jiang *et al.* [151]
	Regulate receptor-like kinase	Hong *et al.* [152]
	ROS accumulation; SIGRX21/SIGRX22 and RBOH	Cui *et al.* [153] Cui *et al.* [154]
	PR-gene expression; miR168a	Hou *et al.* [155]

Table 7.
Other ncRNAs in plant immunity.

of information. Instead, it is based on detecting the 5'end of the small transcripts and recognizing the extension of the transcript where the conserved information of sequenced fragments increases the accuracy [158]. **Table 8** brings together different databases available for the advanced panel of research on ncRNAs.

Recent years have witnessed the advancement in sequencing methodologies, such as long-end sequencing and optical mapping, for more accurate and faster sequencing at affordable rates. Cufflinks, and CIRCexplorer [159, 160] are some of the bioinformatics tools used for the discovery of ncRNA. Molecular approaches, such as cloning and hybridization techniques, were able to detect and characterize ncRNAs, but they came with a lot of false positives. Currently, the most reliable approach for predicting and functionally characterizing ncRNAs are NGS (Next Generation Sequencing) and CRISPR-Cas9 genome editing techniques [161]. The list of databases developed for non-coding RNA and specific ncRNA is given below. (https://rnacentral.org/expert-databases) [162].

9. Conclusion

Non-coding RNAs (ncRNAs) possess little or no protein-coding capacity yet are indeed functional. They make up a huge and significant percentage of eukaryotic transcriptomes. It modulates expression levels at various stages of protein synthesis, playing an important regulative involvement in practically all biological processes. MicroRNAs (miRNAs), small interference RNAs (siRNAs), circular RNAs (circRNAs), and long non-coding RNAs (lncRNAs) are the major non-coding RNAs. These can either operate as long ncRNAs or be converted into small RNAs. They are classed worldwide based on their size, function, and genetic origin. Non-coding modulates its targets via interacting with DNA, RNA, and proteins. These have a role in multiple epigenetic mechanisms controlling phenotypes, as well as the specification of various physiological pathways. MicroRNAs control the level of gene expression by increasing the disintegration of target mRNAs or inhibiting translation. They are involved in many aspects of plant growth and have the power to reconfigure responses to various biotic and abiotic stresses. The modulation of immunological responses in plants has been linked to non-coding RNAs, DNA and RNA methylation, along with

S. No.	Database	Specification
1.	tRNAdb	tRNA sequences and tRNA genes
2.	miRTarBase	microRNA-target interactions database
3.	snoRNA	Archaeal snoRNAs
4.	RiboVision	Ribosomal annotations
5.	Rfam	Collection of non-coding RNA families
6.	piRBase	piRNA
7.	NONCODE	Non-coding RNAs
8.	lncRNAdb	Eukaryotic long non-coding RNAs
9.	GtRNAdb	tRNA gene predictions
10.	5SrRNAdb	5S ribosomal RNAs

Table 8.
Different databases and their specifications.

other epigenetic changes. Regulatory ncRNAs in plants are being highlighted as potential targets for molecular breeding of agricultural trait improved crop plants, such as improved abiotic and biotic stress tolerance, herbicide resistance, yield, enhancement, and plants with amazing nutritional value with prospective high agricultural importance. Non-coding RNAs (ncRNAs) are also observed to work as a defense system against invading viruses as effectors molecules in RNA-mediated gene silencing and are being exploited in agricultural genetic modification. They also act as key moderators in the level of plant immunity and adaptation to different environments. Plant lncRNAs participate in a wide range of biological processes, including regulation of flowering time and morphogenesis of reproductive organs, as well as abiotic and biotic stress responses. Given the discoveries of these ncRNA in the above-discussed processes, be it physical or physiological, they show a new ray of light toward the use of them in crop breeding. In this regard, the areas, especially the quality breeding, stress breeding for abiotic and biotic stresses, have a huge potential. Along with that, looking at their role in changing the flowering and morphogenesis of plants, further research may be carried forward in the direction of their utilization in altering plant growth duration or producing genotypes for off-season breeding. The role of ncRNAs in epigenetics also can be further studied for their exact role in the inheritance pattern of different important traits. Over the last two decades, the research on non-coding RNAs has shown newer insights about their structure, properties, and possible utilities in different fields of life science. Further work is required to be expanded in newer areas to more agriculturally important crops to harness the wonders of ncRNAs.

Author details

Debadatta Panda*, Latha Ananda Lekshmi, Rachel Lissy Vargheese, Nallathambi Premalatha, Mahadevan Kumar and Lakshmanan Mahalingam
Centre for Plant Breeding and Genetics, Tamil Nadu Agricultural University, Coimbatore, Tamil Nadu, India

*Address all correspondence to: debadattapanda555@gmail.com

References

[1] Quan M, Chen J, Zhang D. Exploring the secrets of long non-coding RNAs. International Journal of Molecular Sciences. 2015;**16**(3):5467-5496. DOI: 10.3390/ijms16035467

[2] Taft RJ, Pheasant M, Mattick JS. The relationship between non-protein-coding DNA and eukaryotic complexity. BioEssays. 2007;**29**:288-299. DOI: 10.1002/bies.20544

[3] Shin SY, Shin C. Regulatory non-coding RNAs in plants: Potential gene resources for the improvement of agricultural traits. Plant Biotechnology Reports. 2016;**10**:35-47

[4] Zhou M, Luo H. MicroRNA-mediated gene regulation: Potential applications for plant genetic engineering. Plant Molecular Biology. 2013;**83**:59-75

[5] Cech TR, Steitz JA. The non-coding RNA revolution—Trashing old rules to forge new ones. Cell. 2014;**157**(1):77-94. DOI: 10.1016/j.cell.2014.03.008

[6] Diriba GD. Advances on the application of non-coding RNA in crop improvement. African Journal of Biotechnology. 2021;**20**(11):440-450. DOI: 10.5897/AJB2021.17348

[7] Kresge N, Simoni RD, Hill RL. The discovery of tRNA by Paul C. Zamecnik. Journal of Biological Chemistry. 2005;**280**(40):e37-e39. DOI: 10.1016/S0021-9258(20)79029-0

[8] Sprinzl M, Horn C, Brown M, Ioudovitch A, Steinberg S. Compilation of tRNA sequences and sequences of tRNA genes. Nucleic Acids Research. 1998;**26**:148-153. DOI: 10.1093/nar/26.1.148

[9] Hamilton AJ, Baulcombe DC. A species of small antisense RNA in posttranscriptional gene silencing in plants. Science. 1999;**286**(5441):950-952. DOI: 10.1126/science.286.5441.950

[10] Lee RC, Feinbaum RL, Ambros V. The C. elegans heterochronic gene lin-4 encodes small RNAs with antisense complementarity to lin-14. Cell. 1993;**75**(5):843-854. DOI: 10.1016/0092-8674(93)90529-y

[11] Pachnis V, Belayew A, Tilghman SM. Locus unlinked to alpha-fetoprotein under the control of the murine raf and Rif genes. Proceedings of the National Academy of Sciences. 1984;**81**(17):5523-5527. DOI: 10.1073/pnas.81.17.5523

[12] Fatica A, Bozzoni I. Long non-coding RNAs: New players in cell differentiation and development. Nature Reviews Genetics. 2014;**15**(1):7-21. DOI: 10.1038/nrg3606

[13] Chen HM, Chen LT, Patel K, Li YH, Baulcombe DC, Wu SH. 22-nucleotide RNAs trigger secondary siRNA biogenesis in plants. Proceedings of the National Academy of Sciences. 2010;**107**(34):15269-15274. DOI: 10.1073/pnas.1001738107

[14] Slotkin RK, Vaughn M, Borges F, Tanurdžić M, Becker JD, Feijó JA, et al. Epigenetic reprogramming and small RNA silencing of transposable elements in pollen. Cell. 2009;**136**(3):461-472. DOI: 10.1016/j.cell.2008.12.038

[15] Aravin A, Gaidatzis D, Pfeffer S, Lagos-Quintana M, Landgraf P, Iovino N, et al. A novel class of small RNAs bind to MILI protein in mouse testes. Nature. 2006;**442**(7099):203-207. DOI: 10.1038/nature04916

[16] Morris KV, Mattick JS. The rise of regulatory RNA. Nature Reviews

Genetics. 2014;**15**(6):423-437. DOI: 10.1038/nrg3722

[17] D'Ario M, Griffiths-Jones S, Kim M. Small RNAs: Big impact on plant development. Trends in Plant Science. 2017;**22**(12):1056-1068. DOI: 10.1016/j.tplants.2017.09.009

[18] Brosnan CA, Voinnet O. The long and the short of non-coding RNAs. Current Opinion in Cell Biology. 2009;**21**(3):416-425. DOI: 10.1016/j.ceb.2009.04.001

[19] Voinnet O. Origin, biogenesis, and activity of plant microRNAs. Cell. 2009;**136**(4):669-687

[20] Khraiwesh B, Zhu JK, Zhu J. Role of miRNAs and siRNAs in biotic and abiotic stress responses of plants. Biochimica et Biophysica Acta. 2012;**1819**(2):137-148

[21] Wyman SK, Knouf EC, Parkin RK, Fritz BR, Lin DW, Dennis LM. Post-transcriptional generation of miRNA variants by multiple nucleotidyl transferases contributes to miRNA transcriptome complexity. Genome Research. 2011;**21**:1450-1461

[22] Banerjee A, Roychoudhury A, Krishnamoorthi S. Emerging techniques to decipher microRNAs (miRNAs) and their regulatory role in conferring abiotic stress tolerance of plants. Plant Biotechnology Reports. 2016;**10**(4):185-205

[23] Bologna NG, Voinnet O. The diversity, biogenesis, and activities of endogenous silencing small RNAs in *Arabidopsis*. Annual Review of Plant Biology. 2014;**65**:473-503

[24] Chen X. Small RNAs and their roles in plant development. Annual Review of Cell and Developmental Biology. 2009;**25**:21-44

[25] Fei Q, Xia R, Meyers BC. Phased, secondary, small interfering RNAs in posttranscriptional regulatory networks. Plant Cell. 2013;**25**:2400-2415

[26] Chekanova JA. Long non-coding RNAs and their functions in plants. Current Opinion in Plant Biology. 2015;**27**:207-216

[27] Wu H, Yang L, Chen LL. The diversity of long non-coding RNAs and their generation. Trends in Genetics. 2017;**33**:540-552

[28] Krahn N, Fischer JT, Söll D. Naturally occurring tRNAs with non-canonical structures. Frontiers in Microbiology. 2020;**11**:596914. DOI: 10.3389/fmicb.2020.596914

[29] Gao H, Ayub MJ, Levin MJ, Frank J. The structure of the 80S ribosome from Trypanosoma cruzi reveals unique rRNA components. In: Single-Particle Cryo-Electron Microscopy: The Path Toward Atomic Resolution: Selected Papers of Joachim Frank with Commentaries. 2005. pp. 383-388. DOI: 10.1142/9789813234864_0038

[30] Morais P, Adachi H, Yu YT. Spliceosomal snRNA epitranscriptomics. Frontiers in Genetics. 2021;**12**:652129. DOI: 10.3389/fgene.2021.652129

[31] Ohtani M. Transcriptional regulation of snRNAs and its significance for plant development. Journal of Plant Research. 2017;**130**(1):57-66. DOI: 10.1007/s10265-016-0883-3

[32] Solymosy F, Pollák T. Uridylate-rich small nuclear RNAs (UsnRNAs), their genes and pseudogenes, and UsnRNPs in plants: Structure and function. A comparative approach. CRC Critical Reviews in Plant Sciences. 1993;**12**:275-369. DOI: 10.1080/07352689309701904

[33] Kiss T. Small nucleolar RNAs: An abundant group of non-coding RNAs

with diverse cellular functions. Cell. 2002;**109**(2):145-148. DOI: 10.1016/S0092-8674(02)00718-3

[34] Dragon F, Lemay V, Trahan C. snoRNAs: biogenesis, structure, and function. e LS. 2001

[35] Liang J, Wen J, Huang Z, Chen XP, Zhang BX, Chu L. Small nucleolar RNAs: Insight into their function in cancer. Frontiers in Oncology. 2019;**9**:587. DOI: 10.1038/npg.els.0003813

[36] Carthew RW, Sontheimer EJ. Origins and mechanisms of miRNAs and siRNAs. Cell. 2009;**136**(4):642-655. DOI: 10.1016/j.cell.2009.01.035

[37] Meister G. Argonaute proteins: Functional insights and emerging roles. Nature Reviews Genetics. 2013;**14**(7):447-459. DOI: 10.1038/nrg3462

[38] O'Brien J, Hayder H, Zayed Y, Peng C. Overview of microRNA biogenesis, mechanisms of actions, and circulation. Frontiers in Endocrinology. 2018;**9**:402. DOI: 10.3389/fendo.2018.00402

[39] Tang G. siRNA and miRNA: An insight into RISCs. Trends in Biochemical Sciences. 2005;**30**(2):106-114. DOI: 10.1016/j.tibs.2004.12.007

[40] Han BW, Zamore PD. PiRNAs. Current Biology. 2014;**24**(16):R730-R733. DOI: 10.1016/j.cub.2014.07.037

[41] Yu CY, Kuo HC. The emerging roles and functions of circular RNAs and their generation. Journal of Biomedical Science. 2019;**26**(1):1-2. DOI: 10.1186/s12929-019-0523-z

[42] Zampetaki A, Albrecht A, Steinhofel K. Long non-coding RNA structure and function: Is there a link? Frontiers in Physiology. 2018;**9**:1201. DOI: 10.3389/fphys.2018.01201

[43] Graf J, Kretz M. From structure to function: Route to understanding lncRNA mechanism. BioEssays. 2020;**42**(12):2000027. DOI: 10.1002/bies.202000027

[44] Liu B, Li P, Li X, Liu C, Cao S, Chu C, et al. Loss of function of OsDCL1 affects microRNA accumulation and causes developmental defects in rice. Plant Physiology. 2005;**139**(1):296-305. DOI: 10.1104/pp.105.063420

[45] Nodine MD, Bartel DP. MicroRNAs prevent precocious gene expression and enable pattern formation during plant embryogenesis. Genes & Development. 2010;**24**(23):2678-2692. DOI: 10.1101/gad.1986710

[46] Choi K, Kim J, Müller SY, Oh M, Underwood C, Henderson I, et al. Regulation of microRNA-mediated developmental changes by the SWR1 chromatin remodeling complex. Plant Physiology. 2016;**171**(2):1128-1143. DOI: 10.1104/pp.16.00332

[47] Guo S, Xu Y, Liu H, Mao Z, Zhang C, Ma Y, et al. The interaction between OsMADS57 and OsTB1 modulates rice tillering via DWARF14. Nature Communications. 2013;**4**(1):1-12. DOI: 10.1038

[48] Marin E, Jouannet V, Herz A, Lokerse AS, Weijers D, Vaucheret H, et al. miR390, Arabidopsis TAS3 tasiRNAs, and their AUXIN RESPONSE FACTOR targets define an autoregulatory network quantitatively regulating lateral root growth. The Plant Cell. 2010;**22**(4):1104-1117. DOI: 10.1105/tpc.109.072553

[49] Wu HJ, Wang ZM, Wang M, Wang XJ. Widespread long non-coding RNAs as endogenous target mimics for microRNAs in plants. Plant Physiology. 2013;**161**(4):1875-1884. DOI: 10.1104/pp.113.215962

[50] Xie M, Yu B. siRNA-directed DNA methylation in plants. Current Genomics. 2015;**16**(1):23-31. DOI: 10.2174/1389202915666141128002211

[51] Nuthikattu S, McCue AD, Panda K, Fultz D, DeFraia C, Thomas EN, et al. The initiation of epigenetic silencing of active transposable elements is triggered by RDR6 and 21-22 nucleotide small interfering RNAs. Plant Physiology. 2013;**162**(1):116-131. DOI: 10.1104/pp.113.216481

[52] Katiyar-Agarwal S, Morgan R, Dahlbeck D, Borsani O, Villegas A, Zhu J-K, et al. A pathogen-inducible endogenous siRNA in plant immunity. Proceedings of the National Academy of Sciences. 2006;**103**(47):18002-18007. DOI: 10.1073/pnas.06082581

[53] Borsani O, Zhu J, Verslues PE, Sunkar R, Zhu JK. Endogenous siRNAs derived from a pair of natural cis-antisense transcripts regulate salt tolerance in Arabidopsis. Cell. 29 Dec 2005;**123**(7):1279-1291. DOI: 10.1016/j.cell.2005.11.035

[54] Held MA, Penning B, Brandt AS, Kessans SA, Yong W, Scofield SR, et al. Small-interfering RNAs from natural antisense transcripts derived from a cellulose synthase gene modulate cell wall biosynthesis in barley. Proceedings of the National Academy of Sciences. 2008;**105**(51):20534-20539. DOI: 10.1073/pnas.0809408105

[55] Zhang X, Xia J, Lii YE. et al. Genome-wide analysis of plant nat-siRNAs reveals insights into their distribution, biogenesis and function. Genome Biology. 2012;**13**:R20. DOI: 10.1186/gb-2012-13-3-r20

[56] Wang H, Chung PJ, Liu J, Jang IC, Kean MJ, Xu J, et al. Genome-wide identification of long non-coding natural antisense transcripts and their responses to light in Arabidopsis. Genome Research. 2014;**24**(3):444-453. DOI: 10.1101/gr.165555.113

[57] Ariel F, Romero-Barrios N, Jégu T, Benhamed M, Crespi M. Battles and hijacks: Non-coding transcription in plants. Trends in Plant Science. 2015;**20**(6):362-371. DOI: 10.1016/j.tplants.2015.03.003

[58] Xin M, Wang Y, Yao Y, Song N, Hu Z, Qin D, et al. Identification and characterization of wheat long non-protein coding RNAs responsive to powdery mildew infection and heat stress by using microarray analysis and SBS sequencing. BMC Plant Biology. 2011;**11**(1):1-3. DOI: 10.1186/1471-2229-11-61

[59] Qi X, Xie S, Liu Y, Yi F, Yu J. Genome-wide annotation of genes and non-coding RNAs of foxtail millet in response to simulated drought stress by deep sequencing. Plant Molecular Biology. 2013;**83**(4):459-473

[60] Wang L, Zhao S, Gu C, et al. Deep RNA-Seq uncovers the peach transcriptome landscape. Plant Molecular Biology. 2013;**83**:365-377. DOI: 10.1007/s11103-013-0104-6

[61] Yu X, Yang J, Li X, Liu X, Sun C, Wu F, et al. Global analysis of cis-natural antisense transcripts and their heat-responsive nat-siRNAs in Brassica rapa. BMC Plant Biology. 2013;**13**(1):1-3. DOI: 10.1186/1471-2229-13-208

[62] Shuai P, Liang D, Tang S, Zhang Z, Ye CY, Su Y, et al. Genome-wide identification and functional prediction of novel and drought-responsive lincRNAs in Populus trichocarpa. Journal of Experimental Botany. 2014;**65**(17):4975-4983. DOI: 10.1093/jxb/eru256

[63] Ye CY, Xu H, Shen E, Liu Y, Wang Y, Shen Y, et al. Genome-wide identification

of non-coding RNAs interacted with microRNAs in soybean. Frontiers in Plant Science. 2014;**5**:743. DOI: 10.3389/fpls.2014.00743

[64] Chen J, Quan M, Zhang D. Genome-wide identification of novel long non-coding RNAs in *Populus tomentosa* tension wood, opposite wood and normal wood xylem by RNA-seq. Planta. 2015;**241**(1):125-143. DOI: 10.1007/s00425-014-2168-1

[65] Ding J, Lu Q, Ouyang Y, Mao H, Zhang P, Yao J, et al. A long non-coding RNA regulates photoperiod-sensitive male sterility, an essential component of hybrid rice. Proceedings of the National Academy of Sciences. 2012;**109**(7):2654-2659. DOI: 10.1073/pnas.1121374109

[66] Takeda T, Suwa Y, Suzuki M, Kitano H, Ueguchi-Tanaka M, Ashikari M, et al. The OsTB1 gene negatively regulates lateral branching in rice. The Plant Journal. 2003;**33**(3):513-520. DOI: 10.1046/j.1365-313x.2003.01648.x

[67] Huang X, Qian Q, Liu Z, Sun H, He S, Luo D, et al. Natural variation at the DEP1 locus enhances grain yield in rice. Nature Genetics. 2009;**41**(4):494-497. DOI: 10.1038/ng.352

[68] Zhang F, Luo X, Zhou Y, Xie J. Genome-wide identification of conserved microRNA and their response to drought stress in Dongxiang wild rice (Oryza rufipogon Griff.). Biotechnology Letters. 2016;**38**(4):711-721. DOI: 10.1007/s10529-015-2012-0

[69] Aung K, Lin S-I, Wu C-C, Huang Y-T, Su C-l, Chiou T-J. pho2, a phosphate overaccumulator, is caused by a nonsense mutation in a microRNA399 target gene. Plant Physiology. 2006;**141**(3):1000-1011. DOI: 10.1104/pp.106.078063

[70] Jabnoune M, Secco D, Lecampion C, Robaglia C, Shu Q, Poirier Y. A rice cis-natural antisense RNA acts as a translational enhancer for its cognate mRNA and contributes to phosphate homeostasis and plant fitness. The Plant Cell. 2013;**25**(10):4166-4182. DOI: 10.1105/tpc.113.116251

[71] Campalans A, Kondorosi A, Crespi M. Enod40, a short open reading frame–containing mRNA, induces cytoplasmic localization of a nuclear RNA binding protein in *Medicago truncatula*. The Plant Cell. 2004;**16**(4):1047-1059. DOI: 10.1105/tpc.019406

[72] Sun Q, Liu X, Yang J, Liu W, Du Q, Wang H, et al. MicroRNA528 affects lodging resistance of maize by regulating lignin biosynthesis under nitrogen-luxury conditions. Molecular Plant. 2018;**11**(6):806-814. DOI: 10.1016/j.molp.2018.03.013

[73] Calarco JP, Borges F, Donoghue MT, Van Ex F, Jullien PE, Lopes T, et al. Reprogramming of DNA methylation in pollen guides epigenetic inheritance via small RNA. Cell. 2012;**151**(1):194-205. DOI: 10.1016/j.cell.2012.09.001

[74] Jenny A, Hachet O, Závorszky P, Cyrklaff A, Weston MD, Johnston DS, et al. A translation-independent role of oskar RNA in early Drosophila oogenesis. 2006:2827-2833. DOI: 10.1242/dev.02456

[75] Fan H, Villegas C, Huang A, Wright JA. Suppression of malignancy by the 3′ untranslated regions of ribonucleotide reductase R1 and R2 messenger RNAs. Cancer Research. 1996;**56**(19):4366-4369

[76] Jha UC, Nayyar H, Jha R, Khurshid M, Zhou M, Mantri N, et al. Long non-coding RNAs: Emerging players regulating plant abiotic stress response and adaptation. BMC Plant Biology. 2020;**20**(1):1-20. DOI: 10.1186/s12870-020-02595-x

[77] Liu X, Zhang X, Sun B, Hao L, Liu C, Zhang D, et al. Genome-wide identification and comparative analysis of drought-related microRNAs in two maize inbred lines with contrasting drought tolerance by deep sequencing. PLoS One. 2019;**14**(7):e0219176. DOI: 10.1371/journal.pone.0219176

[78] Cheah BH, Nadarajah K, Divate MD, Wickneswari R. Identification of four functionally important microRNA families with contrasting differential expression profiles between drought-tolerant and susceptible rice leaf at vegetative stage. BMC Genomics. 2015;**16**(1):1-18. DOI: 10.1186/s12864-015-1851-3

[79] Zhou L, Liu Y, Liu Z, Kong D, Duan M, Luo L. Genome-wide identification and analysis of drought-responsive microRNAs in Oryza sativa. Journal of Experimental Botany. 2010;**61**(15):4157-4168. DOI: 10.1093/jxb/erq237

[80] Waterhouse PM, Graham MW, Wang M-B. Virus resistance and gene silencing in plants can be induced by simultaneous expression of sense and antisense RNA. Proceedings of the National Academy of Sciences. 1998;**95**(23):13959-13964. DOI: 10.1073/pnas.95.23.13959

[81] Missiou A, Kalantidis K, Boutla A, Tzortzakaki S, Tabler M, Tsagris M. Generation of transgenic potato plants highly resistant to potato virus Y (PVY) through RNA silencing. Molecular Breeding. 2004;**14**(2):185-197

[82] Kertbundit S, Pongtanom N, Ruanjan P, Chantasingh D, Tanwanchai A, Panyim S, et al. Resistance of transgenic papaya plants to papaya ringspot virus. Biologia Plantarum. 2007;**51**(2):333-339. DOI: 10.1007/s10535-007-0065-1

[83] Pooggin M, Shivaprasad P, Veluthambi K, Hohn T. RNAi targeting of DNA virus in plants. Nature Biotechnology. 2003;**21**(2):131-132. DOI: 10.1038/nbt0203-131b

[84] Powell P, Stark D, Sanders P, Beachy R. Protection against tobacco mosaic virus in transgenic plants that express tobacco mosaic virus antisense RNA. Proceedings of the National Academy of Sciences. 1989;**86**(18):6949-6952. DOI: 10.1073/pnas.86.18.6949

[85] Dalakouras A, Wassenegger M, Dadami E, Ganopoulos I, Pappas ML, Papadopoulou K. Genetically modified organism-free RNA interference: Exogenous application of RNA molecules in plants. Plant Physiology. 2020;**182**(1):38-50. DOI: 10.1104/pp.19.00570

[86] Vanitharani R, Chellappan P, Fauquet CM. Short interfering RNA-mediated interference of gene expression and viral DNA accumulation in cultured plant cells. Proceedings of the National Academy of Sciences. 2003;**100**(16):9632-9636. DOI: 10.1073/pnas.1733874100

[87] Patil BL, Ogwok E, Wagaba H, Mohammed IU, Yadav JS, Bagewadi B, et al. RNAi mediated resistance to diverse isolates belonging to two virus species involved in cassava brown streak disease. Molecular Plant Pathology. 2011;**12**(1):31-41. DOI: 10.1111/j.1364-3703.2010.00650.x

[88] Dunoyer P, Himber C, Ruiz-Ferrer V, Alioua A, Voinnet O. Intra-and intercellular RNA interference in Arabidopsis thaliana requires components of the microRNA and heterochromatic silencing pathways. Nature Genetics. 2007;**39**(7):848-856. DOI: 10.1038/ng2081

[89] Younis A, Siddique MI, Kim C-K, Lim K-B. RNA interference (RNAi) induced gene silencing: A promising approach of hi-tech plant breeding. International Journal of Biological

Sciences. 2014;**10**(10):115. DOI: 10.7150/ijbs.10452

[90] Hernández I, Chacón O, Rodriguez R, Portieles R, López Y, Pujol M, et al. Black shank resistant tobacco by silencing of glutathione S-transferase. Biochemical and Biophysical Research Communications. 2009;**387**:3004. DOI: 10.1016/j.bbrc.2009.07.003

[91] Fitzgerald A, Van Kan JA, Plummer KM. Simultaneous silencing of multiple genes in the apple scab fungus, Venturia inaequalis, by expression of RNA with chimeric inverted repeats. Fungal Genetics and Biology. 2004;**41**(10):963-971. DOI: 10.1016/j.fgb.2004.06.006

[92] Nowara D, Gay A, Lacomme C, Shaw J, Ridout C, Douchkov D, et al. HIGS: Host-induced gene silencing in the obligate biotrophic fungal pathogen Blumeria graminis. The Plant Cell. 2010;**22**(9):3130-3141. DOI: 10.1105/tpc.110.077040

[93] Mao Y-B, Cai W-J, Wang J-W, Hong G-J, Tao X-Y, Wang L-J, et al. Silencing a cotton bollworm P450 monooxygenase gene by plant-mediated RNAi impairs larval tolerance of gossypol. Nature Biotechnology. 2007;**25**(11):1307-1313. DOI: 10.1038/nbt1352

[94] Ghag SB, Shekhawat UKS, Ganapathi TR. Native cell-death genes as candidates for developing wilt resistance in transgenic banana plants. AoB Plants. 2014;**6**. DOI: 10.1093/aobpla/plu037

[95] Baum JA, Bogaert T, Clinton W, Heck GR, Feldmann P, Ilagan O, et al. Control of coleopteran insect pests through RNA interference. Nature Biotechnology. 2007;**25**(11):1322-1326. DOI: 10.1038/nbt1359

[96] Zhu J-Q, Liu S, Ma Y, Zhang J-Q, Qi H-S, Wei Z-J, et al. Improvement of pest resistance in transgenic tobacco plants expressing dsRNA of an insect-associated gene EcR. PLoS One. 2012;**7**(6):e38572. DOI: 10.1371/journal.pone.0038572

[97] Kumar M. RNAi-Mediated Targeting of Acetylcholinesterase Gene of Helicoverpa Armigera for Insect Resistance in Transgenic Tobacco and Tomato: PhD thesis. University of Delhi; 2011. DOI: 10.1007/s11103-015-0414-y

[98] Xiong Y, Zeng H, Zhang Y, Xu D, Qiu D. Silencing the HaHR3 gene by transgenic plant-mediated RNAi to disrupt Helicoverpa armigera development. International Journal of Biological Sciences. 2013;**9**(4):370. DOI: 10.7150/ijbs.5929

[99] Huang Y, Mei M, Mao Z, Lv S, Zhou J, Chen S, et al. Molecular cloning and virus-induced gene silencing of MiASB in the southern root-knot nematode, Meloidogyne incognita. European Journal of Plant Pathology. 2014;**138**(1):181-193

[100] Nawaz-ul-Rehman MS, Mansoor S, Khan AA, Zafar Y, Briddon RW. RNAi-mediated male sterility of tobacco by silencing TA29. Molecular Biotechnology. 2007;**36**(2):159-165. DOI: 10.1007/s12033-007-0025-1

[101] Toppino L, Kooiker M, Lindner M, Dreni L, Rotino GL, Kater MM. Reversible male sterility in eggplant (Solanum melongena L.) by artificial microRNA-mediated silencing of general transcription factor genes. Plant Biotechnology Journal. 2011;**9**(6):684-692. DOI: 10.1111/j.1467-7652.2010.00567.x

[102] Moritoh S, Miki D, Akiyama M, Kawahara M, Izawa T, Maki H, et al. RNAi-mediated silencing of OsGEN-L (OsGEN-like), a new member of the RAD2/XPG nuclease family, causes

male sterility by defect of microspore development in rice. Plant and Cell Physiology. 2005;**46**(5):699-715. DOI: 10.1093/pcp/pci090

[103] Sinha R, Rajam MV. RNAi silencing of three homologues of S-adenosylmethionine decarboxylase gene in tapetal tissue of tomato results in male sterility. Plant Molecular Biology. 2013;**82**(1):169-180. DOI: 10.1007/s11103-013-0051-2

[104] Angaji S, Hedayati SS, Hosein R, Samad S, Shiravi S, Madani S. Application of RNA interference in plants. Plant Omics. 2010;**3**(3):77-84

[105] Khandagale K, Chavhan R, Nadaf A. RNAi-mediated down regulation of BADH2 gene for expression of 2-acetyl-1-pyrroline in non-scented indica rice IR-64 (Oryza sativa L.). 3 Biotech. 2020;**10**(4):145. DOI: 10.1007/s13205-020-2131-8

[106] Liu Q, Singh SP, Green AG. High-stearic and high-oleic cottonseed oils produced by hairpin RNA-mediated post-transcriptional gene silencing. Plant Physiology. 2002;**129**(4):1732-1743. DOI: 10.1104/pp.001933

[107] Eck VJ, Conlin B, Garvin DF, et al. Enhancing beta-carotene content in potato by rnai-mediated silencing of the beta-carotene hydroxylase gene. American Journal of Potato Research. 2007;**84**:331. DOI: 10.1007/BF02986245

[108] Wei S, Li X, Gruber MY, Li R, Zhou R, Zebarjadi A, et al. RNAi-mediated suppression of DET1 alters the levels of carotenoids and sinapate esters in seeds of Brassica napus. Journal of Agricultural and Food Chemistry. 2009;**57**(12):5326-5333. DOI: 10.1021/jf803983w

[109] Davuluri GR, Van Tuinen A, Fraser PD, Manfredonia A, Newman R, Burgess D, et al. Fruit-specific RNAi-mediated suppression of DET1 enhances carotenoid and flavonoid content in tomatoes. Nature Biotechnology. 2005;**23**(7):890-895. DOI: 10.1038/nbt1108

[110] Meena AK, Verma L, Kumhar BL. RNAi, it's mechanism and potential use in crop improvement: A review. International Journal of Pure and Applied Bioscience. 2017;**5**(2):294-311

[111] Sestili F, Janni M, Doherty A, Botticella E, D'Ovidio R, Masci S, et al. Increasing the amylose content of durum wheat through silencing of the SBEIIagenes. BMC Plant Biology. 2010;**10**(1):1-12. DOI: 10.1186/1471-2229-10-144

[112] Li JC, Guo JB, Xu WZ, Ma M. RNA interference-mediated silencing of phytochelatin synthase gene reduce cadmium accumulation in rice seeds. Journal of Integrative Plant Biology. 2007;**49**(7):1032-1037. DOI: DOI/10.1111/j.1672-9072.2007.0047

[113] Shi J, Lang C, Wu X, Liu R, Zheng T, Zhang D, et al. RNAi knockdown of fatty acid elongase1 alters fatty acid composition in Brassica napus. Biochemical and Biophysical Research Communications. 2015;**466**(3):518-522. DOI: 10.1016/j.bbrc.2015.09.062

[114] Gil-Humanes J, Pistón F, Hernando A, Alvarez JB, Shewry PR, Barro F. Silencing of γ-gliadins by RNA interference (RNAi) in bread wheat. Journal of Cereal Science. 2008;**48**(3):565-568. DOI: 10.1016/j.jcs.2008.03.005

[115] Kusaba M, Miyahara K, Iida S, Fukuoka H, Takano T, Sassa H, et al. Low glutelin content1: A dominant mutation that suppresses the glutelin multigene family via RNA silencing in rice. The Plant Cell. 2003;**15**(6):1455-1467. DOI: 10.1105/tpc.011452

[116] Kempe K, Higashi Y, Frick S, Sabarna K, Kutchan TM. RNAi suppression of the morphine biosynthetic gene salAT and evidence of association of pathway enzymes. Phytochemistry. 2009;**70**(5):579-589. DOI: 10.1016/j.phytochem.2009.03.002

[117] Sunilkumar G, Campbell LM, Puckhaber L, Stipanovic RD, Rathore KS. Engineering cottonseed for use in human nutrition by tissue-specific reduction of toxic gossypol. Proceedings of the National Academy of Sciences. 2006;**103**(48):18054-18059. DOI: 10.1073/pnas.0605389103

[118] Gupta A, Pal RK, Rajam MV. Delayed ripening and improved fruit processing quality in tomato by RNAi-mediated silencing of three homologs of 1-aminopropane-1-carboxylate synthase gene. Journal of Plant Physiology. 2013;**170**(11):987-995. DOI: 10.1016/j.jplph.2013.02.003

[119] Luo Z, Zhang J, Li J, Yang C, Wang T, Ouyang B, et al. A STAY-GREEN protein S l SGR 1 regulates lycopene and β-carotene accumulation by interacting directly with S l PSY 1 during ripening processes in tomato. New Phytologist. 2013;**198**(2):442-452. DOI: 10.1111/nph.12175

[120] Elitzur T, Yakir E, Quansah L, Zhangjun F, Vrebalov J, Khayat E, et al. Banana MaMADS transcription factors are necessary for fruit ripening and molecular tools to promote shelf-life and food security. Plant Physiology. 2016;**171**(1):380-391. DOI: 10.1104/pp.15.01866

[121] Sunkar R, Zhu J-K. Novel and stress-regulated microRNAs and other small RNAs from Arabidopsis. The Plant Cell. 2004;**16**(8):2001-2019. DOI: 10.1105/tpc.104.022830

[122] Yao Y, Ni Z, Peng H, Sun F, Xin M, Sunkar R, et al. Non-coding small RNAs responsive to abiotic stress in wheat (Triticum aestivum L.). Functional & Integrative Genomics. 2010;**10**(2):187-190. DOI: 10.1007/s10142-010-0163-6

[123] Wang J, Meng X, Dobrovolskaya OB, Orlov YL, Chen M. Non-coding RNAs and their roles in stress response in plants. Genomics, Proteomics & Bioinformatics. 2017;**15**(5):301-312. DOI: 10.1016/j.gpb.2017.01.007

[124] Zhang W, Han Z, Guo Q, Liu Y, Zheng Y, Wu F, et al. Identification of maize long non-coding RNAs responsive to drought stress. PLoS One. 2014;**9**(6):e98958. DOI: 10.1371/journal.pone.0098958

[125] Yang L, Mu X, Liu C, Cai J, Shi K, Zhu W, et al. Overexpression of potato miR482e enhanced plant sensitivity to Verticillium dahliae infection. Journal of Integrative Plant Biology. 2015;**57**(12):1078-1088. DOI: 10.1111/jipb.12348

[126] Boccara M, Sarazin A, Thiebeauld O, Jay F, Voinnet O, Navarro L, et al. The Arabidopsis miR472-RDR6 silencing pathway modulates PAMP-and effector-triggered immunity through the post-transcriptional control of disease resistance genes. PLoS Pathogens. 2014;**10**(1):e1003883. DOI: 10.1371/journal.ppat.1003883

[127] Soto-Suárez M, Baldrich P, Weigel D, Rubio-Somoza I, San SB. The Arabidopsis miR396 mediates pathogen-associated molecular pattern-triggered immune responses against fungal pathogens. Scientific Reports. 2017;**7**(1):1-14

[128] Park YJ, Lee HJ, Kwak KJ, Lee K, Hong SW, Kang H. MicroRNA400-guided cleavage of pentatricopeptide repeat protein mRNAs renders Arabidopsis thaliana more susceptible to pathogenic bacteria and fungi. Plant and Cell Physiology. 2014;**55**(9):1660-1668. DOI: 10.1093/pcp/pcu096

[129] Nie P, Chen C, Yin Q, Jiang C, Guo J, Zhao H, et al. Function of miR825 and miR825* as negative regulators in Bacillus cereus AR156-elicited systemic resistance to Botrytis cinerea in Arabidopsis thaliana. International Journal of Molecular Sciences. 2019;**20**(20):5032. DOI: 10.3390/ijms20205032

[130] Zhang X, Zhao H, Gao S, Wang W-C, Katiyar-Agarwal S, Huang H-D, et al. Arabidopsis Argonaute 2 regulates innate immunity via miRNA393∗– mediated silencing of a Golgi-localized SNARE gene, MEMB12. Molecular Cell. 2011;**42**(3):356-366. DOI: 10.1016/j.molcel.2011.04.010

[131] Salvador-Guirao R, Baldrich P, Weigel D, Rubio-Somoza I, San SB. The microRNA miR773 is involved in the Arabidopsis immune response to fungal pathogens. Molecular Plant-Microbe Interactions. 2018;**31**(2):249-259. DOI: 10.1094/MPMI-05-17-0108-R

[132] Navarro L, Dunoyer P, Jay F, Arnold B, Dharmasiri N, Estelle M, et al. A plant miRNA contributes to antibacterial resistance by repressing auxin signaling. Science. 2006;**312**(5772):436-439. DOI: 10.1126/science.1126088

[133] Li Y, Zhao S-L, Li J-L, Hu X-H, Wang H, Cao X-L, et al. Osa-miR169 negatively regulates rice immunity against the blast fungus *Magnaporthe oryzae*. Frontiers in Plant Science. 2017;**8**:2. DOI: 10.3389/fpls.2017.00002

[134] Wang Z, Xia Y, Lin S, Wang Y, Guo B, Song X, et al. Osa-miR164a targets Os NAC 60 and negatively regulates rice immunity against the blast fungus *Magnaporthe oryzae*. The Plant Journal. 2018;**95**(4):584-597. DOI: 10.1111/tpj.13972

[135] Liu M, Shi Z, Zhang X, Wang M, Zhang L, Zheng K, et al. Inducible overexpression of ideal plant Architecture1 improves both yield and disease resistance in rice. Nature Plants. 2019;**5**(4):389-400. DOI: 10.1038/s41477-019-0383-2

[136] Wu J, Yang R, Yang Z, Yao S, Zhao S, Wang Y, et al. ROS accumulation and antiviral defense control by microRNA528 in rice. Nature Plants. 2017;**3**(1):1-7. DOI: 10.1038/nplants.2016.203

[137] Yao S, Yang Z, Yang R, Huang Y, Guo G, Kong X, et al. Transcriptional regulation of miR528 by OsSPL9 orchestrates antiviral response in rice. Molecular Plant. 2019;**12**(8):1114-1122. DOI: 10.1016/j.molp.2019.04.010

[138] Campo S, Peris-Peris C, Siré C, Moreno AB, Donaire L, Zytnicki M, et al. Identification of a novel micro RNA (mi RNA) from rice that targets an alternatively spliced transcript of the N ramp6 (natural resistance-associated macrophage protein 6) gene involved in pathogen resistance. New Phytologist. 2013;**199**(1):212-227. DOI: 10.1111/nph.12292

[139] Sánchez-Sanuy F, Peris-Peris C, Tomiyama S, Okada K, Hsing Y-I, San Segundo B, et al. Osa-miR7695 enhances transcriptional priming in defense responses against the rice blast fungus. BMC Plant Biology. 2019;**19**(1):1-16. DOI: 10.1186/s12870-019-2156-5

[140] Li Y, Zhang Q, Zhang J, Wu L, Qi Y, Zhou J-M. Identification of microRNAs involved in pathogen-associated molecular pattern-triggered plant innate immunity. Plant Physiology. 2010;**152**(4):2222-2231. DOI: 10.1104/pp.109.151803

[141] Li F, Pignatta D, Bendix C, Brunkard JO, Cohn MM, Tung J, et al. MicroRNA regulation of plant innate immune receptors. Proceedings of

the National Academy of Sciences. 2012;**109**(5):1790-1795. DOI: 10.1073/pnas.1118282109

[142] Deng Y, Wang J, Tung J, Liu D, Zhou Y, He S, et al. A role for small RNA in regulating innate immunity during plant growth. PLoS Pathogens. 2018;**14**(1):e1006756. DOI: 10.1371/journal.ppat.1006756

[143] de Vries S, Kukuk A, von Dahlen JK, Schnake A, Kloesges T, Rose LE. Expression profiling across wild and cultivated tomatoes supports the relevance of early miR482/2118 suppression for Phytophthora resistance. Proceedings of the Royal Society B: Biological Sciences. 2018;**285**(1873):20172560. DOI: DOI/10.1098/rspb.2017.2560

[144] He X-F, Fang Y-Y, Feng L, Guo H-S. Characterization of conserved and novel microRNAs and their targets, including a TuMV-induced TIR–NBS–LRR class R gene-derived novel miRNA in brassica. FEBS Letters. 2008;**582**(16):2445-2452. DOI: 10.1016/j.febslet.2008.06.011

[145] Zhai J, Jeong D-H, De Paoli E, Park S, Rosen BD, Li Y, et al. MicroRNAs as master regulators of the plant NB-LRR defense gene family via the production of phased, trans-acting siRNAs. Genes & Development. 2011;**25**(23):2540-2553. DOI: 10.1101/gad.177527.111

[146] Cui X, Yan Q, Gan S, Xue D, Dou D, Guo N, et al. Overexpression of gma-miR1510a/b suppresses the expression of a NB-LRR domain gene and reduces resistance to *Phytophthora sojae*. Gene. 2017;**621**:32-39. DOI: 10.1016/j.gene.2017.04.015

[147] Liu J, Cheng X, Liu D, Xu W, Wise R, Shen Q-H. The miR9863 family regulates distinct Mla alleles in barley to attenuate NLR receptor-triggered disease resistance and cell-death signaling. PLoS Genetics. 2014;**10**(12):e1004755. DOI: 10.1371/journal.pgen.1004755

[148] Zou Y, Wang S, Zhou Y, Bai J, Huang G, Liu X, et al. Transcriptional regulation of the immune receptor FLS2 controls the ontogeny of plant innate immunity. The Plant Cell. 2018;**30**(11):2779-2794. DOI: 10.1105/tpc.18.00297

[149] Zhang H, Tao Z, Hong H, Chen Z, Wu C, Li X, et al. Transposon-derived small RNA is responsible for modified function of WRKY45 locus. Nature Plants. 2016;**2**(3):1-8. DOI: 10.1038/nplants.2016.16

[150] Jiang N, Cui J, Shi Y, Yang G, Zhou X, Hou X. Tomato lncRNA23468 functions as a competing endogenous RNA to modulate NBS-LRR genes by decoying miR482b in the tomato-Phytophthora infestans interaction. Horticulture Research. 2019:6-28. DOI: 10.1038/s41438-018-0096-0

[151] Jiang N, Cui J, Hou X, Yang G, Xiao Y, Han L, et al. Sl-lncRNA15492 interacts with Sl-miR482a and affects Solanum lycopersicum immunity against Phytophthora infestans. The Plant Journal. 2020;**103**(4):1561-1574. DOI: 10.1111/tpj.14847

[152] Hong Y-H, Meng J, Zhang M, Luan Y-S. Identification of tomato circular RNAs responsive to Phytophthora infestans. Gene. 2020;**746**:144652. DOI: 10.1016/j.gene.2020.144652

[153] Cui J, Luan Y, Jiang N, Bao H, Meng J. Comparative transcriptome analysis between resistant and susceptible tomato allows the identification of lnc RNA 16397 conferring resistance to Phytophthora infestans by coexpressing glutaredoxin. The Plant Journal. 2017;**89**(3):577-589. DOI: 10.1111/tpj.13408

[154] Cui J, Jiang N, Meng J, Yang G, Liu W, Zhou X, et al. LncRNA33732-respiratory burst oxidase module associated with WRKY1 in tomato-Phytophthora infestans interactions. The Plant Journal. 2019;**97**(5):933-946. DOI: 10.1111/tpj.13408

[155] Hou X, Cui J, Liu W, Jiang N, Zhou X, Qi H, et al. LncRNA39026 enhances tomato resistance to Phytophthora infestans by decoying miR168a and inducing PR gene expression. Phytopathology. 2020;**110**(4):873-880. DOI: 10.1094/PHYTO-12-19-0445-R

[156] Sprinzl M, Hartmann T, Weber J, Blank J, Zeidler R. Compilation of tRNA sequences and sequences of tRNA genes. Nucleic Acids Research. 1989;**17**(Suppl):r1-r172. DOI: 10.1093/nar/17.suppl.r1

[157] Szymanski M, Erdmann VA, Barciszewski J. Non-coding RNAs database (ncRNAdb). Nucleic Acids Research. 2007;**35**(suppl_1):D162-D164. DOI: 10.1093/nar/gkl994

[158] Leonard S, Meyer S, Lacour S, Nasser W, Hommais F, Reverchon S. APERO: A genome-wide approach for identifying bacterial small RNAs from RNA-Seq data. Nucleic Acids Research. 2019;**47**(15):e88. DOI: 10.1093/nar/gkz485

[159] Zhang XO, Wang HB, Zhang Y, Lu X, Chen LL, Yang L. Complementary sequence-mediated exon circularization. Cell. 2014;**159**(1):134-147. DOI: 10.1016/j.cell.2014.09.001

[160] Trapnell C, Roberts A, Goff L, Pertea G, Kim D, Kelley DR, et al. Differential gene and transcript expression analysis of RNA-seq experiments with TopHat and cufflinks. Nature Protocols. 2012;**7**(3):562-578. DOI: 10.1038/nprot.2012.016

[161] Liu D, Mewalal R, Hu R, Tuskan GA, Yang X. New technologies accelerate the exploration of non-coding RNAs in horticultural plants. Horticulture Research. 2017;**4**. DOI: 10.1038/hortres.2017.31

[162] Annonymous. RNAcentral Expert Databases. Available from: https://rnacentral.org/expert-databases 2006. [Accessed: 2022.06.03]

Chapter 6

Pathogenesis-Related Proteins and Their Transgenic Expression for Developing Disease-Resistant Crops: Strategies Progress and Challenges

Anroop Kaur, Sukhpreet Kaur, Ajinder Kaur, Navraj Kaur Sarao and Devender Sharma

Abstract

Various pathogenic microorganisms (such as fungi, bacteria, viruses and nematodes) affect plant viability and productivity. However, plants combat these pathogens by inducing their defense mechanism to sustain their fitness. The aggregation of pathogenesis-related (PR) proteins in response to invading pathogens is a crucial component of a plant's self-defense mechanism. PR proteins induce innate resistance in plants through fungal cell wall disintegration, membrane permeabilization, transcriptional suppression, and ribosome inactivation. Earlier studies have demonstrated their crucial role in determining resistance against phytopathogens, making them a promising candidate for developing disease-resistant crop varieties. Plant genetic engineering is a potential approach for developing disease-resistant transgenic crops by employing several PR genes (thaumatin, osmotin-like proteins, chitinases, glucanases, defensins, thionins, oxalate oxidase, oxalate oxidases like proteins/ germin-like proteins and LTPs). Furthermore, the overexpression of PR proteins enhances the resistance against phytopathogens. As a result, this chapter gives an overview of PR proteins, including their classification, functional characterization, signaling pathways, mode of action and role in defense against various phytopathogens. It also highlights genetic engineering advances in utilizing these genes singly or synergistically against various phytopathogens to impart disease resistance. Various challenges faced with the products of transgenic technology and synergistic expression of different groups of PR proteins were also discussed.

Keywords: biotic stress, pathogen-related proteins, plant genetic engineering, plant defense signaling, disease-resistant crops

1. Introduction

With the rapid expansion in the world population, the area under cultivation has decreased [1]. Moreover, biotic stress has been a significant challenge for farmers since the dawn of agriculture. Global yield loss due to plant pathogens is estimated at 16% [2]. To overcome the economic loss in agricultural production, most research in this field focuses on protecting crops against pathogens, insect pests and nematodes. Crop production and productivity can be enhanced by significant breakthroughs in agricultural practices such as cultural controls, pesticide application, crop rotation, and plant breeding.

On the other hand, Pathogens frequently escape chemicals through strong selection and evolution, resulting in crop loss due to infection. Altering the genetic architecture of crops through breeding programmes is another option for crop protection, but it is a labor-intensive and time-consuming operation. In theory, genetic engineering, which refers to the use of biotechnology to alter an organism's genetic material directly [3], is a potential tool for improving disease resistance. Furthermore, genetic engineering can overcome the limitations of traditional breeding technology, including the introduction/alteration of specific genes with minimum undesirable changes to the rest of the genome; cross-species exchange of genetic material; and introduction of variations/genes into asexually propagated crops like bananas [4]. As a result, research studies have been directed toward the genes that impart long-term resistance to many pests or pathogens and are safe for consumption.

In plants, tolerance and susceptibility to a particular pathogen are determined by a complex interaction of signals and responses corresponding to specific environmental conditions. So, the major difference between resistant and susceptible varieties is the ability to recognize an invading pathogen and further activate host defense mechanisms. Plants have evolved various defense mechanisms, including activating both constitutive and inducible defense responses to combat the diseases. When pathogens are detected, immune receptors in plants recognize specific molecules that signal the activation of effective defense responses. Despite extensive research, details of host defense mechanisms that limit pathogenic infections have yet to be elucidated. The majority of defense responses are characterized by the transcriptional activation of a large number of genes (>1% of the genome), many of which have unknown functions [5, 6]. Pathogen identification activates signaling pathways that result in the formation of reactive oxygen species (ROS), protein kinases, phytohormones, phytoalexins, phenolic compounds and pathogenesis-related (PR) proteins, and eventually a hypersensitive response (HR). Production and accumulation of PR proteins, which are low molecular weight proteins, in plants during pathogen attack is vital [7]. In most plant species, nineteen families of PR proteins (PR-1 to PR-19) have been identified to date. The role of PR proteins in plant defense includes altering the integrity of pathogen and activating other defense pathways through the generation of elicitors.

Due to improvements in transformation techniques and isolation of numerous pathogenesis-related genes, plants can now be engineered to have effective and broad-spectrum resistance against pathogens. The transgenic approaches using PR genes have been proven to be efficient for obtaining pathogen resistance in plants [8, 9]. Several transgenic plants have been developed that offer varying degrees of protection against certain fungal and oomycete diseases.

This chapter overviews the PR proteins, including their classification, activation as defense signaling indicators, and mode of action against the pathogens. It also highlights the success and challenges of the transgenic approach using PR genes for disease resistance.

Families	Type member	Plant source	Gene accession no.	Classes/source	Size (kDa)	Properties	References
PR-1	Tobacco PR-1a	*Nicotiana tabacum*	YOO707		15–17	Antifungal	[14]
PR-2	Tobacco PR-2	*N. tabacum*	M59443.1	Classes III		β-1,3-Glucanase	[14]
				I plant vacuole	~33		[15]
				II, III extracellular proteins	~36		[15]
PR-3	Tobacco P, Q	*N. tabacum*	X77111.1	Classes V	25–30	Chitinase type I, II, IV, V, VI, VII	[16]
				I	~32		[15]
				II	27–28		[15]
				III	28–30		[15]
				IV	28–30		[15]
				V	41–43		[15]
PR-4	Tobacco "R"	*N. tabacum*	NW_015888419.1	Classes II	15–20	Chitinase type I, II	[16]
				I			[15]
				II			[15]
PR-5	Tobacco S	*N. tabacum*	NW_015793016		22–25	Thaumatin, antifungal, osmotin, zeamatin	[16, 17]
PR-6	Tomato inhibitor I	*Solanum lycopersicum*	NW_004196001.1		8	Proteinase inhibitor	[18]
PR-7	Tomato P69	*S. lycopersicum*	NC_015445.2		75	Endoproteinase	[19]
PR-8	Cucumber chitinase	*Cucumis sativus*	NC_026660.1		28	Chitinase type III	[20]
PR-9	Tobacco "lignin-forming peroxidase"	*Solanum* tuberosum	AJ401150		35	Peroxidase	[21]
PR-10	Parsley "PR1"	*Petroselinum crispum*	NC_026940.1	Classes III	17	Ribonuclease-like protein	[22]

Families	Type member	Plant source	Gene accession no.	Classes/source	Size (kDa)	Properties	References
				I	11 to 30		[15]
				II	~60		[15]
				III	~60		[15]
PR-11	Tobacco “class V” chitinase	*N. tabacum*	gi|899,342	—	40	Chitinase, type I	[23]
PR-12	Radish Rs-AFP3	*Raphanus raphanistrum*	NC_025209.1	Class IV	3–5	Defensin	[15, 24]
PR-13	Arabidopsis THI2.1	*Arabidopsis thaliana*	gi|1,181,531	—	5	Thionin	[25]
PR-14	Barley LTP4	*Hordeum vulgare*	gi|1,045,201	—	8.7–9	Lipid-transfer protein	[15, 26]
PR-15	Barley OxOa (germin)	*H. vulgare*	gi|2,266,668	—	20	Oxalate oxidase	[27]
PR-16	Barley OxOLP	*H. vulgare*	gi|1,070,358	—	20	Oxalate oxidase-like	[28]
PR-17	Tobacco PRp27	*N. tabacum*	—	—	27	Antifungal and antiviral	[29]
PR-18	Carbohydrate oxidases	*Helianthus annuns*	AF472608	—	60.9	Carbohydrate oxidases	[30]
PR-19	antimicrobial protein	*Pinus Sylvestris*	AF410954	—	—	antimicrobial protein	[31]

Table 1.
Classification and properties of PR proteins.

2. PR proteins and their classifications

PR proteins are defined as "Proteins encoded by the host plant but induced only in pathogenic or related conditions" [10]. Plant PR proteins were discovered and published for the first time in tobacco plants infected with the tobacco mosaic virus [11] and initially, only PR-1, PR-2, PR-3, PR-4, and PR-5 classes of PR proteins were reported from tobacco plants, but later different PR proteins were found from numerous plants [12]. These low molecular weight proteins (6–43 kDa) are heat stable, protease-resistant and soluble at acidic pH (<3) [13]. PR proteins are currently classified into 19 major families based on their enzymatic activity, biological roles, and amino acid sequences, as indicated in **Table 1** [31, 32]. These include antifungal (PR1), hydrolytic β-1,3- Glucanase (PR2), chitinases (PR 3, 4, 8,11), thaumatin (PR5), proteinase inhibitors (PR6), endo-proteinase (PR7), peroxidase (PR9), ribonuclease-like (PR10), plant defensins (PR12), plant thionins (PR13), lipid transfer proteins (PR14), oxalate oxidase protein family (PR15 and PR16) secretory protein (PR17) and carbohydrate oxidases (PR 18) [7, 32]. A novel antimicrobial protein from *Pinus sylvestris* was isolated and classified as PR19 [31].

3. PR proteins: functional characterization and mode of action

Plants are constantly being challenged by disease-causing organisms that have co-evolved with the evolution of plant hosts' defense mechanisms. Many PR proteins have been shown to possess antifungal, antibacterial, antiviral and antinematode properties [13]. Different PR proteins have a distinct mode of action against the pathogen depending upon the type of pathogen and the activities of the majority of these protein families are known or can be inferred. PR-1 protein, one of the dominant groups of PRs induced by the pathogen, inhibits pathogen growth by binding and sequestration of sterols from the pathogen. Moreover, the programmed cell death is also inhibited by PR1 upon pathogen infection by releasing a defense signal peptide CAPE1 (CAP-derived peptide 1) [33]. Some PR proteins function as hydrolytic enzymes, *viz.* the PR-2 (endo-β-1,3-glucanases) and PR-3, −4, −8 and − 11 (endo-chitinases) [34, 35]. They function as antifungal proteins by catalyzing hydrolytic cleavage of major components of fungal and oomycete cell wall, i.e. β-1,3-glucan (by the breakdown of β-1,3-glucosidic linkages) or chitin (by the breakdown of internal β-1,4-glycoside bonds) respectively, resulting in the breakdown of the fungal cell wall [36, 37]. Different isoforms of glucanases and chitinases are produced depending upon the plant-pathogen interaction.

Thaumatin-like proteins or Osmotin-like proteins such as PR5 inhibit hyphal growth and spore germination by producing transmembrane pores leading to fungal cell leakiness and blocking the function of plasma membrane receptors molecules involved in cAMP/RAS2 signaling pathways. Also, antifungal action has been demonstrated in some family members, predominantly against oomycetes. PR-5 was also demonstrated to exhibit potato cell's defense against *Phytophthora infestans* by forming a cytoplasmic aggregation through an actin-binding complex [38]. Proteinase inhibitors (PIs) such as trypsin inhibitors and serine inhibitors) belonging to PR6 family proteins, implicated in broad-spectrum defense activity, including suppressing pathogenic nematodes, insects and other herbivores, fungi and bacteria [39]. PIs can provide defense against pathogens, decreasing the lyase activity essential for fungal pathogenicity [40], inhibiting the viral replication cycle [41] and restricting the digestive enzyme activity of nematodes and insects, limiting amino acid release [42]. In addition, HyPep (proteinase inhibitor peptide) also causes cell aggregation and pseudo-mycelia development by inhibiting amylase and serine proteinases [43]. Also, PIs can block chitin synthesis in fungal cell walls by inhibiting endogenous trypsin that is essential for chitin synthase, thus inhibiting fungal growth and development [44].

PR-7 is a major protein that has only been examined in tomatoes as an endoproteinase. It is an antifungal auxiliary protein that aids in destroying fungal cell wall proteins, chitinases, and glucanases [45]. The PR-9 family of peroxidases is believed to have a role in plant cell wall strengthening by facilitating lignin deposition in response to microbial invasion [46]. In susceptible wheat varieties, the transcription level of PR9 is considerably reduced after infestation with the aphid-transmitted fusarium virus and hessian flies [47]. This showed that PR9 catalyzes lignin deposition to protect susceptible cultivars from BPH.

The members of PR10 protein families exhibit ribonuclease activity required to inhibit the growth of pathogenic fungi. The antifungal activity of ribonucleases develops due to penetration of the pathogen and the destruction of cellular RNAs due to phosphorylation of PR10. It further leads to plant cell death at the inoculation site, causing apoptosis and the hypersensitivity reaction [48]. These intracellular PRs may be active against viruses due to their ribonuclease activity, although their ability to cleave viral RNA has yet to be shown.

The PR-12 type defensins, PR-13 type thionins, and PR-14 type lipid transfer proteins show antifungal and antibacterial activity, interacting with the target microorganism's biological membrane, leading to altered membrane permeability [49, 50]. Plant defensins are divided into two groups based on the structure of their precursor proteins: class I and class II. Class I defensins have endoplasmic reticulum (ER) signaling sequences along with defensin domains. In contrast, class II defensins contain an additional domain of 27–33 amino acid residues called C-terminal prepropeptide (CTPP) [51]. Due to a lack of signal sequences, class I defensins do not undergo post-translational modification or subcellular targeting. They accumulate in the cell wall and extracellular space directly upon synthesis through the secretory pathway [52]. However, class II defensins undergo proteolysis in the vesicles due to CTPP signal peptides targeting vesicles and releasing mature short peptides. Mature defensins consist of five segments of non-conserved loops, linking α-helices and β-strands to form high-level structures. Differences in the loop sequences confer different functions, including inhibition of protein synthesis, antimicrobial activity, heavy metal tolerance, plant development, and blocking of ion channels [53].

Oxalate oxidases (PR-15 family) and oxalate-oxidase-like proteins (PR-16 family) play an important role in plant defense [54]. These are essential enzymes to produce reactive oxygen species (ROS) during apoplastic oxidative burst [55]. ROS are produced in the apoplast by an enzyme that produces H_2O_2 and CO_2 when it reacts with oxalic acid. Proteolytic enzymes of the PR17 family play an important role in defense against fungi and viruses. PR19 protein binds to fungal cell wall glucans altering cell wall structure, leading to morphological distortion of hyphae [31].

4. PR protein activation as a defense response

Plant cells have evolved to activate and recruit the cellular machinery in response to various stresses to optimally utilize resources and sustain life. Accordingly, plants modulate genes' expression, activating a wide range of plant protectants and defense

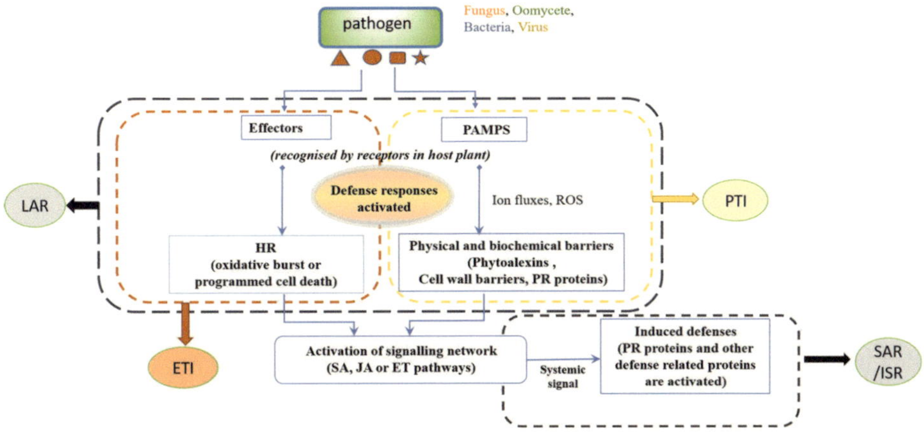

Figure 1.
Overview of the activation of defense response against the pathogen including induction of PR proteins locally as well as systematically.

genes [56]. The pathogenesis-related (PR) protein activation and production are crucial in response to an invading pathogen [57]. While healthy plants may produce a trace amount of PR proteins, they are produced in higher concentrations in response to pathogen attacks, elicitor treatment, wounding, or other stress.

Plants defend themselves against pathogen attacks by employing a variety of defense mechanisms for their survival and fitness [58]. After the pathogen challenge, plants trigger basal defense mechanism, i.e., pattern triggered immunity (PTI), by recognizing the pathogen-associated molecular patterns (PAMPs) and induced defense mechanism, i.e., effector-triggered immunity (ETI) [59]. PTI and ETI are accompanied by a set of preformed defenses (structural and biochemical barriers) and/or induced defense responses (hypersensitive reactions) that usually combat pathogen attacks [60]. Depending upon the plant-pathogen interaction, these defense responses are associated with a coordinated and integrated set of metabolic alterations that lead to induction of systemic acquired resistance (SAR) or induced systemic resistance (ISR) through activation of defense signaling pathways viz., salicylic acid (SA) and jasmonic acid (JA)/ ethylene (ET) respectively. The activation of SA or JA signaling pathway leads to downstream activation and accumulation of PR gene products locally as well as systematically (**Figure 1**). As a result, PR proteins are related to the development of systemic acquired resistance (SAR) or a hypersensitive response (HR) to pathogenic fungi, bacteria, and viruses. Many plant species from many families have been shown to be induced by PRs, implying that PRs have a broad protective effect against biotic stress [57].

5. Role of signaling pathways in PR protein induction

Depending upon the host-pathogen interaction, different signaling systems are activated, producing different sets of PR proteins that provide disease resistance in plants. Basically, pathogens can be categorized into two types depending on the mode of infection: biotrophic and necrotrophic. Based on the type of pathogen, the pathogenic elicitors induce the production of different secondary signals such as ROS, jasmonates, salicylic acid or ethylene, which further induce the expression of different PR genes. Within the plant species, these secondary signals' spatial and temporal production vary depending on pathogen type [61]. Classically, the resistance against biotrophic pathogens is conferred through the salicylic acid (SA) pathway, whereas against necrotrophic pathogens is conferred through activation of jasmonic acid/ ethylene (JA/ET) pathways [62].

In plant-biotrophic pathogen interaction, the SA signaling system induces the expression of signature PR genes related to this pathway *viz* PR1, PR2, PR5, PR8, PR9 and PR10 [63]. The transcription studies (overexpression of PR genes), as well as mutational studies (SA mutants such as *nim1, npr1, sai1, nahG*), have provided evidence of the dependence of these PR genes on SA signaling pathway [64]. SA-mediated defense signaling regulates the expression of the PR genes through binding with and activating the NPR1 (due to conformational changes). Activated NPR1 interacts with transcription factors such as TGACG-binding factor (TGA), thus inducing defense gene expression [65]. However, in plant-necrotrophic pathogen interaction, it has been found through transgenic expression of PR genes as well as JA mutant analysis that the JA/ET signaling pathway induces the expression of PR3, PR4, PR10, PR11, PR12 and PR13 genes [64]. ET signaling pathway induces the expression of PR genes by activating the ETHYLENE RESPONSE FACTOR (ERF) transcription factor through activation of EIN2 and EIN3 proteins. However, in the JA signaling pathway, JAZ (jasmonate ZIM domain) protein is degraded by COL1 (coronatine insensitive 1) mediated 26S proteasome leading to activation of MYC2 transcription factor and hence transcription of JA responsive genes [66]. Furthermore, applying JA or SA hormones (defense hormones) increases the PR genes' transcription level, providing a broad spectrum of resistance [67]. During plant-pathogen interaction, hormonal crosstalks also occur, which can provide novel insights for disease resistance. PR-6 in tomato leaves generated by systemic and jasmonic acid was suppressed by exogenous application of SA. When a pathogen infects tobacco, ethylene may operate downstream of jasmonic acid to activate PR-2 and 3.

The ERF branch's ET/JA-regulated transcription factors are inhibited by the negative regulators of the SA signaling pathway. Also, SA biosynthesis is inhibited upon activation of the ET/JA signaling pathway, depicting these pathways' antagonistic role in defense response [68].

6. Pathogenesis-related proteins (PR-proteins) with their transgenic expression

With the development of modern DNA technology, it is possible to engineer transgenic plants transformed with genes to provide resistance against specific diseases. Recently the transgenic expression of various groups of PR proteins has enhanced the resistance of the transformed plant against several plant pathogens (**Table 2**). PR proteins are found in all organisms and are part of their innate immune systems. They have a wide range of activities, including disrupting fungal cell walls, permeabilizing membranes, inhibiting transcription, and inactivating ribosomes [117]. Genes coding for various PR proteins have been identified, cloned, and expressed in plants, preventing the development of specific diseases and conferring resistance to affected plants. Using modern biotechnology tools, various crops have been engineered to express, or over-express the PR proteins from different sources, such as (i) that are produced during the plant's defense response, (ii) derived from microorganisms or animal cells, (iii) synthetic peptides designed based on sequences of existing antimicrobial compounds [32, 118].

DOI: http://dx.doi.org/10.5772/intechopen.106774

Enzyme	Genes	Source	Target pathogen	Transgenic plant	Transgenic system	Reference
Glucanase	*β-1,3-glucanase*	*Linum usitatissimum*	*Fusarium culmorum*	Potato	*Agrobacterium* -mediated transformation	[69]
	HbGLU	*Hevea brasiliensis*	*Rhizoctonia solani*	Potato	*Agrobacterium*-mediated transformation	[70]
	β-1,3-glucanase II cDNA	*Hordeum vulgare*	*Fusarium graminearum*	Wheat	Particle gun bombardment	[71]
	chi-2, ltp	*Hordeum vulgare, Triticum aestivum*	*Alternaria radicicola* and *Botrytis cinerea*	Carrot	*Agrobacterium*-mediated transformation	[72]
	McCHIT1	*Momordica charantia*	*Magnaporthe grisea* and *Rhizoctonia solani*	Rice	Electroporation	[73]
	OsPR4a–e	*Oryza sativa*	*Magnaporthe grisea*	Rice	*Agrobacterium*-mediated transformation	[74]
	RC7	*Oryza sativa*	*Rhizoctonia solani*	Rice	Biolistic and PEG-mediated transformation system	[75]
	BjCHI1	*Brassica juncea*	*Rhizoctonia solani*	Potato	*Agrobacterium*-mediated transformation	[70]
	chit cDNA	*Hordeum vulgare*	*Fusarium graminearum*	Wheat	biolistic bombardment	[76]
	Chitinase-I	*Oryza sativa*	*Verticillium dahliae and Fusarium oxysporum*	Eggplant	*Agrobacterium*-mediated transformation	[77]
	RC24	*Oryza sativa*	*Puccinia striiformis f.sp. tritici*	Wheat	Particle bombardment	[78]
	rcc2 and *rcg3*	*Oryza sativa*	*Puccinia striiformis f.sp. tritici*	Wheat	*Agrobacterium*-mediated transformation	[79]
	LcCHI2	*Leymus chinensis*	*Pseudomonas tabaci, A. alternata, Exserohilum turcicum, Curvularia lunata*	Maize	*Agrobacterium*-mediated transformation	[80]

Enzyme	Genes	Source	Target pathogen	Transgenic plant	Transgenic system	Reference
Thaumatin	*Thaumatin-likeTaLr19TLP1*	*Triticum aestivum*	*Puccinia triticina*	Wheat	virus-induced gene silencing	[81]
	Tlp	*Triticum aestivum*	*Fusarium graminearum*	Wheat	biolistic transformation	[82]
	Tlp	*Oryza sativa*	*Alternaria solani*	Tomato	*Agrobacterium*-mediated transformation	[83]
	Tlp	*Oryza sativa*	*Rhizoctonia solani*	Rice	Particle bombardment	[84]
	tlp-1	*Hordeum vulgare*	*Fusarium graminearum*	Wheat	*Agrobacterium*-mediated transformation	[71]
	CsTLP	*Camellia sinensis*	*Phytopthora infestans and Macrophomina phaseolina*	Potato	*Agrobacterium* transformation	[85]
	AdTLP	*Arachis diogoi*	*Rhizoctonia solani*	Tobacco	*Agrobacterium*-mediated transformation	[86]
Osmotin-like proteins	*OsOSM1*	*Oryza sativa*	*Rhizoctonia solani*	Rice	*Agrobacterium*-mediated transformation	[87]
	OsmWS	Withania somnifera	*A. solani*	Potato	*Agrobacterium*-mediated transformation	[88]
	JIOsPR10	*Oryza sativa*	*Magnaporthe oryzae*	Rice	*Agrobacterium*-mediated transformation	[89]
Ribonuclease like protein	*GmPRP*	*Glycine max*	*Phytophthora sojae*	Soybean	*Agrobacterium*-mediated transformation	[90]
Ribonuclease inactivating protein	*PAP*	*Phytolacca ameriacana*	Cucumber mosaic virus, Potato virus X, Potato virus Y	Tobacco and Potato	*Agrobacterium*-mediated transformation	[91]
	PAP	*Phytolacca ameriacana*	*Sclerotinia homoeocarpa*	Beet grass	Particle bombardment	[92]
Proteinase inhibitor	*mpi*	*Zea mays*	*Chilo suppressalis*	Rice	Particle bombardment or *Agrobacterium*-mediated transformation	[93]
	cry1B	*Zea mays*	*Chilo suppressalis*	Rice	particle bombardment	[94]

Enzyme	Genes	Source	Target pathogen	Transgenic plant	Transgenic system	Reference
Defensins	*Wasabi*	*Wasabia japonica* L.	*Magnaporthe grisea*	Rice	*Agrobacterium*-mediated transformation	[95]
	Wasabi	*Wasabia japonica* L.	*Botrytis cinerea*	Potato	*Agrobacterium*-mediated transformation	[96]
	MsDef1	*Medicago sativa*	*Fusarium oxysporum f. sp. lycopersici*	Tomato	-	[97]
	MtDef4.2	*Medicago truncatula*	*Puccinia triticina*	Wheat	*Agrobacterium*-mediated transformation	[67]
	RsAFP2	*Raphanus sativus*	Rhizoctonia solani and *Magnaporthe grisea*	Rice	*Agrobacterium*-mediated transformation	[98]
	RsAFP2	*Raphanus sativus*	*Rhizoctonia cerealis, Fusarium graminearum*	Wheat	Biolistic bombardment	[99]
	Wasabi	*Wasabia japonica* L.	*Alternaria solani* and *Fusarium oxysporum*	Melon	*Agrobacterium*-mediated transformation	[100]
	BoDFN	*Brassica oleracea*	Downy Mildew	Wild cabbage	*Agrobacterium*-mediated transformation	[101]
	VrPDF1	*Vigna radiata*	Weevils	mungbean	*Agrobacterium*-mediated transformation	[102]
	TAD1	*Triticum aestivum*	*Typhula ishikariensis, Fusarium graminearum*	Wheat	particle bombardment	[103]
Thionins	*AT1G12660 and AT1G12663*	*A. thaliana*	*R. solani* and *F. oxysporum*	Potato	*Agrobacterium*-mediated transformation	[104]
	Thionin	*Brassica oleracea* var. acephala, *Nasturtium officinale* and *Barbarea vulgaris*	*B. cinerea*	Potato	*Agrobacterium*-mediated transformation	[105]
	α-hordothionin (αHT)	*Hordeum vulgare*	*Ceratocystis fimbriata*	Sweet potato	*Agrobacterium*-mediated transformation	[106]
	Thi2.1	*A. thaliana*	*Fusarium oxysporum*	Tomato	*Agrobacterium*-mediated transformation	[107]

Enzyme	Genes	Source	Target pathogen	Transgenic plant	Transgenic system	Reference
Oxalate Oxidase	*OXO*	*Triticum aestivum*	*Sclerotinia sclerotiorum*	Soybean	*Agrobacterium*-mediated transformation	[108]
	Osoxo4	*Oryza sativa*	*Phytophthora infestans*	Potato	*Agrobacterium*-mediated transformation	[109]
	OXO	*Hordeum vulgare*	*Botrytis cinerea* and *Sclerotinia sclerotiorum*	Tomato	*Agrobacterium*-mediated transformation	[110]
	OXO	*Triticum aestivum*	*Phytophthora infestans*	Potato	*Agrobacterium*-mediated transformation	[111]
Lipid Transfer Proteins	*AtLTP4.4*	*A. thaliana*	*F. graminearum*	Wheat	particle bombardment	[112]
	Ace-AMP1	*Allium cepa*	*Sphaerotheca pannosa var. rosae*, *Blumeria graminis f. sp. tritici* and *Neovossia indica* , *Magnaporthe grisea* and *Rhizoctonia solani*	Wheat and rice	*Agrobacterium*-mediated transformation, microprojectile bombardment, In planta assays	[113–116]
Carbohydrate oxidases	-	*Helianthus Annuus*	*Pectobacterium cartovorum* ssp. *cartovorum*	Tobacco	Electroporation	[30]
Antimicrobial protein	*Sp-AMP*	*Pinus Sylvestris*	*Heterobasidion annosum*	Tobacco	*Agrobacterium*-mediated transformation	[31]

Table 2.
Example of transgenic plants over-expressing PR proteins against plant pathogens.

7. Transgenic plants expressing antifungal activities

Fungi are one of the most harmful phytopathogens, resulting in considerable production losses in most agricultural crops [119]. PR proteins have proven effective in preventing fungal diseases in plants as many of these targets or hydrolyze fungal cell walls, resulting in cell death. PR1, PR2, PR3, PR4, PR5, PR8, PR11, PR12, and PR13 have been identified as plants' most effective antifungal proteins. Transgenic approaches using PR proteins are suitable for developing long-lasting fungal pathogen-resistant crops [64]. Of the various antifungal PR proteins, glucanases and chitinases are most widely used in transgenic technology to provide resistance against fungus.

The transgenic over-expression of glucanase and chitinase genes from different sources has been shown to be effective against pathogens, specifically fungus. It has been reported that overexpression of the tobacco glucanase gene imparted groundnut resistance to *Cercospora arachidicola* and *Aspergillus flavus*, demonstrating that fungal resistance is conferred via *in planta* transformation [120]. Transgenic Arabidopsis plants expressing grapevine *b-1,3-glucanase* (*VvGHF17*) confers resistance to *Colletotrichum higginsianum* and *Botrytis cinerea* [121]. Furthermore, tea with transgenic overexpression of the endo-*1,3-D-glucanase* gene, which expresses a potato glucanase, significantly improved tolerance to the blister blight fungus *Exobasidium vexans* [122]. Recently, oil palm resistance to *G. boninense* was improved by transgenic overexpression of *M. sativa* glucanase (*AGLU1*) [123]. Likewise, transgenic expression of chitinase genes have been reported to be antifungal generated transgenic zoysia grass was generated which overexpressed *Zjchi2* via *Agrobacterium-mediated* transformation and hence showed disease resistance against *Rhizoctonia solani* [124]. Currently, the overexpression of *LcCHI2* gene was identified that increasing the chitinase activity in transgenic tobacco and maize, resulting in improved resistance *to Pseudomonas tabaci*, *Alternaria alternata*, *Exserohilum turcicum*, *Curvularia lunata* [80].

Some other antifungal PR proteins that have been reported to be used in transgenics are thaumatin-like/osmotin-like proteins, defensin-like proteins, thionin, oxalate oxidase and lipid transfer protein. In fungal cells, thaumatin-like proteins are known to form transmembrane pores, whereas osmotin proteins are known to maintain the osmolarity of suitable solutes in cellular compartments [88]. In Arabidopsis thaliana, overexpression of the *TLP29* gene from grape *VqTLP29* improved resistance to powdery mildew and the bacteria *Pseudomonas syringae* [125]. Under *in vitro* conditions, transgenic poplars overexpressing *PeTLP* thaumatin genes showed enhanced resistance to *Marssonina brunnea* [126]. Similarly, in potatoes, overexpression of the osmotin gene (*OsmWS*) conferred resistance to the early blight fungus *A. solani* [88]. Many more transgenic plants have been generated that show increased resistance to phytopathogenic fungi by expressing the TLPs and OLPs as listed in **Table 2**.

The successful developed and characterized transgenic peanut and tobacco plants which overexpress the mustard *defenisn* gene and *Raphanus sativa*, *RsAFP2* gene for fungal resistance respectively [127]. The late leaf spot diseases *Cercospora arachidicola* and *Pheaoisariopsis personata* were more resistant to transgenic peanut plants whereas, *Phytophthora parasitica* pv. nicotianae and *Fusarium moniliforme* resistance was higher in transgenic tobacco plants. Similarly, the *rDrr230a* defensin protein gene suppressed spore germination and growth of both *Fusarium tucumaniae* and *Colletotrichum gossypii* var. cephalosporioides in transgenic *Pichia pastoris* [128]. The antifungal thionin genes (*AT1G12660* and *AT1G12663*) from *A. thaliana* had been used to produce transgenic potato conferring resistance against pathogenic fungi such as *Fusarium solani* and *Fusarium oxysporum* [104]. Furthermore, the overexpression of thionin increased canker resistance and decreased canker bacterial development when transgenic Carrizo plants expressing the modified plant thionin were produced by *Agrobacterium-mediated* transformation [129]. Peanuts with transgenic expression of the oxalate oxidase expressing gene were more resistant to *Sclerotinia* blight [130]. Also, overexpression of oxalate oxidase genes has been developed to increase resistance against *Sclerotinia sclerotiorum* in transgenic Glycine max [108].

Transgenic expression of LTPs has been shown to improve resistance to phytopathogenic fungi in some studies. As an example, antimicrobial protein gene (*Ace-AMP1*) isolated from *Allium cepa* has been overexpressed in both *Triticum aestivum* and *Oryza sativa* through *Agrobacterium-mediated* transformation, microprojectile bombardment, in *planta* assays, conferring resistance against *Sphaerotheca pannosa* var. rosae [113], *Magnaporthe grisea*, *Rhizoctonia solani* and *Xanthomonas oryzae* [116] respectively. Recently, *A. thaliana* LTP overexpressing transgenics has been shown to increase resistance toward pathogens *Plasmodiophora brassicae* and *F. graminearum* [112, 131]. Some other examples of successfully generated transgenic plants with enhanced production of hydrolytic enzymes and resistance against phytopathogenic fungi are given in **Table 2**.

8. Transgenic plant expressing bacterial resistance

Numerous bacterial pathogens causing massive yield losses have been isolated and identified from different agriculturally important crops. Pathogenesis-related proteins are well-known weapons to combat resistance against these bacterial pathogens. Many in-vitro studies have shown the antibacterial properties of many PR proteins *viz* PR10 (Ribonuclease-like proteins), PR12 (defensins), PR13 (thionins) and PR14 (Lipid-transfer protein) [90, 116, 132]. Among these, PR10 shows broad spectrum of antibacterial activity against *P. syringae*, *Agrobacterium tumefaciens*, *A. radiobacter*, *Pseudomonas aureofaciens* and *Serratia marcescens* [90, 133]. Overexpression of lipid transfer protein (PR14) in rice plants showed increased resistance to bacterial as well as fungal pathogens (**Table 2**) [116]. The antibacterial efficacy of additional PR proteins and AMPs against a variety of bacterial diseases in economically significant crops has to be further investigated.

9. Transgenic plant expressing insect resistance

Plants expressing PR genes have been engineered in several experiments, resulting in enhanced pest resistance. The expression of both low and high levels of *MTI-2* was reported by using *Agrobacterium* transformation technique in tobacco and *Arabidopsis* plants leading to resistance against *Spodoptera littoralis* [134]. The wound-inducible expression of a *Bacillus thuringiensis* endotoxin gene which directed significant insecticidal gene expression to protect transgenic rice from *Chilo suppressalis* Walker [94]. Transgenic rice plants were developed by particle bombardment or *Agrobacterium-mediated* transformation of *mpi* gene leading to resistance against *C. suppressalis* (**Table 2**) [93].

10. Transgenic plant expressing viral resistance

Apart from their antifungal or antibacterial effects, PR proteins appear to be a promising candidate gene for producing virus-resistant transgenic crops based on different studies of PR proteins, as given in **Table 2**. Antiviral activities of PR proteins such as defensins, thionins, peroxidase and lipid transfer proteins have been observed *in vitro* [115]. Antiviral activity has also been observed in ribosome-inactivating proteins (RIPs), which suppress translation by enzymatically damaging ribosomes [115]. Plant resistance to plant viruses was improved by a transformation study involving RIPs. In addition, CaPR10 from *Capsicum annuum* has been found to have increased ribonucleolytic activity against the Tobacco mosaic virus (TMV) RNA, allowing it to break viral RNAs [116].

11. Synergistic effect of transgenic PR proteins

In transgenic plants, the synergistic action of two or more PR genes reduces susceptibility to various pathogens. Researchers have reported that *β-1,3-glucanases* and *chitinases* synergistically inhibited the growth of *Fusarium oxysporum* by using *in planta* transformation [135]. Transgenic potato plants co-expressing chitinase (*BjCHI1*) and β-1,3-glucanase (*HbGLU*) suppressed *Rhizoctonia solani* and showed healthier root growth [70]. In another study, transgenes carrying the chitinase gene (*chi11*) and the thaumatin-like protein gene (*tlp*) from rice were introduced by co-bombardment, and overexpression of these antifungal *chi* and *tlp* proteins provided resistance to fungal infections in barley [136]. Likewise, in transgenic carrots, the synergistic action of three different PR-protein genes such as chitinase, β-1,3-glucanase and peroxidase, conferred disease resistance to necrotrophic pathogens namely, *Botrytis cinerea* and *Sclerotinia sclerotiorum* [137]. Amian *et al* [138] reported the development of transgenic pea plants with stable integration of two genes *viz* β-1,3-glucanase (*Hordeum vulgare*) and chitinase gene (*Streptomyces olivaceoviridis*) via *Agrobacterium*-mediated gene transformation and hence produced suppression of fungal spore germination. Chhikara *et al* [139] used *Agrobacterium*-mediated transformation to co-express the barley antifungal genes chitinase and ribosome-inactivating protein in Indian mustard, protecting against Alternaria leaf spot disease. Furthermore, transgenic potato plants expressing *rip30* and *chiA* genes transformed by *A. tumefaciens* strain *GV3101* showed improved resistance to *Rhophitulus solani* [140]. In the case of Oriental melon (*Cucumis melo* Makuwa Group), the fusion of chitinase (*CHI*) and antifungal protein (*AFP*) genes confers enhanced protection against *Rhizoctonia solani* and *Fusarium oxysporum* [141]. Rice plants co-transformed with chitinase (*OsCHI11*) and oxalate oxidase (*OsOXO4*), which are defense-related genes, showed improved resistance to the pathogen that causes sheath blight [142]. Boccardo *et al* [143] suggested co-expression of PR proteins *AP24* and *β-1,3 glucanase* enhanced resistance against *Rhizoctonia solani* in greenhouse conditions and *Peronospora hyoscyami* f.sp. *tabacina* and *phytophthora nicotianae* pathogens in field conditions.

12. Challenges faced by transgenic expression with PR proteins

Since the advent of plant genetic engineering, PR proteins have consistently been the top choice among scientists when creating transgenic plants to increase disease resistance against a variety of diseases. PR proteins expressed either singly or synergistically in transgenic plants can provide broader and more effective disease resistance against different pathogens as described above.

Aside from these successful outcomes, many studies have described the challenges of using PR proteins in transgenic technology. In contrast to the above findings, numerous studies have suggested that the transgenic expression of PR proteins did not lead to increased tolerance to pathogens. Szwacka *et al* [144] reported no relationship between transgenic protein expression level and increased tolerance against the pathogen. Transgenic cucumber plants with stably integrated thaumatin II cDNA under the control of the CAM35S promotor via *Agrobacterium* did not exhibit tolerance to *Pseudoperonospora cubensis*. Moravckova *et al* [145] co-introduced chitinase and glucanase into *Solanum tuberosum* to increase resistance to *R. solani* infection, but hyphal extension assay revealed that transformants did not affect *R. solani* growth in vitro.

Various transgenic plant modifications have been described, with varying degrees of protection against certain fungal and oomycete infections. However, the resulting resistance levels were frequently insufficient for breeding [146]. Furthermore, constitutive expression of PR proteins can lead to the spontaneous production of lesions that look like HR lesions in the absence of a pathogen), which can be an unfavorable outcome [147]. Disease resistance techniques must control specific diseases without affecting crop yield and quality.

Moreover, most researchers have used constitutive promoters to control the expression of PR genes in agricultural plants to enhance resistance, resulting in homology-dependent gene silencing. As a result, unregulated and untimely activation of PR genes or AMPs harms plant growth and development. Human allergenicity is one of the main issues hindering the success of transgenic technology with PR genes. According to the current classification, there are 19 different classes of PR-Proteins, and 8 of them have been confirmed to cause allergic reactions in humans by using *in-silco* approaches. These proteins have been known to trigger allergenic symptoms such as food allergens depending upon their mode of entry into the human body [148], dermatitis, airborne, asthma, airway allergy etc. and if all these allergens have been consumed in greater amount, the gastrointestinal symptoms are also triggered.

13. Conclusion

The goal of this chapter was to review the role of PR-proteins in plant defense and how transgenic expression of PR-proteins in agricultural plants resulted in increased resistance to stresses. Biotic stress has become a significant concern in modern agriculture and many research institutions are actively researching to generate resistant cultivars using PR proteins. PR proteins have become a highlighted topic between scientists because of their effectiveness against biotic agents. Genetic engineering is considered the best way to develop transgenic resistant plants using PR proteins. To increase agronomic characteristics worldwide, new inventions or novel approaches in PR protein transgenic technology are necessary and will continue to improve plant health in the future. Another future concern is that the formation of virulent phytopathogen strains increases as the global climatic change rate increases. So, to cope with such significant obstacles, it is necessary to define and identify novel PR genes functionally. Advances in genomics, transcriptomics, phenomics, proteomics, metabolomics, and ionomics, will substantially aid our understanding of the complex network of PR genes and the interaction of PR proteins with other proteins from plants and pathogens. Therefore, PR proteins could be utilized to develop crop plants more resistant to various stresses. They could also be employed as candidate genes for genetically engineering crop multi-trait factors. Future research is needed to assess the PR transgenic plants' responses to various traits, including biotics, plant development and yield.

Author details

Anroop Kaur[1], Sukhpreet Kaur[1], Ajinder Kaur[1], Navraj Kaur Sarao[1] and Devender Sharma[2]*

1 School of Agricultural Biotechnology, Punjab Agricultural University, Ludhiana, India

2 ICAR-Vivekananda Parvatiya Krishi Anusandhan Sansthan, Almora, India

*Address all correspondence to: devender.kumar1@icar.gov.in

References

[1] Sharma A, Sharma A, Kumar R, Sharama I, Vats AK. PR proteins: Key genes for engineering disease resistance in plants. In: Kumar P, Thakur AK, editors. Crop Improvement: Biotechnological Advances. Boca Raton: CRC press; 2021. pp. 81-98

[2] Ficke A, Cowger C, Bergstrom GC, Brodal G. Understanding yield loss and pathogen biology to improve disease management: Septorianodorum blotch - a case study in wheat. Plant Disease. 2017;**4**:102

[3] Christou P. Plant genetic engineering and agricultural biotechnology. Trends in Biotechnology. 2013;**31**:125-127

[4] Erik A, der Biezen V. Quest for antimicrobial genes to engineer disease-resistant crops. Trends in Plant Science. 2001;**6**(3):0-91

[5] Maleck K et al. The transcriptome of Arabidopsis thaliana during systemic acquired resistance. Nature Genetics. 2000;**26**:403-410

[6] Schenk M et al. Coordinated plant defence responses in Arabidopsis revealed by microarray analysis. Proceedings of the National Academy of Sciences of the United States of America. 2000;**97**:11655-11660

[7] van Loon LC, Rep M, Pieterse CMJ. Significance of inducible Defence-related proteins in infected plants. Annual Review of Phytopathology. 2006;**44**:135-162

[8] Moosa A, Farzand A, Sahi ST, Khan SA. Transgenic expression of antifungal pathogenesis-related proteins against phytopathogenic fungi – 15 years of success. Israel Journal of Plant Sciences. 2017;**65**(1-2):1-17

[9] Nandi AK. Application of antimicrobial proteins and peptides in developing disease-resistant plants. Plant Pathogen Resistance. Biotechnology. 2016;**3**:51-70

[10] Antoniw JF, Pierpoint WS. Purification of a tobacco leaf protein associated with resistance to virus infection [proceedings]. Biochemical Society Transactions. 1978;**6**(1):248-250

[11] Van Loon LC, Van Kammen A. Polyacrylamide disc electrophoresis of the soluble leaf proteins from *Nicotiana tabacum* var. Samsunâ and Samsun NNâ. Virology. 1970;**40**(2):199-211

[12] Breen S, Williams SJ, Outram M. Kobe B and Solomon PS emerging insights into the functions of pathogenesis-related protein 1. Trends in Plant Science. 2017;**22**:871-879

[13] Van Loon LC, Pierpoint WS, Boller T, Conejero V. Recommendations for naming plant pathogenesis-related proteins. Plant Mol. Biol. Reporter. 1994;**12**:245-264

[14] Antoniw JF, Ritter CE, Pierpoint WS, Van Loon LC. Comparison of three pathogenesis-related proteins from plants of two cultivars of tobacco infected with TMV. Journal of General Virology. 1980;**47**:79-87

[15] Selitrennikoff CP. Antifungal Proteins. Applied and Environmental Microbiology. 2001:2883-2894. DOI: 10.1128/AEM.67.7.2883-2894

[16] Van Loon LC. Regulation of changes in proteins and enzymes associated with active defence against virus infection. In: Wood RKS, editor. Active Defence Mechanisms in Plants. Plenum Press: New York; 1982. pp. 247-273

[17] Kereamy A, El-sharkawy I, Ramamoorthy R, Taheri A, Errampalli. Prunus domestica pathogenesis-related protein-5 activates the defence response pathway and enhances the resistance to fungal infection. PLoS One. 2011;**6**:e17973

[18] Green TR, Ryan CA. Wound-induced proteinase inhibitor in plant leaves: A possible defence mechanism against insects. Science. 1972;**175**:776-777

[19] Vera P, Conejero V. Pathogenesis-related proteins of tomato: p-69 as an alkaline endoproteinase. Plant Physiology. 1988;**87**:58-63

[20] Metraux JP, Streit L, Staub T. A pathogenesis-related protein in cucumber is a chitinase. Physiology and Molecular Plant Pathology. 1988;**33**:1-9

[21] Lagrimini LM, Burkhart W, Moyer M, Rothstein S. Molecular cloning of complementary DNA encoding the lignin forming peroxidase from tobacco: Molecular analysis and tissue-specific expression. Proceedings of National Academy of Sciences USA. 1987;**84**:7542-7575

[22] Somssich IE, Schmelzer E, Bollmann J, Hahlbrock K. Rapid activation by fungal elicitor of genes encoding "pathogenesis- related" proteins in cultured parsley cells. Proceedings of National Academy of Sciences USA. 1986;**83**:2427-2430

[23] Melchers LS, Potheker-de Groot MA, Van der Knaap JA, Ponstein AS, Sela-Buurlage MB, Bol JF, et al. A new class of tobacco chitinases homologous to bacterial exo-chitinases displays antifungal activity. Plant Journal. 1994;**5**:469-480

[24] Terras FR, Eggermont K, Kovaleva V, Raikhel NV, Osborn RW, Kester A, et al. Small cysteine-rich antifungal proteins from radish: Their role in host defence. Plant Cell. 1995;**7**:573-588

[25] Epple P, Apel K, Bohlmann H. An Arabidopsis thaliana thionin gene is inducible via a signal transduction pathway different from that for pathogenesis-related proteins. Plant Physiology. 1995;**109**:813-820

[26] Garcia-Olmedo F, Molina A, Segura A, Moreno M. The defensive role of nonspecific lipid-transfer proteins in plants. Trends in Microbiology. 1995;**3**:72-74

[27] Zhang Z, Collinge DB, Thordal-Christensen H. Germin-like oxalate oxidase, a H2O2-producing enzyme, accumulates in barley attacked by the powdery mildew fungus. Plant Journal. 1995;**8**:139-145

[28] Wei YD, Zhang ZG, Andersen CH, Schmelzer E, Gregersen PL, Collinge DB, et al. An epidermis/papilla-specific oxidase-like protein in the defence response of barley attacked by the powdery mildew fungus. Plant Molecular Biology. 1998;**36**:101-112

[29] Okushima Y, Koizumi N, Kusano T, Sano H. Secreted proteins of tobacco cultured BY2 cells: Identification of a new member of pathogenesis-related proteins. Plant Molecular Biology. 2000;**42**:479-488

[30] Custers JHHV, Harrison SJ, Sela-Buurlage MB, Van Deventer E, Lageweg W, Howe PW, et al. Isolation and characterization of a class of carbohydrate oxidases from higher plants, with a role in active defence. The Plant Journal. 2004;**39**:147-160

[31] Sooriyaarachchi S, Jaber E, Covarrubias AS, Ubhayasekera W, Asiegbu FO, Mowbray SL. Expression

and β-glucan binding properties of scots pine (*Pinus sylvestris* L.) antimicrobial protein (Sp-AMP). Plant Molecular Bio. 2011;77:33-45

[32] Ali S, Ganai BA, Kamili AN, et al. Pathogenesis-related proteins and peptides as promising tools for engineering plants with multiple stress tolerance. Microbio. Res. 2018;**212**(13):29-37

[33] Lincoln JE, Sanchez JP, Zumstein K, Gilchrist DG. Plant and animal PR1 family members inhibit programmed cell death and suppress bacterial pathogens in plant tissues. Molecular Plant Pathology. 2018;**19**:2111-2123

[34] Kauffmann S, Legrand M, Geoffroy P, Fritig B. Biological function of `pathogenesis-related' proteins: Four PR proteins of tobacco have 1, 3-b-glucanase activity. EMBO Journal. 1987;**6**:320-3212

[35] Legrand M, Kauffmann S, Geoffroy P, Fritig B. Biological function of pathogenesis-related proteins: Four tobacco pathogenesis-related proteins are chitinases. Proc Natl Acad Sci. 1987;**84**(19):6750-6754

[36] Ebrahim S, Uhsh K, Singh B. Pathogenesis related (PR) proteins in plant defence mechanism. In: Méndez-Vilas A, editor. Science against Microbial Pathogens: Communicating Current Research and Technological Advances. Spain: Formatex Badajoz; 2011. pp. 1043-1054

[37] Patil SV, Jayamohan NS, Kumudini BS. Strategic assessment of multiple plant growth promotion traits for shortlisting of fluorescent pseudomonas spp. and seed priming against ragi blast disease. Plant Growth Regulation. 2016;**80**:47-58

[38] Takemoto D, Furuse K, Doke N, Kawakita K. Identification of chitinase and osmotin-like protein as actin binding proteins in suspension-cultured potato cells. Plant & Cell Physiology. 1997;**38**:441-448

[39] Koiwa H, Bressan RA, Hasegawa PM. Regulation of protease inhibitors and plant defence. Trends in Plant Science. 1997;**2**:379-384

[40] Dunaevskii YE, Tsybina TA, Belyakova GA, Domash VI, Sharpio TP, Zabreiko SA, et al. Proteinase inhibitors as antistress proteins in higher plants. Applied Biochemistry and Microbiology. 2005;**41**:344-348

[41] Gutierrez-Campos R, Torres-Acosta JA, Saucedo-Arias LJ, Gomez-Lim MA. The use of cysteine proteinase inhibitors to engineer resistance against potyviruses in transgenic tobacco plants. Nature Biotechnology. 1999;**17**:1223-1226

[42] Cheng XY, Zhu LL, He GC. Towards understanding of molecular interactions between rice and the brown planthopper. Molecular Plant. 2013;**6**(3):621-634

[43] Vieira Bard GC, Nascimento VV, Ribeiro SFF, Rodrigues R, Perales J, Teixeira-Ferreira A, et al. Characterization of peptides from *Capsicum annuum* hybrid seeds with inhibitory activity against α-amylase, serine proteinases and fungi. The Protein Journal. 2015;**34**:122-129

[44] Lorito M, Peterbauer TC, Hayes CK, Harman GE. Synergistic interaction between fungal cell-wall degrading enzymes and different antifungal compounds enhances inhibition of spore germination. Microbiology. 1994;**140**:623-629

[45] Goldman MHS, Goldman GH. *Trichoderma harzianum* transformant has high extracellular alkaline proteinase

expression during specific mycoparasitic interactions. Genetics and Molecular Biology. 1998;**21**:329-333

[46] Van Loon LC, Van Strien EA. The families of pathogenesis-related proteins, their activities, and comparative analysis of PR-1 type proteins. Physiological and Molecular Plant Pathology. 1999;**55**:85-97

[47] Wu SW, Wang HW, Yang ZD, Kong LR. Expression comparisons of pathogenesis-related (*PR*) genes in wheat in response to infection/infestation by *Fusarium*, yellow dwarf virus (YDV) aphid-transmitted and hessian fly. Journal of Integrative Agriculture. 2014;**13**(5):926-936

[48] He M, Xu Y, Cao J, et al. Subcellular localization and functional analyses of a pr10 protein gene from *Vitis pseudoreticulata* in response to Plasmoparaviticola infection. Protoplasma. 2013;**250**:129-1240

[49] Bohlmann H. The role of thionins in plant protection. Critical Reviews in Plant Sciences. 1994;**13**:1-16

[50] Broekaert WF, Cammue BPA, De Bolle MFC, Thevissen K, De Samblanx GW, Osborn RW. Antimicrobial peptides from plants. Critical Reviews in Plant Sciences. 1997;**16**:297-323

[51] Lay FT, Poon S, McKenna JA, Connelly AA, Barbeta BL, McGinness BS, et al. The C-terminal propeptide of a plant defensin confers cytoprotective and subcellular targeting functions. BMC Plant Biology. 2014;**14**:55-67

[52] Parisi K, Shafee TMA, Quimbar P, van der Weerden NL, Bleackley MR, Anderson MA. The evolution, function and mechanisms of action for plant defensins. Seminars in Cell & Developmental Biology. 2019;**88**:107-118

[53] van der Weerden NL, Anderson MA. Plant defensins: Common fold, multiple functions. Fungal Biology Reviews. 2013;**26**:121-131

[54] Li XC, Liao YY, Leung DWM, Wang HY, Chen BL, Peng XX, et al. Divergent biochemical and enzymatic properties of oxalate oxidase isoforms encoded by four similar genes in rice. Phytochemistry. 2015;**118**:216-223

[55] Muhammad SH, Muhammad J, Jinggui F. Overproduction of ROS: Underlying molecular mechanism of scavenging and redox signaling. In: Biocontrol Agents and Secondary Metabolites. Kidlington, United Kingdom: Woodhead Publishing; 2021

[56] Hammond-Kosack KE. Jones JD Resistance gene-dependent plant defence responses. Plant Cell. 1996;**8**(10):1773

[57] Van Loon LC. Occurrence and properties of plant pathogenesis-related proteins. In: Datta SK, Muthukrishnan S, editors. Pathogenesis-Related Proteins in Plants. Boca Raton: CRC Press LLC; 1999. pp. 1-19

[58] Roux F, Voisin D, Badet TC, Balagué X, Barlet C, Huard-Chauveau, et al. Resistance to phytopathogens e tutti quanti: Placing plant quantitative disease resistance on the map. Molecular Plant Pathology. 2014;**15**:427-432

[59] Jean B, Jean C, Heribert H. Signaling mechanisms in pattern-triggered immunity (PTI). Molecular Plant. 2015;**8**(4):521-539

[60] Van Baarlen P, Van Belkum A, Summerbel RC, Crousl PW, Bart P, et al. Molecular mechanisms of pathogenicity: How do pathogenic microorganisms develop cross-kingdom host jumps? FEMS Micro. Rev. 2007;**3**:239-277

[61] De Vos M, Van Oosten VR, Van Poecke RMP, Van Pelt JA, Pozo MJ, Mueller MJ, et al. Signal signature and transcriptome changes of Arabidopsis during pathogen and insect attack. Molecular Plant-Microbe Interactions. 2005;**18**:923-937

[62] Bari R, Jones JDG. Role of plant hormones in plant defence responses. Plant Molecular Biology. 2009;**69**:473-488

[63] Sudisha J, Sharathchandra RG, Amruthesh KN, Kumar A, Shetty HS. Pathogenesis related proteins in plant defence response. In: Plant defence: Biolog. Cont. Dordrecht: Springer; 2011. pp. 379-403

[64] Ali S, Chandrashekar N, Rawat S, Nayanakantha NMC, Mir ZA, Manoharan A, et al. Isolation and molecular characterization of pathogenesis related PR2 gene and its promoter from *Brassica juncea*. Biologia Plantarum. 2017;**61**:763-773

[65] Fu ZQ, Yan S, Saleh A, Wang W, Ruble J, Oka N, et al. NPR3 and NPR4 are receptors for the immune signal salicylic acid in plants. Nature. 2012;**486**:228-232

[66] Chini A, Fonseca S, Fernández G, Adie B, Chico JM, Lorenzo O, et al. The JAZ family of repressors is the missing link in jasmonate signalling. Nature. 2007;**448**:666-671

[67] Kaur J, Fellers J, Adholeya A, Velivelli SL, El-Mounadi K, Nersesian N, et al. Expression of apoplast- targeted plant defensin MtDef4. 2 confers resistance to leaf rust pathogen *Puccinia triticina* but does not affect mycorrhizal symbiosis in transgenic wheat. Trans Res. 2016;**26**:37-49

[68] Spoel SH, Dong X. Making sense of hormone crosstalk during plant immune responses. Cell Host & Microbe. 2008;**3**:348-351

[69] Wrobel-Kwiatkowska M, Lorenc-Kukula K, Starzycki M, Oszmianski J, Kepczynska E, Szopa J. Expression of *b-1,3-glucanase* in flax causes increased resistance to fungi. Physiological and Molecular Plant Pathology. 2004;**65**:245-256

[70] Chye ML, Zhao KJ, He ZM, Ramalingam S, Fung KL. An agglutinating chitinase with two chitin-binding domains confers fungal protection in transgenic potato. Planta. 2005;**220**:717-730

[71] Mackintosh CA, Lewis J, Radmer LE, Shin S, Heinen SJ, Smith LA, et al. Overexpression of defence response genes in transgenic wheat enhances resistance to Fusarium head blight. Plant Cell Reports. 2007;**26**:479-488

[72] Jayaraj J, Punja ZK. Combined expression of chitinase and lipid transfer protein genes in transgenic carrot plants enhances resistance to foliar fungal pathogens. Plant Cell Reports. 2007;**26**:1539-1546

[73] Li P, Pei Y, Sang X, Ling Y, Yang Z, He G. Transgenic indica rice expressing a bitter melon (*Momordica charantia*) class I chitinase gene (*McCHIT1*) confers enhanced resistance to *Magnaporthe grisea* and *Rhizoctonia solani*. European Journal of Plant Pathology. 2009;**125**:533-543

[74] Wang N, Xiao B, Xiong L. Identification of a cluster of PR4-like genes involved in stress responses in rice. Journal of Plant Physiology. 2011;**168**:2212-2224

[75] Datta K, Tu J, Oliva N, Ona I, Velazhahan R, Mew TW, et al. Enhanced resistance to sheath blight by constitutive expression of infection-related rice chitinase in transgenic elite indica rice cultivars. Plant Science. 2001;**160**:405-414

[76] Shin S, Mackintosh CA, Lewis J, Heinen SJ, Radmer L, Dill-Macky R, et al. Transgenic wheat expressing a barley class II chitinase gene has enhanced resistance against *Fusarium graminearum*. Journal of Experimental Botany. 2008;**59**:2371-2378

[77] Singh D, Haicour R, Sihachakr D, Rajam MV. Expression of rice chitinase gene in transgenic eggplant confers resistance to fungal wilts. Indian Journal of Biotechnology. 2015;**14**:233-240

[78] Huang X, Wang J, Du Z, Zhang C, Li L, Xu Z. Enhanced resistance to stripe rust disease in transgenic wheat expressing the rice chitinase gene RC24. Trans Res. 2013;**22**:939-947

[79] Kovacs G, Sagi L, Jacon G, Arinaitwe G, Busogoro JP, Thiry E, et al. Expression of a rice chitinase gene in transgenic banana ('Gros Michel', AAA genome group) confers resistance to black leaf streak disease. Trans Res. 2013;**22**:117-130

[80] Liu X, Yu Y, Liu Q, Deng S, Jin X, Yin Y, et al. A Na2CO3-responsive chitinase gene from *Leymus chinensis* improves pathogen resistance and saline alkali stress tolerance in transgenic tobacco and maize. Frontiers in Plant Science. 2020;**11**:504

[81] Yanjun Z, Haixia Y, Xuejun W, Jiarui Z, Haiyan Z, Daqun L. Expression analysis and functional characterization of a pathogen-induced thaumatin-like gene in wheat conferring enhanced resistance to *Puccinia triticina*. Journal of Plant Interactions. 2017;**12**:332-339

[82] Anand A, Zhou T, Trick HN, Gill BS, Bockus WW, Muthukrishnan S. Greenhouse and field testing of transgenic wheat plants stably expressing genes for thaumatin-likeprotein, chitinase and glucanase against Fusarium graminearum. Journal of Experimental Botany. 2003;**54**:1101-1111

[83] Radhajeyalakshmi R, Velazhahan R, Balasubramanian P, Doraiswamy S. Overexpression of thaumatin-like protein in transgenic tomato plants confers enhanced resistance to *Alternaria solani*. Arch Phytopathol Plant Prot. 2005;**38**:257-265

[84] Kalpana K, Maruthasalam S, Rajesh T, Poovannan K, Kumar KK, Kokiladevi E, et al. Engineering sheath blight resistance in elite indica rice cultivars using genes encoding defence proteins. Plant Science. 2006;**170**:203-215

[85] Acharya K, Pal AK, Gulati A, Kumar S, Singh AK, Ahuja PS. Overexpression of Camellia sinensis thaumatin-like protein, CsTLP in potato confers enhanced resistance to *Macrophomina phaseolina* and *Phytophthora infestans* infection. Molecular Biotechnology. 2013;**54**:609-622

[86] Singh NK, Kumar KRR, Kumar D, Shukla P, Kirti PB. Characterization of a pathogen induced thaumatin-like protein gene AdTLP from Arachis diogoi, a wild peanut. PLoS One. 2013;**8**:e83963

[87] Xue X, Cao ZX, Zhang XT, Wang Y, Zhang YF, Chen ZX, et al. Overexpression of OsOSM1 enhances resistance to rice sheath blight. Plant Disease. 2016;**100**:1634-1642

[88] Kaur A, Reddy MS, Pati PK, Kumar A. Over-expression of osmotin (*OsmWS*) gene of *Withaniasomnifera* in potato cultivar 'Kufri Chipsona 1'imparts resistance to *Alternaria solani*. Plant Cell Tissue and Organ Culture. 2020;**142**:131-142

[89] Wu SG, Kim KY, Kang JG, Kim SR, Park R, Gupta ST. Overexpression of a pathogenesis-related protein 10

enhances biotic and abiotic stress tolerance in rice. Plant Pathology Journal. 2016;**32**:552

[90] Jiang L, Wu J, Fan S, Li W, Dong L, Cheng Q. Isolation and characterization of a novel pathogenesis-related protein gene (GmPRP) with induced expression in soybean (Glycine max) during infection with phytophthora sojae. PLoS One. 2015;**10**:e0129932

[91] Lodge JK, Kaniewski WK, Tumer NE. Broad-spectrum virus resistance in transgenic plants expressing pokeweed antiviral protein. Proc Natl Acad Sci. USA. 1993;**90**:7089-7093

[92] Dai WS, Bonos Z, Guo W, Meyer P, Day FB. Expression of pokeweed antiviral proteins in creeping bentgrass. Plant Cell Reports. 2003;**21**(5):497-502

[93] Vila L, Quilis J, Meynard D, Breitler JC, Marfà V, Murillo I, et al. Expression of the maize proteinase inhibitor gene in rice plants enhances resistance against the striped stem borer: Effects on larval growth and insect gut proteinases. Plant Biotechnology Journal. 2005;**3**(2):187-202

[94] Breitler JC, Cordero MJ, Royer M, Meynard D, San Segundo B, Guiderdoni E. The −689/+197 region of the maize proteinase inhibitor gene directs high level, wound-inducible expression of the cry1B gene which protects transgenic rice plants from stemborer attack. Molecular Breeding. 2005;7:259-274

[95] Kanzaki H, Nirasawa S, Saitoh H, Ito M, Nishihara M, Terauchi R, et al. Overexpression of the wasabi defensin gene confers enhanced resistance to blast fungus (*Magnaporthe grisea*) in transgenic rice. Theoretical and Applied Genetics. 2002;**105**:809-814

[96] Khan RS, Nishihara M, Yamamura S, Nakamura I, Mii M. Transgenic potatoes expressing wasabi defensin peptide confer partial resistance to gray mold (*Botrytis cinerea*). Plant Biotechnol. 2006;**23**:179-183

[97] Abdallah NA, Shah D, Abbas D, Madkour M. Stable integration and expression of a plant defensin in tomato confers resistance to Fusarium wilt. GM Crops. 2010;**1**:344-350

[98] Jha S, Chattoo BB. Expression of a plant defensin in rice confers resistance to fungal phytopathogens. Trans Res. 2010;**19**:373-384

[99] Li Z, Zhou M, Zhang Z, Ren L, Du L, Zhang B, et al. Expression of a radish defensin in transgenic wheat confers increased resistance to *Fusarium graminearum* and *Rhizoctonia cerealis*. Funct Integ Genom. 2011;**11**:6

[100] Ntui VO, Thirukkumaran G, Azadi P, Khan RS, Nakamura I, Mii M. Stable integration and expression of wasabi defensin gene in "Egusi" melon (*Colocynthis citrullus* L.) confers resistance to *Fusarium wilt* and Alternaria leaf spot. Plant Cell Reports. 2010;**29**:943-954

[101] Jiang M, He CME, Miao LX, Zhang YC. Overexpression of a broccoli defensin gene BoDFN enhances downy mildew resistance. J. Integ. Agric. 2012;**11**:1137-1144

[102] Thao HT, Lan NN, Tuong HM, Thanh NVT, Van Son L, Hoang C. Expression analysis of recombinant Vigna radiata plant defensin 1 protein in transgenic tobacco plants. J. Appl. Biol. Biotechnol. 2017;**5**:70-75

[103] Sasaki K, Kuwabara C, Umeki N, Fujioka M, Saburi W, Matsui H, et al. The cold-induced defensin TAD1

confers resistance against snow mold and Fusarium head blight in transgenic wheat. Journal of Biotechnology. 2016;**228**:3-7

[104] Hammad IA, Abdel-Razik AB, Soliman ER, Tawfik E. Transgenic potato (*Solanumtuberosum*) expressing two antifungal thionin genes confer resistance to Fusarium spp. Journal of Pharmaceutical and Biological Sciences. 2017;**12**:69-79

[105] Hoshikawa K, Ishihara G, Takahashi H, Nakamura I. Enhanced resistance to gray mold (Botrytis cinerea) in transgenic potato plants expressing thionin genes isolated from Brassicaceae species. Plant Biotechnol. 2012;**29**:87-93

[106] Muramoto N, Tanaka T, Shimamura T, Mitsukawa N, Hori E, Koda K, et al. Transgenic sweet potato expressing thionin from barley gives resistance to black rot disease caused by Ceratocystis fimbriata in leaves and storage roots. Plant Cell Reports. 2012;**31**:987-997

[107] Chan YL, Prasad V, Chen KH, Liu PC, Chan MT, Cheng CP. Transgenic tomato plants expressing an Arabidopsis thionin (Thi 2. 1) driven by fruit-inactive promoter battle against phytopathogenic attack. Planta. 2005;**221**:386-393

[108] Yang X, Yang J, Wang Y, He H, Niu L, Guo D, et al. Enhanced resistance to sclerotinia stem rot in transgenic soybean that overexpresses a wheat oxalate oxidase. Transgenic Research. 2019;**28**(1):103-114

[109] Ghosh S, Molla KA, Karmakar S, Datta SK, Datta K. Enhanced resistance to late blight pathogen conferred by expression of rice oxalate oxidase 4 gene in transgenic potato. Plant Cell, Tissue and Organ Culture. 2016;**126**(3):429-437

[110] Walz A, Zingen Sell I, Loeffler M, Sauer M. Expression of an oxalate oxidase gene in tomato and severity of disease caused by *Botrytis cinerea* and *Sclerotinia sclerotiorum*. Plant Pathology. 2008;**57**:453-458

[111] Schneider M, Droz E, Malno P, Chatot C, Bonnel E, Metraux JP. Transgenic potato plants expressing oxalate oxidase have increased resistance to oomycete and bacterial pathogens. Pot Res. 2002;**45**:177-185

[112] McLaughlin JE, Al D, Garcia-Sanchez N, Tyagi J, Trick N, McCormick HN, et al. A lipid transfer protein has antifungal and antioxidant activity and suppresses Fusarium head blight disease and DON accumulation in transgenic wheat. Phytopathology. 2020;**111**:671-683

[113] Li X, Gasic K, Cammue B, Broekaert W, Korban SS. Transgenic rose lines harboring an antimicrobial protein gene, ace-AMP1, demonstrate enhanced resistance to powdery mildew (*Sphaerothecapannosa*). Planta. 2003;**218**:226-232

[114] Roy-Barman S, Sautter C, Chattoo BB. Expression of the lipid transfer protein ace-AMP1 in transgenic wheat enhances antifungal activity and defence responses. Trans Res. 2006;**15**:435-446

[115] Oskar M, Robert N, Anna GJ. Which plant proteins are involved in antiviral Defence? Review on *In vivo* and *In vitro* activities of selected plant proteins against viruses. International Journal of Molecular Sciences. 2017;**18**(11):2300

[116] Patkar RN, Chattoo BB. Transgenic indica rice expressing ns-LTP-like protein shows enhanced resistance to both fungal and bacterial pathogens. Molecular Breeding. 2006;**17**:159-171

[117] van der Biezen EA. Quest for antimicrobial genes to engineer disease-resistant crops. Trends in Plant Science. 2001;**6**:89-91

[118] Montesinos E. Antimicrobial peptides and plant disease control. FEMS Microbiology Letters. 2007;**270**:1-11

[119] Dean R, Van Kan JA, Pretorius ZA, Hammond-Kosack KE, Di Pietro A, Spanu PD, et al. The top 10 fungal pathogens in molecular plant pathology. Molecular Plant Pathology. 2012;**13**:414-430

[120] Sundaresha S, Kumar AM, Rohini S, Math SA, Keshamma E, Chandrashekar SC, et al. Enhanced protection against two major fungal pathogens of groundnut, *Cercosporaarachidicola* and *aspergillus flavus* in transgenic groundnut over-expressing a tobacco *b 1-3 glucanase*. European Journal of Plant Pathology. 2010;**126**:497-508

[121] Fujimori N, Enoki S, Suzuki A, Naznin HA, Shimizu M, Suzuki S. Grape apoplasmic*b-1, 3-glucanase* confers fungal disease resistance in Arabidopsis. Sci Horti. 2016;**200**:105-110

[122] Singh HR, Hazarika P, Agarwala N, Bhattacharyya N, Bhagawati P, Gohain B, et al. Transgenic tea over-expressing *Solanum tuberosum* endo-1, 3-beta-D-glucanase gene conferred resistance against blister blight disease. Plant Molecular Biology Reporter. 2018;**36**(1):107-122

[123] Hanin AN, Parveez GKA, Rasid OA, Masani MYA. Biolistic-mediated oil palm transformation with alfalfa glucanase (AGLU1) and rice chitinase (*RCH10*) genes for increasing oil palm resistance towards *Ganoderma boninense*. Industrial Crops and Products. 2020;**144**:112008

[124] Kang JN, Park MY, Kim WN, Kang HG, Sun HJ, Yang DH, et al. Resistance of transgenic zoysiagrass overexpressing the zoysiagrass class II chitinase gene *Zjchi2* against *Rhizoctonia solaniAG2-2* (IV). Plant Biotechnol. Rep. 2017;**11**:229-238

[125] Xiaoxiao Y, Hengbo Q, Xiuming Z, Chunlei G, Mengnan W, Yuejin W, et al. Analysis of the grape (*Vitis vinifera* L.) thaumatin-like protein (*TLP*) gene family and demonstration that TLP29 contributes to disease resistance. Scientific Reports. 2017;**7**(1):4269

[126] Sun W, Zhou Y, Movahedi A, Wei H, Zhuge Q. Thaumatin-like protein (*Pe-TLP*) acts as a positive factor in transgenic poplars enhanced resistance to spots disease. Physiological and Molecular Plant Pathology. 2020;**112**:101-512

[127] Anuradha TS, Divya K, Jami SK, Kirti PB. Transgenic tobacco and peanut plants expressing a mustard defensin show resistance to fungal pathogens. Plant Cell Reports. 2008;**27**:1777-1786

[128] Lacerda AF, Del Sarto RP, Silva MS, de Vasconcelos EA, Coelho RR, dos Santos VO, et al. The recombinant pea defensin Drr230a is active against impacting soybean and cotton pathogenic fungi from the genera *Fusarium, Colletotrichum and Phakopsora*. Biotechnology. 2016;**6**:1

[129] Hao G, Stover GG. Overexpression of a modified plant thionin enhances disease resistance to citrus canker and huanglongbing (HLB). Frontiers in Plant Science. 2016;**7**:1078

[130] Partridge-Telenko DE, Hu J, Livingstone DM, Shew BB, Phipps PM, Grabau EA. Sclerotinia blight resistance in Virginia-type peanut transformed with a barley oxalate oxidase gene. Phytopathology. 2011;**101**:786-793

[131] Jülke S, Ludwig-Müller J. Response of Arabidopsis thaliana roots with altered lipid transfer protein (*ltp*) gene expression to the clubroot disease and salt stress. Plants. 2016;**5**(1):2

[132] Park CJ, Kim KJ, Shin R, Park JM, Shin YC. Pathogenesis related protein 10 isolated from hot pepper functions as a ribonuclease in an antiviral pathway. The Plant Journal. 2004;**37**:186-198

[133] Xie YR, Chen ZY, Brown RL. Bhatnagar D expression and functional characterization of two pathogenesis-related protein 10 genes from Zea mays. Journal of Plant Physiology. 2010;**167**:121-130

[134] De Leo F, Bonadé-Bottino MA, Ceci LR, Gallerani R, Jouanin L. Opposite effects on Spodoptera littoralis larvae of high and low level of a trypsin proteinase inhibitor in transgenic plants. Plant Physiology. 1998;**118**:997-1004

[135] Jongedijk E, Tigelaar H, van Roekel JSC, Bres-Vloemans SA, Dekker I, van den Elzen PJM, et al. Synergistic activity of chitinases and β-1,3-glucanases enhances fungal resistance in transgenic tomato plants. Euphytica. 1995;**85**:173-180

[136] Tobias DJ, Manoharan M, Pritsch C, Dahleen LS. Co-bombardment, integration and expression of rice chitinase and thaumatin-like protein genes in barley (Hordeum vulgare cv. Conlon). Plant Cell Reports. 2007;**26**:631-639

[137] Wally O, Jayaraj J, Punja Z. Comparative resistance to foliar fungal pathogens in transgenic carrot plants expressing genes encoding for chitinase, b-1, 3-glucanase and peroxidise. European Journal of Plant Pathology. 2009;**123**:31-42

[138] Amian AA, Papenbrock J, Jacobsen HJ, Hassan F. Enhancing transgenic pea (Pisum sativum L.) resistance against fungal diseases through stacking of two antifungal genes (chitinase and glucanase). GM Crops. 2011;**2**:104-109

[139] Chhikara S, Chaudhury D, Dhankher OP, Jaiwal PK. Combined expression of a barley class II chitinase and type I ribosome inactivating protein in transgenic Brassica juncea provides protection against Alternaria brassicae. Plant Cell Tiss Org Cul (PCTOC). 2012;**108**:83-89

[140] M'hamdi M, Chikh-Rouhou H, Boughalleb N, de Galarreta IR. Enhanced resistance to Rhizoctonia solani by combined expression of chitinase and ribosome inactivating protein in transgenic potatoes (Solanum tuberosum L.). Spanish. Journal of Agricultural Research. 2012;**3**:778-785

[141] Bezirganoglu I, Hwang SY, Fang TJ, Shaw JF. Transgenic lines of melon (Cucumis melo L. var. makuwa cv. 'Silver Light') expressing antifungal protein and chitinase genes exhibit enhanced resistance to fungal pathogens. Plant Cell Tiss Org Cult (PCTOC). 2013;**112**:227-237

[142] Karmakar S, Molla KA, Chanda PK, Sarkar SN, Datta SK, Datta K. Green tissue-specific co-expression of chitinase and oxalate oxidase 4 genes in rice for enhanced resistance against sheath blight. Planta. 2016;**243**:115-130

[143] Boccardo NA, Segretin ME, Hernandez I, Mirkin FG, Chacón O, Lopez Y, et al. Expression of pathogenesis-related proteins in transplastomic tobacco plants confers resistance to filamentous pathogens under field trials. Scientific Reports. 2019;**9**(1):2791

[144] Szwacka M, Krzymowska M, Osuch A, Kowalczyk ME, Malepszy S.

Variable properties of transgenic cucumber plants containing the thaumatin II gene from Thaumatococcus daniellii. Act Physiol Planta. 2002;**24**:173-185

[145] Moravckova J, Matusıkova I, Libantova J, Bauer M, Mlynarova LU. Expression of a cucumber class III chitinase and Nicotiana plumbaginifolia class I glucanase genes in trangenic potato plants. Plant Cell Tiss Org Cult (PCTOC). 2004;**79**:161-168

[146] Punja ZK. Genetic engineering of plants to enhance resistance to fungal pathogens: A review of progress and future prospects. Canadian Journal of Plant Pathology. 2001;**23**:216-235

[147] Nishizawa Y et al. Characterization of transgenic rice plants over-expressing the stress-inducible beta-glucanase gene *Gns1*. Plant Molecular Biology. 2003;**51**:143-152

[148] Sinha M et al. Current overview of allergens of plant pathogenesis related protein families. The Scientific World Journal. 2014;**2014**:543195

Section 3

Breeding Practice and Cases

Chapter 7

Recent Advancements in Genetic Improvement of Food Legume Crops

Monika Punia, Lalit Kumar Rolaniya and Ram Lal Jat

Abstract

Legumes are the second-largest source of food after cereals, all over the world, and an essential protein source in the vegetarian diet. These crops remain essential to subsistence production as they have the inherent capacity to survive in an adverse ecosystem and require minimum investment for their management. The increasing challenge of feeding a rapidly growing population places excessive pressure on current food production systems, which can no longer be sustained by traditional plant breeding alone. Therefore, modern breeding methods with increased genetic gains are required to meet the food demand of the increasing population. In the past few decades, the efficiency of legume crop breeding programs has increased considerably using novel and multidisciplinary approaches in breeding programs. A multidisciplinary approach combining conventional plant breeding, mutation breeding, plant biotechnology, and molecular breeding is strategically ideal for production of new and improved crop varieties. This chapter focuses on recent advancements in plant biotechnology, related molecular methods, phenomics, and their application in breeding of legume crops.

Keywords: molecular marker-assisted backcrossing, molecular markers and genomics, food legume crops, genetic improvement, genomic assisted breeding

1. Introduction

Legumes are present in the diet of millions of people worldwide because these crops are associated with nutrition and health. Along with this, also have economic and environmental benefits. These are safe for consumption, relatively inexpensive, readily available, and the preferred food source after cereals. These crops have the ability of symbiotic nitrogen fixation; as a result, they help in the efficient use of fertilizers, lower emission of greenhouse gases and soil health enhancement. This ability of legumes promotes the inclusion of these crops in cultivation systems, thereby contributing to the diversified system and sustainability [1].

Food legumes are divided into two groups: 1) oil seeds and 2) pulses/ grain legumes. Oil seed legumes have high oil/fat content such as soybean and peanuts and pulses are dry seed legumes with low-fat content used as food (moong, urad, lentil, moth, etc.). Grain legumes are grown in both tropical and temperate regions of the world and used with cereals as pulse (dal). These are one of the major sources of income for smallholders who practise sustainable farming.

Today, malnutrition is more pronounced in developing countries due to the increasing population, and the most troubling one is caused by protein deficiency. Plant-based protein sources are the most desirable as they are nutritious, cheap, and easily accessible to poor people. Cereal-based diet system is deficient in protein content and essential amino acid lysine. In contrast, legumes contain protein and lysine amino acids, improving the nutritional status of diets based on cereals. Cereal diets containing legumes are considered one of the greatest therapies for protein caloric deficiency in developing nations [2]. In this way, legumes and cereals complement each other, and they must be eaten in a 35:65 ratio for nutritional balance.

In recent years, more people have substituted animal protein with vegetable protein due to increased health awareness and nutrition. With the increasing demand for vegetable protein, research on food properties and the utilization of indigenous food crops like legumes for protein-rich supplements are significantly increased [3]. These crops are also adapted to adverse climatic circumstances and are resistant to insect pests and disease, so they may be cultivated in arid climates with low or irregular rainfall.

In African and Asian countries, the primary contributors to protein and calories are legumes due to their economic and cultural reasons [3, 4]. Even though legumes crops have several benefits so far production is still less in comparison to cereals and vegetables. Cereal crops clearly overshadow these crops. It is necessary to increase awareness, spread the knowledge among the people, and encourage the farmers to grow legume crops along with cereals to increase their production.

1.1 Constraints in genetic improvement of legumes

Although legumes are a very useful protein source for humans and livestock, the research efforts to increase the productivity of legumes are lesser than the cereals. The poor yield of legumes may be due to growing these crops as subsistence in marginal lands with local varieties that do not tolerate biotic and abiotic stresses. Concerted attempts have been made during recent decades to enhance the yield potential of legumes with conventional methods, but genetic progress is poor compared to cereals [5].

The key challenges facing plant breeders in genetic improvement of legumes are discussed shortly below.

1.2 Genotype and environment interactions

Crops are largely determined by climatic conditions, and even minor changes from optimal conditions can severely affect plant growth and yield. Differential responses of improved cultivar strains are expected in different environments due to unpredictable climatic factors encountered at various sites and/ or years. G x E interactions then become a big challenge for any crop breeding program as they restrict effectiveness of breeding programs and selection responses. Legume crops show phenotypic instability due to environment (70–80%) and genotype x environment (17–27%) interactions for economically important quantitative traits, resulting in variable yield potential. The genotypic effects contribution is very less that is 1.5–7%. As a result, the environment has a crucial role in the stagnation of legume crop progress [6].

1.3 Multiple stress

Legumes are mainly cultivated in rainfed conditions on marginal lands with minimum inputs. In these risk-prone environments, legumes encounter multiple stresses such as various diseases (wilt, rust, Ascochyta blight, powdery mildew), erratic rainfall, prolonged dry spell, extreme temperatures, salinity stress, alkalinity and acidity. Most legumes are susceptible to different stresses, affecting morphological and physiological processes of plants that hamper plant growth. To ensure consistent performance of pulses in these areas, it is essential to develop multiple stress-resistant varieties.

1.4 Limited genetic diversity

The genetic enhancement of crops largely depends on the genetic diversity available in that crop for exploitation. The variability present in legumes for selection is comparatively limited. However, India is rich in available genetic diversity for legume crops, but the production and productivity are poor compared to global production and productivity [7]. To develop new cultivars, breeders use the same germplasm repeatedly in breeding programmes, and the rate of incorporation of new germplasm is less. Extensive use of the same genotypes with common ancestry in breeding programmes is the prime reason for the narrow genetic base of developed cultivars. Thus, the developed varieties are more susceptible to insects, pests, diseases and unpredictable climatic factors.

2. Recent techniques for genetic improvement of food legumes

2.1 Genotyping-by-sequencing of food legumes

The 1st genome sequencing of Arabidopsis (*Arabidopsis thaliana*) was completed in 2000; after this achievement, it has been proved that information about the genome of any crop species is a major and necessary step to the advancement of that crop species (**Table 1**).

Our knowledge of different crop plant traits, including food legumes, has expanded during the past few decades due to advances in plant biotechnology and genomic technologies [27]. Genome sequencing enables crops to be improved on the basis of genomic gains and the selection of genes that possess desirable characteristics that increase the quality and quantity of produce. This also provides detailed information on genome structure and mutagenic changes due to deletions and insertions and discloses the pathways linked to different stress responses. In legumes, *Lotus japonicus* (Japanese trefoil) and Medicago truncatula, these two species with small genome sizes were selected as reference genomes. The genome sequence of the majority of the legume crops is now available; for soybean, groundnut proginator, chickpea, pigeonpea, common bean, and adzuki bean. Completely sequenced legume species (with completed and annotated genomes): *Cajanus cajan* (833 Mb genome), *Cicer arietinum* (738 Mb), *Glycine max* (1,112 Mb), *Lotus japonicus* (472 Mb), *Medicago truncatula* (373 Mb) and *Phaseolus vulgaris*, respectively (588 Mb). These species have between 28,269 and 48,680 genes and 25,640 to 243,067 transcripts, respectively.

In addition to revealing the genome sequences of different legume crops, different research institutes re-sequenced legume germplasm lines because draft genomes are now available, so it is easy to deploy whole genome re-sequencing-based approaches in legumes. This approach will help to learn more about genome architecture,

Crop	Genome sequence	Germplasm lines resequenced
Soybean	• 950 Mb of the 1115 Mb of Glycine max genome; comprised of 46,430 protein-coding genes [8]	• A total of 106 soybean genomes were re-sequenced, representing wild, landrace, and elite lines of the crop. [9] • 89 lines [10] • 286 soybean accessions (14 wild type, 153 landrace and 119 bred accessions) were sequenced; [11]
Pigeonpea	605.78 Mb of the 833.07 Mb pigeonpea genome; total of 48680 genes [12]	• Genome-wide variation in 292 Cajanus accessions, including breeding lines, landraces, and wild species. [13] • 20 Cajanus spp. accessions comprising two wild; species and 18 cultivated species accessions; [14]
Chickpea	~738-Mb whole genome sequence of Kabuli chickpea variety CDC Frontier contains 28,269 protein-coding genes [15]	• 35 chickpea genotypes (parental lines of 16 mapping populations); [16]) • 129 chickpea varieties, comprising 88 desi and 41 kabuli [17]
Groundnut	• 1081 Mb (89%) of Arachis duranensis and 1371 Mb (90%) of A. ipaensis [18] • 50 324 protein-coding genes; A. duranensis var PI475845; [19]	• 11 genotypes including synthetics and their diploid parents [19] • 41 groundnut accessions and wild diploid ancestors [20]
Common bean	• 473 Mb of the 587-Mb genome and genetically anchored 98%; 26 279 protein-coding genes [21]	
Mung bean	• 543 Mb (84.48%) size genome (V. radiata var. radiata VC1973A); 22 427 predicted genes [22]	
Adzuki bean	• 75% of the 612 Mb size genome; 26857protein coding genes [23] • 450 Mb (83% of the genome) sequenced; 50 accessions including 11 wild, 11 semiwild, 17 landraces, and 11 improved varieties; 34 183 predicted genes [24]	
Cowpea	• 36 diverse cowpea accessions [25]	
Pea	3920 Mb of pea cultivar 'Caméor', representing 88% of the estimated pea genome size. Total number of protein-coding genes is 193,976 [26]	42 wild, landrace and cultivars [26]

Table 1.
Summary of genome sequence and resequencing efforts in important legumes.

structural variations, genome evolution, and genome dynamics during domestication. As a result, different genotypes/lines/accessions were chosen for these legumes based on their priority in respective crop improvement programmes.

2.2 Trait mapping from diverse legumes germplasm

Several traits that are agriculturally important are complex. These traits are controlled by many genes and affected by environment and gene-environment interactions.

Over the past few decades, to understand the genetics of complex traits has become a major concern. With the progress made in the area of molecular markers as well as in genomics significant number of QTLs have been found in various crops. In legumes also, several genes/ QTL controlling the target traits have been mapped (**Table 2**). The efficiency and accuracy of breeding practices have been improved significantly with the help of molecular marker-assisted selection of important traits. Further, the mapped gene(s) or QTLs can be introduced individually or pyramided in an improved variety.

Crop	QTL/Gene	Trait	Method	References
Soybean	IDC QTL FAD2-1 and FAD2-2 genes	Iron deficiency chlorosis Oleate biosynthesis	Association mapping Linkage mapping	Wang *et al.* [28] Bachalva *et al.* [29]
Pigeonpea	Hsf genes candidate genes	Heat-response Sterility mosaic disease (SMD), Fusarium wilt (FW) and photoperiod insensitivity	Genome-wide analysis genome-wide association analysis	Maibam *et al.* [30] [13]
Common bean	Co-1–Co-10 10 QTLs/ genes Resistance gene analogs	Resistance to anthracnose Resistance to anthracnose Resistances to different pathogens	Linkage mapping Associations mapping Associations mapping	Kelly and Vallejo [31] Choudhary *et al.* [32]
Cowpea	Candidate genes Hbs-1–Hbs-3 Major QTL	Resistance to root-knot nematodes Heat-induced browning of seed drought tolerance Resistance to root-knot nematodes	QTL mapping QTL mapping QTL mapping	Santos *et al.* [33] Pottorff *et al.* [34] Huynh *et al.* [35]
Pea	nod3 PsMlo PsDREB2A	Hyper nodulation mutation Powdery mildew resistance Drought response	Comparative genomics Comparative genomics Comparative genomics	Bordat *et al.* [36] Mohapatra *et al.* [37] Jovanovic *et al.* [38]
Chickpea	Aquaporins gene family CarERF116 Major QTLs corresponding to flowering time genes (efl-1, efl-3, and efl-4)	Biotic and abiotic stresses Abiotic stress responsive Flowering time	Comprehensive genome-wide analysis Genome-wide association Analysis QTL mapping	Deokar *et al.* [39] Deokar *et al.* [40] Mallikarjuna *et al.* [41]
Adzuki bean	VaAGL, VaPhyE, and VaAP2	Flowering time and pod maturity	QTL mapping	Li *et al.* [42]

Table 2.
QTL mapping in different legume crops.

There are two approaches for marker-trait associations identification in plants: (1) Linkage mapping and (2) Association or linkage disequilibrium (LD) mapping. Linkage mapping is a conventional mapping approach based on genetic recombination between two loci, whereas association mapping is a new approach and based on linkage disequilibrium.

Currently, candidate gene and whole-genome association mapping methods are used in crop plants. As a new approach to conventional linkage analysis, association mapping has the advantages of increased mapping resolution, research speed, and greater allele number. Different loci for iron deficiency chlorosis in soybeans have been effectively mapped using the candidate gene-based method [28].

Similarly, Bachalva *et al.* [29] mapped several candidate genes in soybean involved in oleate biosynthesis and examined their co-segregation with oleate, linolenate quantitative trait loci (QTLs). Whole-genome association mapping has been used in several legume crops; for example- Medicago truncatula, common bean, soybean, chickpea, cowpea, peanut etc. In pigeonpea, 292 accessions were characterized using genome-wide association analysis for the purpose of accelerating genetic gains and identifying associations between several candidate genes and agronomically significant traits [13]. In a diversity panel including 96 Middle American genotypes of common bean, Hoyos-Villegas, Song, and Kelly [43] studied the genetic basis of variation for drought tolerance and related traits, and the GWAS analysis enabled identification of important marker-trait associations for traits related to drought tolerance and candidate genes associated with wilting. Salinity stress, which is intensified by changing climatic conditions, has a negative impact on cowpea at the germination and seedling stages. Ravelombola *et al.* [44] conducted research to identify SNP markers linked to salt tolerance through association mapping.

2.3 Pan-genomes of legume crops

It is clear that a single individual's genome does not adequately represent the diversity of genes in a species. Pangenome assemblies, which capture sequence and structural variation in a species more comprehensively, can be developed as a remedy. Pangenome includes the core complement of genes common to all individuals of the species. The variable genome contributes to species diversity and provides functions that are not essential, but which may. While the availability of reference genomes has significantly supported plant breeding and research, these reference genomes capture only a portion of the species diversity. These reference genomes provide a selective advantage under some circumstances like; biotic and abiotic stress resistance. Tettelin et al. developed the pangenome concept [45] and developed the first-ever pangenome for a bacterial species, Streptococcus agalactiae. The first legume pangenome was generated by sequencing and de novo assembly of seven phylogenetically and geographically representative accessions of the wild relatives of cultivated soybean. The soybean pangenome indicates a faster evolution and greater diversity in dispensable genes than core genes related to adaptation to environmental stresses. Recently, the Pigeonpea pangenome was developed, based on 89 accessions mostly from India and the Philippines. This reveals that in Philippine individuals, there is a substantial genetic variation that is not present in Indian individuals.

Existence of a large number of repeats and several rounds of polyploidy, genome and pangenome assembly in plants is always difficult. The traditional de novo assembly and comparison approach was first used to demonstrate significant genomic differences between individuals. It has the benefit of providing the physical position of genes. During the breeding of certain crops, a decline in genetic variation has been observed, especially associated with the selection of important characteristics. This approach will help identify genes lost during breeding and selection that can be bread back into elite germplasm.

2.4 Mutmap technique in legumes

MutMap is a recently developed method based on whole-genome resequencing of pooled DNA from a segregating population of plants with a useful phenotype. The MutMap scheme to identify rice genes responsible for agronomically significant traits in a rice mutant pool that had been mutagenized by ethyl methane sulfonate (EMS). A recessive mutant from a mutant pool is backcrossed to a wild plant type in MutMap technique. The F1 plant is self-fertilized and the F2 progeny (>100) are screened for plants (>20) with a segregated mutant phenotype. In this method, only a small F2 population (>100 plants) is required for gene identification from crop plants so this is easy to maintain small population in the field. MutMap is particularly useful for identifying genes that control quantitative minor effect phenotypes, which is a challenging thing in crop improvement. This method is being used in the field of legumes to find candidate genes for leaf and plant type mutants in chickpea.

2.5 Genomic assisted breeding

The world's population is rapidly growing and is expected to hit 9 billion by 2050. This massive pressure on population would contribute to a serious food shortage. Pulses in the vegetarian diet are essential sources of proteins, for pulses improvement, extensive breeding programmes have been done through conventional breeding and significantly developed several high-yielding varieties [46]; however, the pace of genetic improvement of pulse crops is very slow. Limited success was achieved through conventional breeding even after continuous and systematic breeding efforts due to several constraints. One of the major constraints on expression of quantitative traits is high G x E interaction leading to slow genetic gain [6]. For strengthening conventional breeding programmes, integration of novel breeding strategies and techniques is required for revolutionary changes. During the last decade, the performance of legume breeding programme has increased significantly, as a result of novel genomic tools and techniques incorporated with the conventional breeding methods. In order to incorporate genomics in breeding, genomics-assisted breeding was suggested, and it has been effective for many traits in cereals and legumes. Genomic-assisted breeding will accelerate the genetic enhancement of pulses which leads to development of cultivars with higher yield and multiple stress tolerant. The availability of molecular markers such as simple sequence repeats and SNPs has enabled the dissection of complex characteristics that limit crop production, In the case of pigeonpea and soybean, genome-wide SNPs focused on resequencing of many germplasm lines were also used to establish marker-trait associations.

2.6 Marker assisted characterization of germplasm

In the twenty-first century, food, water and land are biggest challenges for increasing population. Agricultural activities need to expand, become sustained, and be more adaptive to climate change. To improve sustainability in agricultural systems, new paradigms are required, to explore the genetic potential of the huge although unfortunately underutilized resources of genetic diversity available for different crops. For breeding of climate-resilient varieties, a better understanding of evolutionary genetic variability is essential. Genetic diversity is the precious wealth for any crop improvement programme but due to climatic changes, it is reducing continuously. In last century, 75% decline in genetic diversity was witnessed in farmers' fields and it would further decline by about 20% by 2050.

Besides that, genetic resources can be excellent breeding material to develop superior variety in future breeding programmes. They can also be used in different breeding programs in order to increase the genetic base of cultivated crop varieties.

It has been observed that wild relatives have several desirable characteristics like resistance to biotic and abiotic streses, nutritional characteristics, cleistogamy, photo insensitivity and cytoplasmic male sterility (CMS).

In the past few decades, revolutionary approaches and systems have offered a great wealth of genetic and genomics resources that revolutionized research in both model and crop legumes.

A recent study on chickpea presents evidence of severe domestication bottleneck. Efficiency of cultivated population of chickpea is 100 times lesser than that of wild chickpea (Cicer reticulatum and C. echinospermum). In legume crops, study on landraces and wild relatives are significantly benefited by advanced technologies of genomics, phenotyping and computational biology.

The Vavilov Institute of Plant Genetic Resources (VIR), which houses a special genebank in St. Petersburg, Russia, using a mixture of genomics, computational biology and phenotyping to classify the 147 accessions of chickpea from Turkey and Ethiopia. The combination of high-density genotyping data with historical phenotypic information on these VIR landraces allowed chickpea genomes to enter 'agro islands' or 'domestication islands' that display significant associations with multiple phenotypes. These "genomic gems" have also been identified in chickpeas containing co-adapted and co-localized gene complexes. These are LG4 and LG2 in chickpea containing multiple genes/ QTLs related to drought and disease resistance, respectively. WGRS/RADSeq of 90 Cicer accessions, including cultigens, landraces and wild accessions, previously identified a wide collection of 54 genes on LG3 that could have been targeted during modern breeding efforts to manipulate salient characteristics such as flowering time.

Similarly, a genomic segment with an excess of MTAs for agronomically significant traits was observed on LG9 after re-sequencing of 292 accessions in pigeonpea [13].

In a recent study, to understand the genetic relationships between various lentil species/subspecies, a lentil collection comprised of 467 wild and cultivated genotypes originating from 10 different geographical regions was evaluated. A total of 422, 101 high-confidence SNP markers were identified against the reference lentil genome (cv. CDC Redberry). Phylogenetic analysis clustered the germplasm collection into four groups, namely, Lens culinaris/Lens orientalis, Lens lamottei/Lens odemensis, Lens ervoides, and Lens nigricans. Results of this study indicated that L. nigricans is most distantly related to L. culinaris and major differences were observed in six genomic regions with the largest being on Chromosome 1 (c. 1 Mbp) and further additional structural variants are likely to be identified from genome sequencing studies. In order to improve germplasm and for introgression of novel genes, this will provide further insights into the evolutionary relationship between cultivated and wild lentil germplasm.

Guar (Cyamopsis tetragonoloba (L.) Taub.) is primarily grown as an industrial crop due to its high-quality galactomannan gum used as a thickener, flocculant, emulsion stabiliser and gelling agent. Therefore, the novel set of molecular markers (nSSR) could be adopted as a useful tool to characterize the guar accessions for future breeding programmes.

2.7 Marker assisted backcrossing (MABC)

Research on legumes has greatly benefited from different available molecular markers in crop plants. Association between molecular markers and plant traits in these crops has introduced a novel approach to breeding that is based on the crossing of selected genotypes and selection of suitable progenies based on associated markers/QTL(s) rather than depending solely on phenotypes. Over the past three decades, the advancement and development of molecular marker technologies have been steady, such as low-throughput restriction fragment length polymorphisms (RFLPs) in the 1980s, high-throughput array-based markers in the 2000s and next-generation ultrahigh-throughput sequence-based marker systems in the 2010s. RFLP, RAPD, AFLP and SSR markers are low-throughput marker systems and are also considered past molecular markers. Besides these, next-generation sequencing (NGS) and genotyping by sequencing (GBS) are high and ultrahigh-throughput marker systems. These are based on low-cost and high-throughput sequencing technologies and are considered as present marker systems.

In cereals, so many outstanding achievements of marker-assisted breeding are available, but in legumes, negligence and lack of genomic resources adversely affected their initial establishment in the field of molecular breeding. Now recent advances in pulse genomics have led to the launch of several marker-assisted breeding projects.

RAPD and RFLP markers were used in five wild lentil taxonomic groups to understand their genetic makeup [47]. A genetic linkage map was also constructed in lentils with RAPD, AFLP and RFLP markers [48].

For shielding the varieties against various biotic and abiotic stresses and for ensuring crop productivity; gene mapping, tagging and marker-assisted selection have vital importance. Identifying and deploying molecular markers/QTLs in a desired background would be a priority. Marker-aided selection (MAS) greatly reduce the time and effort required to recover high levels of resistance from the donor and simultaneously recover the genomes of the recurrent parent. It has become more easier to transfer desirable genes/QTLs from wild relatives to existing cultivars due to MAS and transgenics.

Fusarium wilt (FW) and Ascochyta blight (AB) are two major constraints in chickpea (Cicer arietinum L.) production. The most affordable approach for long-term control of ascochyta blight and fusarium wilt in chickpeas is known to be the use of varieties with high resistance levels.

The availability of molecular markers associated with QTL for ascochyta blight resistance provided an opportunity to introgress the traits into adapted chickpea cultivars. backcrossing between moderately resistant donors (CDC Frontier and CDC 425-14) and the adapted varieties (CDC Xena, CDC Leader and FLIP98-135C) resulted in a variety with improved resistance to ascochyta blight [49].

More recently, five resistant lines representing foc2 gene introgressed into the background of Pusa 256 were reported with the help of foreground selection aided by two SSR markers (TA 37 and TA110). Cultivar Vijay was used as a donor of foc2 gene [50]. Annigeri 1 and JG 74 are elite high-yielding desi cultivars of chickpea, in Karnataka and Madhya Pradesh. in recent years, have become susceptible to race 4 of Fusarium wilt (FW).

A widely grown cowpea variety in Africa was improved by adding drought tolerance, striga and root knot nematode resistance QTLs using SNP markers. The major QTL region on LG 8 was introgressed from cultivar V-16 into the bacterial leaf blight susceptible variety C-152 through marker-assisted backcrossing (MABC) [51]. Similarly, By backcrossing resistance to CpMv gene was transferred into variety C-152. Cowpea mosaic virus (CpMV) was responsible for 80–100% yield loss in cowpea. SSR markers were used for linkage map construction and indicated that two markers MA15 and MA 80 were linked to CpMV resistance.

At ICRISAT in hybrid pigeonpea programmes, markers associated with fertility restoration and CMS are being used. This improved the selection efficiency of hybrid breeding and accelerated the breeding work [52]. In addition, a range of markers, including SSRs and SNPs are now available to enable genetic purity testing of pigeonpea hybrids and their parents. Recently, ICRISAT has launched a collaborative effort with ICAR-IIPR and other NARS institutions/universities to accelerate and target the improvement of ruling mega varieties of pigeonpea in India.

In groundnut breeding, the use of molecular markers in backcross breeding programme accelerated selection of recombinant progenies bearing nematode resistance and high oleic acid. Selection for high oleic acid content in groundnut was facilitated by one CAPS marker along with gel-free SNP assay using HybProbe design for the selection of nematode resistance SCAR, SSR and CAPS marker were used.

Recently in peanuts, two ahfad2 alleles from SunOleic 95R were introgressed into ICGV 05141 using marker-assisted selection. Marker-assisted breeding effectively increased oleic acid and oleic to linoleic acid ratio in recombinant lines up to 44% and 30%, respectively as compared to ICGV 05141. Subsequently, In the marker-assisted backcrossing-introgression lines, a 97% increase in oleic acid, and a 92% reduction in linoleic acid content were observed in comparison to the recurrent parent [53].

As opposed to traditional breeding, gene stacking or pyramiding is a useful strategy for transferring multiple desired genes or QTLs from various parents into a single genotype in the shortest possible time (two to three generations). Molecular markers that may be beneficial for marker-assisted selection and gene pyramiding have been identified through genetic linkage analyses and QTL mapping. The most effective and inexpensive means of combating plant diseases is the use of genetic resistance. Gene pyramiding is thus a sensible approach to creating multiple and enduring resistance. Most successful approach in common bean for wide spectrum control of common mosaic virus is to combine I, bc-u, bc-12, bc-22, and bc-3 genes. SCAR marker was used for MAS. In lentils, molecular marker-assisted gene pyramiding was used for resistance to ascochyta blight and anthracnose. In this research, two genes for resistance to ascochyta blight and the gene for anthracnose resistance in lentil breeding lines were pyramided using linked RAPD marker [49].

2.8 Genome editing

Crop plant genome editing is a faster-growing technique for inserting specified changes into the genome precisely and with great accuracy. Genome editing has emerged as an alternative approach to conventional plant breeding, and transgenic (GMO) approaches to improve food legumes and their sustainable production. Instead of spontaneous non-specific changes caused by radiation or chemical mutagenesis, crop researchers have long required mutations at specific sites in the genome. This approach allows for site-specific DNA insertion, deletion, modification, or replacement in a living organism's genome. The plant research community has not been widely involved with earlier SSN-specific (sequence-specific) genome-based editing technologies, because of the complex design and labour-intensive assembly of particular DNA binding protein for each target gene. A relatively new and comparatively easier technique for genome editing is CRISPR (clustered regularly interspaced short palindromic repeats) technique which is based on a simplified version of the bacterial CRISPR-Cas9 antiviral defence system. CRISPR genome editing technique is based on Cas9 protein which is an endonuclease. This endonuclease induces double-strand breaks using a guide RNA that is complementary to a target gene [54]. In order to create mutants for inaccessible genes, CRISPR-Cas9 would be a very useful technique. It can mutate multiple loci and make large deletions, thereby speeding the plant breeding without directly adding any transgene. The sequence-specific nucleases-based plant genome editing has a great potential to develop modified crops which can address the increased global food requirements and sustainable agriculture production. CRISPR/Cas9 was applied first in model legume plants to induce targeted mutagenesis.

A web tool was designed to identify potential CRISPR/Cas9 target sites and also a soybean codon-optimized CRISPR/Cas9 platform to induce mutation at target sites in somatic cells of Glycine max and Medicago truncatula [55]. In a recent study, an efficient CRISPR/Cas9 system was developed for targeted gene mutations in the model legume M. truncatula. A specific sgRNA was designed that targeted medicago phytoene desaturase (MtPDS) gene involved in the carotenoid biosynthesis. Very recently in Cowpea, the representative SNF gene has been effectively disrupted with an efficient CRISPR/Cas9-mediated genome editing. Guide RNAs (gRNAs) for the symbiosis of receptor-like kinase (SYMRK), reached ~67% mutagenesis efficiency in plants with hairy roots, and the formation of nodules in both mutants was totally prevented [56]. Conventional breeding is based on natural genetic variation and rigorous back-crossing systems are needed to incorporate the selected traits into an elite genotype. Unlike conventional breeding techniques, the present diversity does not limit CRISPR because it can directly integrate new mutations. This approach will benefit particularly those crops which have narrow genetic diversity and low variability for traits. Therefore, genome editing can speed up plant breeding programmes by inserting correct and predictable modifications directly in desirable backgrounds. The CRISPR/Cas9 system is especially beneficial because multiple traits can be modified simultaneously.

2.9 Mutation breeding in legumes improvement

The basis for any crop improvement programme is the variations present in the concerned crop. For generation of new variations, mutation is a prerequisite. These mutations are caused by various factors and are broadly divided into two major categories: spontaneous and induced mutations. Natural causes like as ultraviolet (UV) irradiation, reactive oxygen species, and transposable elements may generate spontaneous mutations in nature. On the other hand, physical and chemical mutagens cause artificial mutations. Different mutagenesis techniques have been successfully utilized in molecular plant breeding to study gene functions. The alterations induced can be random or particular to the target. Chemicals and physical mutagens cause random mutations. Unfortunately, random mutagenesis is costly, time-consuming and also difficult to screen desirable mutants from a large, mutated population. In addition to conventional plant breeding and

GMO techniques, targeted mutagenesis has arisen as an alternative for improving crop plants. This approach relies on the use of nucleases that allow for precise double-stranded breaks to occur at certain sites within the genome. The specific methodologies for targeted mutagenesis include PCR-based techniques for in vitro mutation generation and analysis, transposon mutagenesis, RNA interference (RNAi), TILLING (Targeting Induced Local Lesion IN Genomes), and programmed meganucleases [also called homing endonucleases, site-directed nucleases (SDNs) or site-specific nucleases (SSNs)]. TALENs, ZFNs, and CRISPR/Cas9 are frequently used meganucleases.

2.9.1 TILLING

Identifying a mutation in a particular gene and relating this mutation to the phenotypic alteration in the mutant organism is one of the most straightforward ways of determining gene function. TILLING (Targeting induced local lesions in genomes) is a non-transgenic, high throughput, general reverse-genetic strategy which aims to identify SNPs (single nucleotide polymorphisms) and/or INDELS (insertions/deletions) in a gene/gene of interest from a mutagenized population. TILLING has developed a few decades ago as an alternative to insertional mutagenesis in Arabidopsis thaliana. High-throughput TILLING provides a quick and cheap diagnosis method of induced mutations in artificially mutagenized populations. The important feature of TILLING is that it can be applied to any species, regardless of its genome size and ploidy level.

2.9.2 Eco TILLING

EcoTILLING (Ecotype Targeting Induced Local Lesions IN Genomes) is the modification of TILLING, which identifies natural genetic variations in populations in contrast to induced mutations in TILLING. This has been successfully used in animals and plants to discover SNPs and small INDELs. The classical method of Eco-TILLING is based on the enzyme endonuclease (Cel1, Endo1), which cleaves at the point of mutation by detecting mismatches in double-strand DNA. EcoTILLING is convenient for those plant species in detection of natural mutations where chemical mutagenesis is not suitable.

So far, TILLING and EcoTILLING have been implemented in many legume crops. In soybean, Tilling was used to screen more than 40,000 mutant lines and to create novel mutant alleles [57]. In chickpea, TILLING was also used to diagnose mutations in the M2 generation. Recently, in mungbean, five exon residing mutations were identified by TILLING and confirmed the potential role of each mutation in altering mungbean plant architecture to develop an ideal plant type [58].

2.10 Transgenic approaches/genetically modified legumes

Traditional breeding is tedious and success rate of obtaining desirable gene/genes or gene combinations from a large number of crosses is very less. These limitations hamper the desirable changes in crop plants. Therefore, biotechnological approaches are complementary to traditional breeding methods for addressing global food demands. Today we have access to vast gene pools due to new biotechnological approaches, which can be utilized in food crops to add favourable features. In this way, Genetically-modified (GM) crops can contribute to satisfying the food demand by developing varieties which are high yielding, good in quality, nutrition-rich and different kinds of stress-tolerant. Genetically modified crops are plants in which one or more genes have been introduced using genetic engineering techniques to produce desirable traits for agricultural purposes. Genetic engineering facilitates the direct gene transfer not only within the species and between the different plant species but also from unrelated organisms as well as also resolves the problem of linkage drag. Soybean was the first grain legume for which transgenic plants were developed [59]. Glyphosate-resistant soybean was developed by transferring gene derived from Agrobacterium sp. strain CP4, which encodes a glyphosate-tolerant enzyme EPSP synthase. Genetically transformed other legume species have successfully developed glyphosate-resistant lines for example- narrow-leaf lupin (Lupinus angustifolius L.) [60]. This is an easy way of weed control, reduces the cost of production and has a positive impact on the environment. Water stress causes significant yield losses in soybean crops; to resolve this problem, transgenic soybean was developed by transferring a gene encoding an osmotin-like protein extracted from Solanum nigrum var. americanum [61].

Helicoverpa armigera, a food legume insect, causes significant yield losses in pigeonpea. To minimize the losses caused by Helicoverpa armigera; transgenic pigeonpea was developed by transferring two synthetic Bacillus thuringiensis insecticidal crystal protein genes, cry1Ac and Cry2Aa. The transgenic pigeonpea expressed Cry1Ac and Cry2Aa proteins exhibited 80–100% mortality of insect [62]. Chickpea crop often encounter terminal drought stress that affects its production. Desi chickpea variety C235 that has 120 days of crop cycle, a transcription factor DREB1A was transferred and observed better root and shoot partitioning as well as higher transpiration efficiency in transgenic chickpea under drought stress [63]. In storage, cowpea seeds are severely damaged by storage pests (Callosobruchus maculatus and C. chinensis). Introduction of the bean (Phaseolus vulgaris) α-amylase inhibitor-1 (αAI-1) gene into a commercial Indian cowpea cultivar (Pusa Komal) strongly inhibited the development of these insects [64].

Genetic engineering, however, has excellent potential to maximize crop performance coupled with conventional methods, even though there is somewhat risk related to the effects of transgenic crops on the environment and human health. To overcome these risks, each product should be critically examined. Appropriate biosafety and food safety measures should be strictly followed.

2.11 Phenomics

A better understanding of the biological processes is required to increase yield potential and multiple stress tolerance. Any crop for its improvement majorly depends on favourable genetic changes in the crop genome, but the current pace of crop improvement is incapable of meeting future food demands. Therefore, crop improvement requires introducing new approaches for genetic changes in crop plants and their breeding. Marker-assisted breeding/ molecular breeding gives more importance to genotypic information of a crop, but phenotypic information is also equally important. Plant phenotyping is now a bottleneck in advancing crop yield. To enhance the selection efficiency of crop plants, phenotyping is also important, along with genotyping. The rapid and accurate evaluation of the phenotype of breeding lines and different crop populations is required for new variety development.

Phenomics is the investigation of phenomes, which are the collection of phenotypes (physical and biochemical traits) that a given organism may generate during development and in response to environmental effects. Crop phenomics is a multidisciplinary approach which integrates agronomy, life sciences, information science, math and engineering sciences and combines high-performance computing and artificial intelligence technology. This technique provides non-destructive and non-invasive ways of imaging, including colour imaging, near-infrared imaging, far-infrared imaging and fluorescence imaging for different phenomena like; plant structure, biomass, leaf health, for measuring soil and tissue water content, canopy/leaf temperature measurement etc. High-throughput phenotyping has been widely used, offering automated digital analyses of large data samples. The main benefit of high throughput phenomics approaches is the speed at which data can be collected: field data that could take several days to collect using conventional methods can be collected in a matter of hours using several sensors installed on a phenotyping platform. This saves time and allows several observations of a given plant/plot in a single day. These phenomics tools and techniques are making way for crop plant genetic improvement by using the potentiality of genomic resources.

2.12 Rapid generation advancement approaches in legumes

The biggest challenge for breeding higher-yielding and more resilient crops is the inability to complete more generations in lesser time. Generally, legume crops complete one or two breeding cycles in a year, so developing a new variety is time-consuming. Speed breeding is a rapidly emerging method among plant breeders to develop new varieties in a short period of time. This technique greatly enhances breeding and research speed by reducing generation time. Plants are grown in controlled growth chambers or greenhouses with optimal light intensity and quality, as well as specific day length and temperature (22 h light, 22 °C day/17 °C night, and high light intensity), to speed up different physiological functions; especially photosynthesis and flowering, and thus reduce generation time. Under normal glasshouse conditions, 2–3 generations can be produced per year, while speed breeding can produce up to 4–6 generations per year. Chickpea was induced to flower early by Gaur *et al.* [65] using 24-hour photoperiod, which, with the aid of offseason nurseries, allowed the production of three generations per year. Similarly, early and late flowering genotypes of pea, chickpea, faba bean, lentil and lupin were grown by Croser *et al.* [66] in controlled environments under different light spectra (blue and far red-enriched LED lights and metal halide). The time it took for the first seed to germinate was reduced significantly in half, and pollen viability was enhanced. In addition, costs for speed breeding are also reduced by combining this with genomics-based breeding and high-throughput phenotyping.

3. Conclusion and future perspectives

Legumes have a lot of potential to offset the effects of climate change by contributing to sustainable cultivation and expanding the crop spectrum, which is largely controlled by a few major cereal crops. Furthermore, in legumes, considerable advancement has been made in identifying novel genes for useful traits; however, the full potential of legume crops is still unknown. A range of tools and techniques have been provided by advancements in biotechnology, molecular breeding and genomics that can significantly enhance the hereditary gains in legume breeding programs. Genome sequencing allows the improvement of legume crops based on desirable gene selection and now the genome sequence of most legume crops has been sequenced. Marker-assisted breeding also significantly improved the accuracy and efficiency of crop breeding practices. Different loci have been mapped in several legume crops.

The high-throughput genotyping platform would undoubtedly allow for low-cost, large-scale screening of segregating individuals to select suitable genotypes. In the future, it will increase the utility of MAS breeding for legume crops, being productive, and cost-effective. Marker-assisted breeding/ molecular breeding gives more importance to genotypic information of a crop, but phenotypic information is also equally important. This gives the information about phenome of an organism eventually increasing the selection efficiency and reducing the time required for evaluation. CRISPRs allow the development of novel cultivars containing multiple genes in just one generation. Besides these, the availability of the reference genome, combined with high-density genotyping and sequencing assays, opens new possibilities for harnessing genetic variations for climate-resilient traits. These modern techniques are significantly accelerating the pace of legume crop development, ensuring overall food security.

Author details

Monika Punia*, Lalit Kumar Rolaniya and Ram Lal Jat
ICAR-Indian Institute of Pulses Research, Regional Research Centre, Bikaner, Rajasthan, India

*Address all correspondence to: monika.punia@icar.gov.in

References

[1] Bravo L, Siddhuraju P, Saura-Calixto F. Composition of underexploited Indian pulses. Comparison with common legumes. Food Chemistry. 1999;**64**(2):185-192

[2] Machaiah JP, Pednekar MD. Carbohydrate composition of low dose radiation-processed legumes and reduction in flatulence factors. Food Chemistry. 2002;**79**:293

[3] Khattab RY, Arntfield SD. Nutritional quality of legume seeds as affected by some physical treatments 2. Antinutritional factors. LWT-Food Science and echnology. 2009;**42**(6):113-118

[4] Iqbal A, Khalil IA, Ateeq N, Khan MS. Nutritional quality of important food legumes. Food Chemistry. 2006;**97**(2):331-335

[5] Ali M, Kumar S. Advances in Pigeonpea Research. Kanpur, India: Indian Institute of Pulses Research; 2005

[6] Kumar S, Ali M. GE interaction and its breeding implications in pulses. The Botanica. 2006;**56**:31-36

[7] Rana JC, Gautam NK, Gayacharan MS, Yadav R, Kuldeep T, Yadav SK, et al. Genetic resources of pulse crops in India: An overview. Indian Journal of Genetic and Plant Breed. 2016;**76**:420-436

[8] Schmutz J, Cannon SB, Schlueter J, Ma J, Mitros T, Nelson W, et al. Genome sequence of the palaeopolyploid soybean. Nature. 2010;**463**(7278):178-183

[9] Valliyodan B, Qiu D, Patil G, Zeng P, Huang J, Lu D, et al. Landscape of genomic diversity and trait discovery in soybean. Scientific Reports. 2016;**6**(1):1-10

[10] Lee Y-G, Jeong N, Kim JH, Lee K, Kim KH, Pirani A, et al. Development, validation and genetic analysis of a large soybean SNP genotyping array. Plant Journal. 2015;**81**(4):625-636

[11] Zhou L, Wang S-B, Jian J, Geng Q-C, Wen J, Song Q, et al. Identification of domestication-related loci associated with flowering time and seed size in soybean with the RAD-seq genotyping method. Scientific Reports. 2015;**5**(1):1-8

[12] Varshney RK, Chen W, Li Y, Bharti AK, Saxena RK, Schlueter JA, et al. Draft genome sequence of pigeonpea (Cajanus cajan), an orphan legume crop of resource-poor farmers. Nature Biotechnology. 2012;**30**(1):83

[13] Varshney RK, Saxena RK, Upadhyaya HD, Khan AW, Yu Y, Kim C, et al. Whole-genome resequencing of 292 pigeonpea accessions identifies genomic regions associated with domestication and agronomic traits. Nature Genetics. 2017;**49**(7):1082

[14] Kumar V, Khan AW, Saxena RK, Garg V, Varshney RK. First generation HapMap in Cajanus spp. reveals untapped variations in parental lines of mapping 1 populations. Plant Biotechnology Journal. 2016;**14**:1673-1681

[15] Varshney RK, Song C, Saxena RK, Azam S, Yu S, Sharpe AG, et al. Draft genome sequence of chickpea (Cicer arietinum) provides a resource for trait improvement. Nature Biotechnology. 2013;**31**(3):240-246

[16] Thudi M, Chitikineni A, Liu X, He W, Roorkiwal M, Yang W, et al. Recent breeding programs enhanced genetic diversity in both desi and kabuli varieties of chickpea (Cicer arietinum L.). Scientific Reports. 2016;**6**(1):1-10

[17] Thudi M, Khan AW, Kumar V, Gaur PM, Katta K, Garg V, et al. Whole genome re-sequencing reveals genome-wide variations among parental lines of 16 mapping populations in chickpea (Cicer arietinum L.). BMC Plant Biology. 2016;**16**(1):53-64

[18] Bertioli DJ, Cannon SB, Froenicke L, Huang G, Farmer AD, Cannon EKS, et al. The genome sequences of Arachis duranensis and Arachis ipaensis, the diploid ancestors of cultivated peanut. Nature Genetics. 2016;**48**(4):438

[19] Chen X, Li H, Pandey MK, Yang Q, Wang X, Garg V, et al. Draft genome of the peanut A-genome progenitor (Arachis duranensis) provides insights into geocarpy, oil biosynthesis, and allergens. Proceedings of the National Academy of Sciences. 2016;**113**(24):6785-6790

[20] Pandey MK, Agarwal G, Kale SM, Clevenger J, Nayak SN, Sriswathi M, et al. Development and evaluation of a high density genotyping 'Axiom_Arachis' array with 58 K SNPs for accelerating genetics and breeding in groundnut. Scientific Reports. 2017;**7**(1):1-10

[21] Schmutz J, McClean PE, Mamidi S, Wu GA, Cannon SB, Grimwood J, et al. A reference genome for common bean and genome-wide analysis of dual domestications. Nature Genetics. 2014;**46**(7):707-713

[22] Kang YJ, Kim SK, Kim MY, Lestari P, Kim KH, Ha B-K, et al. Genome sequence of mungbean and insights into evolution within Vigna species. Nature Communications. 2014;**5**(1):1

[23] Kang YJ, Satyawan D, Shim S, Lee T, Lee J, Hwang WJ, et al. Draft genome sequence of adzuki bean, Vigna angularis. Scientific Reports. 2015;**5**(1-3):1-8

[24] Yang K, Tian Z, Chen C, Luo L, Zhao B, Wang Z, et al. Genome sequencing of adzuki bean (Vigna angularis) provides insight into high starch and low fat accumulation and domestication. Proceedings of the National Academy of Sciences. 2015;**112**(43):13213-13218

[25] Muñoz-Amatriaín M, Mirebrahim H, Xu P, Wanamaker SI, Luo MC, Alhakami H, et al. Genome resources for climate-resilient cowpea, an essential crop for food security. Plant Journal. 2017;**89**(5):1042-1054

[26] Kreplak J, Madoui M-A, Cápal P, Novák P, Labadie K, Aubert G, et al. A reference genome for pea provides insight into legume genome evolution. Nature Genetics. 2019;**51**(9):1411-1422

[27] Varshney RK, Kudapa H, Pazhamala L, Chitikineni A, Thudi M, Bohra A, et al. Translational genomics in agriculture: Some examples in grain legumes. Critical Reviews in Plant Sciences. 2015;**34**(1-3):169-194

[28] Wang J, Phillip E, Lee R, Goos RJ, Helms T. Association mapping of iron deficiency chlorosis loci in soybean (Glycine max L. Merr.) advanced breeding lines. Theoretical and Applied Genetics. 2008;**116**(6):777

[29] Bachlava E, Dewey RE, Auclair J, Wang S, Burton JW, Cardinal AJ. Mapping genes encoding microsomal ω-6 desaturase enzymes and their cosegregation with QTL affecting oleate content in soybean. Crop Science. 2008;**48**(2):640-650

[30] Maibam A, Tyagi A, Satheesh V, Mahato AK, Jain N, Raje RS, et al. Genome-wide identification and characterization of heat shock factor genes from pigeonpea (Cajanus cajan). Molecular Plant Breeding. 2015;**6**(7):1-11

[31] Kelly JD, Gepts P, Miklas PN, Coyne DP. Tagging and mapping of genes and QTL and molecular marker-assisted selection for traits of economic importance in bean and cowpea. Field Crops Research. 2003;**82**(2-3):135-154

[32] Choudhary N, Bawa V, Paliwal R, Singh B, Bhat MA, Mir JI, et al. Gene/QTL discovery for Anthracnose in common bean (Phaseolus vulgaris L.) from North-western Himalayas. PLoS One. 2018;**13**(2):e0191700

[33] Santos JRP, Ndeve AD, Huynh B-L, Matthews WC, Roberts PA. QTL mapping and transcriptome analysis of cowpea reveals candidate genes for root-knot nematode resistance. PloS One. 2018;**13**:e0189185

[34] Pottorff M, Roberts PA, Close TJ, Lonardi S, Wanamaker S, Ehlers JD. Identification of candidate genes and molecular markers for heat-induced brown discoloration of seed coats in cowpea [Vigna unguiculata (L.) Walp]. BMC Genomics. 2014;**15**(1):1-11

[35] Huynh B-L, Matthews WC, Ehlers JD, Lucas MR, Santos JRP, Ndeve A, et al. A major QTL corresponding to the Rk locus for resistance to root-knot nematodes in cowpea (Vigna unguiculata L. Walp.). Theoretical and Applied Genetics. 2016;**129**(1):87

[36] Bordat A, Savois V, Nicolas M, Salse J, Chauveau A, Bourgeois M, et al. Translational genomics in legumes allowed placing in silico 5460 unigenes on the pea functional map and identified candidate genes in Pisum sativum L. G3: Genes|Genomes|Genetics. 2011;**1**(2):93

[37] Mohapatra C, Chand R, Singh VK, Singh AK, Kushwaha C. Identification and characterisation of Mlo genes in pea (Pisum sativum L.) vis-à-vis validation of Mlo gene-specific markers. Turkish Journal of Biology. 2016;**40**(1):184-195

[38] Jovanovic Z, Stanisavljevic N, Mikic A, Radovic S, Maksimovic V. The expression of drought responsive element binding protein ('DREB2A') related gene from pea ('Pisum sativum'L.) as affected by water stress. Australian Journal of Crop Science. 2013;**7**(10):1590-1596

[39] Deokar AA, Tar'An B. Genome-wide analysis of the aquaporin gene family in chickpea (Cicer arietinum L.). Frontiers in Plant Science. 2016;**7**:1802

[40] Deokar AA, Kondawar V, Kohli D, Aslam M, Jain PK, Karuppayil SM, et al. The CarERF genes in chickpea (Cicer arietinum L.) and the identification of CarERF116 as abiotic stress responsive transcription factor. Functional & Integrative Genomics. 2015;**15**(1):27

[41] Mallikarjuna BP, Samineni S, Thudi M, Sajja SB, Khan AW, Patil A, et al. Molecular mapping of flowering time major genes and QTLs in chickpea (Cicer arietinum L.). Frontiers in Plant Science. 2017;**8**:1140

[42] Li Y, Yang K, Yang W, Chu L, Chen C, Zhao B, et al. Identification of QTL and qualitative trait loci for agronomic traits using SNP markers in the adzuki bean. Frontiers in Plant Science. 2017;**8**:840

[43] Hoyos-Villegas V, Song Q, Kelly JD. Genome-wide association analysis for drought tolerance and associated traits in common bean. The Plant Genome. 2017;**10**(1):512

[44] Ravelombola W, Shi A, Weng Y, Mou B, Motes D, Clark J, et al. Association analysis of salt tolerance in cowpea (Vigna unguiculata (L.) Walp) at germination and seedling stages. Theoretical and Applied Genetics. 2018;**131**:79-81

[45] Tettelin H, Masignani V, Cieslewicz MJ, Donati C, Medini D, Ward NL, et al. Genome analysis of multiple pathogenic isolates of Streptococcus agalactiae: Implications for the microbial "pan-genome". Proceedings of the National Academy of Sciences. 2005;**102**(39):13950-13955

[46] Saxena KB. Heterosis Breeding in Pulses-problems and Prospects. 2002. pp. 61-72

[47] Ferguson ME, Robertson LD, Ford-Lloyd BV, Newbury HJ, Maxted N. Contrasting genetic variation amongst lentil landraces from different geographical origins. Euphytica. 1998;**102**(2):265-273

[48] Eujayl I, Baum M, Powell W, Erskine W, Pehu E. A genetic linkage map of lentil (Lens sp.) based on RAPD and AFLP markers using recombinant inbred lines. Theoretical and Applied Genetics. 1998;**97**(1):83

[49] Tar'an B, Buchwaldt L, Tullu A, Banniza S, Warkentin TD, Vandenberg A. Using molecular markers to pyramid genes for resistance to ascochyta blight and anthracnose in lentil (Lens culinaris Medik). Euphytica. 2003;**134**(2):223-230

[50] Pratap A, Chaturvedi SK, Tomar R, Rajan N, Malviya N, Thudi M, et al. Identification of candidate genes and molecular markers for heat-induced brown discoloration of seed coats in cowpea [Vigna unguiculata (L.) Walp]. Molecular Genetics and Genomics. 2017;**292**(6):1237-1245

[51] Dinesh HB, Lohithaswa HC, Viswanatha KP, Singh P, Rao AM. Identification and marker-assisted introgression of QTL conferring resistance to bacterial leaf blight in cowpea (Vigna unguiculata (L.) Walp.). Plant Breeding. 2016;**135**(4):506

[52] Saxena RK, Patel K, Kumar CVS, Tyagi K, Saxena KB, Varshney RK. Molecular mapping and inheritance of restoration of fertility (Rf) in A4 hybrid system in pigeonpea (Cajanus cajan (L.) Millsp.). Theoretical and Applied Genetics. 2018;**131**:1605

[53] Bera SK, Kamdar JH, Kasundra SV, Patel SV, Jasani MD, Maurya AK, et al. Steady expression of high oleic acid in peanut bred by marker-assisted backcrossing for fatty acid desaturase mutant alleles and its effect on seed germination along with other seedling traits. PloS One. 2019;**14**(12):e0226252

[54] Scheben A, Edwards D. Genome editors take on crops. Science. 2017;**355**(6330):1122-1123

[55] Michno J-M, Wang X, Liu J, Curtin SJ, Kono TJY, Stupar RM. CRISPR/Cas mutagenesis of soybean and Medicago truncatula using a new web-tool and a modified Cas9 enzyme. GM Crops & Food. 2015;**6**(4):243

[56] Ji J, Zhang C, Sun Z, Wang L, Duanmu D, Fan Q. Genome editing in cowpea Vigna unguiculata using CRISPR-Cas9. International Journal of Molecular Sciences. 2019;**20**(10):2471

[57] Anai T. Potential of a mutant-based reverse genetic approach for functional genomics and molecular breeding in soybean. Breeding Science. 2012;**61**(5):462-467

[58] Varadaraju A, Ramadoss BR, Gurunathan S, Balaram R, Ganesamurthy K, Ram SG. TILLING by sequencing (TbyS) reveals mutations in flowering control genes that are associated with altered plant architecture in Mungbean (Vigna radiata (L.) R. Wilczek). Genetic Resources andCrop Evolution. 2021;**68**(3):849-864

[59] Hinchee MAW, Connor-Ward DV, Newell CA, McDonnell RE, Sato SJ, Gasser CS, et al. Production of transgenic soybean plants using Agrobacterium-mediated DNA transfer. Bio/technology. 1988;**6**(8):915

[60] Barker SJ, Si P, Hodgson L, Ferguson-Hunt M, Khentry Y, Krishnamurthy P, et al. Regeneration selection improves transformation efficiency in narrow-leaf lupin. Plant Cell, Tissue and Organ Culture (PCTOC). 2016;**126**(2):219-228

[61] Weber RLM, Wiebke-Strohm B, Bredemeier C, Margis-Pinheiro M, de Brito GG, Rechenmacher C, et al. Expression of an osmotin-like protein from Solanum nigrumconfers drought tolerance in transgenic soybean. BMC Plant Biology. 2014;**14**(1):777

[62] Ghosh G, Ganguly S, Purohit A, Chaudhuri RK, Das S, Chakraborti D. Transgenic pigeonpea events expressing Cry1Ac and Cry2Aa exhibit resistance to Helicoverpa armigera. Plant Cell Reports. 2017;**36**(7):1037-1051

[63] Anbazhagan K, Bhatnagar-Mathur P, Vadez V, Dumbala SR, Kishor PB, Sharma KK. DREB1A overexpression in transgenic chickpea alters key traits influencing plant water budget across water regimes. Plant Cell Reports. 2015;**34**(2):199-210

[64] Solleti SK, Bakshi S, Purkayastha J, Panda SK, Sahoo L. Transgenic cowpea (Vigna unguiculata) seeds expressing a bean α-amylase inhibitor 1 confer resistance to storage pests, bruchid beetles. Plant Cell Reports. 2008;**27**(12):1841-1850

[65] Gaur PM, Samineni S, Gowda CLL, Rao BV. Rapid generation advancement in chickpea. Journal of SAT Agricultural Research. 2007;**3**(1):1037

[66] Croser JS, Pazos-Navarro M, Bennett RG, Tschirren S, Edwards K, Erskine W, et al. Time to flowering of temperate pulses in vivo and generation turnover in vivo–in vitro of narrow-leaf lupin accelerated by low red to far-red ratio and high intensity in the far-red region. Plant Cell, Tissue and Organ Culture (PCTOC). 2016;**127**(3):591

Chapter 8

Soybean Cultivation Technology Innovation and Environmentally Friendly Pest Control in Paddy Fields in South Sulawesi, Indonesia

Abdul Fattah, Elisa Winanda, Salim, Sri Wahyuni Manwan, Idaryani, Wanti Dewayani, Warda, Ayyub Arrahman, Herniwati, Nurjanani, M. Yasin, Sahardi Mulia, M. Basir Nappu, Abdi Negara, Asni Ardjanhar, M. Sudjak Saenong, Amiruddin Syam, Muh Taufik, Sunanto, Andi Yulyani Fadwiwati, Muslimin and Sitti Nuraeni

Abstract

South Sulawesi is one of the centers for soybean development in Indonesia and farmers generally plant it on paddy fields. Soybean cultivation technology innovation in paddy fields in general, farmers use a cropping system without tillage, rice fields after planting rice are directly sprayed with herbicides 3 times and then planted with soybeans. The varieties that are favored and widely planted by farmers in South Sulawesi have large seeds (weighing 100 seeds around 15.0–19.5 g) such as Detap-1, Devon-2, Derap-1, Dega-1, and Dena-2. The spacing used by farmers is 20 × 40 cm with double rows and single-row models. Types of pests that mostly damage soybean plants in South Sulawesi include *Spodoptera litura*, pod borer *Etiella zinckenella*, whitefly *Bemisia tabaci*, aphid *Aphis glycines*, leaf-rolling caterpillar *Lamprosema indicata* Fabricius, caterpillar Helicoverpa *Heliothis armigera*, pod ladybug *Riptortus linearis* F. green *Nezara viridula* L., soybean beetle *Phaedonia inclusa* and grasshopper *Locust migratoria*. To control these pests, farmers combine the use of vegetable insecticides and chemical insecticides.

Keywords: soybean, paddy field, technology, cultivar, variety, pests control, farmer

1. Introduction

South Sulawesi is one of the soybean development centers with a land potential of around 586,492 ha. The average soybean production achieved in South Sulawesi ranges from 1.10 to 1.50 t/ha [1]. This production is still much lower than the potential that is often achieved from research results of 2.0–3.0 t/ha [2]. The low production was due, among other things, to high pest attacks and the use of fertilizers that had not been optimal.

In the cultivation of soybeans in paddy fields after rice, the timeliness of planting greatly determines the success of farming because of the limited time for land preparation. Soybean planting is immediately carried out 2–4 days after rice harvesting with a no-tillage system (zero tillage) because it is associated with soil moisture conditions, in addition to saving energy and production costs. To produce well, soybean crops need to be irrigated 3–4 times, water can come from irrigation networks and groundwater with a pumping system.

According to the Report of the Indonesian Horticultural and Food Crop Protection Agency [3], the area of attack by pod borers, pod suckers, and armyworms for 5 years ranging from 7182 ha. The main pest attack rate on soybeans in 2008 was around 15–35% [4]. Furthermore, the results of research by Fattah and Hamka [5], conducted in Panincong Village (soybean development center in South Sulawesi), the intensity of pod borer attacks (9.59–13.16%), the intensity of pod suckers (6.17–22.55%), and armyworm intensity (8.61–17.26%). Control efforts carried out by farmers generally use insecticides with high doses and a spraying frequency of 1–2 per week. The use of these chemical pesticides has impacts including (1) polluting the environment including pesticide residues, (2) poisoning humans and animals, (3) killing natural enemies and other useful organisms such as bees which are plant pollinators, (4) creating new pests strains that are resistant to pesticides, (5) causing pest resurgence or an increase in pest populations after excessive application of pesticides. To avoid the negative effects of the use of chemical pesticides, the main soybean pest control is directed to the use of bio-pesticides or insecticides.

Biopesticides or insecticides from plant materials are not something new but have been used by farmers for a long time, even at the same time as agriculture itself was born. Farmers in Indonesia are already using vegetable materials as pesticides, including using soursop leaves to control locust pests and rice stem borers. Meanwhile, farmers in India use neem seeds as an insecticide to control insect pests. In addition to vegetable materials which are widely used by farmers as vegetable insecticides, the use of biopesticides is also used, such as the use of NVP from armyworms.

This book is structured to provide information on environmentally friendly soybean cultivation technologies including the use of crop residues such as straw as organic fertilizers and also to provide information about the importance of using natural ingredients in pest control and the use of biopesticides in soybean farming systems carried out by farmers in South Sulawesi.

2. Technology of soybean cultivation at the farmer level

2.1 How to process land

Farmers in South Sulawesi generally plant soybeans in rainfed lowland areas with no-tillage systems. Planting soybeans after rice in rainfed lowlands with a no-tillage system (TOT) has several advantages such as; saving costs, energy, and time. Conversely, if the rainfed lowland paddy field is perfectly processed (OTS), it will be less profitable due to a delay in planting time, in addition to the land losing water because the soil surface is open. Other benefits obtained from soybean planting with a non-tillage system in paddy fields can break the pest cycle, use the remaining fertilizer that is still left in the soil, and make the remaining soybean plants green fertilizers [6].

Rice field preparation is crucial so that soybeans grow and produce well. There are two kinds of rice harvesting models, namely those that harvest rice manually (human labor) and those using a Combine Harvester. The two methods of harvesting cause differences in straw residue in paddy fields.

2.1.1 Land preparation in paddy fields whose solids are harvested manually, using human power

Planting soybeans in paddy fields harvested by paddy using a manual scythe, the rice stems are cut off at the base of the rice stems so that they do not separate the rice stalks that stand on the paddy fields making it easier to grow soybeans (**Figure 1a**). In a non-tillage system to kill grass in paddy fields, farmers use herbicides both contact and systemic. Contact herbicides that are widely used by farmers are herbicides with active ingredients of paraquat dichloride. This herbicide is a full-grown herbicide to control weeds in rainfed lowland fields, while a systemic herbicide that is widely used by farmers is an active ingredient of isopropyl amine glyphosate (Roundup Max 660 SL). Herbicide spraying was carried out on paddy fields.

Rice that has been harvested and straw stumps are cut about 20–30 cm from the ground which aims to prevent the growth of new shoots and facilitate the planting of soybeans. In addition, it also functions to block the seeds of pea fly pests from laying eggs on pieces of seeds so that the dead and attacked plants become reduced. Because soybeans are not resistant to drought and waterlogging, a drainage canal is needed before planting with a distance of 3–5 m and a depth of 20–30 cm. This channel beside flowing water so that it is not flooded also functions for irrigation if the plants experience drought, especially if irrigation water is available. The straw which is still present in paddy fields should be spread over the surface of the land (**Figure 1b**). The results of research in Indonesia show that soybean yields that are planted after paddy fields without tillage are better than those with perfectly cultivated soil because perfectly treated soils can cause evaporation so groundwater supplies are not sufficient for plant growth. In addition, perfect soil processing can cause delays in planting time so that the plants will experience drought in the stage of development and filling of seeds, especially in the dry season. Planting soybeans immediately after harvesting rice, at which time the rainfall has been reduced but still enough for soybean growth [6].

The harvesting system uses Combine Harvester to separate pieces of straw which are about 50–75 cm high so that when planted directly soybeans will be disrupted. In rice fields that are still high in the hay, farmers use two ways to prepare soybeans, namely, some farmers cut back the straw to the base of the stem and some farmers

Figure 1.
Rice field that has been cut with straw: (a) spraying herbicides, (b) making canals with a hoe, and (c) making canals using a hand tractor.

Figure 2.
Former rice plantations were then planted with soybeans: (a) straw stalks, (b) soybean plants mixed with rice stalks, and (c) soybean growth after the rice stalks were removed.

do not cut the straw again but immediately sprayed the herbicide 2–3 times until the straws die and dry like **Figure 2a**. The advantages of soybean plants that are planted between rice stalks are not attacked by peanut fly pests. This is probably caused by the imago of the bean fly being blocked by rice stems when they want to lay their eggs on soybean cotyledons. Another advantage of soybean plants planted between rice straws is straw stems that have been extracted and then immersed in the soil so that the soil becomes fertile and the soybean grows fertile as shown in **Figure 2b** and **c**.

The way farmers grow soybeans in paddy fields with this system can create an environmentally friendly organic farming system because all the remaining rice straw stems are immersed in organic fertilizer. This makes farmers not use chemical fertilizers in their farming systems, moreover, the land used for planting is still fertile because it contains a lot of fertilizer from residues during fertilization in the rice planting period.

Soybean cultivation techniques that are appropriate after paddy is without tillage (TOT), also known as "zero tillage". This technology is appropriately developed in anticipation of the limited workforce in South Sulawesi and at the same time utilizes the remaining availability of groundwater at the time of rice harvesting, especially in areas with simple irrigation or rainfed rice fields. Components of growth and seed yield in soybeans grown with a system without tillage are better than those with perfect tillage systems (**Table 1**). Components of growth such as plant height, number of branches, number of pods, and seed yields were significantly different between systems without tillage with tillage systems. Weaknesses in a perfect tillage system will result in a delay in planting time, so that in areas with a short period of rain it will cause plants to lack water, the plants will experience drought, and seed yield will decrease.

Parameter	Complete tillage	No tillage	t-Hitung
Plant height (cm)	79.80	86.30	*
Number of branches (branches)	3.30	4.00	*
Number of pods	205.00	235.00	*
Weight of 100 seeds (g)	28.80	29.10	Ns
Seed yield (t/ha)	1.90	2.40	*

**Explanation = significant at 5% level.*
Ns = no significant at 5% level.
Source: Idaryani and Yusmasari [7].

Table 1.
Growth and yield of soybean seeds in the complete tillage and zero tillage systems in South Sulawesi. Indonesia. 2015.

2.1.2 Land preparation using perfect tillage carried out by farmers

The process of harvesting rice in paddy fields using combine generally compacts the soil, making it difficult to plant soybeans using the zero tillage method. To fertilize their land, some farmers use tractors. The use of 4-wheel tractors for tillage is only carried out by farmers who have a lot of farming costs. This is because the cost of cultivating the land using a 4-wheel tractor costs Rp. 1,350,000 per hectare. Meanwhile, planting soybeans using the zero tillage method only costs around IDR 450,000–600,000 per hectare.

2.2 How farmers plant soybeans in paddy fields

The remaining straw in paddy fields is spread over the soil surface (**Figure 3b**). This method can inhibit the growth of weeds and reduce the evaporation of groundwater and prevent the attack of fly pests on peanut seeds (**Figure 4**).

2.3 Ways for farmers to control weeds and soil loosening of soybean plants

The way farmers control weeds in soybean cultivation generally uses traditional tools such as hoe. Farmers do not use herbicides in controlling weeds in soybean plantations because they can cause soybean death or leaf. In controlling weeds in soybean cultivation in paddy fields, farmers usually also loosen the soil. Weeds that have been honed are collected and then processed into organic fertilizer (fermentation). The fermented weeds after they become weathered are used for organic fertilizer on soybean plantations (**Figure 5**).

Figure 3.
How to cultivate the land using a tractor (a) and cultivated land (b) in paddy fields where the rice is harvested using a combine.

Figure 4.
How farmers plant soybeans on uncultivated land (a), how farmers grow soybeans on cultivated land (b), and how farmers grow soybeans using soybean planting tools (c), and Soybean planting tool (d).

(a) (b) (c)

Figure 5.
The way farmers control weeds manually and the application of organic fertilizers on soybeans (a), growth of soybeans that had been given organic straw fertilizer at the age of 25 days after planting (b), and growth of soybeans that had been given organic straw fertilizer at the age of 35 days after planting (c).

2.4 Application of organic fertilizer

2.4.1 Application of organic fertilizer from straw fermentation

Composting straw increases the levels of macro and micronutrients, especially Phosphorus (P2O5) and Potassium (K2O), as well as Magnesium (Mg) and Potassium (K) (**Table 1**). The main nutrient elements that need to be added to fertilizing cocoa plants include Nitrogen, Phosphorus, Potassium, and Magnesium [8] (**Figure 6**).

The results of soil analysis after immersion of rice straw into the soil and fertilization showed an increase in the percentage of clay mass in the soil. After most of the material has decomposed in the composting process, the temperature will gradually decrease, and at this time advanced compost maturation occurs, namely the formation of a humus clay complex [9].

2.4.2 Application of liquid organic fertilizer from cow urine (Biourin)

This ceremony uses the means of plant destruction as a vegetable pesticide. However, this is forgotten by subak member farmers in Bali [10]. Biourine is a liquid organic fertilizer derived from the urine of fermented livestock. Fermentation technology is used in processing cow urine into bio urine. This process can cause changes in the properties of materials into simpler molecules so that they are easily absorbed by plants. Based on research conducted by Sutari [11], there was an increase

(a) (b)

Figure 6.
How farmers make fermented straw for organic fertilizer (a) and how farmers apply organic fertilizer from fermented straw (b).

(a) (b) (c)

Figure 7.
Liquid organic fertilizer from cow urine (Biourine) (a), liquid organic fertilizer application on soybean (b), and growth of soybeans that have been given biourine liquid fertilizer (c).

in macronutrient content, micronutrients, and pH in the urine of cattle that had been fermented into bio urine. Cow bio urine can improve plant growth because, cow urine contains elements of N (0.36%), P2O5 (5.589 mg/l), K2O (975.0 mg/l), Ca (25.5 mg/l). and C-organic (0.706%) [12]. Besides that, cow bio urine can improve the physical properties of the soil because beef bio urine is fermented using Azotobacter and Bacillus sp. Biourine contains the hormone Indo Acetate Acid (IAA) of 1197.6 mg/l, while the urine of fresh cattle containing IAA is only 704.26 mg/l. IAA hormone functions as the main auxin in plants [11].

MOL (local micro-organisms) can function as decomposers and also as liquid organic fertilizer. According to Septiana et al. [13], plant residues such as kale, spinach, mustard greens, cabbage, and bamboo shoots can be made into liquid organic fertilizer by adding a biocatalyst. The addition of 60 ml of the biocatalyst is very good for increasing the phosphorus content to 79.26 ppm in the manufacture of liquid organic fertilizer from mustard greens and spinach waste. Baharuddin [14], agricultural waste such as municipal waste, straw, corn waste, sugarcane waste, and livestock manure can be processed using biotechnology to produce liquid organic fertilizer/MOL and biopesticides. The results of research by Suhera et al. [15], microbes as a type of MOL are quite effective for increasing the weathering process in plant residues. Giving Microbat 20% can inhibit *Phytophthora palmivora* by about 50%. Furthermore, it was said that giving Microbat 10% could accelerate weathering (92%) compared to using EM-4 10% (75%). Giving 10 cc/l of Microbat water can increase potato production 30–45% (**Figure 7**).

According to Widhiastuti et al. [16], some other agricultural wastes are good enough to make liquid organic fertilizers such as palm oil mill waste as soil biodiversity fertilizer. The waste can function as organic fertilizer by increasing the physical, chemical, and biodiversity properties of the soil, and increasing the total soil bacteria.

2.5 Types of caterpillars found in soybean planting and how to control them at the farmer level

2.5.1 Spodoptera litura armyworm pest

The pests that attack soybeans in the districts of Maros, Pangkep, Wajo, and Soppeng are armyworms with varying intensities. The intensity of each variety is different. The highest attack intensity of armyworm attacks was found in Detam-2 (16.24%) and Gema (16.29%) varieties in Wajo District and the lowest in Grobogan (10.38%) and Argomulyo (11.25%) varieties, while in The highest attack rate of

Figure 8.
Armyworm larvae (Spodoptera litura on soybeans) (a) and S.litura pests and symptoms of damage to soybean leaves (b).

armyworm pests in Maros was the Burangrang variety (15.26%) and the lowest was the Grobogan variety (9.87%) (**Figure 8**) [17].

Spodoptera litura is an important pest that damages soybean leaves compared to other leaf-damaging pests [18]. Yield losses due to *Spodoptera litura* pest attacks can reach 80%, even puso if not controlled [19]. The rate of yield loss depends on the variety used, the growth phase, and the time of attack [18]. *Spodoptera litura* is known as a polyphagous pest and migratory insect which causes serious damage to soybean crops (**Table 2**) [20].

2.6 How to control soybean pests at the farmer

2.6.1 Use of botanical insecticides

Botanical pesticides are pesticides that are produced from plant parts. Several types of plants can be used as vegetable pesticides: Srikaya seeds (annonacin) which are stomach poison and contact to control aphids, jicama (pchyrrhizid) to control (*Plutella zinckenella*), tuba roots (Derris), *Lantana cedar* (salira), *Fragrant Lemongrass* (Andopogon), Patchouli (*Pogostemon cabilin*), Clove (*Euginia sygium*),

Variety	The intensity of S.litura attack on soybean leaves (%)				Seed yield(t/ha)
	Wajo regency	Maros regency	Pangkep regency	Soppeng regency	
Anjasmoro	14.36	11.24	12.30	10.94	2.71
Detam-2	16.24	12.32	17.01	15.54	2.21
Detam-1	13.24	12.11	16.20	12.53	1.90
Kaba	17.36	11.37	15.43	13.50	1.97
Gepak Kuning	13.29	10.39	10.40	12.30	2.00
Grobogan	10.38	9.87	12.32	8.83	2.10
Argomulyo	11.25	11.95	13.84	10.36	2.10
Gema	16.29	11.32	14.49	14.10	2.18
Burangrang	14.36	15.26	13.85	12.12	2.08

Source: Abdul Fattah et al. [21].

Table 2.
Average intensity of soybean leaf damage due to S.litura pest attack.

Neem (*Azadirachta indica*), tobacco leaves and pork nuts (Kphrosia candida). Of all these vegetable ingredients, pork and neem nuts have the highest ability and are almost comparable to carbaryl insecticides in controlling weevil pests [22].

Neem extract should be sprayed at an early stage of insect development, sprayed on the leaves, and sprinkled on the roots so that it can be absorbed by plants and control insects in the soil [23]. Furthermore, it was said that 50 g of neem seed extract was dissolved in 1 liter of water and added 0.5 ml/l grading agent effectively suppressed mite populations on sweet potatoes with a mortality of 70%. neem 50 g/l water can reduce yield loss of *Maruca testulalis* by 13–45%. The results of research by Sukorini [24], the application of vegetable pesticides from amethyst leaves gives the lowest attack intensity (0.53–0.89%) and the highest on butrowali plants (1.02–1.94%) in cabbage plants clove leaves contain eugenol between 70 and 95% which can kill microorganisms such as *Bacillus subtillis*, *Staphylococcus aureus*, and *Escherichia coli*. In addition, eugenol can also kill or suppress the development of plant pathogens such as *Fusarium oxyspora, Phytopthora capsici, Rhizoctonia solani*, and *Sclerotium rolfii* [25].

Based on their origin, biopesticides are divided into two: Botanical pesticides, which are extracts from certain parts of plants, including leaves, fruit, seeds, and roots, which have toxic properties against certain pests and diseases. Botanical pesticides are generally used to control pests (insecticides) and diseases (bactericidal or fungicide). Several types of plants are capable of controlling pests such as the Meliaceae family (Neem) and the Anonaeceae family (Srikaya seeds and Soursop seeds). The results Indiati, S.W. dan Marwoto [26] that the use of castor seed extract (*Ricinus communis*) as a vegetable antifertility ingredient in field rats with 2 ml/100 g of rat body weight/day given for 5 days, causes infertility in female field rats and has the effect of reducing 64.2–90.70% active sperm temperature in male rats compared to controls. The results of Balitsa Lembang research [27], several types of plants that can be used as vegetable pesticides: sugar apple seeds (annonain) which are stomach and contact poisons for controlling aphis, yam seeds (pchyrrhizid) for controlling (*P. zinckenella*), tuba roots (Derris), *Cypress lantana* (salira), *Fragrant citronella* (Andopogon), patchouli (*Pogostemon cabilin*), cloves (*Euginia sygium*), neem (*Azadirachta indica*), tobacco leaves and pork nuts (*Kphrosia candida*). Of all these vegetable ingredients, pig beans and neem have the highest ability and are almost comparable to carbaryl insecticides in controlling weevils [27].

According to Thamrin et al. [28], extract from the bark of kapayang (*Pangium edule*) can kill the puith stem borer (*Scipopaga innotata*) around 80% after application, while controls using synthetic insecticides (BPMC) have a mortality of around 100%. Furthermore, it is said that rose, papaya, jengkol, lemongrass, noni, pepper, and gadung plants can kill caterpillars (*Plutella xylostella*) around 65–100%. The use of kedondong leaf extract can kill Plusia sp. larvae around 26.7% at 36 hours after infestation (jsi), 66.7% at 48 hrs, and 77.0% at 72 hrs, while the control (chlorpyrifos (control) kills 83, 3% at 36 jsi, 100% at 48 isi. Similarly, Luwa leaf extract (*Ficus glomerata*) can kill plutella caterpillar plants (*Plutella sp*) by about 70%, parang red beans, green severe beans, and soursop are quite effective in killing pariah fruit caterpillars by about 75–80%. Betel leaf contains saponins, flavonoids, and polyphenols, while galangal rhizome contains benzyl benzoate, −methoxycinamal and xanthorhizal which can be used to control neck disease in rice (*Pyricularia oryzae*), and leaf spot disease in peanuts.

Neem (*Azadirachta indica*) contains the active compounds azadirachtin, meliantriol, and salanin. It is in the form of powder from leaves or liquid oil from seeds/fruit. Effectively prevents eating (antifeedant) for insects and prevents insects from approaching plants (repellent) and is systemic. Neem can make insects sterile because it can interfere with hormone production and insect growth. Neem has a spectrum effective for controlling soft-bodied insects (200 species), including grasshoppers, thrips, caterpillars, white butterflies, etc. Besides that, it can also be used to control fungi (fungicides) at a preventive stage, causing fungal spores to fail to germinate. Controlled fungi include powdery mildew, rot, leaf smallpox/scab, leaf rust, and leaf spot. And prevent bacteria in powdery mildew (powdery mildew). Neem extract should be sprayed at an early stage of insect development, sprayed on the leaves, and sprinkled on the roots so that it can be absorbed by plants and control insects in the soil.

Tuba root (*Deris eliptica*) is a compound that has been found, including rotenon. Rotenone can be extracted using ether/acetone to produce 2–4% rotenone resin, made into water concentrate. Rotenon works as a very strong cell poison (insecticide) and as an antifeedant that causes insects to stop eating. Insect death occurs several hours to several days after rotenone exposure. Rotenone can be mixed with pyrethrin/sulfur. Rotenone is a broad-spectrum (non-systemic) contact poison and a stomach poison. Rotenone can be used as a molluscicide (for mollusks), insecticide (for insects), and acaricide (for mites).

Rotenone can be used as a molluscicide (for mollusks), insecticide (for insects), and acaricide (for mites).

The compound tobacco contains is nicotine. It turns out that nicotine is not only toxic to humans but can also be used to poison insects. Dry tobacco leaves contain 2–8% nicotine. Nicotine is a fast-acting nerve poison. Nicotine acts as a contact poison for insects such as caterpillars that destroy leaves, aphids, triphs, and control fungi (fungicides).

In addition to being able to kill plant-disturbing insects, vegetable insecticides can also function as (1) Reference, which repels the presence of insects mainly due to their smell or the substances they contain, (2) Antifidants, causing insects to dislike plants, for example, because they taste bad, (3) Preventing insects from laying eggs and inhibiting the process of hatching eggs, (4) Poisons that can interfere with the nervous system and insect hormones, and (5) Attractants, as attractants for the presence of insects that can be used as a trap plant. Natural ingredients that contain bioactive compounds can be classified into three, namely (1) natural ingredients containing anti-phytopathogenic compounds (agricultural antibiotics), (2) natural ingredients containing compounds that are phytotoxins and plant growth regulators (phytotoxins, plant hormones, and the like) and natural ingredients containing compounds that are active against insects (insect hormones, pheromones, anti-oxidants, repellents, attractants, and insecticides that poison plants) (**Figure 9**).

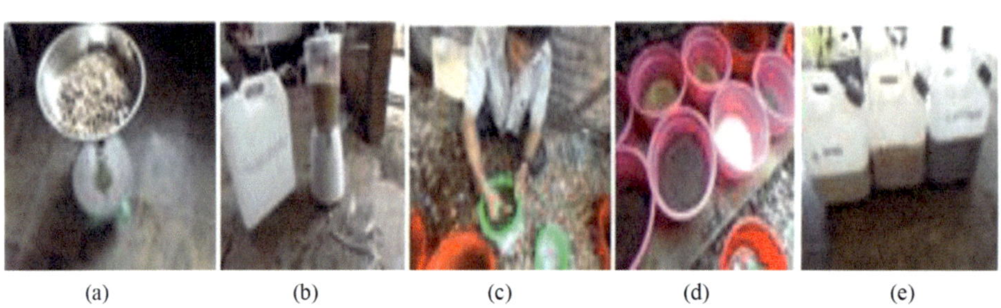

Figure 9.
The process of making vegetable insecticides by farmers (a-c) and the extract results for vegetable insecticides (d).

2.6.2 Use of pathogens

SlNPV propagation was carried out by taking several armyworms instars 4 and 5 which died naturally due to virus infection in soybean plantations with the characteristics of the caterpillars being elongated/expanding, not shrinking when massaged emitting a foul-smelling liquid, and sometimes hanging on the lower surface of the leaves.. The armyworm then made an emulsion using sterile aqua dest. The armyworm NPV emulsion was then diluted with sterile aqua dest and rubbed on the surface of the mulberry leaves. The caterpillars were kept until they died. After death, the armyworms were collected, extracted, mixed with distilled water, and then filtered using nylon gauze to obtain a pure coarse polyhedra suspension. According to Bedjo [29], the use of SLNPV (*Spodoptera liture* Nuclear Polyhedrosis Virus) 150–200 g/ha can kill around 80–100% of the *S liture* armyworm. S LNPV multiplies in its host's cells, so the transmission is through food. Symptoms of SLNPV transmission in armyworms appear 1–3 days after application. Instar-1 caterpillars infected with SLNPV will look milky white. Symptoms in instar caterpillars 3 and 4 will appear brownish white on the abdomen (abdomen), while on the back it is blackish milk brown. If the 5th and 6th instar caterpillars are infected with SLNPV, then at the pupal stage they will rot. In caterpillars that are infected with the SLNPV virus, their feeding activity is reduced, movement is slow, and the body swells due to the replication or multiplication of SLNPV virus particles. The caterpillar's integument usually becomes soft and brittle and easily torn. If the body of the caterpillar is broken, it will emit a very pungent odor. The death of caterpillars infected with this virus in the field is characterized by symptoms of the bodies of the larvae hanging or clinging to leaves or twigs of plants. Armyworm death usually occurs 3–7 days after contracting the virus. Furthermore, Sanjaya et al. [30], stated that a dose of 438 PIB/ml SINPV is sufficient. Effective for killing Instar-5 larvae in armyworms.

Biological pesticides are formulations that contain certain microbes in the form of bacteria, fungi, or viruses that are antagonistic to other microbes (causing plant diseases) or produce certain compounds that are toxic to both insects and nematodes. Some examples of biological insecticides include: (1) Nuclear polyhedrosis virus (NPV), (2) *Beuveria bassiana* (sunflower isolate) which is capable of controlling noncong beetles, the main pest of orchids and ticks on chrysanthemum plants, (3) Bio-PF contains Pf to control wilt, (4) Bio-GL contains (Gliodadium spp) to control soil-borne diseases, and (5) Prima—BAPF contains Bacillus spp. to control root swelling, wilt disease and root rot [31]. To distinguish between armyworms that die due to virus infection and pesticide poisoning in the field, it can be seen the characteristics and differences that arise, namely the death of caterpillars affected by the virus, they tend to elongate or not shrink, whereas if they die from pesticides, they tend to shrink. Larvae that die from the virus, when they are massaged or pricked, tear easily and secrete mucus like pus which smells bad, while caterpillars that are exposed to pesticides do not smell bad [32].

The results of laboratory experiments show that NPV has a high biotic potential, indicated by its level of pathogenicity which is expressed by the LC50 value (the concentration that kills 50% of the population). The LC50SlNPV for the armyworm was 5.4 × 103 polyhedra inclusion bodies (PIBs)/ml [33], while for the pod-eating caterpillar, it was 6 × 103 PIBs/ml [34]. The NPV infection process begins with the ingestion of the polyhedra by the caterpillar with the feed. In the digestive tract, which is alkaline (pH 9.0–10.5), the polyhedra coat dissolves,

(a) (b) (c)

Figure 10.
Ulat yang terinfeksi virus NPV (a), making process and biological insecticides and their application at the farm level (b).

freeing the virions. Virions penetrate the wall of the digestive tract to enter the body cavity, then infect susceptible cells. Virion replication occurs in the cell nucleus. Within 1–2 days after the polyhedra are ingested, the hemolymph which was originally clear turns cloudy. The caterpillar looks greasy, accompanied by swollen integumentary membranes and changes in body color to pale-reddish, especially on the stomach. Its ability to eat decreases, so its growth is slow. The caterpillar tends to crawl to the top of the plant and then dies hanging upside down with the pseudo limbs at the end of the plant. The integument of the dead caterpillar undergoes lysis and disintegration, making it very fragile. Polyhedra. Young caterpillars (instars l-lll) die within 2 days, while old caterpillars (instars IV-VI) in 4–9 days after the polyhedra are ingested [35].

Considering that it is susceptible to sun exposure, especially ultra-violet rays, and the behavior of caterpillars that are active in the evening and at night [36].

The Nuclear Polyhedrosis Virus (NPV) bioinsecticide is one type of pathogenic virus that has the potential as a biological agent in controlling armyworms because it is specific, selective, effective for pests that are resistant to insecticides, and safe for the environment. NPV has been developed in vivo in the Balitkabi laboratory, for biological control of Lepidoptera pests. As a bioinsecticide, the virus can control target insect pests precisely because it is specific, has a fairly high killing ability, is relatively inexpensive, and does not pollute the environment. The results of NPV engineering with carrier materials can maintain NPV virulence so that it can suppress armyworm populations on soybean plants in the field by up to 90% [29] (**Figure 10**).

To distinguish between armyworms that die due to virus infection and pesticide poisoning in the field, it can be seen the characteristics and differences that arise, namely the death of caterpillars affected by the virus, they tend to elongate or not shrink, whereas if they die from pesticides, they tend to shrink. Larvae that die from the virus, when they are massaged or pricked, tear easily and secrete mucus like pus which smells really bad, while caterpillars that are exposed to pesticides do not smell bad [32]. NPV application should be done in the afternoon or evening under favorable weather conditions, considering that it is susceptible to sun exposure, especially ultra-violet rays, and the behavior of caterpillars that are active in the evening and at night [36]. The results of other studies regarding the use of NPV to control armyworms in rice can cause 53% mortality at 3 days after inoculation and 95% at 9 days after inoculation [37].

3. Types of soybean varieties in Indonesia according to the description of new superior varieties of soybeans in Indonesia

Several high-yielding soybean varieties that have been produced by researchers in Indonesia are according to the Description of Soybean Varieties [38].

3.1 Derap-1

Growth type: Determine Flowering age: ± 34 days Ripening age: ± 76 days Hypocotyl color: Purple Epicotyl color: Green Leaf color: Green Flower Color: Purple Hair color: White Pod skin color: Yellow Seed coat color: Yellow Cotyledon color: White Hilum color: Light brown Leaf shape: Round Leaf size: Medium Branching: 2–4 branches/plant number of pods per plant: ±45 podsPlant height: ±59 mourning: Moderate lodging resistantPod break: Slightly resistant to splitting pods Seed size: Large100 seed weight: ±17.62 gramSeed shape: RoundBrightness of seed coat: ShinyYield potential: 3.16 tons/ha of dry beans (at 12% KA) Average yield: ±2.82 tons/ha of dry beans (at 12% KA) Protein content: ±39.17% BKFat content: ±18.10% BKPest resistance: and disease (*Phakopsora pachirhyzi* Syd), sensitive to SMV virus disease, resistant to pod sucking pest (*Riptortus linearis*), resistant to pod borer (*Etiella zinckenella*), and moderately resistant to armyworm (*Spodoptera litura* F.) Breeder: Ayda Kri snawati, M. Muchlish Adie, Apri Sulistyo Researchers: Marida Santi Yudha Ika Bayu, Kurnia Paramita Sari, Erliana Ginting, Joko Susilo Utomo, Eriyanto Yusnawan Technician: Arifin Proposer: Research Institute for Various Beans and Tubers.

3.2 Detap-1

Growth type: Determine Flowering age: ± 35 days Ripe age: ± 78 days Hypocotyl color: Purple Epicotyl color: Green Leaf color: Green Flower Color: Purple Feather color: White Pod skin color: Yellow Seed skin color: Yellow Cotyledon color: White Hilum color: Yellow Leaf shape: Slightly round Leaf size: Medium Branching: 3–6 branches/plant Number of pods per plant: ±51 pods Plant height: ±68.70 cm Bedding: Slightly resistant to lodging Split pods: Resistant to the splitting of pods Seed size: Large 100 seed weight: ±15.37 great shape: Round Yield potential: 3.58 ton/ha Average yield: ±2.70 ton/ha Protein content: ±40.11% PCFat content: ±16.16% PCPest resistance: Resistant to leaf rust disease, sensitive to SMV virus disease, resistant to pod-sucking pests, moderately resistant to pod borers, and sensitive to armyworm pest Breeders: M. Muchlish Adie, Ayda Krisnawati, Gatut Wahyu AS. Researchers: Erliana Ginting, Eryanto Yusnawan, Marida Santi YIB, Kurnia Paramita Sari, Didik Hanowo Technician: Arifin Proposer: Research Institute for Various Nuts and Tubers, Agency for Agricultural Research and Development.

3.3 Deja-1

Growth type: Determinite Flowering age: ±39 days Ripe age: ± 79 days Hypocotyl color: Purple Epicotyl color: Purple Leaf color: Green Flower color: Purple Fur color: Brown Pod skin color: Dark brown Seed coat color: Yellow Cotyledon color: Yellow Hilum color: Light brown Leaf shape: Oval Leaf size: Medium Branching: 3 branch/plant Number of pods per plant: ±36 pods Plant height: ±52.7 cm Creeping: lodging resistant Split pods: Not easily broken Seed size: Medium 100 seed weight: ±12.9 gram Seed shape: Oval Yield potential: 2.87 tons/ha Average yield: ±2 .39 ton/ha Protein content: ±39.6% PCFat content: ±17.3% PCPest resistance: Moderately resistant to armyworm pests, resistant to pod borers, resistant to pod suckers and moderately resistant to leaf rust disease.Remarks: Very tolerant to water stress from 14 days to maturity. Breeders: Purwantoro, Suhartina, Gatut Wahyu A.S., Novita Nugrahaeni and Titik Sundari. Researchers: Abdullah Taufiq, Suh arsono, A. Ghozi Manshuri, Eriyanto Yusnawan, and Kurnia Paramita. Proposers: Research Institute for Various Nuts and Tubers, Agency for Agricultural Research and Development.

3.4 Deja-2

Growth type: Determine Flowering age: ± 37 days Ripe age: ± 80 days Hypocotyl color: Purple Epicotyl color: Purple Leaf color: Green Flower Color: Purple Fur color: Brown Pod skin color: Light Brown Seed coat color: Yellow Cotyledon color: Yellow Hilum color: Brown Leaf shape: Oval Leaf size: Medium Branching: 3 branches/ plant Number of pods per plant: ± 38 pods Plant height: ± 52.3 cm Mortality: lodging resistant Split pods: Pods do not break easily Seed size: Large 100 seed weight: ± 14.8 gram Seed shape: Oval Yield potential: 2.75 ton/ha Average—average yield: ±2.38 ton/ha Protein content: ±37.9% PCFat content: ±17.2% PCPest resistance: Susceptible to armyworm pest, moderately resistant to pod borer, moderately resistant to pod sucker, and slightly resistant to leaf rust disease. Description: Tolerant of water saturation stress from 14 days old until the cooking phase. Breeders: Suhartina, Purwantoro, Gatut Wahyu AS, Novita Nugrahaeni and Titik Sundari. Researcher: Abdullah T aufiq, Suharsono, A. Ghozi Manshuri, Eriyanto Yusnawan, and Kurnia Paramita. Proposers: Agency for Agricultural Research and Development (**Figure 11**).

3.5 Gamasugeng-2

This variety has a determinate growth type, which means that the plant is upright and flowers in unison. Some varieties have an indeterminate growing type, meaning that the plant spreads and the flowers appear gradually. While the semi-determinate type of growth is a plant that has an upright growth type the flowering is not simultaneous or the flowers appear gradually.

The Gamasugeng-2 variety has a flowering age of about 30 days and a ripening age of about 68 days. Hypocotyl purple, epicotyl green, flowers purple, fur brownish white, pod skin brown, seeds yellow, cotyledons green, plant height about 45 cm, number of branches 4–5 stems per plant, yield potential 2.6 t/ha, average seed yield 2.4 t/ha, medium seed size (11.5 g seed weight), 37.4% protein content and 13.2% fat content.

Figure 11.
Appearance of several Indonesian soybean varieties.

Gamasugeng-2 variety, resistant to lodging, and resistant to leaf rust disease (*Phakospscora pachithyzi* Syd), resistant to brown leaf spot/blight (Cercospora), resistant to shoot borer (*Malanagromyza sojae*).

3.6 Gamasugeng-1

The Gamasugeng-1 variety is almost the same as the Gamasugeng-2 variety which were both released in 2013. and is the result of radiation of the Tidar variety at a dose of 200 gray. Determinate growth type with a flowering age of 30 days and maturity of 66 days. The hypocotyl is purple, the epicotyl is green, the flower is purple, the pod skin is brownish white, the seed coat is bright yellow, the cotyledons are green, and the helium is green. The number of branches is 4–5 branches per plant, the number of pods per plant is 52 pods, the plant height is 45 cm, it is resistant to lodging, and the seeds contain 37.6 and 13.2% fat content.

Gamasugeng-1 variety has medium seed size (11.5 g in 100 seed weight), 2.60 t/ha seed yield potential, round seeds, resistant to leaf rust disease (*Phakospscora pachithyzi* Syd), resistant to brown leaf spot/blight (Cercospora), resistant to shoot borer (*Malanagromyza sojae*).

3.7 Detam 4 prida

Determinate growth type, flowering age 36 days, harvest age 76 days, hypocotyl purple color, epicotyl green color, flower purple color, fur color brown, pod skin color brown, seed skin color black, cotyledon color white, helium color white, leaf shape oval (triangular), oval seed shape, plant height 53.2 cm, number of pods per plant 55 pods, moderately tolerant to lodging and moderately tolerant to breaking of pods.

This variety has a medium seed size (11.0 g in 100 seeds), a potential seed yield of 2.90 t, an average seed yield of 2.50 t, the seeds have a protein content of 40.3%, and a fat content of 19.2%.. Moderately resistant to pod-sucking pests, moderately resistant to rust, early maturing and drought tolerant.

3.8 Detam 3 prida

The Detam 3 Prida variety was released in 2013 which was the result of cross-selection between the W9837 and Cikuray lines. Determinate growth type, flowering age 34 days, ripening age 75 days, hypocotyl purple color, epicotyl green color, flower purple color, fur color brown, pod skin color brown, seed skin color black, cotyledon color white, leaf shape oval (triangular), plant height 56.9 cm, somewhat tolerant of lodging, moderately tolerant of pod splitting, oval seed shape and number of pods per plant 51 pods. The potential seed yield per hectare is 3.20 t, while the average seed yield per hectare is 2.90 t, has a medium seed size (100 seeds weigh 11.8 g), the seeds have a protein content of 36.4% and a fat content of 16.7%. This variety is sensitive to pod-sucking pests, sensitive to rust.

3.9 Dering-1

The variety was released in 2012 and is a single cross-product of the Davros x MLG 2984 superior variety. It has a determinate growth type. Flowering age 35 days after planting and harvest age 81 days after planting. Plant height 57 cm, brown fur color, oval leaf shape, purple hypocotyl color, purple epicuticle color, purple flower color, brown pod skin color, yellow seed skin color, helium seed color, dark cotyledon color, resistance to fall, number of branches 3–6 Batang per plant. Having a medium seed size (weight of 100 seeds 10.7 g), the potential yield of seeds is 2.80 t/ha, the average yield of seeds is 2.0 t/ha, and the seeds have a protein content of 34.2% and fat content of 17.1%.

These varieties are resistant to pod borer (*Etiella zinckenella*) and susceptible to armyworm (*Spodoptera litura*), resistant to leaf rust disease (*Phakospscora pachithyzi* Syd), and tolerant of dryness during the reproductive phase.

3.10 Gema

The Gema variety was released in 2011 which was the result of the crossing of Shirome's introduction with the Wilis variety. According to type, varieties have determinate growth types with light brown feathers, purple cotyledon colors, purple hypocotyl colors, green epicuticle colors, and white cotyledon colors. This echo variety has a plant height of 55 cm, a medium size (weighs 100 seeds 11.90 g), a flowering age of 35 days, harvests 73 days, a yield potential of 3.06 t/ha, and an average yield of 2.47 seeds t/ha. Brown pod color, purple flower color, round seed shape, light yellow seed skin color, and brown helium color. The seeds have a protein content of 39.07% and fat content of 19.11%.

The echo variety is sensitive to leaf viruses (CMMV) and moderate rust. Besides that, the variety is also rather sensitive to pod-sucking pests, rather resistant to pod borer, and moderate to armyworm pests.

3.11 Varietas gepak kuning

The variety was released in 2008 which is a selection of local varieties of Gepak Kuning. Determinant growth type, purple hypocotyl color, green epicuticle color, purple flower color, greenish young yellow seed color, old brown pod color, brown stem fur color, tagak branching, and oval leaf shape. Plant height of 55 cm, age of flowering 28 days, age of cooking 73 days, have small seed size (weight of 100 seeds 8.25 g), potential yield of 2.86 t/ha, and the average yield of seeds 2.22 t/ha. In this variety, the seeds have a protein content of 35.38% and fat content of 15.10%.

Gepak Kuning varieties are rather resistant to armyworms, Aphis sp., and leaf scavengers of Phaedonia sp. Adapt well to paddy fields and tegal land both in the rainy season and in the dry season.

3.12 Varietas grobogan

The Grobogan variety was released in 2008 which is a local population purification of Malabar Grobogan. It has a determinate growth type, purple hypocotyl color, purple epicuticle color, brown bark color, purple flower color, brown old pod color, lanceolate leaf shape, and helium brown cocoa color. Plant height 50–60 cm, flowering age 30–32 days, age of cooked pods 76 days, have large seed size (weight of 100 seeds, 18 g), seed yield potential of 3.40 t/ha, and an average yield of seeds 2, 77 t/ha. The seeds have a fat content of 18.4% and a protein content of 43.9%. Adapts well to several different growing environmental conditions. Having pods is not easily broken, and at harvest, the leaves are shed 95–100%.

3.13 Varietas Detam-2

The variety was released in 2008 which was a selection of intra-line crossings introduced 9837 with Wilis. Determinant type of growth, purple hypocotyl color, green epicuticle color, purple flower color, dark brown pod color, purple stem hair color, black seed skin color, helium cocoa color, oval leaf shape, oval shape, and brightness of dull seed skin. Age of flowering is 34 days, cooking age of pods 82 days, plant height is 57 cm, medium size (weight 100 seeds 13.54 g), potential yield of 2.96 t/ha, and yield of seeds 2.46 t/ha. The seeds have a protein content of 45.56% and fat content of 14.83%. The nature of resistance to pests, rather resistant to pod suckers, this variety is sensitive to armyworms. Other properties are rather resistant to drought.

3.14 Varietas Detam-1

Detam-1 variety was released in 2008, which was the result of the selection of crossing lines introduced in 9817 with Kawi. This variety has a determinant growing type, purple hypocotyl color, green epicuticle color, purple flower color, light brown stems, old pods of dark brown skin, black seed skin color, yellow cotyledon color, slightly round leaf shape, skin brightness shiny seeds.

This variety also has a plant height of 58 cm, a flowering age of 35 days, a ripe pod age of 84 days, a large seed size (weight of 100 seeds 14.84 g), a yield potential of 3.45 t/ha, and an average yield of 2 seeds, 51 t/ha, the seeds have a protein content of 45.36% and fat content of 33.06%. The nature of resistance to pests is sensitive to armyworms and somewhat resistant to pod suckers and other properties rather sensitive to drought.

3.15 Varietas anjasmoro

Anjasmoro variety was released in 2001 which was the result of mass selection from the population of the full-fledged Mansusia line. It has purple hypocotyl color, purple epicuticle color, white stem hair color, purple flower color, yellow seed skin color, light brown pod color, and helium color of brownish lining seeds.

This variety also has oval leaves, wide leaf size, determine the growing type, flowering age 35–39 days, aged pod pods 82–92 days, plant height 64–68 cm, number of branches 2–5 branches, has a large seed size (weight of 100 seeds 14.8–15.3 g). The seeds have a protein content of 41.8–42.1%, a content is 17.2–18.6%, and does not hold down. Anjasmoro varieties are moderate to leaf rust, and pods are not easily broken.

3.16 Varietas mahameru

This variety was released in 2001 which was the result of the mass selection of a population of pure Mansuria strains. It has purple hypocotyl color, purple epicuticle color, white stem hair color, purple color, yellow seed skin color, brown pod color, brownish yellow helium color, oval leaf shape, wide leaf size, and determine growth type. Flowering age 36–39 days, and the age of pod pods 83–94 days.

The Mahameru variety has a plant height of 62–64 cm, branches of 2–5 branches, several books 12–15 books, size of seeds (large seeds) with a weight of 100 seeds 16.5–17.0, pods are not easily broken, and resistant fall down The seeds have a protein content of 42.9–44.3%, and fat content of 17.3–18.2%. Another characteristic of this variety is that it is moderate to leaf rust.

3.17 Kaba

This variety was released in 2001 which was the result of a double crossing of 16 elders. It has purple hypocotyl, green epicotyls, yellow cotyledons, brown stems, purple flowers, yellow seed coat, brown pods, brown helium seeds, and oval-shaped seeds..

The Kaba variety has a determinant growing type, a plant height of 66 cm, a flowering age of 35 days, ages of mature pods of 88 days, medium-sized seeds (weight of 100 seeds 10.37 g), seed yields of 2.13 t per ha, have 44.0, 8.0% fat content, resistant to fall, rather resistant to leaf rust disease, has pods that are not easily broken and adaptive to paddy fields.

3.18 Burangrang

This variety released in 1999 originating from natural cross aggregates, taken from farmers' plants in Jember, is the result of pure line selection, three generations of segregation. Burangrang varieties have purple hypocotyls, yellowish brown feathers, purple flowers, yellow seeds, bright helium seeds, oblong-shaped leaves, and pointed edges.

Burangrang variety has determined growing type, the number of branches of 1–2 branches, age of 35 days of flowering, age of pods aged 80–82 days, plant height of 60–70 cm, large size seeds (weight of 100 seeds 16 g), seed yields ranging from 1.6–2.5 t/ha, has a 39% protein content, 20% fat content, not easy to fall, tolerant of leaf rust disease. This variety is suitable for raw materials for soy milk, tempeh, and tofu.

3.19 Argomulyo

This variety was released in 1998 from the introduction of Thailand by PT. Nstle Indonesia in 1998 by the name of Nakhon Sawan 1. Has purple hypocotyl, brown fur, purple flower color, yellow seed coat, bright white helium seed, determined growing type, 35-day flowering age, 80-age harvest 82 days, plant height 40 cm, number of branches per plant 3–4 stems from the main stem, having a large seed size (weight 100 seeds 16.0 g), having seed yield of 1.5–2.0 t/ha, has a protein content of 39.4%, has a fat content of 20.8%, has resistance to falling. In addition, the Argomulyo variety is tolerant of leaf rust and this variety is suitable for soy milk raw materials.

4. Conclusion

The soybean cultivation system carried out by farmers in paddy fields in South Sulawesi using environmentally friendly technology has a quite high potential. This is supported by the availability of organic materials which are quite widely available around farmers' fields such as straw, cow dung, and cow urine (bio urine). These raw materials are processed by farmers into organic fertilizers for use in soybean cultivation to reduce the use of chemical fertilizers. Similarly, in controlling pests and diseases, farmers use raw materials from extracted plants to become vegetable insecticides. This control technique has high potential because the raw materials are widely available in nature, such as areca nut, cashew skin, clove flower, mengkuduh fruit, betel leaf, srikaya leaf, neem seeds, yam seeds, saponins, pine lantana, and tobacco leaves. The advantages of the soybean cultivation technology developed by farmers in South Sulawesi include being more efficient because the raw materials are from nature and do not need to be purchased at high prices such as using chemicals.

Besides that, other advantages of the soybean cultivation technique used by farmers are that it is more environmentally friendly because the organic fertilizers used by farmers are the raw materials from plant residues and livestock manure such as straw and cow urine (bio urine). Likewise, the control of plant pests carried out by farmers in South Sulawesi uses raw materials from plants and biological microorganisms so it is more environmentally friendly than using chemicals. However, there are also drawbacks to the soybean cultivation technology used by farmers, including the time needed for the manufacturing process and the availability of some raw materials used by farmers in nature is limited, so they need to be developed through cultivation, such as neem seeds, areca nut, and sugar apple seeds. In general, the soybean cultivation technology used by farmers in South Sulawesi has greater prospects and opportunities to be developed as a whole in Indonesia, which has a tropical climate.

Author details

Abdul Fattah[1], Elisa Winanda[2], Salim[1], Sri Wahyuni Manwan[3], Idaryani[1], Wanti Dewayani[2], Warda[1], Ayyub Arrahman[1], Herniwati[1], Nurjanani[3], M. Yasin[1*], Sahardi Mulia[1], M. Basir Nappu[1], Abdi Negara[1], Asni Ardjanhar[1], M. Sudjak Saenong[1], Amiruddin Syam[4], Muh Taufik[4], Sunanto[5], Andi Yulyani Fadwiwati[5], Muslimin[5] and Sitti Nuraeni[6]

1 Research Center for Food Crops, Research Organization for Agriculture and Food, National Research and Innovation Agency, Cibinong, Bogor, West Java, Indonesia

2 Research Center for Food Technology and Processing, Research Organization for Agriculture and Food, National Research and Innovation Agency, Cibinong, Bogor, West Java, Indonesia

3 Research Center for Horticultural and Estate Crops, Research Organization for Agriculture and Food, National Research and Innovation Agency, Cibinong, Bogor, West Java, Indonesia

4 Research Center for Macroeconomics and Finance, Research Organization for Governance, Economy, and Community Welfare, South Jakarta, Java, Indonesia

5 Research Center for Behavioral and Circular Economics, Research Organization for Governance, Economy, and Community Welfare, South Jakarta, Java, Indonesia

6 Faculty of Forestry, Hasanuddin University, Tamalandrea, Makassar, Indonesia

*Address all correspondence to: yasin311962@yahoo.com

References

[1] Food Crops Agriculture and Horticulture Office of South Sulawesi Province. Annual Report. Makassar: South Sulawesi Provincial Government; 2008

[2] Research Institute for Nuts and Roots Malang. Annual Report. Malang: Center for Research and Development. Agricultural Research and Development Agency; 2009

[3] Maros Food Crops and Horticulture Protection Center. Annual Report. South Sulawesi Province: Food Crops Agriculture and Horticulture Office; 2010

[4] Maros Food Crops and Horticulture Protection Center. Annual Report. South Sulawesi Province: Food Crops Agriculture and Horticulture Office; 2009

[5] Fattah A, Hamka. The Level of Attack by the Main Soybean Pests in South Sulawesi. Report on the Study Results of the South Sulawesi Agricultural Technology Study Center; 2010

[6] Winardi. Prospects of soybean cultivation in rainfed rice fields and simple irrigated rice fields to increase soybean production in Indonesia. Journal of Agritech. 2014;**16**(2):89-97

[7] Idaryani, Yusmasari. Increasing soybean production through a No-tillage system (TOT) in rain-fed Paddy fields in bone regency, South Sulawesi. Proceedings of the Banjarbaru National Seminar on Agricultural Technology Innovation. 2016:702-710

[8] Wachjar A, Kadarisman L. The effect of the combination of liquid organic fertilizer and inorganic fertilizer and the frequency of their application on the growth of immature cocoa (Theobroma cacao L.) plants. Agronomy Bulletin. 2007;**35**(3):212-216

[9] Isroi. Composting Cocoa Waste. Material Presented at the Coffee and Cocoa Cultivation TOT Training Event at the Coffee and Cocoa Research Center, 2007

[10] Okada T, Tengkano W, Djuwarso T. An outline of soybean pests in Indonesia in faunal aspects. Bogor: Proceedings of the Bogor Food Crops Research Institute Seminar. 1988, 1988:37

[11] Sutari WS. Quality Test of Fermented Biourine with Microbes Derived from Plant Materials on Growth and Yield of Mustard Green Plants. Thesis. Bali: Udayana University; 2010

[12] Kamara IK. Effect of Doses of Vermicompost Fertilizer and Cow Biourine on the Growth and Yield of Peanuts. Thesis. Postgraduate Program. Bali: Udayana University; 2011

[13] Septiana S, Siti Y, Sholikhati U, Putra S. Phosphorus Extracts from Various Types of Simulated Waste for the Manufacture of Liquid Organic Fertilizer. Yogyakarta: National Seminar on Human Resources for Nuclear Technology Yogyakarta; 2009

[14] Baharuddin. The management and utilization of agricultural waste support IPM and environmentally friendly agriculture. In: Proceedings. PEI; 2010

[15] Suhera T, Kuswinanti T, Rosmana A, Baharuddin. Effectiveness of microbial bioactivator in reducing the intensity of pod disease Phytopthora palmivora Bult and in decomposing cocoa pod shell. In: Proceedings. PEI Scientific Seminar and Annual Meeting. Makassar: PFI XX South Sulawesi Regional Commissariat; 2010

[16] Widhiastuti R, Suryanto D, Muklis, Wahyuningsih H. Palm oil mill waste can function as organic fertilizer.

Agricultural Scientific Journal of Kultura. 2006;**41**(1):15

[17] South Sulawesi Agricultural Technology Study Center. Annual report. Makassar: Agricultural Research and Development Agency, Ministry of Agriculture; 2014

[18] Adie MM, Krisnawati A, Mufidah AZ. Degree of resistance of soybean genotypes to armyworm pests. h. 29-36. In: Rahmianna AA et al., editors. Proceedings of the Seminar on Research Results on Various Nuts and Tubers. Competitiveness Improvement and Implementation of Development of Nuts and Tubers Commodities Supports the Achievement of the Four Successes of Agricultural Development. Bogor: Center for Food Crops Research and Development; 2012

[19] Marwoto, Suharsono. Strategy and technological components of armyworm control (Spodoptera litura Fabricius) on soybean plants. Journal of Agricultural Research and Development. 2008;**27**(4):131-136

[20] Djuwarso T, Naito V, Matsuura H, Kikuchi A. The behavior of the moth Spodoptera litura fab. (Lepidoptera: Noctuidae) at night. Research Bulletin. 1986;**3**:35-43

[21] Fattah AS, Sjam ID, Daud VS. Dewi: The type caterpillar of Lepidoptera ordo and control techniques by farmers for soybean in South Sulawesi. Scientific Research Journal (SCIRJ). 2018;**6**(5):1-6

[22] Lembang Vegetable Plant Research Institute. Annual Report. Palembang, Bandung: Agricultural Research and Development Agency; 2009

[23] Indiati SW. Utilization of neem (Azadirachta Indica) as a botanical insecticide for legumes. Tubers with ACIAR-Smallholder Agribusiness Development Initiative (SADI). 2008

[24] Sukorini H. Effect of organic pesticides and spraying intervals on Plutella xylostella on cabbage plants. Journal. Gamma. 2006;**2**(1):11-16

[25] Sumartini. Clove extract to control rust (Phakopsora pachyrhizi) on soybean leaves. In: Linkages Visit for Staff from Provincial Research and Development Organization to the Indonesian Legumes and Tuber Crops Research Institute in Malang. East Java: Collaboration between the Research Institute for Legumes and Tubers with ACIAR-Smallholder Agribusiness Development Initiative (SADI); 2008

[26] Indiati SW d M. Potensi ekstrak biji mimba sebagai insektisida nabati. Buletin Palawija. 2008;**15**:9-14

[27] Vegetable Research Institute. Annual Report. Agricultural Research and Development Agency. 2009

[28] Thamrin M, Asikin S, Muklis, Budiman A. Potential Extracts of Swampland Flora as Vegetable Pesticides. Report of Swampland Agriculture Research Institute. Banjar Baru: Agricultural Research and Development Agency; 2009

[29] Bejo R. Utilization of Spodoptera litura nuclear Polyhedrosis virus (SlNPV) to control armyworm (Spodoptera litura F.) on soybean plants. In: Workshop on Utilization of Nuclear Polyhedrosis Virus (NPV) as a Biological Agent to Control Soybean Leaf-Eating Pests *Spodoptera litura*. 2003

[30] Sanjaya Y, Machmudin D, Kurniawati ND. Histological study of SINPV infection on body weight and peritrophic membrane damage of Spodoptera litura larvae. Nusantara Bioscience Surakarta. 2010;**2**(3):135-140

[31] Ornamental Plant Research Institute. Annual Report. Agricultural Research and Development Agency; 2008

[32] Hariyanto, D., Sri Hartati, and Marwoto. Utilization of NPV (Nuclear Polyhedrosis Virus). 2014. Available from: http://www.agrikencana.com [Accessed: August 15, 20]

[33] Arifin M, Waskito WIS. Susceptibility of soybean armyworm (Spodoptera litura) to nuclear polyhedrosis virus. Seminar on Food Crops Research Results, Center for Research and Development. Sukamandi, 16-18 January 1986. 1 (Palawija). 1986. pp. 74-78

[34] Gotham AAA. Increasing the effectiveness of Helicoverpa armigera (Lepidoptera: Noctuidae) nuclear-polyhedrosis virus with several adjuvants. In: National Seminar on the Role of Entomology in Environmentally Friendly and Economical Pest Control. Bogor: Pel Bogor Branch and the National PHT Program; 1999

[35] Arifin M. Production techniques and utilization of NPV bioinsecticides for controlling armyworms on soybeans. In: Sunihardi et al, editors. Food crops Research and Development Center. PRoceedings of the IV Food Crops Research Symposium: Milestones of Technological Advancement in Food Crop Production, Components, and Packages of Palawija Production Technology. Bogor; 1999. pp. 121-134

[36] Arifin M. Compatibility of SlNPV with HaNPV in the control of armyworms and soybean pod-eating caterpillars. Bogor: Journal of Food Crop Agriculture Research. 2006;**25**(1):65-71

[37] Trisnaningsih, Kartohardjono A. Formulation of nuclear Polyhedrosis virus (NPV) for controlling Rice armyworm (Mythimna separata Walker) in Rice plants. Indonesian Journal of Entomology. Indonesian Entomological Association. 2009;**6**(2):86-94

[38] Description of Superior Varieties of Soybean. Agricultural Research and Development Agency. Ministry of Agriculture; 2016. p. 142

Chapter 9

Breeding of Major Legume Crops through Conventional and Molecular Techniques

Satya Prakash, Suhel Mehandi and Harmeet S. Janeja

Abstract

Legume crops are universally applicable for human and animal food and sustenance because of their relatively high protein and essential amino acid content. Furthermore, they have been linked to sustainable agriculture, noting their ability to bind to atmospheric nitrogen-fixing bacteria. Despite this, several technical limitations of leguminous crops keep their world production far behind that of cereals. This chapter of the book focuses on current developments in breeding and biotechnology of major legume crops. Conventional breeding has primarily set out to recover a number of vegetative and reproductive traits that are associated with different heritability values, which reflect how susceptible each character is to genetic improvement. In conclusion, legume breeding programs using classical breeding methods and biotechnological tools face a promising boost for further application of knowledge and information that may boost their overall production. In plant breeding, the development of improved crop varieties is limited by very long periods of cultivation. Therefore, to increase crop breeding efficiency, they are using new strategies such as high-throughput phenotyping and molecular breeding tools. In this chapter, recent findings on various aspects of crop improvement, plant breeding practices, to explain the development of conventional and molecular techniques.

Keywords: plant breeding, molecular techniques, pigeon pea, speed breeding, conventional breeding

1. Introduction

Legumes are of particular nutritional and economic importance forming part of the diet of millions of people worldwide. Legume seeds in human nutrition are important cause of proteins and peptides, carbohydrates and dietary fibers, and a high-quality source of some micronutrients such as vitamins, fatty acids, folic acid and minerals that have significant health benefits [1].

Different approaches have been used to cut down the period of plant reproductive cycles. Innovative techniques developed in this decade, such as genomic selection, high-throughput plant phenotyping and modern speed breeding, have been shown to speed up plant breeding. Plant genetic engineering also played a precious role in developing crops with desirable quality related traits using gene transformation [2, 3].

Conventional breeding techniques are not adequate for plant genome augmentation to develop new plant varieties. To overcome this hindrance in plant breeding methods, molecular markers have been used for the assortment of superior hybrid lines. Improving plant phenotype for an exact desirable trait involves the artificial selection and breeding of this given trait by the plant breeder. Breeders always promote to use crops with shorter reproductive phase, which permit the production of a number of generations in a single year as well as help in crop rotation generally wheat rice cropping system. In this cropping system summer mung benefited as one extra crop in a year as well as also get better the soil health [4]. Plant breeding combined with genome studies increases the quality of breeding practices and saves time [5]. Research interest in genetically engineered crops has been increasing in legume crops given the fundamental need to ensure food security for the growing whole human population [6].

The use of molecular and conventional plant breeding techniques for many legume crops, as well as the use of genome editing methods, modify and improve required desired plant phenotypes. Moreover, the latent association between these approaches used to formulate the future strategy for crop variety/ hybrid development will also be explored.

2. Conventional breeding

2.1 Germplasm conservation and plant genetic diversity

Prebreeding performance as phenotypic and genetic appraisal of germplasm collections are2 key functions of a breeding program to obtain basic information about the genetic relationships1 amongst accessions, inheritance patterns of some important traits and to select lines for subsequent1 crossing cycles [7]. In this regard, the characterization of germplasm banks of legume crops1 worldwide has been crucial for the development of agriculture because they are the reservoirs of1 genetic diversity [8].

The genetic resources of other legume species are also a primary locus of genes associated with biotic and abiotic resistance and agronomic traits of value to breeders. 1 The genetic diversity of legume species has been described, which has been extremely useful for separating major collections of germplasm1 and genetically identifying different sets of parental lines used in breeding cycles/stages [8–11]. The germplasm of major legume species has shown similar values when they were1 analyzed using micro-satellites (SSRs) markers. This is not surprising because most of the legumes are highly self-pollinate with low and very low out-crossing rate values except pigeon pea [12]. Thus, they1 have tended to display low to moderate genetic variability at intra-population and intra-group. However, most of their inherited variability is spread amongst populations or groups of accessions,1 which is very prominent for breeding purposes.

The technology makes it possible to insert genetic material from unusual sources. It is now possible to insert genetic material from species, families and even kingdoms that may not previously have been sources of genetic material for a specific species, and even to insert custom-engineered genes which do not exist in nature. As a result,

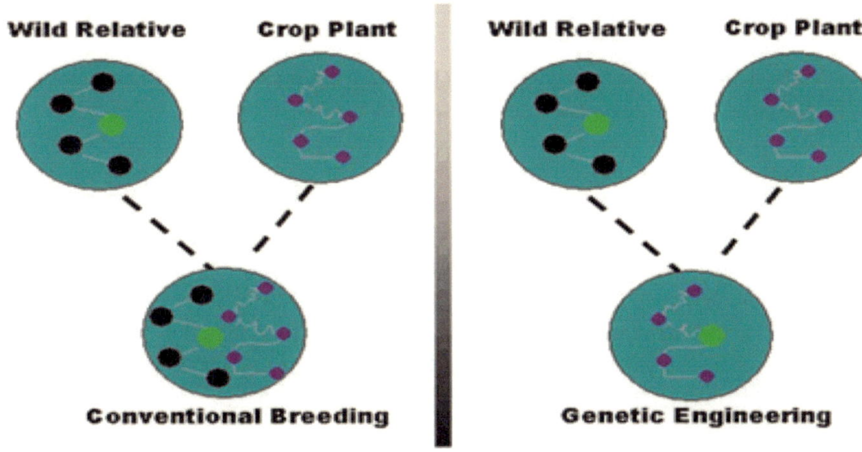

Figure 1.
Comparing conventional breeding and genetic engineering. https://www.isaaa.org/resources/publications/agricultural_biotechnology/download/

we can create what could be considered synthetic life forms, something that cannot be done through conventional breeding (**Figure 1**).

2.2 Characterization of legumes

The legume genotypes have showed significant differences on morphological and phenological1 traits such as pod curvature, days to flowering, hypocotyl color, growth habit, number of nodes, number of flower buds and hundred or thousand seed weight, which is significant for legume breeding [13].

They indicated that traditional breeding approaches have been particularly successful in improving monogenic traits, such as color, size, texture, appearance of some traits, although they are less specific and slow when it comes to quantitative traits, which are controlled by many genes. Are strongly influenced by the environment and are influenced by the environment and genetic interactions [14, 15].

3. Molecular and advance breeding method

3.1 Genomics-assisted breeding

Recent advances in the field of pulsed genomics deserve attention, for example, the discovery of genome-wide genetic markers, high-throughput genotyping and various sequencing platforms, high-density gene linkage1/QTL mapping, and most importantly, whole-genome sequence access. With the genome sequence in hand, there is considerable potential for using whole genome methods for trait mapping1 using correlation studies and selecting desirable genotypes through genomic selection. It is anticipated that GAB will accelerate progress in pulse/legume breeding, leading to rapid expansion of varieties with high yield, high stress tolerance and broad genetic base [16, 17].

3.2 Genetic engineering in legumes

The consequences that may result in the release of Genetically Modified crops (GM crops) in agriculture are a matter of ongoing debate [18]. However, it is logical to technically evaluate the risks1 of utilizing GM crops relative to their benefits and evaluate them with the conventional methods of1 genetic improvement [19]. The most successful case of public information is glyphosate resistant1 transgenic soybean [20], which has been commercializedc for over 20 years [21], and it is1 undoubtedly the most important genetic modification in soybeans [22].

Genetic engineering opens the door for plant breeders to bring together useful genes from a1 variety in one plant [23]. The development of glyphosate resistant variety utilized the CP4 gene from1 *Agrobacterium* spp., which encodes a glyphosate-resistant form of EPSPS, initially introduced in1 soybean [20].

Although gene flow is a legitimate concern of GM soybean [24], trans genes frequently represent gain of function, which might release wild relatives from constraints that limit their fitness1 [25–27]. This was a major breakthrough because no practical resistance to BGMV was known in1 common bean genotypes.

3.3 Modern legume breeding tools

There are many modern breeding tools are available that can speed up the legume breeding1 programme. The Arabidopsis plant model has allowed the study of metabolic and physiological1 processes during plant growth and in responses to biotic and abiotic stress through genome-wide gene expression analysis [28–30].

In parallel, the major version of the complete common bean genome sequence was recently published [31] and the chickpea genome sequence is also available in "The Cool Season 1 Food Legume Genome Database" [32]. References to legume genomes have also opened the door to feature 1 RNA sequencing approaches to conduct global transcriptomic profiling studies and discover new genes and ESTs [33–35]. Much effort has been made to compare genomes between model plant species and legume crops to correctly translate the information obtained [36].

These traits and their beneficial alleles can be introgressed in breeding lines through conventional1 genetic improvement in an easy manner, however, the application of MAS significantly reduces the1 time taken to select for resistant lines [37–39].

3.3.1 Abiotic stress breeding

Stress by low and towering temperatures in legumes can harshly affect plant growth, limiting1 yields and restricting the manufacture of certain regions and in specific periods of the season [40, 41]. Most of legume crops are full-grown in arid to semi-arid climate regions in India, and some countries1 in Africa [42–45].

3.3.2 Breeding for biotic stress

A large wealth of advances in genomic resources of legumes are associated to such as (**1**) Insects1 [46], (2) Fungi [47–50], (4) Bacteria [51, 52], Virus [53, 54] and Nematodes [55] (**Tables 1** and **2**).

Crops	Varieties	Release year	ICAR Institute/ $SAU_{s/}$ Organization centre	Salient features
Chickpea	Pant Kabuli chana-1	2010	Pantnagar	Irrigated condition, semi spreading, late mature, large seeded, medium height.
	Gujarat Junagadh gram 3	2010	Junagarh	Rainfed area, medium plant height, semi erect yellow large seeded, early maturity, resistant to wilt and stunt.
	MNK 1	2011	Gulbarga	Irrigated area, erect plant, seeds are milky white, extra-large seed.
	Raj Vijay Kabuli gram 201	2011	Sehore	Irrigated, desi type, early maturity, resistant to wilt.
	HK 4	2012	Hisar	Irrigated area, large seeds and white color, resistant reaction against wilt.
	JSC 55	2012	Sehore	Late sowing, suitable for sown under irrigated and late sowing condition, resistant to wilt, dry root rot and collar rot.
	GLK 28127	2013	Ludhiana	Irrigated condition, large old seeded variety, tolerant to drought and wilt, good rooting quality.
	NBeG 3	2013	Nandyal	Irrigated condition, long seed old variety, tolerant to drought and wilt and good rooting quality.
	WCGK 2000-2016	2015	Modipuram	Irrigated condition, long seeds, white color, resistant to fusarium wilt.
	Birsa chana 3	2015	BAO, Jharkhand	Normal sown condition, old type, tolerant to gram pod borer, resistant to wilt disease, shattering and lodging.
	GNG 2144	2016	Sri Ganganagar	Irrigated late sown condition, old and medium bold seeded, tolerant to fusarium wilt disease.
	CSJ 515	2016	Durgapura	Irrigated area, resistant to dry root rot, wilt and collar rot, tolerant to Ascochyta blight.
	Indira chana 1	2017	IGKV, Raipur	Rainfed and irrigated area, erect plant, resistant to wilt, primary branches.
	Meera	2017	ARS, Sri Ganganagar (Rajasthan)	Irrigated condition, tolerant to fusarium wilt.
	Pusa 3043	2018	IARI, Pusa (New Delhi)	Timely sown, escaping terminal drought, heat and stresses, resistant to wilt, tolerant to dry root rot, collar rot, stunt, Ascochyta blight.
	GNG 2207	2018	ARS, Sri Ganganagar (Rajasthan)	Timely sown, moderately resistant to fusarium wilt.
	IPC 2006-77	2019	ICAR-IIPR Kanpur (U.P)	Late sown under rice fallow, moderately resistant to wilt, dry root rot and stunt.
	Haryana Chana No 7	2019	CCS HAU, Hisar (Haryana)	Late sown irrigated, tolerant to Helicoverpa armigera.

Crops	Varieties	Release year	ICAR Institute/ $SAU_{s/}$ Organization centre	Salient features
	Pusa Chickpea 10,216	2020	ICAR-IIPR Kanpur (U.P)	Timely sown rainfed, First Marker Assisted Backcross variety of chickpea in India, drought tolerant.
	Pusa Parvati	2020	ICAR-IIPR Kanpur (U.P)	Timely sown irrigated, highly resistant to wilt, dry root rot and stunt.
	Kota Kabuli Channa-2	2021	AU, Kota-ARS, (Rajasthan)	Timely sown, resistant to wilt, dry root rot and collar rot, moderately resistant to stunt disease.
	Pusa Chickpea 20,211	2021	ICAR-IARI & ICRISAT	Rice-based cropping systems, moderately resistant to dry root rot, collar rot, pod borer and stunt.
Pigeon pea				
	TS 3R	2010	ARS Gulbarga	Indeterminate growth, semispreading, white and bold seeded, resistant to fusarium wilt.
	PKV, Tara	2011	PDKV, Akola	Indeterminate growth, semi spreading and tolerant to pod borer & pod fly.
	Rajeev Lochan	2011	IGKV, Raipur	Resistant to wilt & sterility mosaic disease.
	WRG-65	2012	ARS, Warangal	Indeterminate growth, spreading medium, resistant to wilt, tolerant to pod borer.
	Phule T 0012	2012	MPKV, Rahuri	Resistant to fusarium wilt and tolerant to pod borer and pod fly.
	ICPH 2671	2013	RAK College Sehore	Indeterminate growth, medium mature, tolerant to wilt and sterility mosaic.
	BRG 4	2014	UAS, Bangalore	Indeterminate growth, semi determinate, suitable for normal and delayed sowings.
	IPA 203	2014	IIPR, Kanpur	Resistant to sterility mosaic disease, tolerant to fusarium wilt.
	PRG 176	2015	RARS, Palem	Indeterminate growth, suitable to low rainfall conditions.
	ICPH 2740	2015	ICRISAT	Indeterminate growth, semi spreading, resistant to sterility mosaic and wilt.
	GRG 881	2016	ARS Gulbarga	Indeterminate growth, semi spreading, resistant to fusarium wilt, moderately resistant to sterility mosaic disease.
	CORG 8	2016	TNAU Coimbatore	Indeterminate growth, bold seeded, resistant to sterility mosaic disease, tolerant to Helicoverpa armigera and Maruca vitrata.
	LRG 52	2017	RARS, Lam	Indeterminate growth, semi spreading, dark purple pods, brown and large seeded and moderately resistant to wilt.
	BRG 3	2018	UAS Bangalore	Intermediate growth, semi spreading, red flowers, mottled seed, resistant to wilt.

Crops	Varieties	Release year	ICAR Institute/ $SAU_{s/}$ Organization centre	Salient features
	GT 104 (NPMK 15-05)	2018	NAU Navsari	Intermediate growth, semi spreading, long pods cream color.
	WRGE – 93	2019	SAU, Warangal	Moderately resistant to wilt.
	MPV-106	2020	MSSC, Akola	Moderately resistance to wilt diseases, resistant level for sterility mosaic is at par with resistant check variety.
	IPH 15-03	2020	IIPR, Kanpur	Resistant to fusarium wilt, moderately resistant to Phytophthora blight.
	LRG 133-33	2021	RARS, Lam, Guntur	Disease reaction at natural field conditions, resistant to wilt.
Soyabean	NRC-77	2010	DSR, Indore	Resistant to charcoal rot, Rhizoctonia root rot.
	RKS-24	2011	AAU, Kota, Rajasthan	Moderately resistant to bacterial pustule, collar rot and YMV, moderately resistant to girdle beetle, stem fly and defoliators.
	GC-00209-4-1-1	2011	UAS, Bangalore	Vegetable type.
	DSb-1	2012	UAS, Dharwad	Better germination, resistant to rust, tolerant pod shattering.
	SL 744	2012	PAU, Ludhiana	Timely sown irrigated areas, resistant to yellow mosaic virus and soybean mosaic virus.
	PS-19	2013	GBPUA&T Pantnagar	Resistant to major foliar diseases, bacterial pustule and rhizoctonia aerial blight.
	MACS-1188	2013	ARI, Pune	High oil content, early maturity, resistant to pod, shattering and Rhizoctonia aerial blight, bacterial pustules, charcoal rot, stem fly, pod borer, leaf folder, leaf minor and defoliators.
	JS-20-34	2014	JNKVV, Jabalpur	Resistant to charcoal rot, girdle beetle and stem fly.
	MAUS-2 (Pooja)	2014	VNMW Krishi Vidyapeeth, Parbhani	Resistant to bacterial pustule and leafspots, leaf miner, stem fly and blue beetle.
	KPS-344	2015	RRS, MPKV, Sangli	Tolerant to rust, resistant to stem fly, pod borer and leaf roller.
	Pusa 12	2015	IARI, New Delhi	Resistant to YMV, Rhizoctonia aerial blight and bacterial pustules.
	JS 20-69	2016	JNKVV Jabalpur	Resistant to YMV, charcoal rot, bacterial pustules, Alternaria leaf spot, pod blight, Indian bud blight, Target leaf spot.
	VL Bhat 201	2016	VPKAS, Almorah	Highly resistant to frog eye leaf spot, target leaf spot and moderately resistant to pod blight, highly resistant to girdle beetle, moderately resistant to stem fly.
	Raj Soya-24	2017	RVSKVV, Sehore	Resistant to YMV.

Crops	Varieties	Release year	ICAR Institute/ $SAU_{s/}$ Organization centre	Salient features
	Pant Soya 21	2017	GBPUA&T Pantnagar	Resistant to Yellow Mosaic Virus (YMV), SMV & bacterial pustule, tolerant to Rhizoctonia Aerial Blight.
	CG Soya-1	2018	IGKV, Raipur, Chhattisgarh	Resistant to Indian bud blight, Rhizoctonia aerial blight, Myrothecium leaf spot and bacterial pustule disease, moderately resistant to pod blight.
	JS 20-98	2018	JNKVV Jabalpur	Resistant against most dreadful disease ***i.e.,*** charcoal rot, blight, bacterial pustules, leaf spots and insect pests.
	KSD 726	2019	RRS, MPKV Sangli	Resistant to rust, purple seed stain disease, moderately resistance to Stem fly and defoliators.
	VL Soya 89	2019	VPKAS, Almorah	Moderate resistance against frog eye leaf spot and pod blight diseases, moderate resistance against Chauliops and resistant against defoliators.
	Pant Soybean 25	2020	GBPUA&T Pantnagar	Resistant to bacterial pustule & BLB, moderately resistance to RAB, BS & FLS.
	Pant Soybean 26	2020	GBPUA&T Pantnagar	Resistant to bacterial pustule & BLB, moderately resistance to RAB.
Mungbean	Pairy Mung	2010	IGKV, Raipur	Rabi season, tolerant to YMV, resistant to powdery mildew.
	SML 832	2010	PAU, Ludhiana	Spring and Summer term, tolerant to thrips.
	DGGV 2	2012	UAS, Dharwad	Kharif season, resistant to shattering of pods and suitable for mechanical harvesting.
	Shalimar Mung 2	2013	Srinagar centre, SKUASTA	Early maturing, resistant to Cercospora leaf spot, moderate resistant to aphid.
	CO. (Gg) 8	2013	TNAU, Coimbatore	Rainfed area, resistant to YMV.
	SGC 16	2014	RARS, Shilongani, AAU, Assam	Resistant to CLS and YMV.
	BGS 9 (Somnath)	2014	UAS, Raichur	Moderately resistant to PM, bold seed and long pod.
	Pant Mung 8 (PM 9-6)	2016	GBPUAT, Pantnagar	Resistant to MYMV, CLS and PM.
	RMG 975	2016	RARI, Durgapura	Kharif season, moderately tolerant to MYMV and root knot nematode.
	PUSA 1371	2017	IARI, New Delhi.	Kharif season, multiple resistant to MYMV, resistant to root rot, web blight and Anthracnose.
	DDG-1	2017	UAS, Dharwad	Kharif season, resistant to powdery mildew.
	Varsa	2018	IIPR, Kanpur	Kharif season, resistant to MYMV and powdery mildew.

Crops	Varieties	Release year	ICAR Institute/ SAU$_{s/}$ Organization centre	Salient features
	PKV AKM 4	2018	IIPR, Kanpur	Spring and Kharif season, resistant to MYMV, green attractive and medium-large seed.
	VBN 4	2019	NPRC, Vamban	Suitable for all season, moderately resistant to MYMV, ULCV and PM diseases.
	Pant M 9	2019	GBPUA&T, Pantnagar	Suitable for Spring and Kharif seasons, resistant to MYMV.
	KM- 2342	2020	CSAU&T, KANPUR	Suitable for Kharif seasons, resistant to MYMV.
	IPM 312-20	2020	IIPR, Kanpur	Suitable for Spring seasons, moderately resistance to Anthracnose and powdery mildew, resistance to MYMV.
Field Pea	Aman (IPF 5-19)	2010	IIPR, Kanpur	Resistant to PM disease, tolerant to rust, moderately resistance to pod borer and stem fly.
	Gomati	2010	ICAR, Reas. Compl., Agartala	Tolerant to pod borer and stem fly, tolerant to *M. incognita* and *M. Javanica*.
	IPF 4-9	2011	IIPR, Kanpur	Resistant to PM disease, moderately resistant to rust, pod borer and stem fly.
	VL Matar 47	2011	VPKAS, NWPZ/ Uttarakhand	Resistant to wilt, rust and powdery mildew.
	HFP-529	2012	CCS HAU, Hisar	Resistant to rust, PM, Ascochyta blight, mod. Resistant to PB, aphids, leaf miner and stem fly, dwarf type.
	Salimar pea 1	2013	Srinagar centre, SKUAST	Resistant to powdery mildew, moderately resistant to rust, pod borer, high protein content.
	IPFD 10-12	2014	IIPR, Kanpur	Resistant to powdery mildew, dwarf type, green dry seeds.
	HFP 715	2014	CCS HAU	Resistant to powdery mildew, dwarf type plant.
	Punjab-89	2015	PAU, Punjab	Resistant to rust.
	Pant pea 155	2016	GBPUA&T, Pantnagar	Resistant to rust and powdery mildew diseases, tolerant to pod borer pest, dwarf type.
	IPFD-6-3	2016	IIPR, Kanpur	Resistant to powdery mildew, moderately resistant to rust, tendril type.
	Pant pea 250	2017	GBPUAT, Pantnagar	Resistant to powdery mildew, moderately resistant to rust, Ascochyta blight and root rot diseases.
	Pant pea 243	2017	GBPUAT, Pantnagar	Moderately resistant to powdery mildew, rust, Ascochyta blight and root rot diseases.
	IPFD 2014-2	2018	IIPR, Kanpur	Dwarf and early vigor, resistant to pod borer, aphid, leaf miner & nematode.
	TRCP 9	2018	ICAR res. Complex Agartala	Resistant to rust, root knot and rust, dwarf type.

Crops	Varieties	Release year	ICAR Institute/ SAUs/ Organization centre	Salient features
	Kota Matar 1	2020	AU, Borkhera (Kota)	Moderately resistant against powdery mildew, rust and root knot nematodes, ss incidence of pod borer.
	IPFD 12-8 (Aakash)	2020	IIPR, Kanpur	Resistant to powdery mildew and rust disease, moderately resistant to pod borer.
	Pant Pea 250	2021	CCS, HAU	Resistant to powdery mildew, Ascochyta blight and root rot and moderately resistant to rust.
	HFP 1428	2021	CCS, HAU	Resistant to powdery mildew, Ascochyta blight and root rot and moderately resistant to rust.

Table 1.
Central & State released varieties of Pulses in India.

Crops	Traits	QTL/candidate	Reference
Chickpea	Ascochyta blight	QTL-1 and QTL-2	[56]
	heat tolerant	CaLG05 and CaLG06	[57]
	Days to flowering initiation	CaDFI_LS6.1, CaDFI_LS8.1, CaDFI_LS6.1, CaDFI_LS8.1	[58]
	Days to maturity	CaDM_LS1.1, CaDM_LS1.2, CaDM_LS1.3	[58]
	Days to pod initiation	CaDPI_LS7.2, CaDPI_LS7.1, CaDPI_LS6.1, CaDPI_LS6.1, CaDPI_LS6.1, CaDPI_LS1.1	[58]
	Days to pod filling	CaDPF_LS8.1, CaDPF_NS4.2, CaDPF_NS4.1	[58]
	Number of filled pods	CaFP_NS6.1	[58]
	100 seed weight	Ca100SW_LS7.1, Ca100SW_LS1.1, Ca100SW_LS4.1, Ca100SW_LS7.1	[58]
	Seed yield/plant	CaSYPP_LS2.1, CaSYPP_LS6.1, CaSYPP_NS6.2, CaSYPP_NS6.3	[58]
	Biological yield/plant	CaBYPP_LS6.5, CaBYPP_NS6.1, CaBYPP_NS6.1, CaBYPP_LS6.3,	[58]

Crops	Traits	QTL/candidate	Reference
		CaBYPP_LS2.1, CaBYPP_LS6.4, CaBYPP_LS6.5, CaBYPP_NS6.2	
	Harvest index (HI, %)	CaHI_NS5.1, CaHI_NS7.1, CaBYPP_NS6.3, CaHI_LS6.2, CaHI_LS8.1, CaHI_NS7.2, CaHI_NS6.1, CaHI_NS7.1	[58]
Pigeon pea	Plant height	qPH4.1a, qPH5.1	[59]
	Primary branching	qPB4.1a, qPB5.1	[59]
	Secondary branching	qSB5.1	[59]
	Number of pods	qPD3.1, qPD4.1, qPD5.1a	[59]
	Flowering	qFL4.1a, qFL5.1	[59]
	Maturity	qMT4.1, qMT5.1a, qMT10.1	[59]
	Fertility restoration	QTL-RF-1, QTL-RF-2, QTL-RF-3, QTL-RF-4a	[60]
	SMD resistance	qSMD4	[61]
	Determinacy	Six DArT and 19 SNP markers	[62]
	Fusarium wilt	qFW11.1, qFW11.2 and qFW11.3	[63]
Soyabean			
	Seed weight	swHCA2-1, swHCC2-1, swHCD1b-1, swHCA2-2	[64]
	Early-maturing	cl1, cl2, cl3 and cl4,	[65]
	Branch number	*qBN-1*	[66]
	Drought tolerance	qPH2, qPH6, qPH7, qPH17, qPH19-1, qPH19-2, and qPH19-3	[67]
	Plant height	qSWPP2, qSWPP6, qSWPP13, qSWPP17, and qSWPP19	[67]
	Water-logging	Satt160, Satt269, Satt252, Satt485, and Satt385	[68]
Mungbean	Yellow mosaic virus	2015 (qMYMV4-1 and qMYMV5-1), 2016 (qMYMV4-1, qMYMV6-1 and qMYMV10-1)	[69]
	Bruchid resistance	SS1, SS2, and SS3,	[70]
	Days to first flower	Fld2, Fld4.1, Fld4.2, Fld11	[71]
	Days to first pod maturity	Pddm2, Pddm4.1, Pddm4.2	[71]
	Days to harvest	Pddh2, Pddh4.1, Pddh4.2	[71]

Crops	Traits	QTL/candidate	Reference
	100 seed weight (g)	Sd100wt2.1, Sd100wt2.2, Sd100wt4, Sd100wt8, Sd100wt9, Sd100wt11	[71]
	Number of seeds per pod	Sdnppd1.1, Sdnppd1.2	[71]
	Pod length (cm)	Pdl7, Pdl8	[71]
Pea	Bleaching resistance	LG II, LG III, LG IV and LG V	[72]
	Sclerotinia sclerotiorum resistances	PRIL17 and PRIL19	[73]
	Height	Le**	[73]
	Drought adaptation	A6, AA175, AC74, AD57, AB141, AB64, Psblox2, PsAAP2_SNP4, and Dipept IV_SNP1	[74]
	Ascochyta blight	abIII-1, abI-IV-2, abI-IV-2.1 and abI-IV-2.2	[75, 76]

Table 2.
QTL mapping.

4. Conclusions and prospects

The importance of legume crops for the agriculture and the environment is considered1 ancestral; however, the invention and the breeding constraints have led to their current lower relative1 significance compared to cereals. Currently, we are witness and significant boost of legume crops-associated research where abundant studies from conventional breeding to advanced genomics, are1 being carried out and published to address and overcome the various constraints faced by the1 production of legumes. Conventional breeding has made significant contributions to legume genetic1 development, especially by developing lines with superior monogenetic traits such as resistance to1 fungi and insects. Considering the accumulation of published information, it is feasible to forecast1 similar achievements for lentil crop in the near future. MAS in legumes have also shaped opportunity1 for the use of pyramiding approaches and the introgression of quantitative traits for resistance of1 certain diseases.

Furthermore, the dominant entry of legumes to the genomics period has endorsed their breeding programs to adopt new biotechnological and bioinformatics tools such as GWAS, which are hopeful in augment the effectiveness and efficiency of modern breeding techniques.

Legume breeding programs may also consider food superiority parameters as important traits to expand materials of high nutritional and commercial value, meeting the needs of both consumers and the food industry.

The application of genetic modification in legumes shows remarkable progress as reflected in the two successful cases discussed in this article. As the effects of GM crops are still well thought out and public acceptance is not well established, editing tools appear to be more appropriate. Scientist must develop the alternative tools of GM crop.

The strong efforts to accelerate and augment legume breeding and their current global production status show great potential to increase their relative importance in the alimentation and the nourishment of the world population. Presently, cereals exceed pulses as well as legumes 6.7 times in harvested area and 6.0 times in production.

This manufacture boost states a precedent and shows actual possibilities of increasing legume crops production to unexpected levels by genetic advances and improved cultivars, linked to advances in farming technology and agronomic practices.

Acknowledgements

Authors highly thank the department of genetics and plant breeding, School of Agriculture, Lovely professional University, Phagwara, Jalandhar Punjab (India) for providing all necessary facilities to write this book chapter.

Conflict of interest

The authors declare no conflict of interest.

Author details

Satya Prakash*, Suhel Mehandi and Harmeet S. Janeja
Department of Genetics and Plant Breeding, Lovely Professional University, Punjab, India

*Address all correspondence to: prakashsatya056@gmail.com

References

[1] Singh NP, Pratap A. Food legumes for nutritional security and health benefits. In: Singh U, Praharaj SC, Singh SS, Singh PN, editors. Biofortification of Food Crops. New Delhi: Springer; 2011. pp. 41-50

[2] Majid A, Parray GA, Wani SH, Kordostami M, Sofi NR, Waza SA, et al. Genome editing and its necessity in agriculture. International Journal of Current Microbiology and Applied Sciences. 2017;**6**:5435-5443

[3] Mujjassim NE, Mallik M, Rathod NKK, Nitesh SD. Cisgenesis and intragenesis a new tool for conventional plant breeding: A review. Journal of Pharmacognition Phytochemistry. 2019; **8**:2485-2489

[4] Mehandi S, Singh IP, Bohra A, Singh CM. Multivariate analysis in green gram [*Vigna radiata* (L.) Wilczek]. Legume Research. 2015;**38**(6):758-762

[5] Doust A, Diao X. Plant genetics and genomics: Crops and models. Genetics Genome Setaria Series. 2017;**19**:377

[6] Godwin ID, Rutkoski J, Varshney RK, Hickey LT. Technological perspectives for plant breeding. Theoretical and Applied Genetics. 2019;**132**:555-557. DOI: 10.1007/s00122-019-03321-4

[7] Ranalli P, Cubero JI. Bases for genetic improvement of grain legumes. Field Crops Research. 1997;**53**:69-82

[8] Smýkal P, Coyne CJ, Ambrose MJ, et al. Legume crops phylogeny and genetic diversity for science and breeding. Critical Reviews in Plant Sciences. 2015;**34**:43-104

[9] Kumar J, Choudhary AK, Solanki RK, Pratap A. Towards marker-assisted selection in pulses: a review. Plant Breeding. 2011;**130**:297-313

[10] Malik SR, Shabbir G, Zubir M, Iqbal SM, Ali A. Genetic diversity analysis of morpho-genetic traits in Desi chickpea (*Cicer arietinum* L.). International Journal of Agriculture. 2014;**16**:956-960

[11] Rubiales D, Fondevilla S, Chen W, Gentzbittel L, Higgins TJV, Castillejo MA, et al. Achievements and challenges in legume breeding for pest and disease resistance. Critical Reviews in Plant Sciences. 2015;**34**:195-236

[12] Pratap A, Kumar J, editors. Biology and Breeding of Food Legumes. Hyderabad: CABI International; 2011. p. 405

[13] Okii D, Tukamuhabwa P, Odong T, Namayanja A, Mukabaranga J, Paparu P, et al. Morphological diversity of tropical common bean germplasm. African Crop Science Journal. 2014a;**22**: 59-67

[14] Kumar S, Ali M. GE interaction and its breeding implications in pulses. Bottomline. 2006;**56**:31-36

[15] Piepho H, Mohring J. Computing heritability and selection response from unbalanced plant breeding trials. Genetics. 2007;**177**:1881-1888

[16] Gupta PK, Rustgi S, Kulwal PL. Linkage disequilibrium and association studies in higher plants: present status and future prospects. Plant Molecular Biology. 2005;**57**:461-485

[17] Flint F. Mapping quantitative traits and strategies to find quantitative trait genes. Methods. 2011;**53**:163-174

[18] Sanchez MA, Parrott WA. Characterization of scientific studies usually cited as evidence of adverse effects of GM food/feed. Plant Biotechnology Journal. 2017;**15**(10):1227-1234

[19] Schwember AR. An update on genetically modified crops. CiencInvestigAgrar. 2008;**35**:231-250

[20] Padgette SR, Kolacz KH, Delannay X, Re DB, LaVallee BJ, Tinius CN, et al. Development, identification, and characterization of a glyphosate-tolerant soybean line. Crop Science. 1995;**35**:1451-1461

[21] James C. 20th anniversary (1996 to 2015) of the global commercialization of biotech crops and biotech crop highlights in 2015. ISAAA briefs no 51. The international service for the acquisition of agri-biotech applications, Ithaca, NY, USA. 2015. Available from: http://www. isaaa.org /resources/ publications/briefs/ 51/ default.asp [Accessed 20 June 2016]

[22] Arruda SCC, Barbosa HS, Azevedo RA, Arruda MAZ. Comparative studies focusing on transgenic through cp4EPSPS gene and non-transgenic soybean plants: an analysis of protein species and enzymes. Journal of Proteomics. 2013. DOI: 10.1016/j. jprot.2013.05.039

[23] Slater A, Scott NW, Fowler MR. Plant Biotechnology: The Genetic Manipulation of Plants. Oxford: Oxford University Press; 2003

[24] Duke SO. Taking stock of herbicide-resistant crops ten years after introduction. Pest Management Science. 2005;**61**:211-218

[25] Stewart CN et al. Transgene introgression form genetically modified crops to their wild relatives. Nature Reviews. Genetics. 2003;**4**:806-817

[26] Aragao FJL, Faria JC. First transgenic geminivirus-resistant plant in the field. Nature Biotechnology. 2009;**27**:1086-1088

[27] Aragao FLJ, Nogueria OPL, Tinocoa MLP, Faria JC. Molecular characterization of the first commercial transgenic common bean immune to the Bean golden mosaic virus. Journal of Biotechnology. 2013;**166**:42-50

[28] Scheible WR, Morcuende R, Czechowski T, et al. Genome-wide reprogramming of primary and secondary metabolism, protein synthesis, cellular growth processes, and the regulatory infrastructure of Arabidopsis in response to nitrogen. Plant Physiology. 2004;**136**:2483-2499

[29] Dijk VK, Ding Y, Malkaram S, Riethoven JJM, Liu R, Yang J, et al. Dynamic changes in genome-wide histone H3 lysine 4 methylation patterns in response to dehydration stress in *Arabidopsis thaliana*. BMC Plant Biology. 2010;**10**:1-12

[30] Gupta PK, Kumar J, Mir RR, Kumar A. Marker assisted selection as a component of conventional plant breeding. Plant Breeding Review. 2010; **33**:145-217

[31] Schmutz J, McClean PE, Mamidi S, et al. A reference genome for common bean and genome-wide analysis of dual domestications. Nature Genetics. 2014; **46**:707-713

[32] Main D, Cheng CH, Ficklin SP, Sanad M, Jung S, Lee T, et al. The cool season food legume database: an integrated resource for basic, translational and applied research. In: International Plant and Animal Genome Conference. San Diego, CA, USA; 2014

[33] Yang SS, Tu ZJ, Cheung F, Xu WW, Lamb JFS, Jung HJG, et al. Using RNA-

Seq for gene identification, polymorphism detection and transcript profiling in two alfalfa genotypes with divergent cell wall composition in stems. BMC Genomics. 2011;**3**:56-64. DOI: 10.1186/1471-2164-12-199

[34] Jones SI, Vodkin LO. Using RNA-Seq to profile soybean seed development from fertilization to maturity. PLoS One. 2013;**5**:72-85. DOI: 10.1371/journal.pone.0059270

[35] Singh VK, Garg R, Jain M. A global view of transcriptome dynamics during flower development in chickpea by deep sequencing. Plant Biotechnology Journal. 2013;**11**:691-701

[36] Zhu H, Choi HK, Cook DR, Shoemaker RC. Bridging model and crop legumes through comparative genomics. Plant Physiology. 2005;**137**:1189-1196

[37] Cruz-Izquierdo S, Avila CM, Satovic Z, Palomino C, Gutierrez N, Ellwood SR, et al. Comparative genomics to bridge *Vicia faba* with model and closely-related legume species: stability of QTLs for flowering and yield-related traits. Theoretical Application Genetics. 2012;**125**:1767-1782

[38] Castro P, Rubio J, Madrid E, Fernández-Romero MD, Millán T, Gil J. Efficiency of marker-assisted selection for Ascochyta blight in chickpea. The Journal of Agricultural Science. 2015;**153**:56-67

[39] Varshney RK, Thudi M, Nayak SN, Gaur PM, Kashiwagi J, Krishnamurthy L, et al. Genetic dissection of drought tolerance in chickpea (*C. arietinum* L.). Theoretical Application Genetic. 2014; **127**:445-462

[40] Gusta LV, Wisniewski M. Understanding plant cold hardiness: An opinion. Physiologia Plantarum. 2013;**147**: 4-14

[41] Klein A, Houtin H, Rond C, Marget P, Jacquin F, Boucherot K, et al. QTL analysis of frost damage in pea suggests different mechanisms involved in frost tolerance. Theoretical Application Genetics. 2014;**127**: 1319-1330

[42] Rubio MC, Bustos-Sanmamed P, Clemente MR, Becana M. Effects of salt stress on the expression of antioxidant genes and proteins in the model legume Lotus japonicus. The New Phytologist. 2009;**181**:851-859

[43] Golldack D, Li C, Mohan H, Probst N. Tolerance to drought and salt stress in plants: Unraveling the signaling networks. Frontiers in Plant Science. 2014;**3**:112-119. DOI: 10.3389/fpls.2014.00151

[44] Abid G, Muhovski Y, Mingeot D, Watillon B, Toussaint A, Mergeai G, et al. Identification and characterization of drought stress responsive genes in faba bean (*V. faba* L.) by suppression subtractive hybridization. Plant Cell, Tissue and Organ Culture. 2014;**121**: 367-379

[45] Yang W, Wang M, Yue A, Wu J, Li S, Li G, et al. QTLs and epistasis for drought-tolerant physiological index in soybean (*Glycine max* L.) across different environments. Caryologia. 2014;**67**:72-78

[46] Huynh BL, Ehlers JD, Ndeve A, Wanamaker S, Lucas MR, Close TJ, et al. Genetic mapping and legume synteny of aphid resistance in African cowpea (*Vigna unguiculata* L. Walp.) grown in California. Molecular Breeding. 2015;**2**: 432-439. DOI: 10.1007/s11032-015-0254-0

[47] Carrillo E, Satovic Z, Aubert G, Boucherot K, Rubiales D, Fondevilla S. Identification of quantitative trait loci

and candidate genes for specific cellular resistance responses against Didymellapinodes in pea. Plant Cell Reports. 2014;**33**:1133-1145

[48] Anderson J, Akond M, Kassem MA, Meksem K, Kantartzi SK. Quantitative trait loci underlying resistance to sudden death syndrome (SDS) in MD96–5722 by 'Spencer' recombinant inbred line population of soybean. 3 Biotech. 2014;**4**: 65-71. DOI: 10.1007/s13205-014-0211-3

[49] Patil BS, Ravikumar RL, Bhat JS, Soregaon CD. Molecular mapping of QTLs for resistance to early and late Fusarium wilt in chickpea. Czech Journal of Genetics and Plant Breeding. 2014;**50**: 171-176

[50] Pottorff MO, Li GJ, Ehlers JD. Genetic mapping, synteny, and physical location of two loci for Fusarium oxysporum f. sptracheiphilum race 4 resistance in cowpea [*V. unguiculata* (L.) Walp]. Molecular Breeding. 2014;**33**: 779-791

[51] Agbicodo EM, Fatokun CA, Bandyopadhyay R, Wydra K, Diop NN, Muchero W, et al. Identification of markers associated with bacterial blight resistance loci in cowpea [*V. unguiculata* (L.) Walp.]. Euphytica. 2010;**175**:215-226

[52] Trabanco N, Asensio-Manzanera MC, Pérez-Vega E, Ibeas A, Campa A, Ferreira JJ. Identification of quantitative trait loci involved in the response of common bean to Pseudomonas syringaepv. Molecular Breeding. 2014; **33**:577-558

[53] Blair MW, Rodriguez LM, Pedraza F, Morales F, Beebe S. Genetic mapping of the bean golden yellow mosaic geminivirus resistance gene bgm-1 and linkage with potyvirus resistance in common bean (*Phaseolus vulgaris* L.). Theoretical Applications Genetics. 2007; **114**:261-271

[54] Hema M, Sreenivasulu P, Patil BL, Kumar PL, Reddy DVR. Tropical food legumes: virus diseases of economic importance and their control. Advances in Virus Research. 2014;**90**:431-505

[55] Abdelmajid KM, Ramos L, Hyten D, Bond J, Bendahmane A, Arelli PR, et al. Quantitative trait loci (QTL) that underlie scn resistance in soybean [*G. max* (L.) Merr.] PI438489B by 'Hamilton' recombinant inbred line (RIL) population. Atlas Journal of Plant Biology. 2014;**1**:29-38

[56] Santra DK, Tekeoglu M, Ratnaparkhe M, Kaiser JW, M. F. J. Identification and Mapping of QTLs Conferring Resistance to Ascochyta Blight in Chickpea. Crop Science. 2000; **40**:1606-1612

[57] Paul PJ, Samineni S, Thudi M, Sajja SB, Rathore A, Das RR, et al. Molecular mapping of QTLs for heat tolerance in chickpea. International Journal of Molecular Science. 2018;**19**: 2166

[58] Jha UC, Nayyar H, Palakarthi R, Jha R, Valluri V, Bajaj P, et al. Major QTLs and potential candidate genes for heat stress tolerance identified in Chickpea (*C. arietinum* L.). Frontier in Plant Science. 2021;**12**:655103

[59] Kumawat G, Raje RS, Bhutani S, Pal JK, Mithra ASVCR, Gaikwad K, et al. Molecular mapping of QTLs for plant type and earliness traits in pigeon pea (Cajanus cajan L. Millsp.). BMC Genetics. 2012;**13**:84

[60] Bohra A, Saxena RK, Gnanesh BN, Saxena KB, Byregowda M, Rathore A, et al. An intra-specific consensus genetic map of pigeon pea [*C. cajan* (L.)

Millspaugh] derived from six mapping populations. Theoretical Applications Genetics. 2012;**125**:1325-1338

[61] Gnanesh BN, Bohra A, Sharma M, Byregowda M, Pande S, Wesley V, et al. Genetic mapping and quantitative trait locus analysis of resistance to sterility mosaic disease in pigeon pea [*C. cajan* (L.) Millsp.]. Field Crops Research. 2011; **123**:53-61

[62] Mir RR, Saxena RK, Saxena KB, Upadhyaya HD, Kilian A, Cook DR, et al. Whole-genome scanning for mapping determinacy in pigeon pea (Cajanus spp.). Plant Breeding. 2013;**132**: 472-478

[63] Sexena RK, Singh VK, Kale SM, Tathineni R, Parupalli S, Kumar V, et al. (2017) Construction of genotyping-by-sequencing based high-density genetic maps and QTL mapping for fusarium wilt resistance in Pigeonpea. Scientific Reports. 2017;7:1911

[64] Han Y, Li D, Zhu D, Li H, Li X, Teng W, et al. QTL analysis of soybean seed weight across multi-genetic backgrounds and environments. Theoretical and Applied Genetics. 2012; **2012**(125):671-683

[65] Khan AN, Githiri MS, Benitez ER, Abe J, Kawasakis S, Hayashi T, et al. QTL analysis of cleistogamy in soybean. Journal of Biotechnology. 2008;**117**(4): 479-487

[66] Lamlom SF, Zhang Y, Su B, Wu H, Zhang X, Fu J, et al. Map-based cloning of a novel QTL qBN-1 influencing branch number in soybean (G. max L.). The Crop Journal. 2020;**2020**:793-801

[67] Ren H, Han J, Wang X, Zhang B, Gao H, Hong H, et al. QTL mapping of drought tolerance traits in soybean with SLAF sequencing. The crop Journal. 2020;**2020**(8):977-989

[68] Cornelious B, Chen P, Chen Y, de Leon N, Shannon JG, Wang D. Identification of QTLs underlying water-logging tolerance in soybean. Molecular Breeding. 2005;**2005**(16):103-112

[69] Mathivathana MK, Murukarthick J, Karthikeyan A, Jang W, Dhasarathan M, Jagadeeshselvam N, et al. Journal of Applied Genetics. 2019:56-62. DOI: 10.1007/s13353-019-00506-x

[70] Mei L, Cheng XZ, Wang SH, Wang LX, Liu CY, Sun L, et al. Relationship between bruchid resistance and seed mass in mungbean based on QTL analysis. Genome. 2009;**52**:589-596

[71] Kajonphol T, Chontira SC, Somta P, Toojinda T, Srinives P. SSR map construction and quantitative traits loci (QTL) identification of Major agronomic traits in mungbean (*V. radiata* L.). SABRAO Journal of Breeding and Genetics. 2012;**44**(1):71-86

[72] Ubayasena L, Bett K, An BT, Vijayan P, Warkentin T. Genetic control and QTL analysis of cotyledon bleaching resistance in green field pea (*Pisum sativum* L.). Genome. 2010;**53**(5): 346-359

[73] Mahini RA, Kumar A, Elias EM, Fiedler JD, Porter LD, McPhee KE. Analysis and Identification of QTL for Resistance to Sclerotinia sclerotiorum in Pea (P. sativum L.). Frontiers in Genetics. 2020;**11**:587968. DOI: 10.3389/fgene.2020.587968

[74] Garcia RL, Prats E, Fondevilla S, Satovic Z, Rubiales D. Quantitative trait loci associated to drought adaptation in pea (*P. sativum* L.). Plant Molecular Biology Reporter. 2015;**33**:1768-1778. DOI: 10.1007/s11105-015-0872-z

[75] Jha AB, Gali KK, Taran B, Warkentin TD. Fine mapping of QTLs for ascochyta blight resistance in pea using heterogeneous inbred families. Frontier in Plant Science. 2017:118-126. DOI: 10.3389/fpls.2017.00765

[76] Vange T, Moses OE. Studies on genetic characteristics of pigeon pea germplasm at Otobi, Benue state of Nigeria. WJAS. 2009;5:714-719

Chapter 10

Soybean Molecular Design Breeding

Xianzhong Feng and Muhammad Hafeez Ullah Khan

Abstract

Soybean is a globally important crop being rich source of edible oil and protein. Traditional phenotypic-based breeding procedures have contributed significantly to the development of several improved soybean varieties. In this context, molecular breeding technology, is seen as a viable way to address the issues and providing great opportunities to accelerate the process of soybean breeding. Hence, marker-assisted breeding (MAB) has been greatly applied in the soybean breeding to accelerate the improved soybean cultivars, transgenic breeding technology achieves great success in the soybean production. New genomics approaches and the development of genome editing technologies have increased soybean genetic diversity in its germplasm and have created new possibility to make precise genes modifications to controlling essential agronomic traits in an elite background Besides, the establishment of genotype driven phenotypic design breeding model has become a great challenge for soybean molecular breeding in the future. These approaches have the potential to expand the practical utility of molecular design breeding and speed up the germplasm and breeding materials in soybeans. This chapter goes into great detail about how current advances in genomics and phenomics can increase the efficiency and potential of MAB, transgenic technology, molecular design breeding and gene editing technology in soybean improvement.

Keywords: soybean, molecular design breeding, phenomics, genomics, genome editing

1. Introduction

Around 5000 years ago, cultivated soybean (*Glycine max* [L.] Merr.) had been domesticated from wild soybean (*Glycine soja Sieb*. & *Zucc*.). This crop has originated in China, and it spreads gradually around the different parts of the world [1]. Soybeans is now one of the most economically important oilseed and biodiesel crops, as well as a major source of protein and oil for human and animal consumption [2]. Early soybean breeding relied primarily on farmers selecting preferred seeds from the planted population. Artificial hybridization has been used since the early 1900s. In the 1940s, North American breeding programs published the first modern soybean cultivar developed through hybridization [3, 4]. Artificial hybridization became more commonly used in soybean breeding, after that it is investigated that artificial hybridization dramatically enlarged the genetic basis of established lines and increased soybean adaption as well as productivity [5]. Soybean is largely crushed into soy oil and meal, and it can be found in a variety of edible and nonedible goods, including cooking oil, animal grains, vegan food, and milk, as well as biodiesel and other industrial applications. Soybean oil is the most widely used cooking oil in the world, second only to palm oil [6].

The major objective of the most plant breeding projects/programs in soybean is to increase the yield and quality [2]. However, in the field of plant breeding, measuring primary traits such as yield or quality, which are mostly complex quantitative traits in a large breeding populations with thousands of genotypes, is time-consuming and labor-intensive [7, 8]. Due to genetic and environmental influences, breeding for yield is recognized to be a highly complicated and nonlinear process [9]. To this end, plant breeders can efficiently identify the promising lines at early growth stages using secondary traits for selection (e.g., yield component traits and reflectance bands), which are strongly correlated with the primary trait [10, 11].

The recent advances in sequencing technology have triggered a data boom in the biology field, propelling molecular biology into a stunning postgenomic *era*. From structural characterization to functional analysis, genomic research has progressed [12]. Despite the fact that genomic mapping, bioinformatics prediction, and other technologies aid in inferring gene function; however, any theory in life science requires ultimate confirmation. This inference is required for genetic transformation and vice versa; and it appears to be a powerful tool in functional genomics. Transgenic breeding is other important approach used to introduce genetic changes for specific plant traits. This method has been successfully used to increase crop productivity, production of biofuels, improve food quality and plant resistance against severe environmental conditions by breaking species limits. Furthermore, the implementation of genome-editing tools such as CRISPR/Cas9 relies on transformation procedures, demonstrating the necessity and importance of this technology.

Marker-assisted selection (MAS) has speeded up the breeding process especially in the production of disease and insect pest-resistant cultivars [13]. Linkage and physical maps are created using various types of genetic markers [14, 15]. Consensus Map 4.0 was created to combine known genetic and physical maps [16]. Large numbers of quantitative trait loci (QTLs) associated with different crop traits have been identified in soybean using genetic markers. However, the efficiency and precision of QTL location were restricted by limited number of molecular markers and their uneven distribution. To this end, the advances in the high-throughput genotyping and phenomics have greatly enhanced the precision and resolution in the gene mapping [17, 18]. Although, the advances in high-throughput genotyping were significant to alleviate the challenges in the plant breeding [7, 19, 20], but the advances in the high-throughput field phenotyping, is far lagging behind the genomics. Hence, the phenomics is a major bottleneck in current breeding programs [19].

The mechanism of genome editing technology is to introduce double-strand breaks (DSBs) within the genome at targeted sites using sequence specific artificial nucleases (SSNs), which are then repaired using nonhomologous end joining (NHEJ) or homologous recombination repair (HR) mechanisms, resulting in targeted mutagenesis by adding, removing, or replacing DNA bases [21]. Zin finger nucleases (ZFNs), transcription activator-like effector nucleases (TALENs), and clustered regularly interspaced short palindromic repeats/CRISPR-associated protein (CRISPR/Cas) are the most common SSNs at the moment [22, 23]. Despite their early development, ZFNs and TALENs are complex and expensive, which has limited their use. Since its inception, the CRISPR/Cas system has gained popularity in biological science due to its simplicity. The CRISPR/Cas9 system is the most well-known and has been increasingly used in the crop plants in the last few years [24]. The genome editing toolset has been broadened after the CRISPR/Cas9 system by selecting Cas9 orthologs and created variations [25–27]. Dead *Cas9* and *Cas9* nickases are two of them, and they have been employed extensively in base editing, gene expression regulation, epigenome editing, cell imaging, and other domains [28].

2. Marker-assisted breeding in soybean

Marker-assisted breeding (MAB), also known as molecular-assisted breeding, is the use of molecular tools, primarily DNA markers, in conjunction with linkage maps and genomics to change and enhance plant as well as animal characteristics using genotypic tests [29]. The term MAB is used to explain the various novel strategies including MAS, marker-assisted backcrossing (MABC), marker-assisted recurrent selection (MARS), and genome-wide selection (GWS) or genomic selection (GS) [30]. MAB is recognized as a unique technique and a potent methodology for agricultural plant genetic modification, and it has been widely applied in a variety of crop species to date [29, 31].

Classical markers and DNA markers are two types of genetic markers used in plant breeding [32, 33]. Morphological, biochemical, and cytological indicators are examples of traditional markers. Random amplified polymorphic DNA (RAPD), amplified fragment length polymorphism (AFLP), microsatellites or simple sequence repeats (SSRs), restriction fragment length polymorphism (RFLP), and single-nucleotide polymorphism (SNP) are all examples of DNA markers (SNP). Marker-assisted breeding (MAB) is the most promising of the many applications of DNA markers in plant science for cultivar creation. MAB has huge potential to increase conventional plant breeding efficiency and precision by using DNA markers that are firmly related to critical genes or loci [33]. Several allele-specific functional markers for essential soybean features such as blooming and maturity, pod dehiscence, aroma, salt tolerance, soybean cyst nematode oleic acid content, raffinose content, and Kunitz trypsin inhibitor have recently been discovered [33, 34]. Phytic acid content, glycinin, conglycinin concentration, fragrance, and lipoxygenase were also discovered as strongly connected markers for seed nutritional value, which could help with the selection of novel varieties that are free of antinutritional chemicals [33].

MAB allows selection of plant features (that are expressed late in the plants genotype) at the seedling stage based on the genotypic data; hence, reducing the time it takes to identify the phenotypic of a single plant. MAS can swiftly remove unwanted genotypes for features that are displayed at later developmental stages. This trait is very significant and valuable for backcrossing as well as recurrent selection breeding programs, which require crossing with or between chosen individuals [17, 30]. MAB is unaffected by the environment, allowing selection to take place in any setting (e.g., greenhouses and off-season nurseries). This is particularly useful for improving qualities that are only expressed in the presence of favorable environmental circumstances, such as disease/pest resistance as well as stress tolerance [30]. MAS is based on reliable markers that are strongly connected to the QTLs associated with particular trait of interest and is more effective and efficient than phenotypic selection for low-heritability traits that are highly influenced by the environment (PS). In the heterozygous state, MAB utilizes the codominant markers (e.g., SSR and SNP) to allow effective selection of recessive alleles of desirable features. To detect quality controlled by recessive alleles, no selfing or test-crossing is required; hence, MAB may save time and speed up breeding process [29].

The MAB method significantly accelerates the accurate and efficient introgression of targeted genes into recipient varieties, as well as the recovery of the recurrent parent genetic background. With just two backcrosses ($BC_2F_{2:3}$), marker-assisted background selection in wheat was able to transfer *Yr15*, a stripe rust resistance gene in a recurrent variety and recover 97% of the genetic background of the recurrent parent, whereas phenotypic selection could only recover 82% of the background in BC_4F_7 plants [35, 36]. In this case, the MAB successfully saves the time it takes to obtain advanced breeding lines in half when compared with traditional approaches.

MAS and MABC have been frequently used to increase disease resistance and other relatively basic qualities [33, 37, 38]. MAB has been used successfully in a few soybean breeding programs to introduce single genic as well as polygenic traits into the desired genetic background (**Table 1**). Moreover, MAB has been proven to be effective in improving quantitative features that contribute to soybean nutritional value, such as seed protein content and oil quality. MAB for seed protein content (SPC) in soybeans using SSR markers yielded up to 9% transgressive segregation in the trait after two cycles [48, 49].

3. Transgenic breeding in soybean improvement

Transgenic soybeans are one of the few vegetable-based foods that contain all nine necessary amino acids. As a result, the transgenic soybean has grown in importance as a human and animal protein source, with 85% of its production going to animal feed, and the rest is going to direct human consumption [55]. Transgenic crops have been embraced by key soybean-growing countries such as the United States, Brazil,

Target trait	Gene/locus	Type of marker	References
Resistance to soybean mosaic virus	*Rsv1*, *Rsv3*, and *Rsv4*	SSR	[39, 40]
Resistance to soybean mosaic virus	R_{SC4}, R_{SC8}, and R_{SC14Q}, SC_7	SSR	[41, 42]
High oleic acid content	*FAD2–1A*, *FAD2–1B*	Gene-specific Simple Probe	[43–45]
Grain yield	Yield QTL	SSR	[46]
Resistance to soybean cyst Nematode	*rhg1*, *Rhg4*	SSR	[47]
Seed protein content (SPC)	(QTL *Prot-08-1*)	SSR	[48, 49]
Salt tolerance	*GmSALT3*	SSR	[50–52].
Elimination of Kunitz trypsin inhibitor (kti)	*Ti3*	SSR	[33, 34]
Elimination of off-flavor and improvement of seed Longevity	*lox2*	lox2 specific	[53]
Low raffinose family oligosaccharides content	*RS3*	Gene-specific Simple Probe	[54]

Table 1.
Details of marker-assisted breeding conducted for improvement of soybean for various traits.

and Argentina, and they now account for about 85–95% of total soybean in terms of crop harvested area. The major markets for genetically modified agricultural seeds are North America and South America, which together account for more than 90% of the global GM seed industry. Nearly 85–95% of the soybean crop grown in North and South America is genetically modified. Demand for America's produced corn and soybean produce from other countries (particularly China) is a major factor in determining planted acreage and seed demand [56].

The landmark products of transgenic soybean's genetic composition allow it to be used for a wide range of purposes, which keeps it in high demand. Initially, manufacturers only wanted to use transgenics to grow more soy at a low cost in order to meet this demand, as well as to fix any problems in the growing process. But eventually it was discovered that soybean can be genetically modified to contain healthier components or even focus on one aspect of the soybean to produce in larger quantities. The first and second generations of genetically modified (GM) foods were named after these periods. The benefits of first-generation GM foods were oriented toward the manufacturing process and companies, whereas the second generation of GM foods offers a variety of advantages and added value for the consumer, including improved nutritional composition or even therapeutic effects [57]. The main and important landmark products of soybean are Roundup ready soybean, Generic GMO soybean, and genetic modification in soybean to improve soybean oil. Roundup Ready soybeans (the original variety was also known as GTS 40–3-2 (OECD UI: MON-04032-6)) are a series of glyphosate-resistant soybean cultivars developed by Monsanto. Glyphosate is a herbicide that kills plants by interfering with the production of phenylalanine, tyrosine, and tryptophan, which are all necessary amino acids. These amino acids are referred as "essential" since only plants and microorganisms can produce them, and mammals are dependent on the plants for these amino acids [58].

Soybean transgenic technology is an essential tool for validating the soybean gene function. Soybean genetic transformation has been explored for over two decades, but progress has been slow and inefficient, which is why some studies used Arabidopsis instead of soybean for functional validation. Several transformation systems have been developed, including shoot meristems [59], hypocotyls ([60], embryo [61], immature cotyledons, half-seed explants [62, 63], and cotyledonary nodes [64]. Agrobacterium-mediated cotyledonary node (CN) soybean transformation is currently widely employed due to its ease of usage, reproducibility, quantity of copies of foreign DNA, and low cost of experimentation [63, 65]. The overall average efficiency of transformation was 3.8–8.7% [63, 64]. Recently, the average transformation efficiency of soybean had been improved to 18.7% [66]. However, it is still lower than the 23% as reported in rice [67] and more than 30% found in maize [68].

Seed sterilization and germination followed by Agrobacterium infection, cocultivation of soybean explants and Agrobacterium, shoot induction, shoot elongation, rooting, and finally, moving the plants to pots containing soil are the different steps involved in the general transformation process. Many factors affect the transformation efficiency at the above different steps. For example, the soybean genotypes used in the transformation are the initial effector. The transformation efficiency and regeneration rate of 20 soybean varieties have been studied, and it was reported that transformation efficiency varied greatly (0.31–4.59%) among the different genotypes [69]. Second, all Agrobacterium concentrations, soybean explant status, Agrobacterium suspension medium, and cocultivation time will affect the infection efficiency during the Agrobacterium infection process, which is one of the most significant processes. Explant regeneration is another important factor in affecting transformation efficiency. Plant hormone has been found to have a vital function in promoting explant regeneration, and the right dose could boost efficiency [64]. Previously research studies have showed that the combination of L-glutamine and L-asparagine in culture media increases the transformation efficiency by inhibiting *GmPRs* expression [70].

4. Soybean molecular design breeding

For molecular design breeding, the breeders could design the superior genotypes with particular breeding objectives based on the molecular networks regulating the agronomic traits. The breeding process can be simulated and optimized "in silico" before going to the field, which enhanced breeding accuracy and efficiency. Recent advances in genomics and phenomics have opened up new possibilities for more efficient molecular design breeding [71, 72]. Several soybean databases have been created for genomes, transcriptomics, proteomics, and germplasm analysis. SoyBase is a USDA-ARS database for genetics and genomics [73, 74]; SoyTEdb is a database of transposable elements [75]; SoyNet is a database for cofunctional networks [76]; SoyProDB is a database for seed proteins [77]; SoyPro [78]. These databases are quite useful for soybean molecular design breeding, which could provide the multiple levels of soybeans (**Table 2**).

4.1 Applications of genome selection in soybean design breeding

The genomics-assisted breeding (GAB) is one of important tools for molecular design breeding, which has allowed for higher genetic gain for complex traits at a lower cost, but it requires a molecular understanding of the trait [86]. MAS and GS are the two basic techniques used in GAB [87]. MAS is dependent on the presence of markers linked to the trait of interest, which can be discovered by linkage mapping or genome-wide association studies (GWASs). Many previous studies have shown that MAS may be successfully used in soybean by adding significant genes and large-effect QTLs for many attributes [88, 89]. Minor genes, on the other hand, control the majority of inheritance in complex characteristics, but they have never been studied because of the limitation of MAS [90]. Furthermore, the influence of the environment, epistatic interactions, and the effect of genetic background have made breeding complex traits extremely difficult. As a result, plant breeders have concluded that MAS is not an appropriate method for breeding complex plant characteristics [91].

GS uses the entire genome-wide marker profile of breeding lines to predict the genomics- estimated breeding value (GEBV) using several models, preventing the loss of a significant percentage of variation dictated by modest impact QTLs/genes [90]. However, precise genotyping and phenotyping analyses are required for accurate detection of marker-trait relationships and determination of GEBV, which determines the effectiveness of GAB. Manual low-throughput phenotyping and genotyping frequently result in the identification of false positives or negatives [37]. In this sense, high-throughput genotyping and phenotyping based on next-generation sequencing (NGS) enables for successful MAS and GS, as well as greater molecular design breeding programs success [92, 93]. The availability of high-throughput NGS-based genotyping methods has significantly speeded up the gene identification

Database	URL	Description	References
Soybean gene expression atlas	http://www.soybase.org/soyseq	A database of soybean 14 tissues specific gene expression	[79]
Soybean cDNA sequenced	http://digbio.missouriemi/soybean_atlas	A cDNA database of soybean developmental tissues specifically in root hair and meristem	[80]
SoyNet	http://www.inetbio.org/soynet	A database for network-based functional predictions	[76]
Soybean transcriptome data	http://venanciogroup.uuen.br/cgi-bin/gmax_atlas/index.cgi	A database of 1298 publicly available soybean transcriptome	[81]
Proteomics of oilseeds	http://oilseedproteomics.missurm.edu	Expression profile data for proteomics research on soybean and other oilseeds plants	[82]
Soybean Proteome Database	http://proteome.dc.affrc.go.jpSjpSoyb/	SPD a database of soybean proteomics	[83, 84]
SoyBase	http://www.soybase.org	A database of soybean genetics and genomics	[73, 74]
SoyTEdb	http://www.soytedb.org	A database of soybean transposable elements	[75]
SoyProDB	http://bioinformatics.towson.ede/Soybean_Seed_Proteins_2D_Gel_DB/Home.aspx	A database for soybean seed proteins	[77]
SoyProLow	http://bioinformatics.towson.ede/Soybean_low_abundance_prprotei_2D_Gel_DB/Gel1.aspx	A database for soybean low abundant proteins	[78]
SoyKB	http://soykb.org	A database of soybean translational genomics and for soybean molecular breeding	[85]

Table 2.
Resources and databases of soybean.

and GS, particularly in agricultural plants with bigger and more complex genomes, such as soybeans [87]. In this regard, phenomics and genomics are equally important for accurate gene identification and the development of a GS model to quantify the breeding population's GEBV (BP). Consequently, integrating these methodologies with suitable genetic diversity, soil and meteorological data, analytical tools, and databases, new varieties with improved yield, quality, and stress tolerance might be developed quickly [91].

MAS has not yielded satisfactory results in soybean for minor genes that contribute only a modest amount of obvious phenotypic variation for the complex trait [48, 87]. Most economically important soybean traits, including as yield, oil and protein content, and stress tolerance, are complex in nature, with modest effect genes controlling the majority of phenotyping variance for these traits [94]. The GS develops a prediction model by combining marker profile and phenotypic data from the training population, which is then used to estimate the GEBV of all BP individuals [95, 96]. Cross-validation on subsets of the training population is used to assess the accuracy of the prediction model before using it to select individuals from BP [87]. Following successful validation, this model can be used to select desirable plants from the BP based on GEBVs estimated solely from marker/genotypic data; hence, only genotypic data are utilized to predict the phenotypic performance of BP individuals [97]. The main benefit of GAB is that genotypic data collected at an early stage of plant development (such as seedling) can be utilized to predict phenotypic performance in mature individuals. As a result, it can significantly reduce the amount of time, money, and labor required for broad phenotypic examination across many habitats and years [98]. GAB also allows higher number of breeding selection cycles and genetic gain per unit time [87].

4.2 NGS-based genotyping for soybean design breeding

In recent decades, the total dependence on phenotypic selection has gradually shifted to a greater use of genotypic-based approaches for plant selection, facilitated by NGS-based genotyping platforms [99–101]. NGS technology has boosted throughput, speeed of genome-wide genotyping, and cost-effectiveness [109] (**Table 3**). NGS-based genotyping technologies have tremendously aided in enhancing the resolution of gene mapping and tagging the gene/QTL extremely closer to the neighboring maker. In GWAS analysis, for example, the use of NGS has made it feasible to genotype huge populations of plants with a greater density of markers than was previously possible, which directly leads to better mapping resolution [110, 111]. In GWAS analysis, using a varied and big population allows for the discovery of more recombination break sites, which aids in the identification of candidate genes with greater precision [112]. Many studies have used NGS-based genotyping for GWAS analysis in soybean for various traits, and these studies have shown significant success in identifying candidate genes for specific traits of interest. For instance, previous study used the RAD-seq method to find a candidate gene underlying the main QTL controlling flooding tolerance in soybean [113]. Many other studies have shown that NGS-based genotyping facilitated candidate gene identification in areas such as nitrogen fixation [114], soybean plant height and primary branches [115], agronomic traits [116], disease resistance [117], and protein content [118]. The NGS-based WGRS has greatly improved the power of bulk segregant analysis (BSA) and its modified techniques, and it is now extensively employed in a variety of plant species, including soybean. For example, in another study, WGRS was employed to resequence different DNA pools

Trait	SNP genotyping platform	Number of SNPs	Population size	Training population	Prediction accuracy	Model	References
Yield and related traits	SNP Chip	2647	483	483	0.26–0.81	RR-BLUP	[97]
Amino acid contents	GBS	23,279	249	199	0.25–0.61	RR-BLUP	[100]
Chlorophyll content	SNP Chip	4089	172	100	0.31–0.76	RR-BLUP	[102, 103]
Grain yield	GBS	2395	139	55	0.64	G-BLUP	[104]
Yield-related traits	SNP Chip	4947	324	324	0.56	eBLUP	[99]
Yield-related trait	SNP Chip	5361	235	—	0.47–0.86	RR-BLUP	[105]
Yield and protein content	SNP Chip	4141	1248	252–492	0.55–0.62	G-BLUP	[106]
Yield	GBS	3000	227	227	0.6	RR-BLUP and Bayesian Models	[107]
Seed weight	SNP Chip	31,045	309	97–197	0.75–0.87	RR-BLUP	[108]
Nematode resistance	SNP Chip	3782	234	117–201	0.43–0.48	gBLUP	[102]

Table 3.
Tools of design breeding in soybean using high-throughput SNP genotyping platforms.

in the BSA study, and they discovered two significant genes that regulate cotyledon color in soybeans at the same time. Furthermore, many studies have used NGS-based techniques in the BSA approach to identify candidate genes for various soybean traits, including soybean mosaic virus [119], charcoal rot resistance [120], flowering time [121], phytophthora resistance [122], and powdery mildew resistance [123].

5. Genome editing provides the powerful tool for soybean design breeding

Genome editing has emerged as more powerful approach for functional study and molecular design breeding compared with traditional genetics approaches, namely mutagenesis, transgenic RNAi, or overexpression in obtaining plant cultivars with predictable and desired traits [124, 125]. CRISPR (clustered regularly interspaced short palindromic repeat)/Cas (CRISPR-associated) is one of the most efficient genome editing systems and has been widely used in various plant species [126]. In 2015, the first knockout and DNA homology-directed recombination (HDR) in soybean plant was successfully obtained using CRISPR/Cas9 technology [127]. Du et al. discovered that altering the AtU6–26 promoter of the CRISPR/Cas9 system to the GmU6-16g-1 promoter might considerably improve the efficacy of targeted mutagenesis in soybean [128]. Nearly 75% of the genes in soybean are duplicated, thus knocking out a single gene usually does not result in a mutant phenotype. It is critical to create a CRISPR/Cas9 system that can edit several homologous genes at the same time to obtain the desired phenotype. The CRISPR/Cas9 system that can achieve multiplex mutagenesis with better efficiency was established by refining the phases of vector synthesis, sgRNA assessment, pooled transformation, and sgRNA identification [129]. Naturally, single-nucleotide polymorphism (SNP) variants account for a major portion of phenotypic variability in agronomic traits, in addition to alleles induced by loss-of-function mutations. When a gene function is disrupted by utilizing a gene-editing technique, mostly it results in undesirable phenotype that is difficult to optimize for agronomic trait improvement [27]. As a result, in molecular breeding, generating point mutations at specific locations impacting crucial agronomic properties is extremely valuable [130]. The CRISPR/Cas9 system has recently been used to develop "base editing," which changes single bases into another without the use of DNA DSBs or a donor template [131]. Cai et al. successfully used the technique to generate the point mutants of *GmFT2a* and *GmFT4* in soybean [132].

Currently, CRISPR/Cas9 is being frequently used in soybean functional studies [128, 133, 134]. For example, to identify the genes that are responsible for flowering time, frame shift mutations created by CRISPR/Cas9 revealed that *GmFT2a* functions primarily during short day (SD), while *GmFT5a* has more substantial impacts under long day (LD) [132, 135, 136]. Similarly, knockout of *GmPRR37* by the CRISPR system suggested that it can repress flowering under LD [137]. Male sterility was seen in two CRISPR/Cas9 gene-editing mutants of *Glyma.13G114200*, showing that it was the casual gene *GmMS1* responsible for male sterility [138–140].

5.1 The advantages of genome editing for soybean molecular design breeding

A key issue and important research goal for soybean researchers since the completion of the soybean genome sequencing project is to elucidate the function of 46–56 thousand identified genes [129]. Transgenic technology is a great tool for functional genomic research and crop genetic enhancement, but it has certain drawbacks when used in soybeans. *Agrobacterium-mediated transformation* and particle bombardment have been widely utilized to make transgenic soybean plants in recent decades. When considering its easy technique, low cost, single or low copy number of insertions, and relatively infrequent rearrangement, *Agrobacterium-mediated transformation* is a superior alternative [141]. However, because soybean is a refractory crop as far as the transformation and regeneration are considered, no sustained soybean genetic transformation has yet been developed, regardless of whatever technique is used. Furthermore, the efficacy of Agrobacterium-mediated transformation is affected by soybean tissue, cultivar, or species [142], resulting in a restricted number of soybeans types that may be enhanced directly by genetic modification. Transgenic technology is used to investigate gene function by integrating foreign DNA sequences into the plant genome, which results in either overexpression or silence of the target gene [142]. As a result of potential risks such as unintended gene insertions, endogenous gene disruption, and unpredictable gene expression that arise during transformation [127], transgenic plants frequently cause potential bio-safety issues and are subjected to regulatory restrictions on genetically modified organisms (GMOs).

Furthermore, being a diploid that developed from palaeotetraploid, soybean has a highly duplicated genome, with around 75% of projected genes possess multiple copies, resulting in substantial genetic redundancy and complicating the elucidation of soybean gene function. On the one hand, conventional random mutagenesis methods (physical, chemical mutagenesis, or T-DNA insertion) make it difficult to link genotype and phenotype because the loss of one homolog can be fully compensated by redundant homologous copies [143]. On the other hand, RNA-silencing-based technology frequently silences the entire gene family and is difficult to silence a single gene copy [144], and due to partial gene product depletion, the phenotype may be unstable [145]. To accelerate soybean gene function and breeding research, more accurate and efficient genetic engineering technology is required. Current genome editing technology has provided opportunity to overcome the aforesaid difficulties, at least in part. For starters, genome editing technology can not only alter a single gene without impacting other members of the gene family, but it can also edit many genes of interest at the same time using a single transformation, making it ideal for soybean and other polyploid crop studies [126].

Secondly, unlike transgenic technology, genome editing technology incorporates sequences that are genome editing components rather than foreign genes of interest into plant genomes. These genome editing components may be deleted once the target gene has been edited to yield transgene-free mutants, which are safer to employ in breeding and easier to commercialize under tight GMO rules. For example, in the United States, a gene-edited soybean oil with a different fatty acid profile was recently released (https://calyxt.com/first-commercial-sale-of-calyxt-high-oleic-soybean-oil-on-the-usmarket/).

Finally, because the soybean genetic transformation is inefficient, one possible approach is to do direct genome editing without transformation or tissue culture. The standard procedure is to edit a variety with a high transformation efficiency to enhance one or more characteristics and then utilize the modified plant as a donor, which contains editing components. By hybridization with donor plants and subsequent backcrossing, the modified target gene or genome editing components can be introgressed into elite lines resistant to transformation. In conclusion, genome editing technology is a strong tool that has a lot of potential for speeding up soybean breeding [137, 146].

5.2 Applications of genome editing in soybean improvement

Hairy-root transformation mediated by *Agrobacterium rhizogenes* is a simple and rapid technique for studying soybean gene function, and it only takes a few weeks to get transgenic hairy root [142, 147]. When ZFNs, TALENs, and CRISPR/Cas9 systems were initially employed in soybean, researchers preferred to use the hairy-root transformation technique to quickly assess the efficiency of these genome editing tools before undertaking the time-consuming soybean whole-plant transformation [142, 147]. As a consequence, ZFNs technology delivered the first example of genome editing in soybean in 2011, producing heritable mutations in two homologous *DICER-LIKE genes,* namely *DCL4a* and *DCL4b* [144]. In 2015, five research groups successfully evaluated the mutation efficiency of the CRISPR/Cas9 system in soybean endogenous or exogenous genes in hairy roots, establishing a precedent for using CRISPR/Cas technology to study soybean genes [127, 148, 149]. One of these teams used NHEJ to accomplish about 76% targeted mutagenesis, and they also used HR to create mutant plants with targeted gene integration at the target location, as well as a chlorsulfuron-resistant soybean with a mutated acetolactate synthase1 (ALS1) gene [127]. All of these early successes have highlighted the genome editing systems' remarkable potential to develop useful features in the near term through focused gene alterations.

Scientists evaluated the effectiveness of TALENs and CRISPR/ Cas9 in editing two phytoene desaturase genes in hairy roots, namely *GmPDS11* and *GmPDS18* [128]. The results demonstrated that CRISPR/Cas9, particularly CRISPR/Cas9 employing the GmU6-16g-1 promoter, was far more effective than TALENs at concurrently targeting two alleles. Another study was carried out to examine a variety of GmU6 promoters in soybean hairy roots and *Arabidopsis thaliana* to determine which ones were best for driving sgRNA production and discovered that GmU6-8 and GmU6-10 promoters had high activity, which improved editing effectiveness [150]. Both results are beneficial in the development of an effective CRISPR/Cas9 system for use in soybean research. In addition, the CRISPR/Cas9 method in combination with the hairy-root transformation technique has been used to edit and explore soybean storage protein genes [134], and the candidate gene governing nodulation specificity [151].

NHEJ has generally been used to delete target genes by tiny insertions or deletions, whereas HR is primarily utilized to replace or integrate targeted genes [152]. Using dual-sgRNA to cleave two neighboring loci on the same chromosome, two research studies recently showed that massive genomic deletions might be generated in soybeans [135, 153]. Due to the two editing chances, the dual-sgRNA design can not only boost gene mutation rates, but also construct substantial fragment deletion to ensure that the target gene is completely eliminated. Other fields of research will benefit from the substantial loss, such as understanding the role of regulatory elements or noncoding genes. Unlike prior HR-mediated donor integration, one study claimed that two large multigene donors (7.1 kb and 16.2 kb) were inserted into a target genomic location of soybean utilizing NHEJ and ZFNs, with donor heritability verified in T1 progeny plants [152]. This study found that NHEJ might be used instead of HR to induce accurate insertions of numerous transgenes in soybeans while avoiding the drawbacks of inefficient HR [154].

In addition to the abovementioned developments in genome editing systems to permit better application in soybean research, several researchers have used genome editing technology and entire plant transformation to investigate gene function or enhance agronomic features. The single and double mutants of the soybean genes, namely *DCL1a* and *DCL1b* using ZFNs-based mutagenesis techniques were created to investigate their roles in the soybean miRNA system [143]. Only the *dcl1a/dcl1b* double mutants showed a morphological phenotype, which was characterized by shrunken and shriveled seed as well as seedling developmental abnormalities, whereas both *dcl1a* and *dcl1b* single mutants showed a normal phenotype, suggesting that *GmDCL1* homologs have functional redundancy. Besides, the two homologous genes, namely *GmPPD1* and *GmPPD2*, coding for Arabidopsis PEAPOD orthologs were edited with a single sgRNA [155]. As a result, double mutants with frame shift mutations displayed a severe phenotype of developmental abnormalities in leaf and pod development.

Previous results revealed that homozygous mutants produced by CRISPR/Cas9-mediated mutagenesis in *FLOWERING LOCUS T2a (GmFT2a)* delayed soybean flowering [133] and male sterile soybean lines [156]. They also used the CRISPR/Cas9 system to alter *GmFT5a* and then crossed it with *ft2a* mutants to get *ft2aft5a* double mutants. This double mutant bloomed 31.3 days later than wild-type plants and produced more pods as well as seeds under short day circumstances [136]. The researchers employed a multiplex genome editing method based on CRISPR/Cas9 to change four *SQUAMOSA PROMOTER BINDING PROTEIN-LIKE9* (*GmSPL9*) genes and generated several soybean mutants with different mutated locus combinations. These mutants revealed altered node number on the main stem and branch number [157]. CRISPR/Cas9 technology has also enhanced soybean seed oil profile [158], offensive beany flavor of soybean seed product [159], and isoflavone content and resistance to soybean mosaic virus [160]. The experiments mentioned above showed that genome editing technology has a lot of potential for improving soybeans.

Despite the fact that CRISPR/Cas9 technology has become the standard for genome editing, NGG PAM's fundamental restriction limits its use in highly precise genomic areas. ZFNs and TALENs have crucial tools as a complement to the CRISPR/Cas9 system since their target range is infinite. The combined usage of these systems will aid the soybean breeding and functional genomics projects. For example, a group of genes were mutated that encode the main machinery proteins involved in small RNA processing in soybean and *Medicago truncatula* using CRISPR/Cas9 and TALENs [161]. Together with the previously reported mutants induced by ZFNs, the resultant mutant plants established a collection of mutant resources for future studies of short RNA biology in legume plants [143].

Many scientists used CRISPR/Cas9-based high-throughput mutagenesis to create a genome-wide mutant library or mutant collection related to certain functions for pooled functional screen, which provided a wealth of resource in mammalian cell, rice, and tomato [162–164]. The success of creating these CRISPR libraries, however, is strongly reliant on a reliable genetic transformation mechanism. The creation of a CRISPR library in soybean is not only a massive task, but it is also extremely important for research and breeding. A first soybean CRISPR library targeting over 100 candidate genes was recently developed. A collection of mutant soybean lines was also created utilizing an enhanced process at several important steps, demonstrating the feasibility and usefulness of using CRISPR/Cas9 technology to execute large-scale mutagenesis in soybeans [129]. Following the CRISPR/Cas9 system, new genomic editing tools such as the CRISPR/Cas12a system, BE systems, and other CRISPR/Cas variants have been developed. However, to our knowledge, most of these have yet to be employed in soybean, with the exception of one work [165] that used Cas12a-RNP in soybean protoplasts to induce gene editing. RNP-based genome editing may not be the best option for soybean research due to a lack of updated information on successful protoplast regeneration (**Figure 1**).

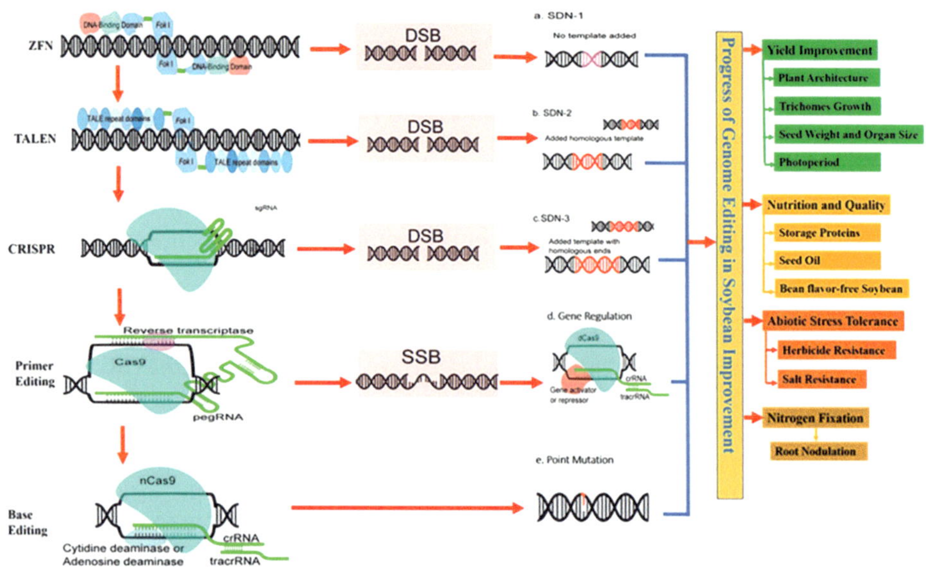

Figure 1.
Genome editing platforms and editing outcomes. Each editing platform (arrow) and its outcomes (rectangular) are coded with the same color. ZFN, zinc-finger nuclease; TALEN, transcription activator-like effector nuclease; CRISPR, clustered regulatory interspaced short palindromic repeat; DSB, double-strand breaks; SSB, single-strand breaks; outcomes of GE created by site-directed nucleases (SDN) include: SDN1 – The approach involves DNA breaks repair through DNA repair mechanisms in the host cellular without using an added repair template; SDN2 – The approach involves the break repair via HR using an added homologous repair template; and SDN3 – The approach involves DNA break repair via either HDR or NHEJ pathway using an added DNA template containing no homologous sequences but with homologous ends: Progress and application of genome editing in soybean improvement.

6. Conclusion and future prospective

Traditional breeding has made important contributions to soybean improvement and the generation of soybean varieties with enhanced yield, quality, and tolerance to numerous stressors throughout the last century. Traditional crop development technologies, on the other hand, are unable to keep up with the world's rapidly rising population and climate change [166, 167]. Reduced generation time allows for faster soybean breeding, which may be accomplished by the quick creation of homozygous lines employing the doubled-haploid (DH) production process. The creation of a high-throughput DH production program in soybean would be tremendously beneficial in achieving the crop's targeted genetic gain. Soybean androgenesis, root development, and unusual shoot induction have all shown slight advancements. However, there is currently no efficient, repeatable way for producing doubled haploids in soybean. One of the primary impediments to the development of a commercial DH production procedure in soybeans may be the tissue's resistance to *in vitro* regeneration [168, 169]. To achieve a sustainable yield, it is necessary to identify genetic resources in the form of water-stress-resistant soybean genotypes and genomics-assisted water-stress mitigation approaches. Several techniques, such as QTL mapping, genome-wide association mapping, and comparative transcriptome analyses, are being used to determine the genetic basis for water-stress tolerance in soybean [170, 171].

Recent advances in NGS-based genotyping technologies and powerful computational pipelines have significantly reduced the cost of WGS/WGRS, allowing the discovery, sequencing, and genotyping of hundreds of thousands of markers in one step. For large-scale marker identification, NGS-marker technologies based on reduced representation sequencing are the ideal solution, especially for the huge and complex soybean genome. These NGS-based marker approaches represent the soybean's partial genome, and they can even be used without a reference sequence. RAD-seq (or its variations) and GBS are two NGS technologies that have previously proven to be efficient and effective procedures for GAB and have been widely employed for GS investigations in various agricultural plants. Furthermore, the NGS has enabled the fabrication of high-density SNP chips for HTG in soybean. The low cost, genome-wide marker coverage, better speed and throughput, and higher marker density of NGS-marker technologies have allowed geneticists to explore the inheritance of numerous traits at the nucleotide level accuracy.

On the other hand, GS employs a number of markers spread over the entire genome to forecast the breeding value of a breeding line for selection. GS can quantify Mendelian sampling without phenotyping the entire population thanks to genome-wide dense markers. It shortens cycles to save time while also increasing genetic gain per unit of time. GS was compared with traditional phenotypic selection in soybean to see if it has any advantages in terms of accuracy and time savings.

GE technologies, particularly CRISPR-based systems, have advanced quickly, with the majority of them being implemented to give effective tools for soybean improvement. If this technique is properly implemented in plant breeding programs, a recent field trial of high oleic soybean employing TALENs has indicated the bright future of soybean improvement. Currently, the discovery of more GE target genes associated with agronomically important traits, the adoption of newly developed GE technologies, the simplification and renovation of editing reagent delivery, and the improvement of target mutant recovery method in soybean will improve editing outcomes, save time, and lower product development costs. The development of GE products will be aided by the cost-effective preparation of intellectual property for GE technologies, as well as breeders' and farmers' comprehension of GE-related government regulation. In several countries, transgene-free or DNA-free edited plants are considered nongenetically modified events, making GE soybean production easier. In future, more applications of "base editing" for single genes or several genes at once would substantially aid functional research and molecular design breeding in soybean.

Next-generation GAB in agricultural plants has been enabled by recent advances in crop phenomics and genomics, which have provided several high-throughput platforms, as well as statistical approaches and computational tools. When these modern technologies are integrated, they can precisely and accurately identify genes/QTLs, as well as their beneficial usage in soybean breeding [172, 173]. Despite the fact that high-throughput SNP genotyping technologies have completely revolutionized marker application in soybean breeding, they have enabled research groups to apply GWAS and GS for soybean improvement on a regular basis. These marker technologies, however, must be paired with HTP to produce meaningful genetic gain from complex features in order to reap the full benefits of genomic investigations. So far, only a few studies involving the use of both HTP and HTG in soybean have been reported. This is because large-scale field-based HTP has a greater cost. New advances in crop phenotyping technology are expected to make HTP more inexpensive for commercial application in soybean breeding projects in the near future. This would undoubtedly increase the scope of germplasm assessment and facilitate the development of better soybean cultivars. WGRS-based genotyping will become increasingly viable and cost-effective as the cost of DNA sequencing falls. Sequencing-based genotyping employing genome-reduction methods such as GBS and RAD-sequencing appears to be more cost-efficient for breeding-based applications such as GS at the moment. Since the cultivated soybean has a limited genetic basis, genome editing and TILLING can be used to produce a variety of changes in these orthologs, from knock-down to knockout alleles. For quick deployment of these alleles in breeding programs, it should be combined with the speed breeding facility.

The ZFNs, TALENs, CRISPR/Cas9, CRISPR/Cas12a, BEs, and other CRISPR/Cas variations provide a robust genome editing toolkit that will aid future functional genomic and genetic improvement studies in soybeans and other plants. CRISPR/Cas9 technology may become the preferred method for soybean breeding due to its efficiency, multiplex editing, and high-throughput mutagenesis capabilities, as well as its maturity. With the progress of additional genome editing methods, however, soybean genome editing will become more versatile. Despite the fact that substantial effort may be required to employ these techniques, given the enormous potential of genome editing and the economic importance of soybean, we anticipate that these issues will be resolved in the near future.

Acknowledgements

This research was funded by the Zhejiang Lab (Grant No. 2021PE0AC04) and Yazhou Bay Seed Lab (Grant No. B21HJ0101).

Conflict of interest

On behalf of all authors, the corresponding author states that there is no conflict of interest.

Author details

Xianzhong Feng* and Muhammad Hafeez Ullah Khan
Key Laboratory of Soybean Molecular Design Breeding, Northeast Institute of Geography and Agroecology, Chinese Academy of Sciences, Changchun, China

*Address all correspondence to: fengxianzhong@iga.ac.cn

References

[1] Wilson RF. Soybean: Market driven research needs. In: Genetics and Genomics of Soybean. Springer; 2008. pp. 3-15

[2] Hartman GL, West ED, Herman TK. Crops that feed the world 2. Soybean worldwide production, use, and constraints caused by pathogens and pests. Food Security. 2011;**3**:5-17

[3] Rincker K, Nelson R, Specht J, Sleper D, Cary T, Cianzio SR, et al. Genetic improvement of US soybean in maturity groups II, III, and IV. Crop Science. 2014;**54**:1419-1432

[4] Wolfgang G, An YC. Genetic separation of southern and northern soybean breeding programs in North America and their associated allelic variation at four maturity loci. Molecular Breeding. 2017;**37**:1-9

[5] Anderson EJ, Ali ML, Beavis WD, Chen P, Clemente TE, Diers BW, et al. Soybean [*Glycine max* (L.) Merr.] breeding: History, improvement, production and future opportunities. In: Advances in Plant Breeding Strategies: Legumes. Springer; 2019. pp. 431-516

[6] Didinger C, Foster MT, Bunning M, Thompson HJJDB. Pulses: Production P, nutrition. Nutrition and human health benefits of dry beans and other pulses. Dry beans pulses: Production, processing. Nutrition. 2022:481-504

[7] Araus JL, Cairns JE. Field high-throughput phenotyping: The new crop breeding frontier. Trends in Plant Science. 2014;**19**:52-61

[8] Xiong Q, Tang G, Zhong L, He H, Chen X. Response to nitrogen deficiency and compensation on physiological characteristics, yield formation, and nitrogen utilization of rice. Frontiers in Plant Science. 2018;**9**:1075

[9] Collins NC, Tardieu F, Tuberosa R. Quantitative trait loci and crop performance under abiotic stress: Where do we stand? Plant Physiology. 2008;**147**:469-486

[10] Jin J, Liu X, Wang G, Mi L, Shen Z, Chen X, et al. Agronomic and physiological contributions to the yield improvement of soybean cultivars released from 1950 to 2006 in Northeast China. Field Crops Research. 2010;**115**:116-123

[11] Montesinos-López OA, Montesinos-López A, Crossa J, de Los Campos G, Alvarado G, Suchismita M, et al. Predicting grain yield using canopy hyperspectral reflectance in wheat breeding data. Plant Methods. 2017;**13**:1-23

[12] Medini D, Serruto D, Parkhill J, Relman DA, Donati C, Moxon R, et al. Microbiology in the post-genomic era. Nature Review Microbiology. 2008;**6**:419-430

[13] Li MW, Wang Z, Jiang B, Kaga A, Wong FL, Zhang G, et al. Impacts of genomic research on soybean improvement in East Asia. Theoretical and Applied Genetics. 2020;**133**:1655-1678

[14] Chan C, Qi X, Li MW, Wong FL, Lam HM. Recent developments of genomic research in soybean. Journal of Genetics and Genomics. 2012;**39**:317-324

[15] Song QJ, Marek LF, Shoemaker RC, Lark KG, Concibido VC, Delannay X, et al. A new integrated genetic linkage

map of the soybean. Theoretical and Applied Genetics. 2004;**109**:122-128

[16] Hyten DL, Cannon SB, Song Q, Weeks N, Fickus EW, Shoemaker RC, et al. High-throughput SNP discovery through deep resequencing of a reduced representation library to anchor and orient scaffolds in the soybean whole genome sequence. BMC Genomics. 2010;**11**:1-8

[17] Song X, Wei H, Cheng W, Yang S, Zhao Y, Li X, et al. Development of INDEL markers for genetic mapping based on whole genome resequencing in soybean. G3: Genes, Genomes, Genetics. 2015;**5**:2793-2799

[18] Bhat JA, Yu D, Bohra A, Ganie SA, Varshney RK. Features and applications of haplotypes in crop breeding. Communications Biology. 2021;**4**:1-12

[19] Araus JL, Kefauver SC, Zaman-Allah M, Olsen MS, Cairns JE. Translating high-throughput phenotyping into genetic gain. Trends in Plant Science. 2018;**23**:451-466

[20] Tardieu F, Cabrera-Bosquet L, Pridmore T, Bennett M. Plant Phenomics, from sensors to knowledge. Current Biololgy. 2017;**27**:R770-RR83

[21] Pacher M, Puchta H. From classical mutagenesis to nuclease-based breeding–directing natural DNA repair for a natural end-product. The Plant Journal, Wiley Online Library. 2017:819-833

[22] Cong L, Ran FA, Cox D, Lin S, Barretto R, Habib N, et al. Multiplex genome engineering using CRISPR/Cas systems. Science. 2013;**339**:819-823

[23] Khan MHU, Khan SU, Muhammad A, Hu L, Yang Y, Fan C. Induced mutation and epigenetics modification in plants for crop improvement by targeting CRISPR/Cas9 technology. Joournal of Cellular Physiology. 2018;**233**:4578-4594

[24] Chen K, Gao C. Targeted genome modification technologies and their applications in crop improvements. Plant Cell Reports. 2014;**33**:575-583

[25] Cebrian-Serrano A, Davies B. CRISPR-Cas orthologues and variants: Optimizing the repertoire, specificity and delivery of genome engineering tools. Mammalian Genome. 2017;**28**:247-261

[26] Zetsche B, Gootenberg JS, Abudayyeh OO, Slaymaker IM, Makarova KS, Essletzbichler P, et al. Cpf1 is a single RNA-guided endonuclease of a class 2 CRISPR-Cas system. Cell. 2015;**163**:759-771

[27] Wang T, Xun H, Wang W, Ding X, Tian H, Hussain S, et al. Mutation of *GmAITR* genes by CRISPR/Cas9 genome editing results in enhanced salinity stress tolerance in soybean. Frontiers in Plant Science. 2021:2752

[28] Moradpour M, Abdulah SNA. CRISPR/dC as9 platforms in plants: Strategies and applications beyond genome editing. Plant Biotechnology Journal. 2020;**18**:32-44

[29] Jiang GL. Molecular markers and marker-assisted breeding in plants. In: Plant Breeding from Laboratories to Fields. Vol. 3. London, UK: IntechOpen; 2013. pp. 45-83

[30] Ribaut JM, de Vicente MC, Delannay X. Molecular breeding in developing countries: Challenges and perspectives. Current Opinion in Plant Biology. 2010;**13**:213-218

[31] Moose SP, Mumm RH. Molecular plant breeding as the foundation for 21st century crop improvement. Plant Physiology. 2008;**147**:969-977

[32] Xu Y. Molecular Plant Breeding. CABI; 2010

[33] Mundhe S, Patil R, Oak M, Jaybhay S. Accelerating soybean improvement through genomics-assisted breeding. In: Accelerated Plant Breeding. Vol. 4. Springer; 2022. pp. 41-62

[34] Kumawat G, Gupta S, Ratnaparkhe MB, Maranna S, Satpute GK. QTL omics in soybean: A way forward for translational genomics and breeding. Frontiers in Plant Science. 2016;7:1852

[35] Randhawa HS, Mutti JS, Kidwell K, Morris CF, Chen X, Gill KS. Rapid and targeted introgression of genes into popular wheat cultivars using marker-assisted background selection. PLoS One. 2009;**4**:e5752

[36] Badri J, Lakshmidevi G, JaiVidhya L, Prasad MS, Laha GS, Lakshmi VJ, et al. Multiparent-derived, Markerassisted introgression lines of the elite Indian rice cultivar,'Krishna Hamsa' show resistance against bacterial blight and blast and tolerance to drought. Plants. 2022;**11**:622

[37] Tuberosa R. Marker-assisted breeding in crops. In: Encyclopedia of Sustainability Science Technology. 2012. pp. 6425-6448

[38] Swarup S, Cargill EJ, Crosby K, Flagel L, Kniskern J, Glenn KC. Genetic diversity is indispensable for plant breeding to improve crops. Crop Science. 2021;**61**:839-852

[39] Shi A, Chen P, Li D, Zheng C, Zhang B, Hou A. Pyramiding multiple genes for resistance to soybean mosaic virus in soybean using molecular markers. Molecular Breeding. 2009;**23**:113-124

[40] Widyasari K, Tran P-T, Shin J, Son H, Kim K-H. Overexpression of purple acid phosphatase *GmPAP2*. 1 confers resistance to soybean mosaic virus in a susceptible soybean cultivar. Journal of Experimental Botany. 2022;**73**:1623-1642

[41] Wang D, Lin Z, Kai L, Ying M, Wang L, Yang Y, et al. Marker-assisted pyramiding of soybean resistance genes RSC4, RSC8, and RSC14Q to soybean mosaic virus. Journal of Integrative Agriculture. 2017;**16**:2413-2420

[42] Zhang Y, Song J, Wang L, Yang M, Hu K, Li W, et al. Identifying quantitative trait loci and candidate genes conferring resistance to soybean mosaic virus SC7 by quantitative trait loci-sequencing in soybean. Frontiers in Plant Science. 2022;**13**

[43] Pham AT, Lee JD, Shannon JG, Bilyeu KD. Mutant alleles of *FAD2-1A* and *FAD2-1B* combine to produce soybeans with the high oleic acid seed oil trait. BMC Plant Biology. 2010;**10**:1-13

[44] Zhang H, Jiang H, Hu Z, Song Q, An YC. Development of a versatile resource for post-genomic research through consolidating and characterizing 1500 diverse wild and cultivated soybean genomes. BMC Genomics. 2022;**23**:1-13

[45] Pham AT, Lee JD, Shannon JG, Bilyeu KD. A novel *FAD2-1A* allele in a soybean plant introduction offers an alternate means to produce soybean seed oil with 85% oleic acid content. Theoretical Applied Genetics. 2011;**123**:793-802

[46] Sebastian S, Streit L, Stephens P, Thompson J, Hedges B, Fabrizius M, et al. Context-specific marker-assisted selection for improved grain yield in elite soybean populations. Crop Science. 2010;**50**:1196-1206

[47] Santana FA, Silva MFD, Guimarães JKF, Ferreira MFDS,

Pereira WD, Piovesan ND, et al. Marker-assisted selection strategies for developing resistant soybean plants to cyst nematode. Crop Breeding Applied Biotechnology. 2014;**14**:180-186

[48] Zhang YH, Liu MF, He JB, Wang YF, Xing GN, Li Y, et al. Marker-assisted breeding for transgressive seed protein content in soybean [*Glycine max* (L.) Merr.]. Theoretical Applied Genetics. 2015;**128**:1061-1072

[49] Zhang S, Du H, Ma Y, Li H, Kan G, Yu D. Linkage and association study discovered loci and candidate genes for glycinin and β-conglycinin in soybean (*Glycine max L. Merr*.). Theoretical Applied Genetics. 2021;**134**: 1201-1215

[50] Liu Y, Yu L, Qu Y, Chen J, Liu X, Hong H, et al. *GmSALT3*, which confers improved soybean salt tolerance in the field, increases leaf Cl^- exclusion prior to Na^+ exclusion but does not improve early vigor under salinity. Frontiers in Plant Science. 2016;**7**:1485

[51] Akanbi M, Egbichi I, Ludidi N. Comparative analysis of responses to field salinity stress in contrasting soybean accessions highlights NaCl exclusion in leaves as a key mechanism for salinity stress tolerance. Journal of Oasis Agriculture and Sustainable Development. 2021:19-24

[52] Karahara I, Horie T. Functions and structure of roots and their contributions to salinity tolerance in plants. Breeding Science. 2021;**71**:89-108

[53] Rawal R, Kumar V, Rani A, Gokhale SM. Genetic elimination of off-flavour generating lipoxygenase2 gene of soybean through marker assisted backcrossing and its effect on seed longevity. Plant Breeding Biotechenolgy. 2020;**8**:163-173

[54] Hagely KB, Jo H, Kim JH, Hudson KA, Bilyeu K. Molecular-assisted breeding for improved carbohydrate profiles in soybean seed. Theoretical Applied Genetics. 2020;**133**:1189-1200

[55] Qin P, Wang T, Luo Y. A review on plant-based proteins from soybean: Health benefits and soy product development. Journal of agriculture. Food Research. 2022:100265

[56] Baghbani-Arani A, Poureisa M, Alekajbaf H, Borz-Abad RK, Khodadadi-Dashtaki K. Current status of genetically modified crops in Iran and the world: Overview of production and consumption challenges. 2021. DOI: https://doi.org/10.21203/rs.3.rs-141395/v1

[57] Celec P, Kukučková M, Renczésová V, Natarajan S, Pálffy R, Gardlík R, et al. Biological and biomedical aspects of genetically modified food. Biomedicine Pharmacotherapy. 2005;**59**:531-540

[58] Zhang M, Liu S, Wang Z, Yuan Y, Zhang Z, Liang Q, et al. Progress in soybean functional genomics over the past decade. Plant Biotechnology Journal. 2022;**20**:256

[59] Rech EL, Vianna GR, Aragao FJ. High-efficiency transformation by biolistics of soybean, common bean and cotton transgenic plants. Nature Protocols. 2008;**3**:410-418

[60] Wang G, Xu Y. Hypocotyl-based Agrobacterium-mediated transformation of soybean (*Glycine max*) and application for RNA interference. Plant Cell Reports. 2008;**27**:1177-1184

[61] Trick H, Finer J. Sonication-assisted Agrobacterium-mediated transformation of soybean [*Glycine max (L.) Merr*] embryogenic suspension culture tissue. Plant Cell Reports. 1998;**17**:482-488

[62] Liu HK, Yang C, Wei ZM. Efficient Agrobacterium tumefaciens-mediated transformation of soybeans using an embryonic tip regeneration system. Planta. 2004;**219**:1042-1049

[63] Paz MM, Martinez JC, Kalvig AB, Fonger TM, Wang K. Improved cotyledonary node method using an alternative explant derived from mature seed for efficient Agrobacterium-mediated soybean transformation. Plant Cell Reports. 2006;**25**:206-213

[64] Li S, Cong Y, Liu Y, Wang T, Shuai Q, Chen N, et al. Optimization of Agrobacterium-mediated transformation in soybean. Frontiers in Plant Science. 2017;**8**:246

[65] Somers DA, Samac DA, Olhoft PM. Recent advances in legume transformation. Plant Physiology. 2003;**131**:892-899

[66] Pareddy D, Chennareddy S, Anthony G, Sardesai N, Mall T, Minnicks T, et al. Improved soybean transformation for efficient and high throughput transgenic production. Transgenic Research. 2020;**29**:267-281

[67] Ge X, Chu Z, Lin Y, Wang S. A tissue culture system for different germplasms of indica rice. Plant Cell Reports. 2006;**25**:392-402

[68] Yang A, He C, Zhang K. Improvement of Agrobacterium-mediated transformation of embryogenic calluses from maize elite inbred lines. In Vitro Cellular Developmental Biology-Plant. 2006;**42**:215-219

[69] Song ZY, Tian JL, Fu WZ, Li L, Lu LH, Zhou L, et al. Screening Chinese soybean genotypes for Agrobacterium-mediated genetic transformation suitability. Journal of Zhejiang University SCIENCE B. 2013;**14**:289-298

[70] Chen L, Cai Y, Liu X, Yao W, Guo C, Sun S, et al. Improvement of soybean *Agrobacterium-mediated transformation* efficiency by adding glutamine and asparagine into the culture media. International Journal of Molecular Sciences. 2018;**19**:3039

[71] Mir RR, Reynolds M, Pinto F, Khan MA, Bhat MA. High-throughput phenotyping for crop improvement in the genomics era. Plant Science. 2019;**282**:60-72

[72] van Bezouw RF, Keurentjes JJ, Harbinson J, Aarts MG. Converging phenomics and genomics to study natural variation in plant photosynthetic efficiency. The Plant Journal. 2019;**97**:112-133

[73] Grant D, Nelson RT, Cannon SB, Shoemaker RC. SoyBase, the USDA-ARS soybean genetics and genomics database. Nucleic Acids Research. 2010;**38**:D843-D8D6

[74] Brown AV, Conners SI, Huang W, Wilkey AP, Grant D, Weeks NT, et al. A new decade and new data at SoyBase, the USDA-ARS soybean genetics and genomics database. Nucleic Acids Research. 2021;**49**:D1496-DD501

[75] Du J, Grant D, Tian Z, Nelson RT, Zhu L, Shoemaker RC, et al. SoyTEdb: A comprehensive database of transposable elements in the soybean genome. BMC Genomics. 2010;**11**:1-7

[76] Kim E, Hwang S, Lee I. SoyNet: A database of co-functional networks for soybean *Glycine max*. Nucleic Acids Research. 2017;**45**:D1082-D10D9

[77] Tavakolan M, Alkharouf NW, Khan FH, Natarajan S. SoyProDB: A database for the identification of soybean seed proteins. Bioinformation. 2013;**9**:165

[78] Tavakolan M, Alkharouf NW, Matthews BF, Natarajan SS. SoyProLow: A protein database enriched in low abundant soybean proteins. Bioinformation. 2014;**10**:599

[79] Severin AJ, Woody JL, Bolon Y-T, Joseph B, Diers BW, Farmer AD, et al. RNA-Seq atlas of *Glycine max*: A guide to the soybean transcriptome. BMC Plant Biology. 2010;**10**:1-16

[80] Libault M, Zhang XC, Govindarajulu M, Qiu J, Ong YT, Brechenmacher L, et al. A member of the highly conserved FWL (tomato FW2. 2-like) gene family is essential for soybean nodule organogenesis. The Plant Journal. 2010;**62**:852-864

[81] Machado FB, Moharana KC, Almeida-Silva F, Gazara RK, Pedrosa-Silva F, Coelho FS, et al. Systematic analysis of 1298 RNA-Seq samples and construction of a comprehensive soybean (Glycine max) expression atlas. The Plant Journal. 2020;**103**:1894-1909

[82] Agrawal GK, Hajduch M, Graham K, Thelen JJ. In-depth investigation of the soybean seed-filling proteome and comparison with a parallel study of rapeseed. Plant Physiology. 2008;**148**:504-518

[83] Ohyanagi H, Sakata K, Komatsu S. Soybean proteome database 2012: Update on the comprehensive data repository for soybean proteomics. Frontiers in Plant Science. 2012;**3**:110

[84] Komatsu S, Wang X, Yin X, Nanjo Y, Ohyanagi H, Sakata K. Integration of gel-based and gel-free proteomic data for functional analysis of proteins through soybean proteome database. Journal of Proteomics. 2017;**163**:52-66

[85] Joshi T, Fitzpatrick MR, Chen S, Liu Y, Zhang H, Endacott RZ, et al. Soybean knowledge base (SoyKB): A web resource for integration of soybean translational genomics and molecular breeding. Nucleic Acids Research. 2014;**42**:D1245-D1D52

[86] Steiner B, Michel S, Maccaferri M, Lemmens M, Tuberosa R, Buerstmayr H. Exploring and exploiting the genetic variation of fusarium head blight resistance for genomic-assisted breeding in the elite durum wheat gene pool. Theoretical Applied Genetics. 2019;**132**:969-988

[87] Bhat JA, Ali S, Salgotra RK, Mir ZA, Dutta S, Jadon V, et al. Genomic selection in the era of next generation sequencing for complex traits in plant breeding. Frontiers in Genetics. 2016;**7**:221

[88] Gupta PK, Balyan HS, Gahlaut V. QTL analysis for drought tolerance in wheat: Present status and future possibilities. Agronomy. 2017;**7**:5

[89] Saghai Maroof M, Jeong SC, Gunduz I, Tucker D, Buss G, Tolin S. Pyramiding of soybean mosaic virus resistance genes by marker-assisted selection. Crop Science. 2008;**48**:517-526

[90] Spindel J, Begum H, Akdemir D, Virk P, Collard B, Redona E, et al. Genomic selection and association mapping in rice (*Oryza sativa*): Effect of trait genetic architecture, training population composition, marker number and statistical model on accuracy of rice genomic selection in elite, tropical rice breeding lines. PLoS Genetics. 2015;**11**:e1004982

[91] Xu K, Zhang X-M, Chen H, Zhang C, Zhu J, Cheng Z, et al. Fine-tuning florigen increases field yield through improving photosynthesis

in soybean. Frontiers. Plant Science. 2021;**12**

[92] Cobb JN, Juma RU, Biswas PS, Arbelaez JD, Rutkoski J, Atlin G, et al. Enhancing the rate of genetic gain in public-sector plant breeding programs: Lessons from the breeder's equation. Theoretical Applied Genetics. 2019;**132**:627-645

[93] Khan A, Sovero V, Gemenet D. Genome-assisted breeding for drought resistance. Current Genomics. 2016;**17**:330-342

[94] Hina A, Cao Y, Song S, Li S, Sharmin RA, Elattar MA, et al. High-resolution mapping in two RIL populations refines major "QTL hotspot" regions for seed size and shape in soybean (*Glycine max* L.). International Journal of Molecular Sciences. 2020;**21**:1040

[95] Gao H, Wang Y, Li W, Gu Y, Lai Y, Bi Y, et al. Transcriptomic comparison reveals genetic variation potentially underlying seed developmental evolution of soybeans. Journal of Experimental Botany. 2018;**69**:5089-5104

[96] Yin L, Zhang H, Zhou X, Yuan X, Zhao S, Li X, et al. KAML: Improving genomic prediction accuracy of complex traits using machine learning determined parameters. Genome Biology. 2020;**21**:1-22

[97] Stewart-Brown BB, Song Q, Vaughn JN, Li Z. Genomic selection for yield and seed composition traits within an applied soybean breeding program. G3: Genes, Genomes, Genetics. 2019;**9**:2253-2265

[98] Bohra A, Saxena K, Varshney RK, Saxena RK. Genomics-assisted breeding for pigeonpea improvement. Theoretical Applied Genetics. 2020;**133**:1721-1737

[99] Matei G, Woyann LG, Milioli AS, de Bem OI, Zdziarski AD, Zanella R, et al. Genomic selection in soybean: Accuracy and time gain in relation to phenotypic selection. Journal Molecular Breeding. 2018;**38**:1-13

[100] Qin J, Shi A, Song Q, Li S, Wang F, Cao Y, et al. Genome wide association study and genomic selection of amino acid concentrations in soybean seeds. Frontiers in Plant Science. 2019:1445

[101] Ravelombola WS, Qin J, Shi A, Nice L, Bao Y, Lorenz A, et al. Genome-wide association study and genomic selection for tolerance of soybean biomass to soybean *cyst* nematode infestation. PLoS One. 2020;**15**: e0235089

[102] Ravelombola WS, Qin J, Shi A, Nice L, Bao Y, Lorenz A, et al. Genome-wide association study and genomic selection for soybean chlorophyll content associated with soybean *cyst* nematode tolerance. BMC Genomics. 2019;**20**:1-18

[103] Shi A, Gepts P, Song Q, Xiong H, Michaels TE, Chen S. Genome-wide association study and genomic prediction for soybean cyst nematode resistance in USDA common bean (*Phaseolus vulgaris*) core collection. Frontiers in Plant Science. 2021;**12**

[104] Jarquín D, Howard R, Graef G, Lorenz A. Response surface analysis of genomic prediction accuracy values using quality control covariates in soybean. Evolutionary Bioinformatics. 2019;**15**:1176934319831307

[105] Ma Y, Reif JC, Jiang Y, Wen Z, Wang D, Liu Z, et al. Potential of marker selection to increase prediction accuracy of genomic selection in soybean (*Glycine max* L.). Molecular Breeding. 2016;**36**:1-10

[106] Duhnen A, Gras A, Teyssèdre S, Romestant M, Claustres B, Daydé J, et al. Genomic selection for yield and seed protein content in soybean: A study of breeding program data and assessment of prediction accuracy. Crop Science. 2017;**57**:1325-1337

[107] Đorđević V, Ćeran M, Miladinović J, Balešević-Tubić S, Petrović K, Miladinov Z, et al. Exploring the performance of genomic prediction models for soybean yield using different validation approaches. Molecular Breeding. 2019;**39**:1-15

[108] Zhang J, Song Q, Cregan PB, Jiang GL. Genome-wide association study, genomic prediction and marker-assisted selection for seed weight in soybean (*Glycine max*). Theoretical Applied Genetics. 2016;**129**:117-130

[109] Getachew T. Pulse crops production opportunities, challenges and its value chain in Ethiopia: A review article. Journal of Environment Earth Science. 2019;**9**:20-29

[110] Jing Y, Zhao X, Wang J, Teng W, Qiu L, Han Y, et al. Identification of the genomic region underlying seed weight per plant in soybean (*Glycine max L. Merr.*) via high-throughput single-nucleotide polymorphisms and a genome-wide association study. Frontiers in Plant Science. 2018:1392

[111] Li D, Zhao X, Han Y, Li W, Xie F. Genome-wide association mapping for seed protein and oil contents using a large panel of soybean accessions. Genomics. 2019;**111**:90-95

[112] Alqudah AM, Sallam A, Baenziger PS, Börner A. GWAS: Fast-forwarding gene identification and characterization in temperate cereals: Lessons from barley. A review. Journal of Advanced Research. 2020;**22**:119-135

[113] Hu D, Kan G, Hu W, Li Y, Hao D, Li X, et al. Identification of loci and candidate genes responsible for pod dehiscence in soybean via genome-wide association analysis across multiple environments. Frontiers in Plant Science. 2019:811

[114] Torkamaneh D, Chalifour F-P, Beauchamp CJ, Agrama H, Boahen S, Maaroufi H, et al. Genome-wide association analyses reveal the genetic basis of biomass accumulation under symbiotic nitrogen fixation in African soybean. Theoretical Applied Genetics. 2020;**133**:665-676

[115] Borah J, Singode A, Talukdar A, Yadav RR, Sarma RN. Genome-wide association studies (GWAS) reveal candidate genes for plant height and number of primary branches in soybean [*Glycine max (L.) Merr.*]. Indian Journal of Genetics Plant Breeding. 2018;**78**:460-469

[116] Sonah H, O'Donoughue L, Cober E, Rajcan I, Belzile F. Identification of loci governing eight agronomic traits using a GBS-GWAS approach and validation by QTL mapping in soybean. Plant Biotechnology Journal. 2015;**13**:211-221

[117] Zhao X, Han Y, Li Y, Liu D, Sun M, Zhao Y, et al. Loci and candidate gene identification for resistance to Sclerotinia sclerotiorum in soybean (*Glycine max* L. Merr.) via association and linkage maps. The Plant Journal. 2015;**82**:245-255

[118] Sui M, Jing Y, Li H, Zhan Y, Luo J, Teng W, et al. Identification of loci and candidate genes analyses for tocopherol concentration of soybean seed. Frontiers in Plant Science. 2020:1366

[119] Yang Q, Jin H, Yu X, Fu X, Zhi H, Yuan F. Rapid identification of soybean resistance genes to soybean mosaic virus by SLAF-seq bulked segregant analysis.

Plant Molecular Biology Reporter. 2020;**38**:666-675

[120] da Silva MP, Zaccaron AZ, Bluhm BH, Rupe JC, Wood L, Mozzoni LA, et al. Bulked segregant analysis using next-generation sequencing for identification of genetic loci for charcoal rot resistance in soybean. Physiological Molecular Plant Pathology. 2020;**109**:101440

[121] Watanabe S, Tsukamoto C, Oshita T, Yamada T, Anai T, Kaga A. Identification of quantitative trait loci for flowering time by a combination of restriction site–associated DNA sequencing and bulked segregant analysis in soybean. Breeding Science. 2017:17013

[122] Cheng Y, Ma Q, Ren H, Xia Q, Song E, Tan Z, et al. Fine mapping of a Phytophthora-resistance gene RpsWY in soybean (*Glycine max* L.) by high-throughput genome-wide sequencing. Theoretical Applied Genetics. 2017;**130**:1041-1051

[123] Jiang B, Li M, Cheng Y, Cai Z, Ma Q, Jiang Z, et al. Genetic mapping of powdery mildew resistance genes in soybean by high-throughput genome-wide sequencing. Theoretical Applied Genetics. 2019;**132**:1833-1845

[124] Gao C. The future of CRISPR technologies in agriculture. Nature Reviews Molecular Cell Biology. 2018;**19**:275-276

[125] Gao C. Genome engineering for crop improvement and future agriculture. Cell. 2021;**184**:1621-1635

[126] Bao A, Zhang C, Huang Y, Chen H, Zhou X, Cao D. Genome editing technology and application in soybean improvement. Oil Crop Science. 2020;**5**:31-40

[127] Li Z, Liu Z-B, Xing A, Moon BP, Koellhoffer JP, Huang L, et al. Cas9-guide RNA directed genome editing in soybean. Plant Physiology. 2015;**169**:960-970

[128] Du H, Zeng X, Zhao M, Cui X, Wang Q, Yang H, et al. Efficient targeted mutagenesis in soybean by TALENs and CRISPR/Cas9. Journal of Biotechnology. 2016;**217**:90-97

[129] Bai M, Yuan J, Kuang H, Gong P, Li S, Zhang Z, et al. Generation of a multiplex mutagenesis population via pooled CRISPR-Cas9 in soyabean. Plant Biotechnology Journal. 2020;**18**:721-731

[130] Mishra R, Joshi RK, Zhao K. Base editing in crops: Current advances, limitations and future implications. Plant Biotechnology Journal. 2020;**18**:20-31

[131] Komor AC, Kim YB, Packer MS, Zuris JA, Liu DR. Programmable editing of a target base in genomic DNA without double-stranded DNA cleavage. Nature. 2016;**533**:420-424

[132] Cai Y, Chen L, Zhang Y, Yuan S, Su Q, Sun S, et al. Target base editing in soybean using a modified CRISPR/Cas9 system. Plant Biotechnology Journal. 2020;**18**:1996

[133] Cai Y, Chen L, Liu X, Guo C, Sun S, Wu C, et al. CRISPR/Cas9-mediated targeted mutagenesis of *GmFT2a* delays flowering time in soybean. Plant Biotechnology Journal. 2018;**16**:176-185

[134] Li C, Nguyen V, Liu J, Fu W, Chen C, Yu K, et al. Mutagenesis of seed storage protein genes in soybean using CRISPR/Cas9. BMC Research Notes. 2019;**12**:1-7

[135] Cai Y, Chen L, Sun S, Wu C, Yao W, Jiang B, et al. CRISPR/Cas9-mediated deletion of large genomic fragments in soybean. International Journal of Molecular Sciences. 2018;**19**:3835

[136] Cai Y, Wang L, Chen L, Wu T, Liu L, Sun S, et al. Mutagenesis of *GmFT2a* and *GmFT5a* mediated by CRISPR/Cas9 contributes for expanding the regional adaptability of soybean. Plant Biotechnology Journal. 2020;**18**:298-309

[137] Wang L, Sun S, Wu T, Liu L, Sun X, Cai Y, et al. Natural variation and CRISPR/Cas9-mediated mutation in *GmPRR37* affect photoperiodic flowering and contribute to regional adaptation of soybean. Plant Biotechnology Journal. 2020;**18**:1869-1881

[138] Fang X, Sun X, Yang X, Li Q, Lin C, Xu J, et al. MS1 is essential for male fertility by regulating the microsporocyte cell plate expansion in soybean. Science China Life Sciences. 2021;**64**:1533-1545

[139] Jiang B, Chen L, Yang C, Wu T, Yuan S, Wu C, et al. The cloning and CRISPR/Cas9-mediated mutagenesis of a male sterility gene MS1 of soybean. Plant Biotechnology Journal. 2021;**19**:1098

[140] Nadeem M, Chen A, Hong H, Li D, Li J, Zhao D, et al. GmMs1 encodes a kinesin-like protein essential for male fertility in soybean (*Glycine max L.*). Journal of Integrative Plant Biology. 2021;**63**:1054-1064

[141] Li S, Cong Y, Liu Y, Wang T, Shuai Q, Chen N, et al. Optimization of Agrobacterium-mediated transformation in soybean. Frontiers in Plant Science. 2017;**8**:246

[142] Homrich MS, Wiebke-Strohm B, Weber RLM, Bodanese-Zanettini MH. Soybean genetic transformation: A valuable tool for the functional study of genes and the production of agronomically improved plants. Genetics Molecular Biology. 2012;**35**:998-1010

[143] Curtin SJ, Michno JM, Campbell BW, Gil-Humanes J, Mathioni SM, Hammond R, et al. MicroRNA maturation and microRNA target gene expression regulation are severely disrupted in soybean dicer-like1 double mutants. G3: Genes, Genomes, Genetics. 2016;**6**:423-433

[144] Curtin SJ, Zhang F, Sander JD, Haun WJ, Starker C, Baltes NJ, et al. Targeted mutagenesis of duplicated genes in soybean with zinc-finger nucleases. Plant Physiology. 2011;**156**:466-473

[145] Shalem O, Sanjana NE, Hartenian E, Shi X, Scott DA, Mikkelsen TS, et al. Genome-scale CRISPR-Cas9 knockout screening in human cells. Science. 2014;**343**:84-87

[146] Kelliher T, Starr D, Su X, Tang G, Chen Z, Carter J, et al. One-step genome editing of elite crop germplasm during haploid induction. Nature Biotechnology. 2019;**37**:287-292

[147] Rahman SU, McCoy E, Raza G, Ali Z, Mansoor S, Amin I. Improvement of soybean; a way forward transition from genetic engineering to new plant breeding technologies. Molecular Biotechnology. 2022:1-19

[148] Cai Y, Chen L, Liu X, Sun S, Wu C, Jiang B, et al. CRISPR/Cas9-mediated genome editing in soybean hairy roots. PLoS One. 2015;**10**:e0136064

[149] Jacobs TB, LaFayette PR, Schmitz RJ, Parrott WA. Targeted genome modifications in soybean with CRISPR/Cas9. BMC Biotechnology. 2015;**15**:1-10

[150] Di YH, Sun XJ, Hu Z, Jiang QY, Song GH, Zhang B, et al. Enhancing the CRISPR/Cas9 system based on multiple GmU6 promoters in soybean. Biochemical Biophysical Research Communications. 2019;**519**:819-823

[151] Tang F, Yang S, Liu J, Zhu H. Rj4, a gene controlling nodulation specificity in soybeans, encodes a thaumatin-like protein but not the one previously reported. Plant Physiology. 2016;**170**:26-32

[152] Bonawitz ND, Ainley WM, Itaya A, Chennareddy SR, Cicak T, Effinger K, et al. Zinc finger nuclease-mediated targeting of multiple transgenes to an endogenous soybean genomic locus via non-homologous end joining. Plant Biotechnology Journal. 2019;**17**:750-761

[153] Do PT, Nguyen CX, Bui HT, Tran LT, Stacey G, Gillman JD, et al. Demonstration of highly efficient dual gRNA CRISPR/Cas9 editing of the homeologous *GmFAD2-1A* and *GmFAD2-1B* genes to yield a high oleic, low linoleic and α-linolenic acid phenotype in soybean. BMC Plant Biology. 2019;**19**:1-14

[154] Khan MHU, Khan SU, Muhammad A, Hu L, Yang Y, Fan C. Induced mutation and epigenetics modification in plants for crop improvement by targeting CRISPR/Cas9 technology. Journal of Cellular Physiology. 2018;**233**:4578-4594

[155] Kanazashi Y, Hirose A, Takahashi I, Mikami M, Endo M, Hirose S, et al. Simultaneous site-directed mutagenesis of duplicated loci in soybean using a single guide RNA. Plant Cell Reports. 2018;**37**:553-563

[156] Chen X, Yang S, Zhang Y, Zhu X, Yang X, Zhang C, et al. Generation of male sterile soybean lines with the CRISPR/Cas9 system. The Crop Journal. 2021;**9**:1270-1277

[157] Bao A, Chen H, Chen L, Chen S, Hao Q, Guo W, et al. CRISPR/Cas9-mediated targeted mutagenesis of *GmSPL9* genes alters plant architecture in soybean. BMC Plant Biology. 2019;**19**:1-12

[158] Al Amin N, Ahmad N, Wu N, Pu X, Ma T, Du Y, et al. CRISPR-Cas9 mediated targeted disruption of FAD2-2 microsomal omega-6 desaturase in soybean (*Glycine max.* L). BMC Biotechnology. 2019;**19**:1-10

[159] Wang J, Kuang H, Zhang Z, Yang Y, Yan L, Zhang M, et al. Generation of seed lipoxygenase-free soybean using CRISPR-Cas9. The Crop Journal. 2020;**8**:432-439

[160] Zhang P, Du H, Wang J, Pu Y, Yang C, Yan R, et al. Multiplex CRISPR/Cas9-mediated metabolic engineering increases soybean isoflavone content and resistance to soybean mosaic virus. Plant Biotechnology Journal. 2020;**18**:1384-1395

[161] Curtin SJ, Xiong Y, Michno JM, Campbell BW, Stec AO, Čermák T, et al. CRISPR/Cas9 and talen s generate heritable mutations for genes involved in small rna processing of glycine max and medicago truncatula. Plant Biotechnology Journal. 2018;**16**:1125-1137

[162] Jacobs TB, Zhang N, Patel D, Martin GB. Generation of a collection of mutant tomato lines using pooled CRISPR libraries. Plant Physiology. 2017;**174**:2023-2037

[163] Lu Y, Ye X, Guo R, Huang J, Wang W, Tang J, et al. Genome-wide targeted mutagenesis in rice using the CRISPR/Cas9 system. Molecular Plant. 2017;**10**:1242-1245

[164] Meng X, Yu H, Zhang Y, Zhuang F, Song X, Gao S, et al. Construction of a genome-wide mutant library in rice using CRISPR/Cas9. Molecular Plant. 2017;**10**:1238-1241

[165] Kim H, Kim S-T, Ryu J, Kang BC, Kim J-S, Kim S-G. CRISPR/Cpf1-mediated DNA-free plant genome

editing. Nature Communications. 2017;**8**:1-7

[166] Gosal SS, Wani SH, Kang MS. Biotechnology and drought tolerance. Journal of Crop Improvement. 2009;**23**:19-54

[167] Détain A, Bhowmik P, Leborgne-Castel N, Ochatt S. Latest biotechnology tools and targets for improving abiotic stress tolerance in protein legumes. Environmental. Experimental Botany. 2022:104824

[168] Croser J, Lülsdorf M, Davies P, Clarke H, Bayliss K, Mallikarjuna N, et al. Toward doubled haploid production in the Fabaceae: Progress, constraints, and opportunities. Critical Reviews in Plant Sciences. 2006;**25**:139-157

[169] Hale B, Ferrie AM, Chellamma S, Samuel JP, Phillips GC. Androgenesis-based doubled Haploidy: Past, present, and future perspectives. Frontiers in Plant Science. 2022:2944

[170] Valliyodan B, Ye H, Song L, Murphy M, Shannon JG, Nguyen HT. Genetic diversity and genomic strategies for improving drought and waterlogging tolerance in soybeans. Journal of Experimental Botany. 2017;**68**:1835-1849

[171] Ratnaparkhe MB, Satpute GK, Kumawat G, Chandra S, Kamble VG, Kavishwar R, et al. Genomic designing for abiotic stress tolerant soybean. In: Genomic Designing for Abiotic Stress Resistant Oilseed Crops. Springer; 2022. pp. 1-73

[172] Patil G, Do T, Vuong TD, Valliyodan B, Lee J-D, Chaudhary J, et al. Genomic-assisted haplotype analysis and the development of high-throughput SNP markers for salinity tolerance in soybean. Scientific Reports. 2016;**6**:1-13

[173] Moreira FF, Oliveira HR, Volenec JJ, Rainey KM, Brito LF. Integrating high-throughput phenotyping and statistical genomic methods to genetically improve longitudinal traits in crops. Frontiers in Plant Science. 2020;**11**:681

Chapter 11

Accelerated Generation of Elite Inbreds in Maize Using Doubled Haploid Technology

Suman Dutta, Vignesh Muthusamy, Rajkumar U. Zunjare and Firoz Hossain

Abstract

The creation of homozygous parental lines for hybrid development is one of the key components of commercial maize breeding programs. It usually takes up to 6 to 7 generations of selfing to obtain homozygous inbreds from the initial cross using the conventional pedigree method. Using doubled haploid (DH) method, concurrent fixation of all the genes covering entire chromosomes is possible within a single generation. For generation of DH lines, haploids are generated first by several means such as *in-vitro* method using tissue culture technique and *in-vivo* method using the haploid inducer (HI) lines. Of which, tissue culture-based methods have shown little promise for large-scale DH production as it needs good infrastructures and technical requirements. In contrast, inducer-based method provides more optimistic solutions for large-scale DH lines production. Due to its rapidity, DH technology is now being adopted in many countries including India for reducing the breeding cycle.

Keywords: doubled haploid, homozygous line, maize, and haploid inducer, inbreds

1. Introduction

Maize breeding strategies rely heavily on the creation of homozygous parental lines for hybrid breeding. Using the traditional pedigree approach, it might take up to 7 generations of selfing to achieve homozygous inbreds from the first cross (**Figure 1**). In this context, because of its economic and logistical practicality, the creation of doubled haploid (DH) has received a lot of attention for varietal development in the last two decades [1]. The DH approach allows for simultaneous fixation of all genes across complete chromosomes in a single generation [2]. Haploids are initially created by a variety of methods, including an *in-vitro* way utilizing tissue culture techniques and an *in-vivo* method employing haploid inducer (HI) lines. On the other hand, Tissue culture-based approaches have not shown much promise for large-scale DH production since they necessitate adequate infrastructure and technical requirements. In contrast, because of its viability for large-scale DH line production, the inducer-based technique is deemed more hopeful and cost-effective [3]. According to the parent from which the haploids are being formed, inducer-based haploids may be divided into maternal haploid and

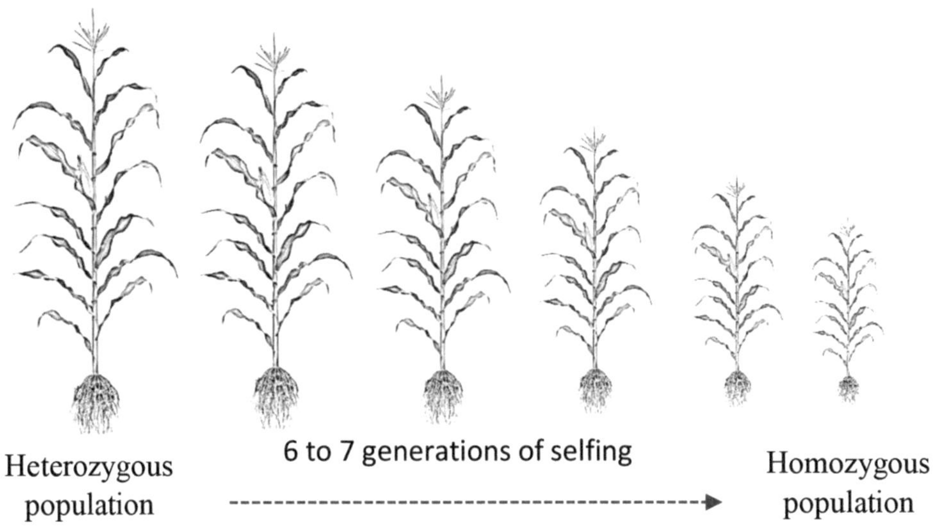

Figure 1.
Conventional breeding method for obtaining inbred lines.

paternal haploid [1]. With the exception of their cytoplasm, the genetic constitutions of both forms of haploids are identical. Once haploids are created, their genetic constitutions can be duplicated in a single step to produce homozygous DH lines. DH technique has recently been implemented in commercial maize breeding programs in various countries because of its speed in decreasing the breeding cycle.

2. Genetic factors involved in maize haploid induction

The inheritance of haploid induction rate (HIR) has been extensively investigated during the last two decades. *In-vivo* haploid induction in maize is accomplished using three approaches (**Figure 2**). The first method uses a mutation originating from 'Stock 6' to induce maternal haploid induction, and it is commonly employed in commercial maize breeding programs [3]. On chromosome 1, a pollen-specific phospholipase-A gene known as *Matrilineal* (*MTL*) was discovered with a 4-bp insertion in the final exon of the gene, causing a flame shift mutation and premature stop codon [4–6]. When coupled with the *MTL* gene in homozygous recessive form (*mtl/mtl*), a second mutant gene called *ZmDMP* (encoding DUF679 domain membrane protein) on chromosome 9 increases the haploid induction rate [7]. As a result, mutations in the *MTL* and *ZmDMP* genes are required for maize maternal haploid induction. The second strategy, on the other hand, entails a mutation in the *indeterminate gametophyte1* (*ig1*) gene, which was identified in Wisconsin-23 (W23) inbred for paternal haploid induction [8]. It was previously recognized that ig1 on chromosome 3 encodes a lateral organ boundary (LOB)-domain protein, which is part of a wide family of transcription factors important for plant lateral organ development [9]. Because the underlying determinants for both procedures (maternal and paternal) were different, the mechanisms of haploid induction in both ways differed greatly in the commercial breeding programs. Aside from these two key changes, induction of haploids in maize is influenced by several variables, including

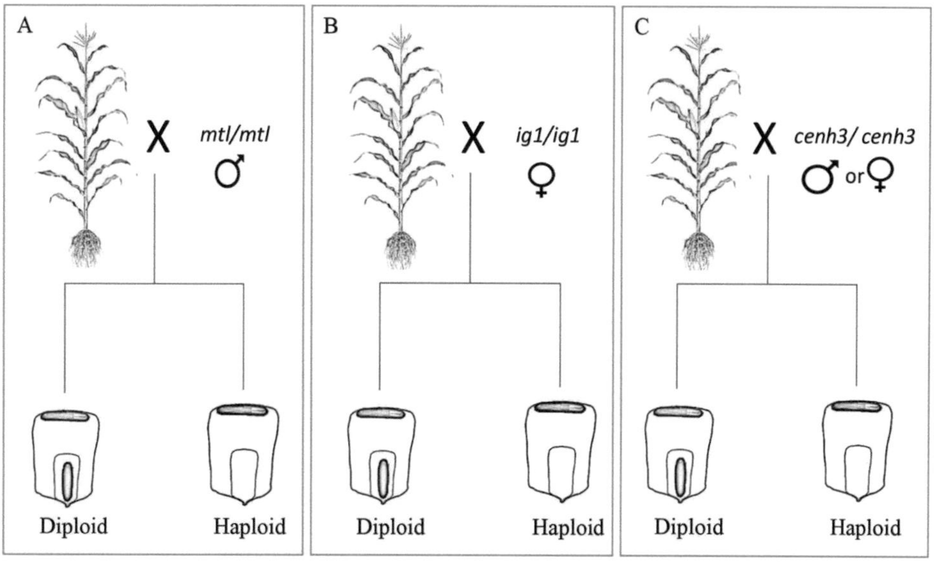

Figure 2.
Production of haploids using in-vivo induction system. A: Maternal haploid induction using mtl *mutant; B: Paternal haploid induction using* ig1 *mutant; C: Both maternal and paternal haploid induction using* cenh3 *mutant.*

the maternal genetic background of the donor germplasms and the environment in which the induction crosses are made [1, 10, 11]. Another approach for generating both maternal and paternal haploids from a single inducer is *CENH3*-based haploidization [12]. *CENH3* gene codes a histone H3 protein variant found in centromeric nucleosomes that have two primary domains. In CENH3 protein, the N terminal tail domain has little in common with regular histone H3, but the C terminal histone fold region has a lot in common with normal histones [13]. At the homozygous stage, a mutation in *CENH3* is fatal because chromosomes fail to segregate to poles during cell division due to the lack of functional centromeres [14].

3. Haploids identification after induction crosses

During an induction cross, haploids appear at a frequency of ~10% depending on the HIR, while the remaining 90% of seeds are diploid with no utility [3]. As a result, distinguishing haploids from diploid offspring at the seed, seedling, or mature plant stage is critical. Reducing the number of progenies would be helpful since it would lower the cost of developing DH lines. Different morphological and molecular indicators can be included in the inducer genotypes utilized in the DH development process [15, 16]. The dominant genetic marker produced in the seed or seedling stage can be included in mother haploid inducers, allowing haploids formed from induction crosses to be differentiated [1]. In most maize breeding programs across the world, haploid inducer with *R1-nj* (*Navajo*) allele is commonly utilized for haploid identification [1, 15]. The purple color of the scutellum of the embryo and the aleurone layer of the endosperm in diploid seeds influence the *Navajo* phenotype. Anthocyanin, on the other hand, is exclusively found in the endosperm of haploid seeds, not in the embryo. *R1-nj* marker makes it easier to distinguish haploids from diploids based on visual inspection due to the different colors of the embryo [17, 18]. The existence of a dominant inhibitor allele *C1-I* in tropical elite inbred lines inhibits anthocyanin production on seeds, which is a fundamental restriction of the *R1-nj* marker-based method [16]. When identifying haploids exclusively based on the *R1-nj* marker is challenging, the *Pl1* gene is employed as an alternative [19]. Pl1 gene induces light-independent anthocyanin synthesis in seedling roots, allowing haploids and diploids to be identified that were previously misclassified by the *R1-nj* marker. The *Pl1* gene, on the other hand, can frequently lead to misinterpretation due to the formation of red roots in seedlings after exposure to sunlight [19]. Furthermore, in the adult stage, recessive phenotypic mutations such as *liguleless* can be utilized to identify haploids [1].

Because of the numerous drawbacks of phenotypic morphological markers, multiple attempts have been made to use genetic markers based on the xenia effect of high oil content for haploid identification [20]. The use of a haploid inducer with a high oil content would be advantageous since the high oil marker is not genotype-dependent, allowing it to be applied to practically all genotypes, including landraces and wild cousins like teosinte [1]. As a result, the genes that cause high oil content may be targeted in order to create inducers with high oil content. The effectiveness of the oil-based identification technique, on the other hand, is dependent on a large difference in oil content between source germplasm and inducer, since a little difference would result in a higher number of false positives and false negatives [21, 22]. Automating the process of haploid identification would be a cost-effective and practical solution since it would considerably cut the cost of wages for those participating in the haploid identification process [23]. Several mechanical approaches have been altered based on *R1-nj* marker expression on embryo and endosperm employing multispectral, hyperspectral, and fluorescence imaging techniques (**Figure 3**). In this case, an imagining-based automated approach powered by machine learning and deep learning understanding might

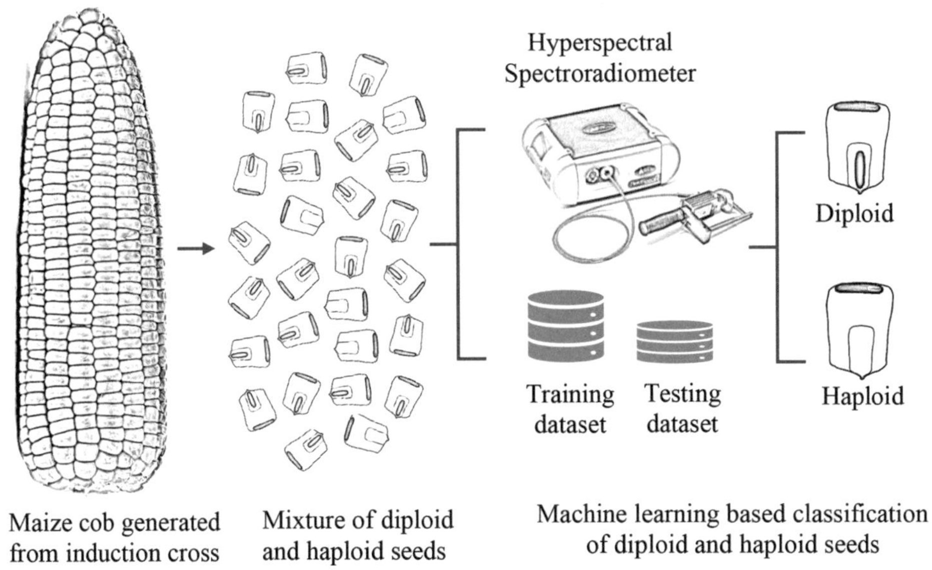

Figure 3.
Development of automation system based on hyperspectral spectroradiometer using a machine learning algorithm.

be a feasible option since it decreases the time and effort required to identify haploids [23, 24]. As a result, numerous approaches for haploid identification employing *in-vivo* HI are available. The main task is to create an HI with a mix of appropriate markers that can solve all of the problems associated with the identifying procedure. No marker method can give a universal strategy that is applicable to all germplasm [1]. As a result, developing an inducer with a proper marker system for the desired breeding program is critical before starting any DH programs. As a result, it is often recommended to create a set of HI with distinct marker systems appropriate for different types of source germplasms.

4. Chromosome doubling of identified haploids

The next objective is to create DH lines from the haploids after a successful induction cross using an appropriate inducer [3]. Haploids are normally infertile since they only have one copy of each chromosome, thus they must be chromosomally duplicated. Chemicals that prevent haploid seedlings from mitotically duplicating are used to achieve artificial chromosomal duplication (**Figure 4**). Colchicine is the chemical of choice in DH pipelines for artificial chromosomal doubling [1, 3]. Initially, haploid seeds are recognized using any of the markers and then germinated on paper towels until the coleoptiles reach a length of 2 cm. Before submersion in colchicine, the coleoptile tip is cut off, and the seedlings are rinsed out under tap water. The seedlings were then placed in trays filled with peat moss and kept at room temperature until they reached the three-leaf stage. Viable seedlings were then transplanted to a DH nursery field with suitable row-to-row and plant-to-plant spacing [3].

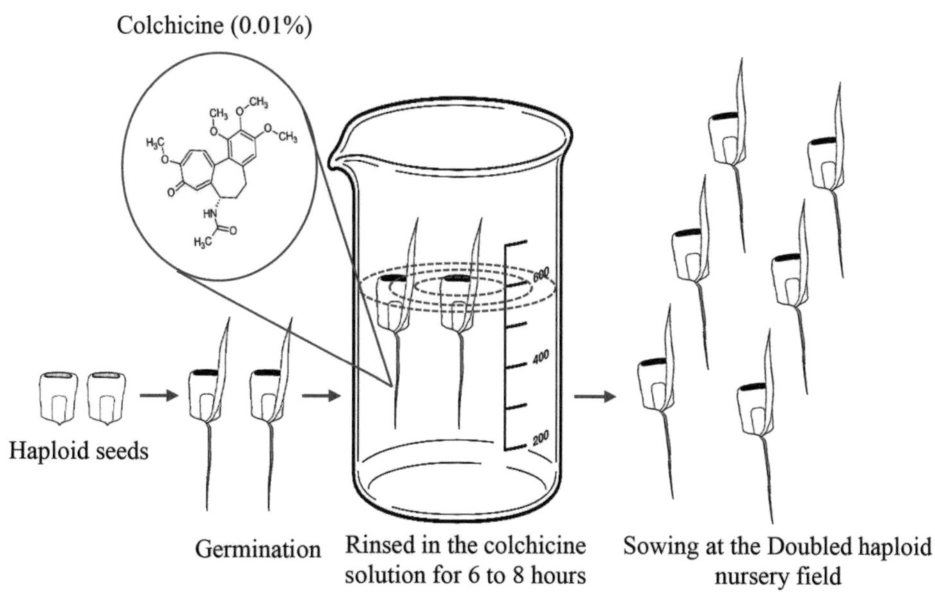

Figure 4.
Pipeline for doubled haploid production following colchicine treatment.

5. Application of doubled haploid technology

Visual selection based on classic pedigree breeding methods within segregating populations for numerous generations is the most common strategy for inbred development in maize [25]. Recurrent selection is another method for improving the breeding population mean by recombining superior progeny after selection [26]. However, using these approaches takes longer to obtain the appropriate amount of homozygosity. In comparison to other approaches, the DH method may achieve homozygosity in a single generation (**Figure 5**). As a result, DH production would be a feasible alternative to conventional methods for rapidly generating homozygous lines [3]. Because they are 100 percent homozygous, DH lines meet all of the DUS (Distinctness, Uniformity, and Stability) requirements for varietal development [1]. The DH population may also be used to gain knowledge on the genetic architecture of complex characteristics through breeding. Because the DH population is made up entirely of additive genetic variation due to homozygosity at all loci, the selection response is substantially higher than in other segregating populations [27, 28]. Additionally, DH breeding can be used with a marker-assisted backcrossing program to transfer the favorable allele of the concerned trait through either phenotypic or marker-assisted procedures, or a mix of both, by omitting the self-pollination stages at the end of the program [29]. Many commercial breeding programs have recently combined DH technology with genomic selection to improve genetic gain, especially for characteristics governed by a large number of QTL with low heritability [30]. Individual haploid plants are genotyped to find superior haploids, followed by self-pollination to establish homozygous lines using genomic selection [1]. Recently, after genetic alterations of three essential genes involved in meiotic recombination (*REC8*, *PAIR1*, and *OSD1* genes) and a single gene involved in haploid induction (*MTL*),

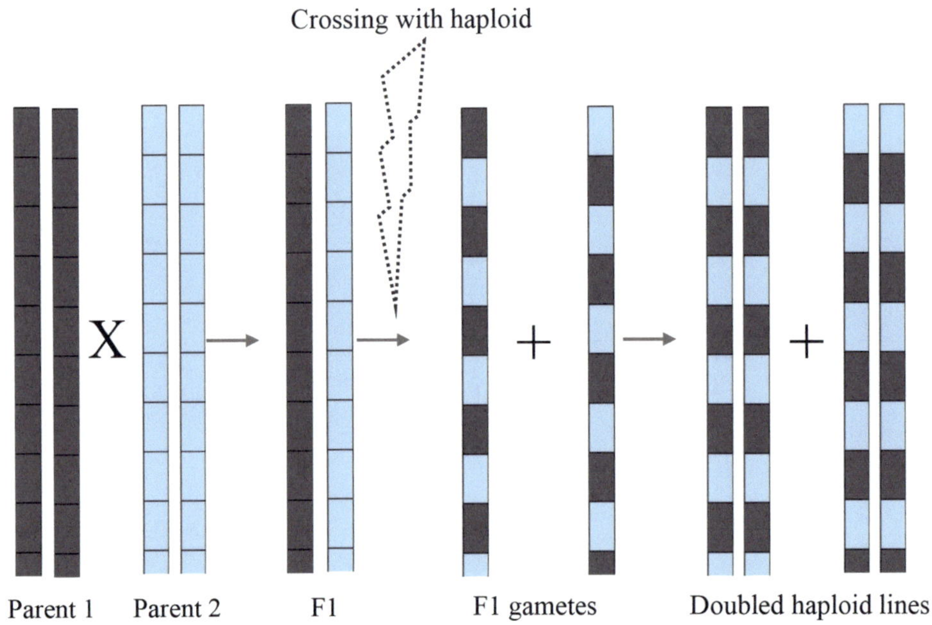

Figure 5.
Chromosome constitution of various stages during homozygous lines development in a single generation.

parthenogenesis was designed to fix heterosis [31–34]. Therefore, there are plenty of opportunities to combine these strategies in order to boost breeding yield.

6. Conclusion

In summary, DH technology has revolutionized commercial maize breeding programs by offering economic viability. This technology can also be integrated with the other major crops of economic importance to fast-track their breeding program. Integration of recent biotechnological approaches in the DH program further enhances the output of the breeding cycle per unit of time.

Acknowledgements

The participant is also thankful to the Human Resource Development Group (HRDG) division of the Council of Scientific & Industrial Research (CSIR), New Delhi, India for the Junior Research Fellowship (File No.: 09/083(0383)/2019-EMR-I) to pursue his Ph.D. program. The participant also shows his sincere gratitude to Vignesh Muthusamy, Rajkumar U Zunjare, and Firoz Hossain for correcting the manuscript.

Conflict of interest

The authors declare no conflict of interest.

Author details

Suman Dutta, Vignesh Muthusamy, Rajkumar U. Zunjare and Firoz Hossain*
Division of Genetics, ICAR-Indian Agricultural Research Institute, New Delhi, India

*Address all correspondence to: fh_gpb@yahoo.com

References

[1] Chaikam V, Molenaar W, Melchinger AE, Boddupalli PM. Doubled haploid technology for line development in maize: technical advances and prospects. Theoretical and Applied Genetics. 2019:**132**:1-17

[2] Kalinowska K, Chamas S, Unkel K, Demidov D, Lermontova I, Dresselhaus T, et al. State-of-the-art and novel developments of *in vivo* haploid technologies. Theoretical and Applied Genetics. 2019;**132**(3):593-605

[3] Prasanna BM, Chaikam V, Mahuku G. Doubled haploid technology in maize breeding: theory and practice. Mexico: CIMMYT Publication Repository. 2012. p. 50

[4] Kelliher T, Starr D, Richbourg L, Chintamanani S, Delzer B, Nuccio ML, et al. *MATRILINEAL*, a sperm-specific phospholipase, triggers maize haploid induction. Nature. 2017;**542**(7639):105-109

[5] Gilles LM, Khaled A, Laffaire JB, Chaignon S, Gendrot G, Laplaige J, et al. Loss of pollen-specific phospholipase *NOT LIKE DAD* triggers gynogenesis in maize. The EMBO Journal. 2017;**36**(6):707-717

[6] Liu C, Li X, Meng D, Zhong Y, Chen C, Dong X, et al. A 4-bp insertion at *ZmPLA1* encoding a putative phospholipase a generates haploid induction in maize. Molecular Plant. 2017;**10**(3):520-522

[7] Zhong Y, Liu C, Qi X, Jiao Y, Wang D, Wang Y, et al. Mutation of *ZmDMP* enhances haploid induction in maize. Nature Plants. 2019;**5**(6):575-580

[8] Kermicle JL. Androgenesis conditioned by a mutation in maize. Science. 1969;**166**(3911):1422-1424

[9] Evans MM. The indeterminate gametophyte1 gene of maize encodes a LOB domain protein required for embryo sac and leaf development. The Plant Cell. 2007;**19**(1):46-62

[10] Prigge V, Sanchez C, Dhillon BS, Schipprack W, Araus JL, Bänziger M, et al. Doubled haploids in tropical maize: I. effects of inducers and source germplasm on *in vivo* haploid induction rates. Crop Science. 2011;**51**(4):1498-1506

[11] De La Fuente GN, Frei UK, Trampe B, Nettleton D, Zhang W, Lubberstedt T. A diallel analysis of a maize donor population response to *in vivo* maternal haploid induction: I. Inducibility. Crop Science. 2018;**58**(5):1830-1837

[12] Ravi M, Chan SW. Haploid plants produced by centromere-mediated genome elimination. Nature. 2010;**464**(7288):615-618

[13] Watts A, Singh SK, Bhadouria J, Naresh V, Bishoyi AK, Geetha KA, et al. Brassica juncea lines with substituted chimeric GFP-CENH3 give haploid and aneuploid progenies on crossing with other lines. Frontiers in Plant Science. 2019;**2017**:7

[14] Ravi M, Marimuthu MPA, Tan EH, Maheshwari S, Henry IM, Marin-Rodriguez B, et al. A haploid genetics toolbox for *Arabidopsis thaliana*. Nature Communications. 2014;**5**(1): 1-8

[15] Melchinger AE, Schipprack W, Würschum T, Chen S, Technow F. Rapid and accurate identification of *in vivo* induced haploid seeds based on oil content in maize. Scientific Reports. 2013;**3**:2129

[16] Chaikam V, Nair SK, Babu R, Martinez L, Tejomurtula J, Boddupalli PM. Analysis of effectiveness of *R1-nj* anthocyanin marker for *in vivo* haploid identification in maize and molecular markers for predicting the inhibition of *R1-nj* expression. Theoretical and Applied Genetics. 2015;**128**(1):159-171

[17] Chase SS, Nanda DK. Screening for monoploids of maize by use of a purple embryo marker. Maize Genetics Cooperation Newsletter. 1965;**39**:59-60

[18] Greenblatt IM, Bock M. A commercially desirable procedure for detection of monoploids in maize. Journal of Heredity. 1967;**58**(1):9-13

[19] Hake S. In: Bennetzen JL, editor. Handbook of Maize. New York: Springer; 2009. pp. 693-713

[20] Chen SJ, Song TM. Identification haploid with high oil xenia effect in maize. Acta Agronomica Sinica. 2003;**29**(4):587-590

[21] Rotarenco VA, Kirtoca IH, Jacota AG. Using oil content to identify kernels with haploid embryos. Maize Genetics Cooperation Newsletter. 2007;**81**:11

[22] Melchinger AE, Schipprack W, Friedrich Utz H, Mirdita V. *In vivo* haploid induction in maize: Identification of haploid seeds by their oil content. Crop Science. 2014;**54**(4):1497-1504

[23] Wang XY, Liao WX, An D, Wei YG. Maize Haploid Identification via LSTM-CNN and Hyperspectral Imaging Technology. 24 May 2018 arXiv: 1805.09105 [cs.CV]

[24] Boote BW, Freppon DJ, De La Fuente GN, Lubberstedt T, Nikolau BJ, Smith EA. Haploid differentiation in maize kernels based on fluorescence imaging. Plant Breeding. 2016;**135**(4):439-445

[25] Chase SS. Production of homozygous diploids of maize from Monoploids 1. Agronomy Journal. 1952;**44**(5):263-267

[26] Hallauer AR, Carena MJ, Miranda Filho JD. Quantitative genetics in maize breeding New York: Springer Science & Business Media. 2010;**6**

[27] Choo TM. Doubled haploids for estimating mean and variance of recombination values. Genetics. 1981;**97**(1):165-172

[28] Snape JW, Simpson E. The genetical expectations of doubled haploid lines derived from different filial generations. Theoretical and Applied Genetics. 1981;**60**(2):123-128

[29] Lubberstedt T, Frei UK. Application of doubled haploids for target gene fixation in backcross programmes of maize. Plant breeding. 2012;**131**(3):449-452

[30] Andorf C, Beavis WD, Hufford M, Smith S, Suza WP, Wang K, et al. Technological advances in maize breeding: Past, present and future. Theoretical and Applied Genetics. 2019;**132**(3):817-849

[31] Khanday I, Skinner D, Yang B, Mercier R, Sundaresan V. A male-expressed rice embryogenic trigger redirected for asexual propagation through seeds. Nature. 2019;**565**(7737):91

[32] Marimuthu MP, Jolivet S, Ravi M, Pereira L, Davda JN, Cromer L, et al. Synthetic clonal reproduction through seeds. Science. 2011;**331**(6019):876-876

[33] Wang B, Zhu L, Zhao B, Zhao Y, Xie Y, Zheng Z, et al. Development of a haploid-inducer mediated genome

editing system for accelerating maize breeding. Molecular Plant. 2019;**12**(4):597-602

[34] Yao L, Zhang Y, Liu C, Liu Y, Wang Y, Liang D, et al. *OsMATL* mutation induces haploid seed formation in indica rice. Nature Plants. 2018;**4**(8):530

Chapter 12

Accelerating Breeding for Drought Tolerance in Sorghum (*Sorghum bicolor*): An Integrated Approach

John Charles Aru, Scovia Adikini, Sam Omaria, Francis Okiasi, William Esuma, Ronald Kakeeto, Moses Kasule Faiso, Michael Adrogu Ugen and Eric Manyaza

Abstract

Sorghum (Sorghum bicolor) is the main food crop for people living in marginal areas. They are faced with a number of production challenges including; drought, insect pests, diseases, soil fertility and striga weeds. To adapt to current and future stresses, there is a dire need to develop tolerant cultivars using multistress lines and varieties from wide genetic backgrounds. Toward better integrated approaches; we conducted participatory field screening in hot spot locations for drought, striga weed and major leaf spot fungal diseases on the 20 lines making mini-core sorghum germplasm. Lines carrying key traits of resistance to stresses have been recycled into the breeding program. The study also identified biochemical traits that could potentially be used as surrogate traits for the selection of tolerant genetic resources with improved yields. Nuclear male fertile crosses have been derived for exploiting differences in the cytoplasm for enhancing resistance. It also integrated variability in phytochemicals and cytoplasmic resistance to develop multi-parent sorghum lines and populations possessing potentially favorable adaptive alleles. In conclusion; unique traits and breeding strategies for sorghum adapted to the dry lowlands have been identified to lay a foundation for a modernized and market-oriented sorghum breeding program to the advantage.

Keywords: integrated, multi-parent, multi-stress, *Sorghum bicolor*, participatory evaluation

1. Introduction

Sorghum (*Sorghum bicolor*) is the main food crop for people living in marginal areas. They are faced with a number of production challenges including; drought, insect pests, diseases, soil fertility and weeds such as striga. Of recent, drought has been the major production challenge limiting sorghum production in Uganda by preventing improved cultivars from expressing their full genetic potential. Three mechanisms namely; drought escape, drought avoidance and drought tolerance are involved in drought resistance. Functional drought resistance categories are based on unique morphological, physiological and biochemical traits working together with genetic factors determined under growth chamber/controlled screening. Nevertheless, morphological and physiological characters show different types of inheritance pattern (i.e., monogenic and polygenic) and different gene actions (additive and non-additive). This implies that the heritability of drought resistance from different genotypes is not consistent so cannot be relied on. The breeding procedure commonly practiced for handling segregating generations affects the rate of genetic progress that can be made under stress that is, yield. Therefore drought resistance is best selected using secondary traits which depict the interaction between water stress, weather variables and the plant [1]. For example, the stay green color is a secondary trait resulting from biochemical adaptation to water stress leading to a change in chlorophyll content. There are genotypes expressing various degrees of stay green traits identified from screening experiments [2–5]. For that reason, a large amount of genetic variability has been reported among sorghum germplasm for their reaction to drought stress necessary for setting a breeding scheme. Therefore innovative breeding as an aspect of forward-looking approach is critical while exploiting wide genetic variation of relevant plant characteristics with farmers who support the plant breeding industry. The approach is inspired by [6, 7], who have described steps and methodologies involved in setting priorities in the breeding program/breeding scheme. They argue that the views of farmers need to be considered to come with a deeper understanding about how yield and quality can be increased within the local production system and its specific risk management strategies. Therefore, incorporated resistance to drought must improve positively agronomic characteristics as well as the quantity and quality of harvested products. To address the availability of suitable varieties in the long term, the sorghum improvement program in Uganda has considered the following integrated approaches, in line with the early stages of product profile development under the dry lowland agroecology. Research objective 1: Participatory exploration of sorghum breeding targets. Research objective 2: Identification of genotypes with broad adaptation. Research objective 3: Profile elite germplasm on the basis of phytochemical defense compounds to exploit major factors of the evolution of the crop. Research objective 4: To test pollen fertility restoration of selected parents in cytoplasmic male sterile (CMS) background, to exploit heterosis for productivity and resistance to stresses in the derived lines and populations.

1.1 Participatory exploration of drought tolerance breeding targets in sorghum

The approach was through engaging farmers using a system approach in product profile development, to ensure practical actions by beneficiaries exposed to drought stresses are contextualized. This will contribute to harnessing opportunities and generate context-based technologies and innovations for dry lowland agroecologies of Uganda. This chapter compares farmers' sorghum traits and breeding targets by breeders/scientists.

1.2 Materials and methods

1.2.1 Sorghum varieties and breeding process

The current six popular sorghum varieties in Uganda were developed between 2011 and 2017 and are phenotypically distinct. There exists a number of other elite cultivars from the breeding program following six year breeding program to release. That is; from germplasm evaluation through initial screening, making crosses, advancing populations (F_1–F_5) as single plants selections under the pedigree system. This is followed by stages of preliminary yield trials (PYT) and advanced yield trials (AYT) in designated places. It is then followed by adaptation trials, national performance tests and release. Yield improvements currently have stagnated and there is a need to replace the current varieties released five years ago but suitable to dry lowland agroecology. Therefore participatory evaluation of potential candidate cultivars for replacement was done in 2017 B (short rain season), to set better targets for innovative breeding and selection approaches. This was to enhance the germplasm base for key traits that stabilize production and could increase genetic gains in farmers' fields (**Table 1**). This summarizes the priority traits related to drought tolerance and drought escape given by the two groups of sorghum farmers, during the field day organized for the Teso and the Lango farmers of Uganda in the 2017 B season. These groups of farmers are from different sorghum production systems. The participants were encouraged to screen the on-station field experiment block and pick up the eye-catching ones for in-depth discussion. The scores and ranks for each group were summarized into four categories. The prioritization of drought tolerance and escape were mentioned along with the description of traits that will contribute to the stability of production (**Table 1**). Specific ranking based on value for use (**Table 2**) and ranking based on gender following focus group discussions in the field (**Table 3**).

Priority	Plant traits	Explanation/Description	Breeders related views
High priority	Robustness	**Both groups rated high1** Robustness is associated with (Vigor, Leafy, stout stem, plant health, seed yield, less lodging). The traits are indicative of high tolerance to drought or can escape drought	Biological-yield Selection based on selection-index Heterosis/genetic distance of parents Leaf area/stay green Node diameter Tillers
Medium-High priority	Large seeds /Yield (**Table 2**) Panicle-width panicle length	**One group rated high1 and other medium2** Large seeds are attractive and marketable	Harvest-index Threshing percentage Seed size(weight) Good agronomic practices
Medium-Low priority	Uniform Early maturity Non-senescent	**One group rated high1 and other medium2**	-Genotype by environment interaction -Production areas -Agroecology -Certified seed
Low-priority	Plant color/Seed color	**One group rated high1 and the other low4**	Anthocyanin/alkaloid Content Stem sugar Agromorphological characters Quality flour

Table 1.
Prioritization of identified selection traits by two participatory groups. Each group consisted of a random mix of six researchers and 30 farmers male and female. The farmers ranked groups of perceived drought tolerance traits from 1 to 4 and rated from 1 to 4 in order of priority.

Genotype	Yld	Good for food	Disease and pest resist	Local Mrkt	Attractive seed color	Early maturity	Bird damage	Average	Rank
SESO1	4	3	5	5	3	1	4	**3.57**	8th
IESV24029SH XICSB479-1	3	3	2	3	4	3	3	**3.00**	7th
NAROSORGH 2	1	1	1	1	1	2	1	**1.14**	**1st**
NAROSORGH3	1	3	3	4	3	2	3	**2.71**	6th
SESO3	3	1	4	1	1	1	2	**1.86**	4th
IESV24029SH	1	1	1	1	1	3	1	**1.33**	**3rd**
SEREDO × SRN 39H2-2-1	1	2	2	2	2	2	2	**1.86**	4th
IESV92207DL	1	5	1	5	3	4	5	**3.43**	10th
IESV92034DLSEL2	2	5	1	5	3	4	5	**3.57**	9th
KAK7780	1	1	1	1	1	2	2	**1.29**	**2nd**
Value	**18**	**25**	**21**	**28**	**22**	**24**	**28**		
Rank	**1st**	5th	**2nd**	6th	**3rd**	4th	6th		

Brief explanation of the results:

Low numbers are most preferred and therefore i.e., 1 is the best and 5 is the worst.

Highly valued traits are yield, disease and pest resistance and attractive seed color.

Best overall candidate varieties for release are; KAK77880 and IESV24029SH.

Yld = Yield, method described by [7], most farmer participants associated grain color with variety, hence able to describe agromorphological differences.

Table 2.
Matrix ranking of traits by farmers (based on value for use), by pairwise wise ranking method.

Genotype	Females N = 15	Males N = 15	Average	Rank
SESO1	8	10	9	9th
IESV24029SH XICSB479-1	6	7	6.5	6th
NAROSORGH 2	4.5	1	2.8	4th
NAROSORGH3	9	6	7.5	8th
SESO3	1	5	3	**3rd**
IESV24029SH	3	2	2.5	**1st**
SEREDO × SRN 39H2-2-1	4.5	4	4.3	5th
IESV92207DL	7	8.5	7.7	7th
IESV92034DLSEL2	10	8.5	9.7	10th
KAK7780	2	3	2.5	**1st**
Highly ranked candidate varieties; KAK-7780, IESV24029SH, SESO3.				

Table 3.
Ranking based on gender following focus group discussions in the field.

1.3 Results and discussion

High-valued traits considered for drought tolerance and escape by farmers are high plant vigor and robustness to nurture the seeds of high quality and hence high yield. These traits were identified from three genotypes; KAK-7780, IESV 24029SH and SESO3. Further description of the robustness trait takes care of plant health and associated responses to biotic and abiotic stresses (pests, diseases, drought and soil fertility). Additionally, it dictates the maturity and uniformity of the crop and hence was rated highly by participating farmers. Drought is a complicated trait to screen; robustness and vigor show good nutrition and good germination. This is how farmers quickly demonstrate yield potential. Furthermore, farmers emphasized morphological traits that breeders do not systematically select for. Breeders, on the other hand, rely on quantitative measurements but are aware of the heritability of specific traits and the influence of genotypes by the environment on plants growing under optimal conditions. Therefore, observed selection goals of farmers and breeders fit into the description of [8], who stated that science takes a reductionist approach toward breeding that is, to reduce the number of variables considered but study them in detail. Farmer's positive selection in favor of healthy, good germinating, vigoros, large panicles and large seeds demonstrate their ability to select stable genetic responses. These selection traits have actual advantages on yield under rain-fed conditions. Number of factors have been found to be associated with resistance in sorghum and include; seedling vigor, glossiness, morphological and biochemical characteristics. Any condition such as drought, low fertility and temperature makes the plant susceptible to attack by insect pests and diseases [9]. Besides involving farmers in target setting and selection, it is important to consider the selection environment. Selection is more efficient when the correlation between selection and the target environment is high and increases selection efficiency for direct and indirect selection for broader or specific adaptation [10]. This study recommends selection indices that will help in selecting seedlings at a very early stage without losing important information. The farmers and scientists preferred selections made from advanced lines to be planted in regional adaptation trials to analyze genotype by environment interaction to prove the need for more varieties to be released with certainty. This will help define clusters of target environments based on differences in environmental parameters, production systems, and farmer preferences [11]. From the participatory study; drought resistance is best measured in terms of robustness (vigor) and plant health traits as priority traits when designing product profiles (**Table 2**).

1.4 Sectional conclusion

Several relevant traits are often considered simultaneously in plant breeding, particularly when selection is done by farmers. The relationship among them determines breeding strategies and response to selection (study 2). There is a need to develop an index which when applied to sorghum seedlings during the first eight weeks of development in the field (at anthesis), will indicate the best progenies and best plants within the progeny. The possible vigor parameters associated with drought tolerance include; Plant height, node diameter, basal stem sugar, internode length, Leaf area, leaf dry weight and stay green from where correlations can be calculated from non-senescent genotypes. Some of these traits identified can reasonably be bred depending on their level of expression. Therefore, it is a need to conduct multi-location trials to help to monitor levels and expression of adaptive associated traits in plants that will be incorporated into the model while taking account of environmental variance in the expression of traits.

2. Determination of yield stability of rain-fed sorghum and GGE biplot analysis of multi-environment trials

2.1 Materials and methods

Drought, yield and biotic stress traits are quantitatively inherited and highly environmentally interactive and require multi-environment testing to accurately characterize them. Therefore integrating the yield and stability of genotypes tested in very unpredictable environments is an important breeding strategy to identify superior sorghum genotypes for the rain-fed areas of Uganda. It helps to determine the existence of different mega-environments for maximizing genetic gains. This was achieved by testing 20 elite genotypes identified in study 1, in four locations via GGE (genotype + genotype-by-environment) biplot analysis. Data from two seasons (2019b–2020b) of rain-fed sorghum were used in this study. The materials consist of elite germplasm; landraces, advanced breeding lines and popular cultivars within crop improvement programs and introductions which contribute to variation and differentiation. **Table 4** is a representative sample of all the diversity in the large collection and would facilitate the enhanced use of sorghum germplasm in the breeding programs for major economic and production traits. They were selected together with farmers as having the best attributes for food and niche markets. The information on genotype-environment interaction was a useful supplement for classifying genotypes. The nature of genotype-environment interaction helps in the development suitable procedures for selection and the nature of stability for vital characters. Meanwhile, multivariate analysis was used to offer a more complete examination of data by looking at all possible independent variables and their relationships to one another. Finally, levels of similarity were computed as percentages among genotypes for key variables useful for planning crosses.

Entry/ NPT	Genotype	Origin	Pedigree	Traits
1	NAROSORGH 3	Uganda	IS8193 × SEREDO	Midge resistance
2	NAROSORGH 4	U.S.A-Perdue	GE17/1/2003A	Striga + Early
3	SILA	Kenya	SILA	Malting
4	NTJ2	Ethiopia	NTJ2	Malting + Grain size
5	IESV24029SH	Ethiopia	GADAM × IS8193	Food grain
6	IESV92172DL	Ethiopia	IESV92172DL	Drought + Dwarf
7	ASERECA 13-3-1	Sudan	GADAM	Striga, Drought
8	ASERECA 15-2-1	Sudan	GADAM	Striga
9	IESV24029SH × ICSB479-1	Uganda	IESV24029SH × ICSB479-1	Stemborer resistance + Male sterility
10	KAK-7780	Kenya	Landrace	Grain quality (food), drought
11	IESV142001	Ethiopia	IESV142001	Grain yield
12	ICSV142012	India	ICSV142012	Grain Yield
13	SEREDO X SESO1	Uganda	SEREDO X SESO1	Striga + grain Yield
14	IESV92207DL	Ethiopia	IESV92207DL	Drought
15	IESV92034DLSEL2	Ethiopia	IESV92034DLSEL2	Drought
16	ICSL71052	India	ICSL71052	Yield
17	EPURIPURI	Uganda	Tegemeo	Grain Yield
18	SESO2	Uganda	SRN39	Striga + grain yield
19	IESV23007DL	Ethiopia	IESV23007DL	Drought + grain yield
20	SEREDO	Uganda	SEREDO	Drought + striga

Table 4.
Genetic materials.

2.2 GGE analysis

In GGE biplot analysis, the first two principal components (PC1 and PC2), derived by subjecting the environment-centered yield bi-plot (**Figure 1**), to singular value decomposition (SVD). Principal component (PC1) was significant and location accounted for 17% of the total sums of squares. Genotype by environment variation was greater in the two seasons confirming that food productivity is threatened by environmental variables. The small yield variation due to location is relevant to cultivar evaluation. The yield obtained from across environments selected the following genotypes as they combine yield and stability and this should be considered during genotype selection. Genotypes with the above mean performance were; IESV 92207DL, IESV92024SH × ICSB 497-1, on the basis of the

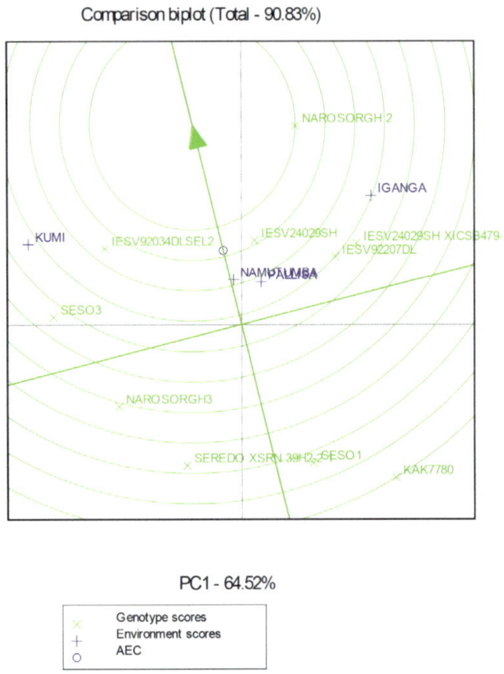

Figure 1.
Environment-focused comparison bi-plot shows the performance and stability of genotypes.

ATC (Average Tester coordinate X-axis) or (Average Tester coordinate Y –axis, the stability axis). The research identified two mega environments for sorghum under rain-fed areas. This has several implications for future breeding and genotype evaluations of sorghum that is, warm-dry ecology (Kumi) and sub-humid environments; (Namutumba, Pallisa and Iganga). The closer an environment is to this virtual environment (ATC axis); the better it is as a test environment [11]. Thus Pallisa and Namutumba are relatively favorable test environments and most representative and as well discriminative, suitable for multiple stress evaluation with a yield above 2000 kg/ha. Kumi was a most discriminative environment, probably due to the high level of striga from low fertile sandy soils which enhance water stress, hence strong genotype by environment interaction. The large part of the genotype × environment interaction was also indicated by a positive correlation between different yield components (**Figure 2**).

2.3 Cluster dendrograms based on grain yield

A dendogram with clusters was created from 12 elite lines based on their level of similarity in grain yield/ha (**Figure 3**). This is important for planning a crossing program; for example, using IESV24029SH × ICSB 497 and KAK-7780 could have a good combination of favorable alleles. They were also among the best four highly ranked genotypes from Additive Main effects and Multiplicative Interaction (AMMI) analysis (**Table 5**). Genotype KAK-7780 was selected from a landrace population that could possess both major and minor gene systems for stress protection which contributes to yield.

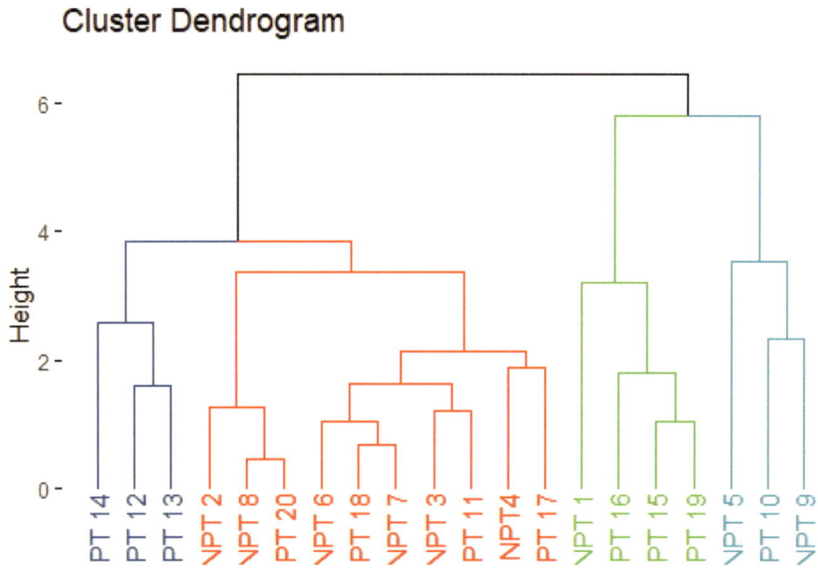

Figure 2.
Dendogram showing how sorghum founder lines can be clustered together in groups on the basis of important biochemical composition.

```
** Levels   100.0 90.0 80.0 70.0 60.0 50.0 (based on similarity)

IESV24029SH              1 ..........
IESV24029SH XICSB479-1   2. )
IESV92207DL              4.) ........).
NAROSORGH 2              6 ..............) ...........
KAK7780                  5 ................. )
SESO1                    9 .................) ........).
IESV92034DLSEL2          3 ..................... )
NAROSORGH3               7 ...........)      )
SEREDO XSRN 39H2-2-1     8 ...........).)        )
SESO3                    10 ..............) .....).......)...........
```

Figure 3.
Dendrogram with clusters based on level of similarity (%) in grain yield (Kgm/ha).

Number	Environment	Mean	Score	1	2	3	4
1	IGANGA	2517	24.63	NPT 21	NPT 9	NPT 14	NPT 5
4	PALLISA	2125	5.71	NPT 21	NPT 23	NPT 14	NPT 22
3	NAMUTUMBA	2664	1.26	NPT 15	NPT 5	NPT 14	NPT 9
2	KUMI	2612	−31.61	NPT 23	NPT15	NPT 1	NPT 5

NPT 1 = NAROSORGH3, NPT 5 = IESV24029SH, NPT9 = IESV24029SH × ICSB479-1, NPT22 = SESO1. NPT14 = IESV92207DL. NPT15 = IESV92034DLSEL2, NPT21 = NAROSORGH 2, NPT23 = SESO3.

Table 5.
AMMI ranking of the best four selections.

2.4 Cluster dendogram based on plant height

Using IESV92034DLSEL2, KAK-7780, NAROSORGH 2 and IESV92207DL will be very useful for improved yield components. These differences could have resulted from differences in biomass production and seed weights and hence they are the best ranking candidates for rain-fed areas (**Figure 4**). Since the genotype cannot restrict their heights significantly across environments, it implies resistance operates against major sorghum production challenges (drought, disease and pests) contributing to stability. More research needs to be conducted to unravel the underlying principles for plant-stress interaction with respect to plant height so that they can be incorporated into breeding.

2.5 Correlation between traits

Genetic variability for yield and seedling vigor components exits and they include; seed size, days to flowering (maturity), plant height, panicle length, panicle width, pest and disease resistance. They are a result of active physiological processes driven by active translocation and they influence dry matter accumulation. Plant height is strongly positively correlated to panicle length (PL) and moderately to days to flowering (DF) and hundred seed weight (HSWT). On the other hand, stem borer (SB) resistance is moderately positively correlated to stay green (SC), plant height (PHT) and days to flowering (**Figure 2**), where the correlation is low (R = 0.2) or negative, or both, little progress can be made. For example, the relationship between hundred seed weight and stay green could be due to competition for sink source relationship. So grain yield may not be true determinant for drought tolerance [12].

This could be because hundred seed weight is calculated after seed cleaning. For breeding purposes, it is therefore important to compare hundred seed weight (seed size) among genotypes with respect to the standard check/commercial varieties. The data should be interpreted based on physiological time of maturity, growth patterns, dry matter accumulation, partitioning of sink and genetic differences. Although many traits have been studied for their use in breeding for drought resistance, there is a general consensus among breeders that only a few of them can be recommended for use in practical breeding programs at this time. The study identified maturity (50% flowering), plant height, stay green [13].

```
      **     Levels              100.0        90.0         80.0         70.0         60.0         50.0

ASERECA15-2-1                                                                          1          ........
ICSV142002                                                                           3       .....    )
NAROSORGH1                                                                       11        .....)     )
IESV24029SH                                                             6          ..           )     )
NTJ2                                                                              13         ..)..)   )
IS2262                                                                                9        .....)..)..
SEREDO        XSRN39H2-2-1                                                 14             ...........)........
ICSL71052                                                                          2         ........  )
ICSV92172DL                                                            4          ........)...........)................
IES24029-SH         XICSB        479-1                                    5              .....      )
IESV92038/2SH                                  7     .....)..                                       )
IESV92207DL                                      8     ........)                                    )
NAROSORGH2                                          12     ........).....                           )
KAK     7780                                                       10       ..............).......................)..
```

Figure 4.
Level of similarity (%) based on plant height.

2.6 Sectional conclusion

It is possible to select stable genotypes combining the number of yield components since genetic merit for these traits exist. For example, tall plants of a height of above 170 cm (centimeters) can be used to select improved grain yield components. Considering the strong weak negative correlation between turscicum leaf blight (TLB) infection with plant height and days to flowering, implies delayed flowering is associated with decreased diseased levels and is common among tall plants. The key traits are plant height, stay grain and days to 50% flowering which are most likely to improve the rate of genetic gains for drought tolerance. There is a need to exploit information gained from correlated traits and pedigree for the selection of parents with multiple traits [14, 15].

3. Determination of grain quality profile and phytochemical content of elite lines

3.1 Introduction

Adaptation to drought is a result of biochemical adaptation to water stress leading to a change in chlorophyll content, production of antioxidant scavenging enzymes, increase in proline content, production of secondary metabolites such as; alkaloids, terpenes, flavonoids, mevalonic acid, shikimic acid among others [4, 5]. Drought like other environmental changes can bring about marked differences in the defense chemistry of the plant. These qualitative traits (secondary metabolites) also control the aroma, taste and acceptability of products and can be integrated into the breeding pipeline at the priority setting and trait discovery stage. Variations in phytochemicals could be used to broaden the genetic base in the three current gene pools; food, feed, fodder through introgression to generate desirable genetic complexes (linkage groups). From breeder point of view, these phytochemicals can be grouped as valuable or useful and sometimes negative hampering the application of germplasm in a breeding program.

3.2 Materials and methods

The study evaluated the hypothesis that alkaloid content reduce in advanced generations as result of selection. They are however associated with serious side effects on products at high levels. This was investigated among the 20 breeding materials in study 2. The genotypes included; (7 cultivars, 4 progeny lines, 8 Varieties and 1 Landrace populations) making minicore-germplasm. There are from different gene pools and breeding history. Biochemical analysis was carried out at the National Crops Resources Research Institute (NACRRI) Bio-Nutrition Laboratory in 2019. Principles and methods of biometrical designs were applied according to the protocol developed by [16]. Absorbance was read at wavelength 470 nm. The variation with respect to chemical composition in genotypes was attributed to genetic differences. Canonical correlation analysis was used between sets of independent variables for data interpretation and cluster analysis for grouping genotypes.

3.3 Variability for biochemical contents in grain sorghum

The multivariate analysis utilized all the variations of traits in generic way to group genotypes with similar sets of traits and quantify the importance of various traits in grouping/clustering genotypes. Results revealed statistically significant ($P < 0.001$) differences among the sorghum accessions profiled for biochemical contents, demonstrating the influence of genotype with respect to checks (**Table 6**). The subset of sorghum minicore germplasm has been categorized into four clusters. This could be the influence of physiological and biochemical processes in modifying the plant in response to abiotic and biotic stresses [4, 17]. Cluster 1 in red (**Figure 5**), with the largest number of genotypes displayed more scope for selection against water stress as supported by positive and significant correlations (**Figure 6**). The coefficient of genotypic variance was above 80% indicative of substantial genetic diversity and prospects for improvement through chemical selection (**Table 6**). Low levels of phytochemicals of less than 4 mg/100 g were consistent as all released varieties were clustered together with checks supporting the hypothesis (**Figure 5**). Polyphenols were positively correlated among themselves but negatively associated to the levels of carbohydrates. The level of tannins was important in establishing groups and contributed a lot to the total variability among the accessions. Landraces and their derived lines and hybrids clustered together, hence exploitation of this material require a lot of chemical selection due to probably strong linkages with the wild (i.e., NPT 10, NPT 5 and NPT 9). Improvement of genetic gain for these plant chemical defense compounds might be possible through hybridization.

Entry	Flavs		Carbh		Color		Pheno		Amylose		Tannins	
NPT4	0.172	A	1.337	Cd	1.182	h	0.039	E	0.5817	b	0.2747	F
NPT 8	0.183	A	1.944	I	0.152	a	0.00500	A	0.737	h	0.225	E
NPT 6	0.207	B	1.773	H	1.432	j	0.00633	A	0.6037	c	0.0273	A
NPT 20	0.221	C	1.973	I	0.289	b	0.01367	Bc	0.7297	h	0.0677	Bc
NPT 18	0.237	D	1.735	H	1.008	f	0.024	D	0.6693	e	0.086	C
NPT 2	0.242	D	1.893	I	0.745	e	0.01533	C	0.733	h	0.025	A
NPT 12	0.265	E	1.928	I	1.915	m	0.077	G	0.8775	j	1.1453	I
NPT 7	0.29	F	1.776	H	1.105	g	0.00333	A	0.692	f	0.4253	G
NPT 14	0.326	G	1.763	H	1.334	i	0.16183	I	0.631	d	1.2475	J
NPT 13	0.332	G	1.693	Gh	1.5	jk	0.01067	B	0.9717	k	0.9553	H
NPT 16	0.35	H	1.261	Bc	1.355	i	0.15217	H	1.0715	m	1.1215	I
NPT 15	0.356	Hi	1.079	A	1.646	l	0.0695	F	0.9782	k	1.8665	Lm
NPT 3	0.358	Hi	1.488	Ef	1.191	h	0.014	Bc	0.5567	a	0.122	D
NPT 17	0.361	Hi	1.514	F	0.594	d	0.00667	A	0.7363	h	0.02	A
NPT 19	0.365	I	1.269	Bc	1.551	k	0.07133	F	0.8882	j	1.3975	K
NPT 11	0.395	J	1.611	G	1.648	l	0.00533	A	0.7103	g	0.0387	Ab
NPT 1	0.436	K	1.226	B	1.983	m	0.2915	J	0.8608	i	1.8372	L
NPT 10	0.562	L	1.496	Ef	0.506	c	0.0145	Bc	1.0525	l	0.9725	H
NPT 5	0.724	M	1.974	I	1.306	i	0.02267	D	1.5667	n	1.9015	M
NPT 9	0.884	N	1.402	De	0.749	e	0.01217	Bc	1.0787	m	1.9598	N

Checks NPT 3 = SILA, NPT 17 = EPURIPURI a, b, c, d, e, f, g, h, I, j, k, l, m, n = Mean separation.

Table 6.
Variability for biochemical contents in grain sorghum in Serere in 2017 based on observed values of absorption spectrophotometer.

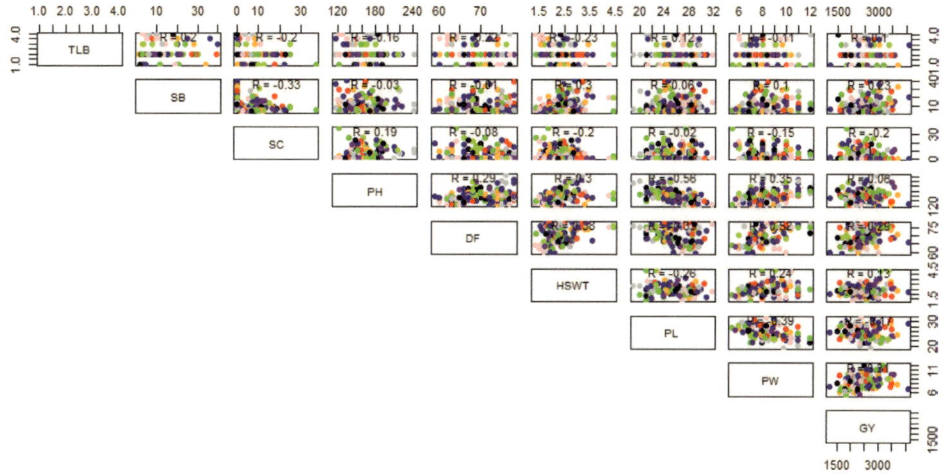

Figure 5.
Correlation among key trait.

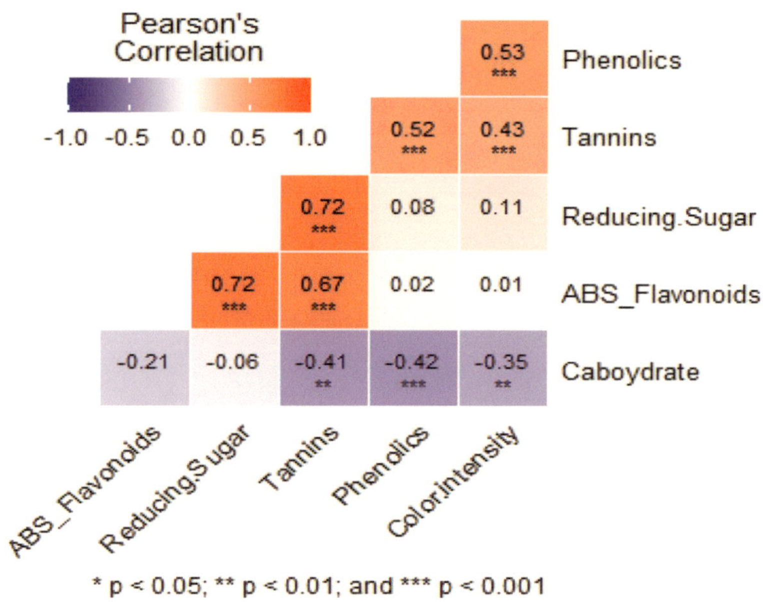

Figure 6.
Correlation among variables in grain sorghum in Serere in 2017 with their corresponding coefficient values and probabilities potential.

3.4 Correlated response to selection and indirect selection

A high level of carbohydrates was negatively correlated to the levels of polyphenols. Therefore, selection for large seed size of ≥3 gms/100 seed could reduce the concentration of polyphenols relative to the increase in water and carbohydrate (starch) content in the seed. In nature, the association among desirable traits can be negative as for the case of maize, (e.g., increasing grain yield is associated with lower protein content [18]. The variation for biochemical traits was represented by the two-dimensional scatter diagram that accounted for 70% of the variance. Genotypes; NPT15 (IESV92034DLSEL2) and NPT 19 (IESV23007DL), were plotted in the upper right quadrat. Meanwhile genotypes; NPT 5 (IESV24029SH), NPT 9 (IESV24029SH × ICSB 479) and NPT10 (KAK-7780) are intermediate occupying the lower right quadrant (**Figure 7**). The differences could reflect breeding, selection history and complex interrelationships between ecological factors important for parental selections for multiple traits [14, 15]. Diversity among genotypes has been categorized into groups of similar characteristics that can be used for designing optimized crossing strategies. Released varieties clustered together support the argument that selection and hybridization among themselves have taken place (**Figure 5**). The key traits that are most likely to improve the rate of genetic gains for grain quality traits are levels of tannins and carbohydrates since they are negatively correlated.

3.5 Sectional conclusion and recommendations

Chemical factors such as high carbohydrate content, less intensive color and tannin content are good attributes for better quality products. This study showed the value of exploiting the information in correlated traits that will contribute toward improving the accuracy of breeding values of the products such as bread and malt quality. Accurate selection of sorghum breeding lines can accelerate annual genetic gain for these correlated traits when used to generate an optimized crossing design

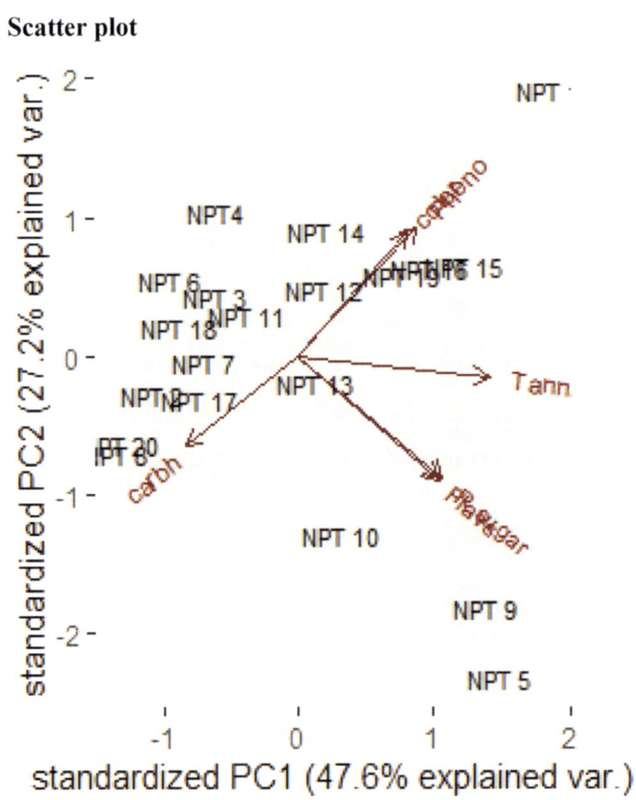

Figure 7.
Scatter plot showing the best selections on the upper right hand.

(study 4); where there is a high and positive correlation between secondary traits such as color and target traits such as tannins and phenolic, then greater selection intensities can be applied to the secondary trait during screening in big populations. Positive results will be expected when the information is integrated with the pedigrees during breeding. Furthermore innovative research is important into processes that mitigate ant nutritional factors while enhancing bio-availability of proteins, amylose starch among the high tannin genotypes in the utilization of such materials in feed, food and beer value chains.

4. Pollen fertility restoration in diverse cytoplasmic sterility lines for improved populations

Superimposed on the major forces of evolution is variation due to the interaction of the mitochondria (cytoplasm) and nuclear genes in mediating resistance to some major stresses and physiological processes in the plant that needs to be exploited to enhance genetic variation. This material is useful in developing hybrids, inbred lines and gene pools based on combining ability for specific traits (components of grain yield). The study identified testers and restorers for extraction of lines from improved populations for yield-related traits such as; seed size, plant height panicle length, panicle width and resistance to stem borers.

4.1 Population development

The experiment was designed to make improvements in quantitative traits to maximize as much as possible the additive effects, maternal, as well as to gather genes with complementary dominant and epistatic effects in genotype. The hypothesis tested was that Hybrids derived from inbred lines (A-lines) with complementary heterotic groups have superior performance. (**Table 7**). A two-way cross hybrids were generated from seven pollinator testers of (*Sorghum bicolor*) with drought tolerant backgrounds mated to A_2 cytoplasmic sterile lines using an appropriate model [17]. The lines used were selected from previous studies (1–3) perceived to be containing favorable alleles for the prioritized traits of; robustness (vigor), large seeds, disease resistance and high threshing percentage. Data was analyzed using NCII model appropriate to line × tester crossing design with two reps in the 2022 A season. The variance between the testers was subdivided into variance within cytosteriles and that due to interaction.

4.2 Combining ability

Testers varied significantly for all characters (**Tables 7** and **8**). The differential behavior of the genotypes was reflected in general combining ability (GCA) effects. Variance for GCA is greater than the variance for specific combining ability (SCA) implying large additive gene effects and over-dominance effects. There is a possibility of deriving superior lines from such populations containing balanced cumulative effects of genes. Lines with highly significant positive GCA effect such as DINKIMASH-17 for grain yield could be useful for contributing favorable alleles for breeding for improved grain yield under drought conditions. Female inbred lines; ICSA 12, CK60A, P 9518A and ICSA 90001 and the male lines GE30, KAK-7780, NAROSORGH4, IESV98038/2SH and DINKIMASH-17, were selected for their desirable GCA effects for agronomic traits.

SOURCE	DF	PHT	PANL	PWD	100 GNW	GRN YLD
Hybrids	21	251.5**	102.5**	18.8**	18.6**	1029.9**
Lines	5	7408***	296***	38.**	39.8**	2708**
Testers	8	4820***	166***	104.2**	62.9**	1589.9*
Between genotypes with A-lines	6	8641***	382***	147.6**	142.1***	654.1 ns
A-Line × Testers	30	636.3***	30.9***	2.9***	6.6 ns	408.3 ns
Hybrids × Replication	42	129**	30.9***	0.7***	5.5 ns	423.5.5**
Error	360	178.6	7.4	0.45	1.88	152.1
Var GCA		78.31	0.95	0.33	0.43	8.21
var SCA		46.35	0.4	0.07	0.01	0.51
VarGCA:Var SCA		**1:0.59**	**1:0.42**	**1:0.21**	**1: 0.65**	**1:0.06**

*SCA = specific combining ability, GCA = General combining ability, Var = variance, ***=significant at 0.01 probability, **p = 0.1, *p = 0.05, ns = not significant, DF = degrees of freedom, PHT = Plant height, PANL = Panicle Length, GNW = Grain weight, GRNYLD = Grain yield per hectare.*

Table 7.
Analysis of variance for key yield components.

Parents	PHT	PANL	PWD	100 GNW	GRN YLD
ICSA 12 × DINKIMASH-17	7.92**	0.8**	0.5 ns	0.5**	2.94**
ICSA 1630139 × DINKIMASH-17	5.44**	0.31**	0.35*	0.14**	0.3.07**
ICSA 1630139 × IESV92038/2SH	7.25	1.24	0.19*	0.61*	1.05
P9518A × NAROSORGH4	−6.75	1.29.	0.63*	0.89*	1.87
ICSA9 × GE30/1/1-2-2	11.24*	1.29*	0.1*	0.55*	1.87*
ICSA9001 × IESV92021DL-1-9-2-1	26.55*	1.04	0.17	7.42	0.44*
ISCA 6 × IESV 92038/2SH -2-1	0.87*	2.24	1.06**	0.47**	0.5*
ICSA8 × GE30/1/2013A-1-1-1-3	3.36	0.53	0.47	0.52	0.77
P9518A × NAROSORGH4	11.38*	3.24*	0.91	34.64*	2.68
ICSA × KAK-7780-1-5-1-3	2.34*	2;46**	0.37	6.72**	3.57*

*Probability, **p = 0.1, *p = 0.05, ns = not significant, PHT = Plant height, PANL = Panicle Length, GNW = Grain weight, GRNYLD = Grain yield per hectare.*

Table 8.
Estimates of general combining ability for top 10 ranking genotypes.

Test cross F_1-derived F_2 families have been developed by exploiting differences in the cytoplasm (**Table 8**). The identified lines have a positive effect on fertility restoration. The effect of genes carried was likely large enough to influence the full seed set of the panicles among many families. The Partial sterility observed among some $F1_S$ does not present a serious problem because the F_1 may be either selfed/backcrossed to recurrent parent Populations. This permits the best exploration of the intricate assortment of both major genes and genes with small effects for the traits under consideration [19]. The breeding products from this study are important in enhancing sorghum germplasm base by contributing favorable alleles for expressing vigor of yield and cytoplasmic pest and disease-mediated resistance [20].

5. Perspectives: Toward breeding to for drought resistance

Drought resistance is a complex trait controlled by many linked genes. Therefore, the probability of selecting drought-tolerant lines increases measurably as the percentage of adapted genotypes in gene combination increases. This progressively will increase the rate of genetic gains. The account below gives a summary of components (insights) of a breeding strategy for drought resistance for a practical breeding program.

5.1 Identification of important secondary traits from multi-locational trials

Analysis of genotype-by environmental interaction (GEI) was carried out in selected elite breeding material using nine characters including grain yield. The study has highlighted important secondary traits well expressed under field screening conditions because of good interaction of genotype × environment with other weather variables. Such traits; stay green, plant height, vigor (size), day to flowering, disease and pest response. The secondary traits can be used to increase progress made with primary traits (functional resistance) that is determined under growth chamber conditions. Therefore toward better integrated approaches, field screening of elite sorghum lines was carried out in hot spot locations for key sorghum production constraints; that is, drought, striga, pests and leaf spot diseases. This is because incorporated resistance to drought must have a positive effect on agronomic attributes as well as the quantity and quality of harvested products (study 1). Therefore the research captured a wide genetic variation of relevant plant characters to include in designing a breeding scheme. The GEI was approached through variance components, regression and multivariate methods (study 2 and study 3). Through these analyses, genotypes and environments were grouped, and stable genotypes were identified and ranked. The GEI for yield was attributed to various traits and genotypic correlations determined between plant height, panicle length, panicle width, days to 50% flowering, hundred grain weight, grain yield/ha, pest and disease response. The correlations ranged from negative, low, moderate, and positive and this guided selection. These analyses helped to design optimal crossing design (study 4).

5.2 Development of multi-parent population using cytoplasmic male sterile system (CMS)

The selected good parents from study 2 and study 3 were evaluated to determine their effectiveness in a breeding scheme through line × tester (North Carolina II) mating design. General combining ability for agronomic characteristics and full fertility restoration ability was tested under A_2 form of CMS which is easily transmissible to the progeny. The use of such derived progenies helped to exploit diverse nuclear backgrounds to enhance stability. The use of CMS-derived crosses/progenies has improved; grain sets, hundred seed weight, germination capacity and vigor which are components of drought resistance. This will permit selection for drought resistance but guarantee essential gene recombination necessary for the stability and adaptability of breeding lines.

Acknowledgements

The study was supported by the National Agricultural Research Organization (NARO), International Crop Research Institute for Semi-Arid Tropics I (CRISAT) under the Accelerated Variety Improvement and Seed Delivery of legumes and cereals in Africa (AVISA) project, National Semiarid Resources Research Institute (NaSARRI), And the National Crops Resources Research Institute (NACRRI)-BIO NUTRITION LAB.

Conflict of interest

None.

Author details

John Charles Aru[1*], Scovia Adikini[1], Sam Omaria[1], Francis Okiasi[1], William Esuma[1], Ronald Kakeeto[1], Moses Kasule Faiso[1], Michael Adrogu Ugen[1] and Eric Manyaza[2]

1 National Semi-Arid Resources Research Institute, Soroti, Uganda

2 International Crop Research Institute for Semi-Arid Tropics (ESA Regional Hub), Nairobi, Kenya

*Address all correspondence to: arucharles40@yahoo.com

References

[1] Edmeades GO, Bolaños J, Chapman SC. Value of secondary traits in selecting for drought tolerance in tropical maize. In: Edmeades GO, Bänziger M, Mickelson HR, Peña-Valdivia CB, editors. Developing Drought and Low-N Tolerant Maize. Proceedings of a Symposium; 25-29 March 1996. El Batan, México: CIMMYT. 1997

[2] Duncan RR, Bockholt AJ, Miller FR. Descriptive comparison of senescent and n genotypes. Agronomy Journal. 1981;**73**:849-853

[3] Rosenow DT, Clark LE. Drought and lodging resistance for a quality sorghum crop. In: Proceedings of the 5thAnnual Corn and Sorghum Industry Research Conference. Scientific Research Publisher(SRP). 1995. pp. 82-97

[4] Tiwari A, Rastogi A, Singh V, Arunachalam A. Water stress effects on nutritional values and relative water content of barnyard and finger millet crops. International Journal on Agricultural Sciences. 2019;**10**:23-28

[5] Tiwari A, Rastogi A, Singh V, Arunachalam A. Effect of water stress on oxidative damage and antioxidant enzyme activity in finger millet and barnyard millet. Indian Journal of Hill Farming. 2020;**33**:36-45

[6] Schubotz D. Participatory Action Research. Queens University Belfastsa, SAGE publisher; 2019. DOI: 10.4135/9781526421036840298

[7] Weltzien E, Smith ME, Meitzner LS, Sperling L. Technical and institutional issues in participatory plant breeding-from the perspective of formal plant breeding. In: A Global Analysis of Issues, Results and Current Experience. PPB Monograph. Cali, Columbia: PRGA Program Coordination Office, CIAT; 2003

[8] Vernocy R, Withshrestha P, Song Y, Humphries S. Towards new roles, responsibilities and rules: The case of participatory plant breeding. In: Ceccarelli S, Guimaraes EP, Eel Zen E, editors. Plant Breedin and Farmer Participation. 2009. pp. 613-628

[9] Taneja SL, Leuschaner K. Resistance and screening mechanisms of resistance in sorghum. In: Proceedings of International Sorghum Entomology Workshop; Patancheru. USA: TEXAS A&M University College; 1981. pp. 115-127

[10] Atlin GN, Baker RJ, McRae KB, Lu X. Selection response in subdivided target regions. Crop Science. 2000;**40**:7-13. DOI: 10.2135/cropsci2000.4017

[11] Ceccarelli S. Positive interpretation of genotype by environment interactions in relation to sustainability and biodiversity. In: Cooper M, Hammer GL, editors. Plant Adaptation and Crop Improvement. Wallingford, UK; ICRISAT: Patancheru, India; and IRRI: Manila, The Philippines: CAB International; 1996. pp. 467-486

[12] Austin RB, Bingham J, Blackwell RD, Evans LT, Ford MA, Morgan CL, et al. Genetic improvement in winter wheat yield since 1900 and associated physiological changes. Journal of Agricultural Science (Cambridge). 1980;**94**:675-689

[13] Atlin G. Improving drought tolerance by selecting for yield. In: Fischer KS, Lafitte R, Fukai S, Atlin G, Hardy B, editors. Breeding Rice for Drought-Prone Environments. Los Baños, Philippines: IRRI. 2003. pp. 14-22. Available from: http://www.knowledgebank.irri.org/drought/ [Accessed: January 10, 2023]

[14] Bauer AM, Léon J. Multiple-trait breeding values for parental selection in self-pollinating crops. Theoretical and Applied Genetics. 2008;**116**:235-242

[15] Piepho HP, Möhring J, Melchinger AE, Büchse A. BLUP for phenotypic selection in plant breeding and variety testing. Euphytica. 2007;**161**:209-228

[16] AOAC. Official Methods of Analysis. Washington, DC: Association of Official Agricultural Chemists; 1970

[17] Assefa T, Zeleke H, Afriye T, Otyama P. Line × tester analysis of tropical high land maize (Zea mays L.) inbred lines top crossed with three east African maize populations. American. Journal of Plant Sciences. 2017;**8**(2):126-136. DOI: 10.4236/ajps.2017.82010

[18] Duvick DN, Cassman KG. Post green revolution trends in yield potential of temperate maize in the north central United States. Nov 1999;**39**(6):1622-1630

[19] Weyhrich R, Lamkey KR. Responses to seven methods of recurrent selection in the BS11 maize population. Crop Science. 1998;**38**(2):308-321. DOI: 10.2135/cropsci1998.0011183X003800020005x

[20] Sharma HC, Abraham CV, Vidyasagar P, Stenhouse JW. Gene action for resistance in sorghum to midge, Contarinia sorghicola. Crop Science. 1996;**36**:259-265

Chapter 13

New Perspectives in Grapevine (*Vitis* spp.) Breeding

Arif Atak

Abstract

Many grape varieties or genotypes of *Vitis* species are grown for different purposes in various parts of the world. However, despite a large number of cultivars, there is a demand for different grape cultivars due to changing consumer expectations. Grapevine breeding programs are carried out by scientists in different countries in order to meet these expectations. Breeding studies, which used to take a long time with traditional crossbreeding methods, have become studies that achieve the desired results in a much shorter time with the development of molecular methods and biotechnology. One of the most important developments in grapevine breeding is that the relevant gene regions in hybrid populations developed from breeding programs can be identified in a very short time. In recent years, the demand for cultivars that are more resistant or tolerant to biotic and abiotic stress conditions has increased, and for this purpose, there has been a significant increase in breeding studies on cultivars and rootstocks that are resistant or tolerant to different stress conditions. Considering the current breeding programs, genetically manipulated new cultivars with desired characteristics and interspecies hybrid cultivars will soon become the main study subjects of grapevine breeding programs.

Keywords: *Vitis* spp., breeding, molecular methods, tissue culture, biotic and abiotic stress

1. Introduction

Grapevine is one of the most widely grown crops in the world and covers about seven million hectares. According to the data of 2020, approximately 78 million tons of grapes are obtained from this area. The countries with the highest production, respectively, are China, Italy, Spain, France, and USA. Half of the world's total grape production is made in these five countries [1]. Grapevine cultivation is in high commercial demand on a global scale due to its high yield and different consumption patterns. In 2020, grapes were the world's 289th most traded product, with a total trade of $11B. Between 2019 and 2020 the exports of grapes grew by 2.27%, from $10.8B to $11B. Trade in grapes represent 0.066% of the total world trade. In 2020, the top exporters of grapes were Chile ($1.18B), China ($1.09B), United States ($1.02B), Peru ($1.01B), and Italy ($831M). In 2020, the top importers of grapes were United States ($1.36B), Germany ($943M), China ($817M), Netherlands ($816M), and United Kingdom ($812M). The countries with the highest import tariffs for grapes are

Turkmenistan (100%), India (73.9%), Iran (55%), and Turkey (54.6%) [2]. Grapes are the world's third most valuable horticultural crop (after potatoes and tomatoes). Cultivation of grapes for fruit and wine began at least 7000 years ago in the Near East, and over the millennia, thousands of cultivars have been developed and selected for different purposes. Nowadays, grapes are used to produce diverse consumer products including wine, table grapes, raisins, grape juice concentrate, and distillate for various industrial uses as well as making fortified wine and brandy. While wild Vitis species are very valuable to breeders, new cultivars developed from particular different breeding programs are important for grape growers to sell their quality products at high prices. With the rapid change in consumer preferences, different government policies, increased awareness of human/environmental health, global warming, and some other factors, it has become more important for researchers to better examine and understand the grapevine genome and, as a result, to develop new varieties that will meet all these expectations with the help of modern methods. Grapevine is not only an economically valuable species but also a highly preferred model for both cultivation and breeding studies due to its genetic characteristics [3]. In addition, for many countries, the culture of viticulture is a cultural heritage that has great meaning. For all these reasons, the grapevine plant (*Vitis* spp.) is among the most important plant species in which research and investment are made [4].

Grapevines (*Vitis* spp.) are members of the Vitaceae and include two subgenera, *Euvitis* (38 chromosomes) and *Muscadinia* (40 chromosomes), with about 60 species in total. The Vitaceae family is the most important agricultural species in the genus *Vitis*. Especially the varieties belonging to the *V. vinifera* species are the most widely cultivated species all over the world and dominate the markets. However, this species has some disadvantages of its own. Especially, the cultivars of this species are highly susceptible to biotic and abiotic stress factors. As a result of this susceptibility, there are significant quality losses [5].

In each country, grape production can be done for different purposes. Grapes are grown for wine, table, raisins, juice, jam, concentrate, seed oils, and other purposes. According to these different growing purposes, market demands, and expectations are changing rapidly and, in this case, the demand for new grape cultivars increases [6]. Controlled grape breeding is thought to start almost 200 years ago. Henri and Louis Bouschet de Bernard are believed to have begun generating hybrids between "Teinturier du cher" and "Aramon" cultivars in 1824 in southern France [7]. The birth of modern grape breeding is connected with the arrival of North American diseases (downy mildew, powdery mildew, and black rot) and insects (mainly phylloxera) to Europe. These diseases and pests caused substantial losses on the highly susceptible *V. vinifera* vines in European vineyards.

Several major progress in viticulture and grapevine breeding occurred as a result of the epidemics spreading through Europe in the late nineteenth and early twentieth centuries. Especially, the advent of rootstock breeding as an effective and immediate means to control phylloxera. Wild vines (*V. riparia* and *V. rupestris*) from North America were first imported to be used as rootstocks and provided *phylloxera* resistance [8].

Although, the cultivars obtained as a result of the hybridization of *V. labrusca, V. aestivalis* and *V. vinifera* species with each other and with *V. vinifera*, they started to be used for different purposes both in America and Europe in the following years. But the hybrids obtained from crosses made especially the hybrids obtained from crosses with wild species did not receive much demand in wine production due to their intense aromas [9].

First grapevine breeding studies started with wine grapes, and later on, at the end of the nineteenth century, also table grapes were included in these breeding studies. With the phylloxera pest affecting vines in the European continent, studies on rootstock breeding started at the same time (late nineteenth century). In the twentieth century, different institutions and organizations in the USA (such as the University, the private sector and the USDA) started breeding studies on table grapes and many new grape cultivars with superior characteristics were developed as a result of these studies. These cultivars, which were developed as a result of breeding programs in the USA and grown by grape growers, led to the development of the table grape industry all over the world. Afterwards, grape breeding programs started rapidly in different countries (Far East, South America, Europe, Turkey, Israel, Argentina, and many other countries) [10]. The main purpose of these breeding studies is to obtain new cultivars that will meet the expectations, taking into account the changing consumer demands, as soon as possible by using the opportunities provided by technology. Especially in recent years, demand for large berry, seedless and high-yielding cultivars that are resistant to different biotic and abiotic stress conditions for table grapes has increased and more emphasis has been given to these issues in breeding studies. Among the wine grapes, cultivars that are resistant to different biotic and abiotic stress conditions and have the desired wine quality criteria have also become prominent selection criteria in breeding studies. Most of the commonly grown *Vitis* species and cultivars are not resistant to biological agents (virus, bacteria, fungi, nematode, and others) that significantly affect yield and quality. The main causes of diseases that cause the most important losses in vineyards are viruses, fungi, and microorganisms such as fungi, oomycetes, and so on. A remarkable resistant cultivar has not yet been detected in economically important cultivars of *V. vinifera* [11]. Compared to other perennial plants, the number of viruses infecting grapevines is quite high [12]. Abiotic stress factors that are effective in grapevine include especially water availability, temperature, and light. In recent years, breeding studies have focused on these issues [13]. Most grapevine breeding programs were initially publicly funded (such as Universities and Research Institutes), but nowadays different private sector companies are also involved in these breeding programs. Since many of the new cultivars are protected by intellectual property rights, growers or organizations that want to grow them must first agree with the right holders. It is also seen that growers who want to grow these new cultivars have come together and created a new model in recent years. Because of this new "breeder club system", many countries and companies have started their own special breeding programs. Many grape varieties developed as a result of breeding studies are protected by strict breeder rights in many countries. Grape producers or companies that want to grow these varieties must first negotiate with the people or organizations that have the breeder's rights. In addition, in recent years, a system has started to spread for the cultivation of newly bred cultivars which have breeder's right in a limited area and with a club system to guarantee certain quality conditions. In addition to this, today companies, grower unions, or cooperatives sometimes cooperate with some government organizations for these breeding programs and try to jointly develop new grape varieties by financing their breeding programs. In recent years, especially in table grape breeding, this system is increasingly in demand and is spreading all over the world [14, 15]. It is among the important problems of breeding studies to reach correct results by analyzing many data obtained in the field, especially in grape breeding programs. After the integration of smart agriculture models into breeding studies, complex data that takes a lot of time can be obtained in a much shorter time and with high accuracy [16–18]. The challenges are given by large field sizes with thousands of plants that need to be phenotyped today by laborious, manual, and subjective classification methods. For this purpose, scientists in Germany have developed systems that take images of different growth stages of plants and process them. The images taken in this system can be processed according to different scenarios and adapted for extreme conditions. These six different RGB images can be used in breeding studies with very high accuracy results in terms of applicability and transferability [19].

In this review, it is aimed to inform scientists from different fields who are interested in grapevine breeding by summarizing the remarkable techniques, methods, and developments in grape breeding studies carried out for different purposes in recent years.

2. New advances in grapevine breeding studies

2.1 Genomic and transgenic researches

Grapevine growing areas are increasing worldwide due to the understanding that grapes and grape products are beneficial for human health. Biotechnology research is increasingly playing a role in improving the yield and quality of grapes. Grapevine breeding and genetics researches increased after the 1950s and spread all over the world. Molecular markers have facilitated research in *Vitis* genetics. It is now possible to map the grapevine genome and to create unique DNA profiles for each genotype. The first plant linkage maps were based on visually scored morphological markers, isozymes, and DNA-based markers, which are virtually limited in number were used to create densely saturated maps. However, in recent years, much more information has been gained about the grapevine genome for breeding studies with much more sensitive SNP and genome sequencing applications. Nowadays, the results of different *Vitis* spp. genome sequencing has led to more innovative and targeted studies in grape breeding studies. Especially in parallel with the developments in biotechnology, it has become possible to obtain different transgenic vines with these innovative approaches. Significant progress has been made in the development of transgenic vines with the development of gene regions and markers associated with desired traits, the development of transformation systems, the use of genetic engineering against biotic and abiotic stress conditions, and the improvement of grape quality characteristics by identifying flavor and aroma components. Considering the results obtained from these studies; While some of them offer positive results for different *Vitis* species and varieties by providing direct application, it has been determined that some of them are far from increasing the quality conditions of the grapes as desired in breeding studies [10]. Very important progress has been made thanks to the Vitisgen1 and Vitisgen2 projects initiated with the participation of different institutions in the USA in order to determine different characteristics in the grapevine genome and to use these important characteristics related markers in breeding programs. The goals of these projects are to develop novel methods to improve production efficiency and profitability long-term throughout the table grape, raisin, and wine industry, such as through plant breeding and genomics. Also, they aimed to identify and address threats from diseases and insect pests, and develop novel methods to improve resistance to these pests and diseases. Projects (VitisGen1 and VitisGen2) are multidisciplinary, collaborative projects focused on decreasing the time, effort, and cost involved in developing the next generation of grapes. Incorporating cutting-edge genomics technology and socioeconomic research into the traditional grape breeding process will speed up the ability to identify important genes related to consumer-valued traits like disease resistance, low-temperature tolerance, and enhanced fruit quality. Identifying these genes will help grape breeding programs from around the world to more rapidly develop new grape varieties that will appeal to a wide range of consumers, while also addressing grape grower and producer needs [20]. Genetic transformations offer many innovative solutions for grape breeding studies. They can be used successfully to transfer regions associated with particularly important traits to traditional varieties. However, researchers still lack the desired success rate in identifying high-throughput regeneration protocols. Positive results were obtained in the development of transgenic plants obtained by transferring different characteristics to some important grape cultivars [21, 22]. Embryogenic tissues are mostly preferred in transformation studies due to their morphogenetic competence. Various explant sources, such as leaves [23] and anthers [24] have been studied under inducing conditions to explore the possibilities of obtaining somatic embryos. In the initial transformation studies, leaf tissues of different *V. vinifera* cultivars [25] and different rootstocks [26–28] were studied, resulting in non-regenerating transgenic callus. Somatic embryogenesis has been used by different researchers in micro propagation and genetic transformation studies of various woody perennial plant species. However, in these studies, it has been reported that the efficiency of somatic embryo induction is generally very low and the success rate varies depending on the developmental stages of the explant [29]. Among the genetic transformation studies carried out with grapevines, the most successful results were obtained from Agrobacterium-mediated transformation coupled with proembryonic masses together with somatic embryos. Studies have reported that some factors (explant source, culture medium, cultivar/genotype, culture medium, and others) affect the efficiency of grapevine initial inducing embryogenic callus and adventitious buds for plant regeneration. Photos of the embryogenesis development of different grapevine (*V. vinifera* L.) organs are given in **Figure 1**. Until today, several grapevine cultivars have been transformed with genes associated with various functions by biolistic bombardment, Agrobacterium-mediated transformation, and transgenic grapevine lines have been obtained using established regeneration systems. However, a healthy plant regeneration is affected by many factors. Especially explant source, variety/genotype, and environment are the most important ones. Furthermore, the selection and use of acceptor materials, cell density, bacterial strain, selectable markers, and selection methods also affect conversion efficiency. In many of the studies, it has been reported that the regeneration capacity of rootstock varieties in organogenesis and somatic embryogenesis is higher than hybrids and varieties belonging to the *V. vinifera* species [30].

Badouin et al. [31] generated a high-quality de novo reference genome for *V. sylvestris*, onto which they map whole-genome re-sequencing data of a cross to locate the sex locus. They described the genomic and evolutionary characterization of the sex locus of wild and cultivated grapevines, providing a coherent model of sex determination in the latter and for the transition from dioecy (separate sexes) to hermaphroditism during domestication.

With the help of RNA sequencing, one of the next-generation sequencing systems developed in recent years, short readings of cDNA sequences that can be quantified absolutely can be made by aligning them with reference sequences [32]. With the development of this technology, it has become possible for many researchers working on grape breeding to conduct important studies on the grapevine genome. In particular, the RNA sequencing technique has been widely used to identify single nucleotide

Figure 1.
Photographs of various regeneration systems during somatic embryogenesis of Vitis vinifera cv. "Thompson seedless" grape [30].

polymorphisms and new cultivar-specific transcripts, also splicing variants [33–35]. As a result of the latest advances in plant biotechnology, full-length cDNA sequencing readings can now be made with much higher accuracy. In addition to all these, it is seen that this developed system is also used to accurately detect alternative transcripts that play a role in different biological processes and stress responses [34, 36–39]. While the reference genome was needed before the full-length cDNA sequencing technique, it is no longer needed thanks to this technique. Thus, it has become possible to obtain healthy information about many traits related to plant breeding, in a much shorter time.

Integration of data banks with the results obtained in genomic studies is of great importance in terms of using the obtained results in breeding studies. In particular, studies on the grapevine genome in different countries and the collection of existing data in different databases continue. Standing out as the most comprehensive of these studies, VitisGDB provides the most comprehensive information of *Vitis* genomic data. The *Vitis* genome and genetic database (VitisGDB) is an integrated

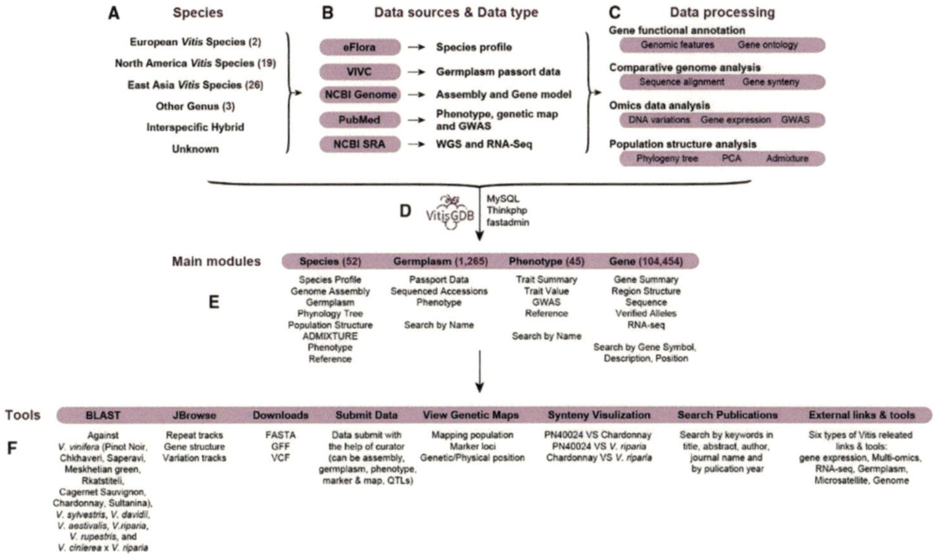

Figure 2.
Schematic VitisGDB platform. (A) Species information for species, (B) data type and source, (C) data processing explanation (D) framework of VitisGDB, (E) main modules, (F) VitisGDB overview [4].

genomic resource and a global web of *Vitis* resources. This platform contains up-to-date genomic data and very important information for Vitis agronomy, breeding, and genomic development studies. VitisGDB is a platform created to make vine research results widely available. As genomic data from sequencing studies become clear and available, they are constantly updated on VitisGBD and presented to relevant researchers VitisGDB provides long-term support to grapevine researchers on a variety of grapevine genomics issues [4]. Schematic information about the working system of this platform is given in **Figure 2**.

Transferable DNA markers are of great importance for the success of breeding and genetic studies. Although grapevine breeders have been using the disease resistance-related alleles of closely related species for many years, it has been reported that the interspecies transmission rates of the current Vitis markers are quite low. Zhu et al. [40] in their study with the Vitis core genome of 40 accessions, they were able to identify PCR primary binding sites of conserved nuclei with high information content surrounding polymorphic haplotypes. Researchers developed markers (2000 rhAmpSeq) as PCR multiplexes from target sites and confirmed this in four biparental populations, also increased the transferability to a very high rate.

2.2 Polyploidy and embryo recovery researches

Polyploidy refers to the condition in which a diploid organism has an excess of chromosomes as a result of the addition of one or more sets of chromosomes. A general classification of polyploids is made as allopolyploids, auto-polyploids, and segmental allopolyploids [41]. In order to obtain polyploid structures, triploid or tetraploid new genotypes were tried to be obtained with different applications and mutations by increasing the chromosome numbers in grapevine and other plant species. However, a high success rate could not be mentioned in these techniques, and sequence-specific mutations were mostly dependent on chance [42].

Somatic embryogenesis is one of the methods preferred by many researchers for the micro propagation of different plant species. It is also used for the removal of many phytopathogens that have infected different plant organs under in vitro conditions [43]. However, abnormalities may be encountered during somatic embryogenesis due to some somaclonal variations. These mutations are desirable in some cases as they sometimes allow the formation of polyploid forms. In particular, induced polyploidy and natural polyploidy are frequently used to obtain new genotypes for polyploidy breeding studies. Because it is reported that new polyploid genotypes have more resistant structures against many biotic and abiotic stress conditions. It is known that plants with polyploid form have some advantages compared to plants with diploid form. Especially polyploid plants are among the most important advantages with their ability to tolerate harmful mutations, larger structures, high heterozygous, and heterozygous structures [44–46]. Polyploid seedless cultivars are obtained in polyploid structure due to some errors during meiosis, and this is common in grapevines. It also allows the reproduction of hybrid genotypes in a sterile structure by doubling the genome [47, 48].

One of the most used methods to provide polyploidy in plants is the application of colchicine to the apical meristem, and it is reported that it is not very effective in applications made on grapevines [49]. Many studies have been carried out to obtain polyploid genotypes by using different species and cultivars belonging to the *Vitis* species. It has been tried to obtain tetraploid genotypes, especially in American grape rootstocks (*Vitis rupestris* and *Vitis riparia*) and *Muscadinia* species [50]. In some studies, on the somatic embryoids and polyploidization of shoot tips of seedless grape cultivars, it has been reported that non-chimeric autotetraploid plants can be obtained [40]. Different studies on the effect of colchicine on proembryogenic cells have reported that it can lead to the regeneration of true utopolyploids [51, 52]. In addition, researchers have identified and published the most effective protocols for the induction of polyploidy [53]. Recently, a significant relationship has been found between the frequency of polyploidy detected in the meristem tissues of plants grown under in vitro conditions and the number of chloroplasts in the stomata of grape somaclones. However, it has been reported that there is a reverse relationship or correlation between the frequency of polyploidy and the stomata number in the leaf area [54]. Cross-breeding between cultivars/genotypes with different ploidy is one of the effective methods to create new germplasms. Most of the triploid fruit plants are sterile and their fruit is seedless.

Triploid breeding researchers have presented a new method for seedless grape breeding as it allows high sterilization and obtaining of parthenocarpic fruits, and ultimately facilitated the achievement of desired results [55]. Seedless is generally desired by breeders, and it has become a more important issue, especially for table grape breeding researchers in recent years. Because the demand for seedless grape cultivars is much higher than the seedless cultivars [56, 57]. However, there exists some mating obstacles in crosses between diploid and tetraploid grape cultivars. Embryo rescue or embryo recovery technique has been used with increasing success rate in recent years to overcome these obstacles. The embryo rescue technique may prevent the early-stage abortion of triploid young embryo, so triploid plants can be produced [58]. The majority of the studies on grape embryo rescue involved studies using seedless or early ripening grape cultivars as the female parent and cross-breeding studies, also the research of a cross between subgenus. There are few studies on embryo rescue from an interspecific cross between diploid and tetraploid grape species. There was a very limited number of studies on cross-breeding and embryo recovery between diploid and tetraploid grape cultivars, including interspecies, but these studies have begun to increase with the techniques and technologies developed in recent years [55, 59–61]. Different studies have been carried out to obtain triploid and tetraploid new genotypes that have larger berry and seedless from different *Vitis* species. It has been stated that the tetraploid grapes obtained as a result of colchicine application have weak vegetative growth, low resistance to cold, and also not at the desired level in terms of yield. Although artificial tetraploids have these disadvantages, they have been successfully used in breeding studies where different *Vitis* species (especially *Vitis rotundifolia*, *Vitis vinifera,* and *Vitis labrusca*) are crossed with each other. In these studies, especially the tetraploid "Kyoho" cultivar (4x = 76), which is an interspecies hybrid, was widely used. Apart from "Kyoho" cultivar, Osuzu (3X) and King Dela (3X) cultivars were also obtained as triploid cultivars as a result of breeding studies. Intensive breeding studies are still continuing on polyploidy breeding, especially in the Far East [62].

Triploid genotypes usually have strong plant formation with seedless berry [63]. Researchers reported that some superior hybrid genotypes were obtained in triploid breeding studies carried out on grapevines. It has been reported that especially larger berry formation is frequently seen in triploid individuals. For this reason, it is seen that both natural and artificial polyploidy studies have increased in recent years [64, 65]. Despite all these studies, the commercial use of polyploid genotypes is still far from the desired levels. Polyploidization can change some phenotypes in plants, but without affecting the appearance of many of the fundamental characteristics of the cultivars. Due to these advantages, polyploidy breeding studies allow the development of some important characteristics (such as quality, yield, and resistance to stress conditions). Researchers still cannot fully explain the genetic and physiological mechanisms affected in the plant as a result of polyploidy. It can also increase the adaptation of artificial tetraploid (4x) grapevine rootstocks to the conditions of biotic and abiotic stresses in *Vitis* spp. Studies on synthetic polyploidy in plant species are still very limited. Due to the high adaptability of polyploid cultivars to different stress conditions, it has become an important study subject in breeding programs. In addition, the importance of polyploid cultivars and rootstocks for sustainable agriculture and their use in production has begun to increase [66].

Sequence-specific nucleases that generate double-stranded DNA breaks in targeted genes are the most important parts of site-specific genome editing in some plants. Induction of knockout mutations to inactivate undesirable features in genome editing has become the preferred method in many plant species in recent years. Different applications of sequence-specific nucleases have come to be used as robust tools for introducing functional mutations in many polyploid species, including grapes. The main approach here to utilize knowledge of biological mechanisms for targeted induction of double-stranded DNA breaks and their error-prone repair. Moreover, these regions may allow very specific changes at designated genome loci [41].

2.3 Biotic stress researches

The main phytopathogenic organisms that cause biotic stress in vines are organisms such as bacteria, nematodes, fungi, oomycetes, and viruses, which cause different infections in the vine and adversely affect many of their functions. All these pathogens get what they need for growth and reproduction from the host plant. Plant pathogens are divided into three different classes based on their infection strategy.

In this classification, the differences of the pathogens according to their feeding patterns and the necrosis they form in the plants are based [67].

Fungal diseases are among the most important biotic stress factors in grapes. Among the fungal diseases, downy mildew (*P. viticola*), powdery mildew (*E. necator*) and botrytis (*B. cinerea*) are the most common and most damaging diseases. Most of the grape cultivars consumed in different ways (such as table grapes, wine grapes, and raisins) belong to *V. vinifera* species whose gene source is Eurasia. This species is mostly preferred because of its unique taste, aroma, and better fruit quality. However, this species has a very high sensitivity to fungal pathogens despite these superior properties, and therefore, grape growers apply very intense fungicides for quality cultivation [68, 69].

Since this intensive fungicide application poses a great risk for both human and environmental health, grape breeding studies have focused on breeding more resistant varieties to diseases that will not need such intensive spraying in recent years. In grape breeding programs, many breeding studies are carried out by researchers in different countries to determine gene regions that are resistant to these diseases and to develop new wine and table varieties that carry these gene regions [13].

Although many of the North American origin wild *Vitis* species show varying levels of resistance to powdery mildew, unfortunately, the fruit quality is not at the desired level. These species are used as a very important genetic resource as a natural resistance source in grapevine breeding programs. To date, many resistant species have been described within these species. These species include *Vitis V. aestivalis, V. cinerea, V. riparia, V. berlandieri, V. labrusca* and *Muscadinia rotundifolia* [70, 71]. However, some cultivars such as "Dzhandzhal kara" and "Kishmish vatkana" belong to the *V. vinifera* species, which are known to be susceptible to diseases, were found to be resistant [72, 73].

As a result of revealing the characteristics and related gene regions related to resistance in grape breeding studies, much more successful results have been obtained in breeding programs. To date, some resistance loci related to fungal diseases have been identified and their mapping has been done. In recent years, not only fungal diseases but also numerous genetic loci associated with a particular phenotype have been identified in the grapevine. Regions associated with these diseases are very important in grapevine breeding studies as they are determined with the help of marker-assisted selection (MAS) and provide a great advantage in achieving results. Gene regions associated with many traits in grapevines have been reported by the Vitis International Variety Catalog [74].

Overall, the table reports potential gene regions found to be related to 20 different traits. In particular, the table includes loci and alleles associated with downy and powdery mildew diseases. Eight sites associated with non-mildew diseases, five with metabolites, five refer to morphology traits, and four with phenology. This table is updated regularly to provide accurate access to loci and markers associated with many of the commercial traits and stress factors required by grapevine research. It also helps the researchers in correct naming and following the same systematic (**Figures 3**–**5**).

The resistance of hybrid genotypes obtained from different grape breeding studies against downy and powdery mildew diseases has been compared in several studies [6, 73, 76–80]. According to the studies conducted by the researchers so far, 31 genomic regions have been associated with downy mildew resistance (Rpv loci) and 13 with powdery mildew disease resistance (Run/Ren loci) (**Figure 3**). In order to determine the presence of these loci in hybrid genotypes, marker-assisted selection (MAS) studies have been successfully performed [76–78, 81–84].

Trait	Symbol	Chr	Position on chr [Mb]	Associated marker	Parent 1	Parent 2	Population size	Genotype of origin	Original species of trait
Resistance									
Agrobacterium spec	*Rcg1*	15	7.1	UDV015	Kunbarát	Sárfehér	272	Kunbarát	*V. amurensis*
			9.3	9M3-3					
Colletotrichum gloeosporioides (Glomerella cingulata)	*CgR1*	14	4.1	np19345	Cabernet Sauvignon	Shuang Hong	91	Shuang Hong	*V. amurensis*
Daktulosphaira vitifoliae	*Rdv1*	13		GF13_1	GF.V3125	Börner	188	Börner	*V. cinerea*
			21.9	GF13_9					
				GF13-1	GF.V3125	Börner	188	Börner	*V. cinerea*
			21.5	GF13-7					
Daktulosphaira vitifoliae	*Rdv2*	14	4.9	S14_4921219	V. cinerea C2-50	Riesling	90	V. cinerea C2-50	*V. cinerea*
Daktulosphaira vitifoliae	*Rdv3*	14	5.0	S14_5049399	MN1264	MN1246	125	MN1264	
Daktulosphaira vitifoliae	*Rdv4*	10			MN1264	MN1246	125	MN1246	
Daktulosphaira vitifoliae	*Rdv5*	5			MN1264	MN1246	125		
Daktulosphaira vitifoliae	*Rdv6*	7			VRH8771	Cabernet Sauvignon	135	VRH8771	*M. rotundifolia*
Daktulosphaira vitifoliae	*Rdv7*	3	5.5	3_5494608	VRH8771	Cabernet Sauvignon	135	VRH8771	*M. rotundifolia*
Daktulosphaira vitifoliae	*Rdv8*	10			VRH8771	Cabernet Sauvignon	135	VRH8771	*M. rotundifolia*
Diaporthe ampelina (Phomopsis viticola)	*Rda1*	15	19.6	S15_19560016	Chardonnay	V. cinerea B9	148	*V. cinerea* B9	*V. cinerea*
				S15_19591538	Horizon	V. cinerea B9	162	*V. cinerea* B9	*V. cinerea*
			19.3	S15_19300044	Horizon	Illinois 547-1	366	Illinois 547-1	*V. cinerea*
Diaporthe ampelina (Phomopsis viticola)	*Rda2*	7	1.2	VVMD7	Horizon	*V. cinerea* B9	162	Horizon	
			1.8	VrZAG62					
			3.1	VVIb22					
			3.1	S7_3127368					
			1.9	S7_1912889	Horizon	Illinois 547-1	366	Horizon	
Erysiphe necator	*Ren1*	13		UDV020	Nimrang	Kishmish vatkana	310	Kishmish vatkana	*V. vinifera*
			18.4	VMC9h4-2					
			18.4	VMCNg4e10.1					
Erysiphe necator	*Ren2*	14	26.9	CS25	Horizon	Illinois 547-1	58	Illinois 547-1	
Erysiphe necator	*Ren3*	15	7.1	UDV015b	Regent	Lemberger	153	Regent	
			10.9	VVIv67					
				ScORA7-760	Regent	Lemberger	152	Regent	
			4.9	VChr15CenGen02	Regent	RedGlobe	206	Regent	
			10.9	GF15-28 / VVIv67	GF GA-47-42	Villard blanc	151		
			9.3	GF15-42	Regent	Lemberger	132	Regent	
Erysiphe necator	*Ren4*	18	26.9	VMC7f2	C166-043	F8909-08	42	C166-043	*V. romanetii*
			26.9	SNPs	C87-41	B70-57	57	C87-41	*V. romanetii*
Erysiphe necator	*Ren5*	14	4.8	VMC9c1	Regale	Regale	191	Regale	*M. rotundifolia*
Erysiphe necator	*Ren6*	9	8.6	PN9-057	F2-35	V. piasezkii (DVIT2027)	277	V. piasezkii (DVIT2027)	*V. piasezkii*
			9.1	PN9-068					
Erysiphe necator	*Ren7*	19	0.2	VVIp17.1	F2-35	V. piasezkii (DVIT2027)	277	V. piasezkii (DVIT2027)	*V. piasezkii*
			0.9	VMC9a2.1					
Erysiphe necator	*Ren8*	18	13.2	UDV117	GF GA-47-42	Villard blanc	151		
				SPS_P_SNP632GF					
Erysiphe necator	*Ren9*	15	1.4	CenGen6	Regent	Lemberger	133	Regent	
Erysiphe necator	*Ren10*	2	17.9	S2_17854965	MN1264	MN1214	147	Seyval blanc	
		2		Haploblock validation	MN1264	MN1246	125		
Erysiphe necator	*Ren11*								

(a)

Trait	Symbol	Chr	Position on chr [Mb]	Associated marker	Parent 1	Parent 2	Population size	Genotype of origin	Original species of trait
Erysiphe (Uncinula) necator	*Run1*	12	13.1	VMC4f3.1	VRH3082-1-42	Cabernet Sauvignon	161	VRH3082-1-42	*M. rotundifolia*
			20.4	VMC8g9					
			16.4	49MRP1.P2	VRH3082-1-42	Cabernet Sauvignon	2575		
			16.8	CB33.54	VRH3176-21-11	Cabernet Sauvignon	722		
					VRH3161-6-4	Cabernet Sauvignon	110		
					BC1 M. rotundifolia	Syrah	139		
Erysiphe (Uncinula) necator	*Run2.1*	18	26.9	VMC7f2	JB81-107-11	Chenin Blanc	97	Magnolia	M. rotundifolia
			20.9	VMCNg1e3					
			23.4	VVIn16	JB81-107-11	Tokay	47		
			26.9	VMC7f2					
			26.9	VMC7f2	A90-71	Flame Seedless	80		
Erysiphe (Uncinula) necator	*Run2.2*	18	26.9	VMC7f2	e2-9	Malaga Rosada	255	Trayshed	*M. rotundifolia*
Erysiphe necator	*Sen1*	9	13.6 - 18.0	S8_19258484	*V. rupestris* B38	Chardonnay	85	Chardonnay	*V. vinifera*
Guignardia bidwellii	*Rgb1*	14	26.7	GF14-42	GF.V3125	Börner	202	Börner	
Guignardia bidwellii	*Rgb2*	16	15.3	VChr16c	GF.V312	Börner	202	Börner	
Meloidogyne javanica (root knot nematode)	*MjR1*	18	31.2 34.0	S18_31160355 S18_33954011	C2-50	Riesling	90	C2-50	*V. cinerea*
Pierce's disease (*Xylella fastidiosa*)	*Pdr1*	14	25.3	VMCNg3h8	*V. rupestris*	*V. arizonica*	181		*V. arizonica*
			26.6	VVIn64					
			26.1	UDV095					
Plasmopara viticola	*Rpv1*	12	10.3	VVIb32	Syrah	28-8-78		28-8-78	*M. rotundifolia*
Plasmopara viticola	*Rpv2*	18			Cabernet Sauvignon	8624	129	8624	*M. rotundifolia*
Plasmopara viticola	*Rpv3*	18		UDV112	Regent	Lemberger	153	Regent	
			24.9	UDV305	Chardonnay	Bianca	116	Bianca	
			26.9	VMC7f2					
			26.9	VMC7f2	Regent	RedGlobe	206	Regent	
	Rpv3.1 (=*Rpv3*$^{299-279}$)		24.9 26.1	UDV305 UDV737				Seibel 4614	*V. rupestris*
			25.9 26.9	GF18-06 GF18-08	GF GA-47-42	Villard blanc	151	Villard blanc	*V. rupestris*
	Rpv3.2 (=*Rpv3*$^{null-297}$)		24.9 26.1	UDV305 UDV737				Munson (Jaeger 70)	*V. rupestris* or *V. lincecumii*
			25.9 26.9	GF18-06 GF18-08	GF.GA-47-42	Villard blanc	151	GF.GA-47-42	*V. rupestris* or *V. lincecumii*
	Rpv3.3 (=*Rpv3*$^{null-271}$)		24.9 26.1	UDV305 UDV737				Noah	*V. labrusca* or *V. riparia*
			23.4 26.1	VVIN16 UDV737	Merzling	Teroldego	136	Merzling	
	Rpv3$^{321-312}$		24.9 26.1	UDV305 UDV737				Noah	*V. labrusca* or *V. riparia*
	Rpv3$^{299-299}$		24.9 26.1	UDV305 UDV737				*V. rupestris* Ganzin	*V. rupestris*
	Rpv3$^{299-314}$		24.9 26.1	UDV305 UDV737				*V. rupestris* Ganzin	*V. rupestris*
	Rpv3$^{null-297}$		24.9 26.1	UDV305 UDV737				Bayard (Couderc 28-112)	*V. rupestris* or *V. labrusca*
Plasmopara viticola	*Rpv4*	4	4.7	VMC7h3	Regent	Lemberger	153	Regent	
			5.2	VMCNg2e1					
Plasmopara viticola	*Rpv5*	9	4.0	VVIo52b	Cabernet Sauvignon	Gloire de Montpellier	138	Gloire de Montpellier	*V. riparia*
Plasmopara viticola	*Rpv6*	12	20.4	VMC8g9	Cabernet Sauvignon	Gloire de Montpellier	138	Gloire de Montpellier	*V. riparia*
Plasmopara viticola	*Rpv7*	7	11.4	UDV097	Chardonnay	Bianca	116	Bianca	
Plasmopara viticola	*Rpv8*	14	6.6	Chr14V015	V. amurensis Ruprecht	V. amurensis Ruprecht	232	V. amurensis Ruprecht	*V. amurensis*
Plasmopara viticola	*Rpv9*	7	16.6	CCoAOMT	Moscato Bianco	V. riparia W63	174	V. riparia W63	*V. riparia*

(b)

Trait	Symbol	Chr	Position on chr [Mb]	Associated marker	Parent 1	Parent 2	Population size	Genotype of origin	Original species of trait
Plasmopara viticola	*Rpv10*	9	3.7	GF09-46	GF GA-52-42	Solaris	256	Solaris	*V. amurensis*
Plasmopara viticola	*Rpv11*	5	4.5	VVMD27	Regent	Lemberger	153	Regent	
				CS1E104J11F	Chardonnay	Bianca	116	Chardonnay	
			4.1	VCHR05C	GF GA-52-42	Solaris	256	Solaris	
Plasmopara viticola	*Rpv12*	14	8.0	UDV014	99-1-48	Pinot noir	180	99-1-48	*V. amurensis*
			9.3	UDV304	Cabernet Sauvignan	20/3		20/3	*V. amurensis*
				rgvvia180					
			10.1	UDV370					
Plasmopara viticola	*Rpv13*	12	10.0	VMC1g3.2	Moscato Bianco	V. riparia W63	174	V. riparia W63	*V. riparia*
Plasmopara viticola	*Rpv14*	5	20.2	GF05-13	GF.V3125	Börner	202	Börner	*V. cinerea*
Plasmopara viticola	*Rpv15*	18			V. piasezkii (DVIT2027)	F2-35	94	V. piasezkii (DVIT2027)	*V. piasezkii*
Plasmopara viticola	*Rpv16*								
Plasmopara viticola	*Rpv17*	8	11.7		V. rupestris B38	Horizon	163	Horizon	
Plasmopara viticola	*Rpv18*	11	15.4		V. rupestris B38	Horizon	163	Horizon	
Plasmopara viticola	*Rpv19*	14	29.5		V. rupestris B38	Horizon	163	V. rupestris B38	*V. rupestris*
Plasmopara viticola	*Rpv20*	6	0.9		Horizon	V. cinerea B9	152	Horizon	
Plasmopara viticola	*Rpv21*	7	2.1		Horizon	V. cinerea B9	152	Horizon	
Plasmopara viticola	*Rpv22*	2	2.1		Cabernet Sauvignan	Shuanghong	91	Shuanghong	*V. amurensis*
			3.5						
Plasmopara viticola	*Rpv23*	15	9.9		Cabernet Sauvignan	Shuanghong	91	Shuanghong	*V. amurensis*
			13.9						
Plasmopara viticola	*Rpv24*	18			Cabernet Sauvignan	Shuanghong	91	Shuanghong	*V. amurensis*
Plasmopara viticola	*Rpv25*	15	3.0	Marker561375	Red Globe	Shuangyou	149	Shuangyou	*V. amurensis*
			3.9	Marker549779					
Plasmopara viticola	*Rpv26*	15	14.7	Marker525926	Red Globe	Shuangyou	149	Shuangyou	*V. amurensis*
			15.0	Marker526446					
Plasmopara viticola	*Rpv27*	18	24.6	VVCS1H077H16R1-1	Norton	Cabernet Sauvignon	182	Norton	*V. aestivalis*
			26.0	UDV737					
Plasmopara viticola	*Rpv28*	10	1.2	VVIH01	V. rupestris B38	V. riparia HP-1	294	V. rupestris B38	*V. rupestris*
			1.3	UDV-073					
Plasmopara viticola	*Rpv29*	14	21.6	chr14_21613512_C_T	Mgaloblishvili	Mgaloblishvili	132	Mgaloblishvili	*V. vinifera*
Plasmopara viticola	*Rpv30*	3	16.2	cn_C_T_chr3_16229046	Mgaloblishvili	Mgaloblishvili	132	Mgaloblishvili	*V. vinifera*
Plasmopara viticola	*Rpv31*	16	21.3	li_T_C_chr16_21398409	Mgaloblishvili	Mgaloblishvili	132	Mgaloblishvili	*V. vinifera*
Xiphinema index	*XiR1*	19	20.9	VMC5a10	*V. rupestris*	*V. arizonica*	185		*V. arizonica*
			20.9	1N2R3b					
				M4F3R					
Xiphinema index	*XiR2*	9		VVBX-A-06	VRH8771	Cabernet Sauvignon	135	VRH8771	
Xiphinema index	*XiR3*	10		SC8-03	VRH8771	Cabernet Sauvignon	135	VRH8771	
Xiphinema index	*XiR4*	18	29.1	UDV108	VRH8771	Cabernet Sauvignon	135	VRH8771	

(c)

Figure 3.
Table of loci for diseases and pests traits in grapevine relevant for breeding and genetics (details in www.vivc.de/data) [75].

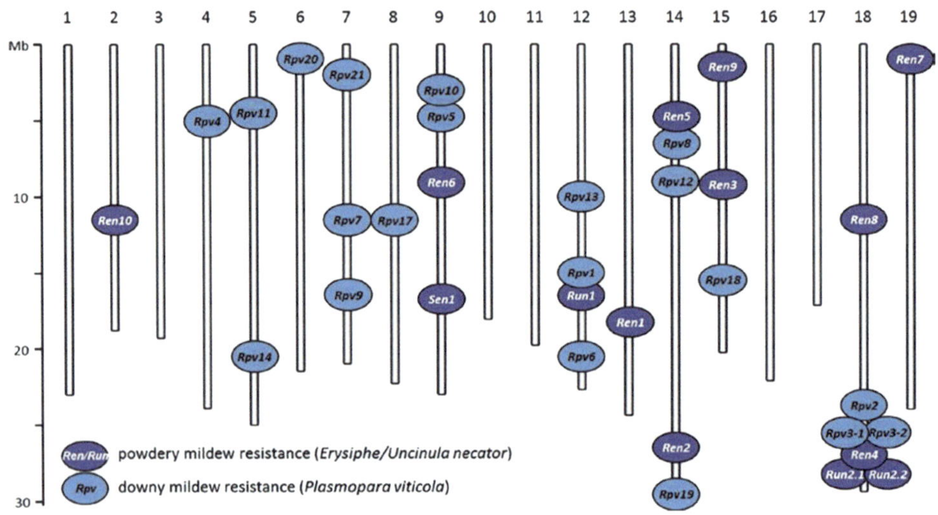

Figure 4.
Downy mildew (P. viticola) and grapevine powdery (E. necator) resistance loci position in the genome. Scale is in megabases (Mb) [74].

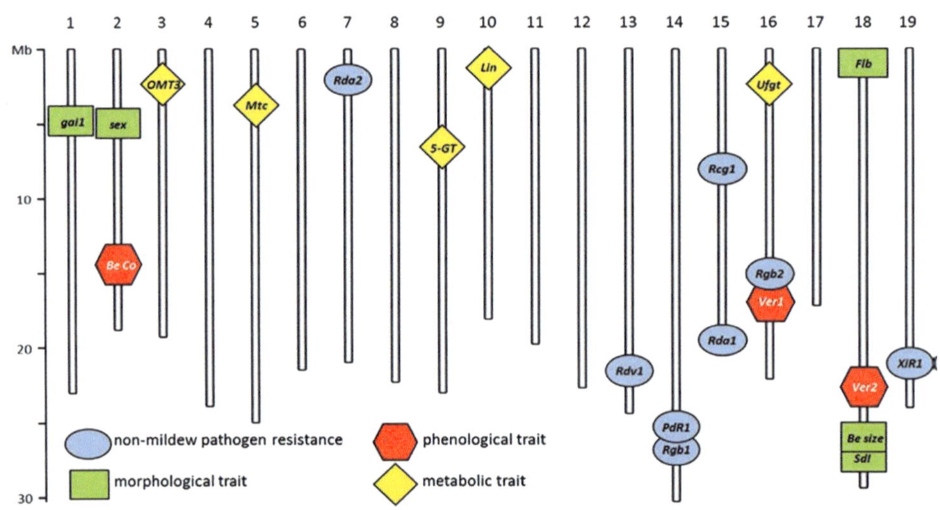

Figure 5.
Genomic positions of morphological, phenological, metabolic trait loci, non-mildew disease, and pest resistances. Scale is in megabases (Mb) [74].

However, the presence of these gene regions alone often does not prove that the variety is resistant to diseases. In addition, the resistance of the genotypes should be tested in the field and under controlled greenhouse/laboratory conditions [85–90]. New downy and powdery mildew resistant cultivar "Regent" was obtained in Germany, its pedigree includes American species carrying Ren3, Ren9, Rpv3, Rpv4, and Rpv11 [91–93]. In recent years, new resistant loci have been identified in different species, especially in relation to resistance to fungal diseases from biotic stress factors. In one of these studies, [94] discovered the REN11 locus from *Vitis aestivalis* for stable resistance to grape powdery mildew.

After determining the gene regions associated with resistance and thus the resistant genotypes, another problem may be encountered in resistance breeding studies. New pathogen races may break this resistance. In order to solve this problem that may arise, one of the methods that breeders usually resort to is to try to collect more than one gene region associated with resistance into new genotypes. Thus, the resistance of the new genotypes is further increased. For example, even if the vine plant is infected with a new virus, it limits the development of the pathogen and shows more resistance against it. Marker-assisted gene pyramid applications have been a highly preferred application by researchers in grape breeding studies in recent years. With the use of molecular markers in grape breeding programs, genotypes that may exhibit the same phenotype in appearance but carry more than one resistance gene in their genomes can be determined [75, 95, 96]. In recent years, breeders and pathologists have worked together to achieve significant success in grape breeding, especially in studies related to resistance. In one of them, with the VitisGen project carried out in partnership with different organizations in the USA, they collected different isolates against powdery mildew disease and identified complementary resistance loci sets to evaluate the phenotypic and genetic resistance gene stacks against them [97].

It has been reported as a result of studies that different chemicals have important effects on defense mechanisms in plants also grapevines. Among them, ethylene, jasmonic acid, and salicylic acid are the most important ones. These chemicals can act synergistically or vice versa, depending on the pathogen. While jasmonic acid and ethylene help plants to defend themselves against necrotrophic pathogens, defense against biotrophic pathogens is mediated by salicylic acid, unlike them. Cultivars of *V. vinifera* species have very low resistance to many pathogens of fungal origin. This is probably related to their insufficient defense systems against these pathogens. However, although studies have been carried out with the details of this defense system in recent years, more information is needed regarding the interaction of cultivars with grapevine diseases. In recent years, researchers have been working intensively on these interactions to obtain useful information for breeders, especially in parallel with the advances made in molecular methods. In one of the studies conducted for this purpose, it was reported that the modulation of chloroplast-associated lipids in the first hours of interaction with downy mildew is important for the protection of photosynthetic machinery and for the biosynthesis of jasmonic acid [67].

Cavaco et al. [98] identified subtilisin-like proteases as strong resistance-associated candidates. The relationship between fungal diseases and phenolic components has become increasingly important in recent years. Researchers evaluated changes in total phenolics, total antioxidant activity, and phenolic compounds in different *Vitis* species and genotypes. After fungal diseases, it has been reported that there is an increase in the amount of total phenolics, total antioxidant activity, and some phenolic compounds [99–101]. Resistance mechanism of *V. vinifera* cv. "Mgaloblishvili", which was grown in Georgia and resistant to downy mildew, was investigated by Ricciardi et al. [102] Researchers explained the disease resistance of the cultivar with low disease density, low sporulation, damaged mycelium, production of antimicrobial compounds such as volatile organic compounds (VOCs) whose activity on the pathogen was evaluated by leaf experiments. These results contain data that can assist and accelerate future resistance breeding programs.

Chibutaru et al. [103] first examined the reaction of mono-locus resistant genotypes against downy mildew after the first and second infection, and also evaluated the pyramid resistance genotypes. Researchers especially investigated different metabolites (not stilbenes and stilbenoids), which accumulate significantly in resistant and susceptible genotypes as a result of disease infection and can be used as potential resistance-related markers. Also, they investigated whether these metabolites could be markers of infection. In their study, it was aimed to provide a better understanding of the different resistance mechanisms of hybrid-pathogen interaction that can affect different *Vitis* species and to find previously undetected resistance biomarkers. As a result of their studies, they determined the components that increased after downy mildew and could be used as biomarkers.

The development of highly reproducible genetic engineering methods for grapevine rootstocks, cultivars, and genotypes now allows the identification, screening, and/or introduction of grapevine-derived genes related to desirable traits, such as disease or pest resistance. It has been reported that genetically modified grapevines constitutively expressing rice chitinase genes have been screened for the responses of pathogenesis-related proteins to fungal pathogen infection, and show increased resistance to powdery mildew disease. As a result of studies, it has been revealed that other grape-derived genes such as polygalacturonase inhibitor protein and other lytic peptides increase resistance to fungal diseases [104, 105].

Grapevine breeding programs have been started in order to develop new resistant hybrid genotypes against powdery mildew and downy mildew diseases in different countries [106]. In one of these, Ruiz-García et al. [107] evaluated the degree of phenotypic resistance or susceptibility for downy and powdery mildew of 28 new genotypes obtained from crosses between "Monastrell" and "Regent". In particular, three genotypes from the hybrid population showed strong combined resistance, and they could be used as a very important source of resistance parents in future breeding studies in terms of both powdery mildew and downy mildew. As a result of their study, they reported that multi-resistant lines provide very valuable material for obtaining resistant genotypes and help to characterize the molecular basis of downy and powdery mildew resistance.

Wild grapevine species are widely recognized as an important source of resistance or tolerance genes for diseases and environmental stresses. Recent studies revealed partial resistance to powdery mildew (*Erysiphe necator*) in *V. sylvestris* from Central Asia. Lukšić et al. [108] investigated the resistance of in situ *V. sylvestris* seedlings collected from different regions of Croatia against powdery mildew. Ninety-one in situ individuals and 67 *V. sylvestris* seedlings were evaluated for Powdery mildew resistance according to OIV 455 descriptor. Three SSR markers (SC47-18, SC8-071-0014, and UDV-124) linked to Powdery mildew resistance locus *Ren1* were used to decipher allelic structure. As a result, they determined that there were varying numbers of resistant genotypes in individuals in different *V. sylvestris* populations. Thus, in their study, powdery mildew resistance was proved for the first time in the germplasm of *V. sylvestris* in the eastern Adriatic region.

In order to increase the resistance to different biotic and abiotic stress conditions in grapevine, research programs have increased primarily on the determination of the responsible gene regions and then the introgression of these regions into susceptible cultivars or the mutation of the genes that cause the susceptibility in recent years. Sometimes the resistance obtained as a result of mutation of the genes can provide a longer-term protection. Especially in breeding studies, genotypes with genes containing resistance are selected as parents and it is aimed to transfer these characteristics to new genotypes. According to Pirello et al. [109] used Arabidopsis as a model in their study, worked with resistant mutants, and investigated the effectiveness of DMR6 and DLOs genes that could confer downy mildew resistance in grapevines.

By examining the relationships between genes and the links between the VviDLO1, VviDMR6-1, and VviDMR6-2 gene groups, they reported that they are associated with genes sensitive to pathogenesis. In particular, the researchers concluded that the VviDMR6-1 region may be a candidate that can be used to produce resistant cultivars by gene editing.

2.4 Abiotic stress researches in grapevine breeding

As a result of climate change affecting the whole world, the development of new grape genotypes with high adaptability to abiotic stress conditions has become a more important priority in recent years. Successful programs are carried out for sustainable viticulture, with the aim of grapevine breeding studies and the transfer of many resistance-related genes in wild grapevine species to new genotypes through interspecies cross-breeding. Different studies are being conducted to identify these alleles in the grapevine genome and understand how they can be used to manipulate phenotypes. The diversity of abiotic constraints (heat stress, drought, salinity, mineral deficiency, etc.) and their timing, duration, and intensity must be taken into account. It is seen that topics such as the type of factors causing abiotic stress (extreme temperatures, salt stress, excessive water, heavy metals, and others), duration, and intensities are taken into account in these studies. It is necessary to clearly identify these by thoroughly examining the characteristics and sensitive development stages underlying the adaptation of the vine plant to different stress factors. Targeted traits are often quite complex and under the control of various genetic mechanisms. Especially in the last decade, various researches on grapevine genome (sequencing, genetics, phenotype development, modeling) and functional characterization of related genes) and significant results have been obtained by carrying out successful projects. In the light of recent developments in grape physiology and genome, molecular mechanisms related to adaptation processes to changing climatic conditions and the gene regions controlling them are explained. The physiology of the vine is actually quite complex, and this complex mechanism is polygenically controlled. However, in recent years, very important information has been obtained about new grapevine genotypes that are more tolerant/resistant to different abiotic stress conditions. Responses to extreme temperatures, heavy metals, droughts, and some other stress conditions have been extensively studied by different researchers in order to obtain new more compatible hybrid genotypes [110].

Grapevine (*Vitis* spp.) can adapt well to environments under different stress conditions. It has even been reported that moderate abiotic stress has a positive effect on the quality of grape products such as wine. In addition, as a result of the high variation in *Vitis vinifera* cultivars and their combination with different rootstocks, it becomes possible for grapevine cultivation in different ecology and soil types [111]. Grapevine plants are considered drought resistant when compared with other horticultural plant species. However, most of the time, as a result of insufficient irrigation, significant decreases in yield and quality can be observed [112, 113]. By utilizing the wide variation in drought tolerance within the Vitis species in breeding studies, new varieties more tolerant to the desired drought can be developed [114–117]. There is very limited information about the mechanisms related to drought tolerance, but promising studies have been carried out in recent years. In order to develop longer-term sustainable irrigation programs for drought, which has become a growing problem in many parts of the world, new drought-resistant/tolerant varieties must be developed.

In a study investigating the response of grapevine plants to temperature, Luchare et al. [118] studied the effects of increases in temperature on carbon balance using microvine mutants. The grapevine plants under controlled conditions studied in detail the photosynthesis, respiration, and carbon allocation in different parts of the plants at different temperature ranges. As a result of the study, they reported that especially net photosynthesis decreased after peaking at 25–30°C, and that respiration at night increased steadily with the increase in temperature. In addition, a less favorable carbon balance was formed at higher temperatures compared to lower temperatures. In another similar study, it was reported that although organogenesis and leaf area was stimulated by high temperature, there was a decrease in carbon balance [119].

In case of exposure to high temperatures, there can be significant changes in the amount of compounds in the content of grapes, especially aroma compounds [120]. For example, in a study with "Gewürztraminer" × "Riesling" hybrids, it was observed that high temperature had different effects on geraniol content linalool and linalool content in grape berries. While geraniol content increased with the high temperature in all genotypes used in the study, linalool content decreased. This was interpreted as the regulatory pathways for the deposition of the components were different for both components. This also showed that high temperatures increased the complexity of the quality control parameters [121].

Roots are vine organs that play an important role in biotic stress factors. Despite the fact that they are less studied due to their underground location, there has been a significant increase in the studies on roots in recent years due to the understanding that roots have an important role in resistance against many stress conditions. Since most of the vines are grafted, rootstocks can be used especially for mineral deficiency/toxicity or drought tolerance, and there is a chance to choose the best scion × rootstock combinations for different soil/climate types [122].

One of the most important problems in mineral nutrition and roots is that genetic variation cannot be fully characterized and the lack of extensive investigation of the genetic architecture of mineral nutrition-related traits. Important studies have been published on limestone tolerance [123] and recently salt tolerance [124] Since rootstocks are often derived from interspecific crosses, it should reveal the extent to which different species possess the relevant alleles that define and differentiate their feeding activities. These alleles can be used successfully in breeding studies to grow highly efficient rootstocks. As a result of the evaluation of the variability between rootstocks from different genetic backgrounds, important data can be provided to achieve the desired goals. In one of these studies, it was reported that rootstocks with *V. riparia* among the parents had a lower phosphorus concentration than the others when their petioles were evaluated [110, 125].

With a full understanding of the conditions that cause abiotic stress, successful results can be obtained by applying the effects of this stress on plants and fruits to breeding studies with integrated approaches of ecophysiological and genetic modeling [126, 127]. With plant modeling, many complex traits can be divided into simpler traits, and as a result, complex traits adapt to the environment and growing environments much more stably with simple genetic control. For the grapevine, researchers are working on the details of the different models. In recent years, important studies have been carried out to describe the physiological and genetic mechanisms of grapevine responses to important abiotic stress conditions. These studies, the number of which has increased in recent years. The increasing development of modern phenotyping and genotyping tools with approaches in molecular physiology, modeling, ecophysiology, and genetics, and their integration with each other, show how knowledge on this subject can be further increased. With the help of these extensive studies, new components related to regulatory pathways have been found and the genetic structure of important traits has been analyzed. As a result, these data provided very important information for the breeding of advanced cultivars and rootstocks in the fight against different diseases and their agents. Despite all these developments and advances in technology, much remains to be done in order to describe the responses of plants to abiotic stress factors and to fully understand how to adapt to these extreme climates. There is still a large gap between the phenotypes and genotypes of newly developed cultivars. We are far from the desirable level of fully understanding and responding to the interactions of plants, both with the environment and with their own structures [110, 128].

2.5 Seedlessness researches in grapevine breeding

Seedless grape cultivars constitute a very important part of table grape production. Especially in recent years, consumers have been demanding more seedless cultivars. This situation has led to the start of studies in many countries for the breeding of high quality, larger berry, long storage life, high yielding, and relatively more disease tolerant seedless cultivars. These breeding programs are carried out by state institutions, private sector, and grower [129].

It is known that the seedless grape cultivars have two different types (parthenocarpic and stenospermocarpic seedless). In stenospermocarpic seedless genotypes, the embryo fails to develop shortly after fertilization during seed development, and such cultivars are used as parents in breeding to obtain seedless genotypes with larger berry size [130]. Berries of parthenocarpic grapes have rather a small berry size that develops without fertilization. For this reason, the embryo rescue/recovery technique is widely used together with conventional breeding methods to obtain new seedless grape varieties.

In traditional hybridization studies on the breeding of seedless grape varieties, seedless parents are used as the father (pollinator) and the seed parent is used as the mother. However, the seedlessness rate in the genotypes obtained from these crosses varies between 0% and 49% depending on the parent combination [131]. As a result of the abortive embryos of stenospermocarpic vine cultivars to continue their development in tissue culture, seedless x seedless hybridizations have been possible in traditional hybridization studies. This application (embryo rescue technique) increased the seedless rate observed in seedless x seedless hybrids in F1 plants between 16.7% and 92% depending on the parent combination. Has changed. For this reason, the embryo rescue technique is widely used together with traditional breeding methods to obtain new seedless grape cultivars [132–136]. Success rate in embryo recovery studies depends on the genotype of the parents [132], sampling time, and composition of the culture medium [137]. Embryo forming capacity and germination of embryos of hybrid genotypes may differ according to both their male and female parents [133]. Seed trace of stenospermocarpic grape cultivars can be in 3 different sizes (small, medium, and large). Generally, those with a larger seed trace have a higher rate of transformation into a living plant with the embryo rescue method [138].

Also, in another study, Li et al. [56] conducted studies to obtain seedless, disease-resistant and high-quality grape cultivars by using the embryo recovery method and reported that the sampling time has a very significant effect on the development and recovery of the embryo. The genetic structure of seedlessness in grapes has been studied by different researchers. Finally, a model of three recessive genes (independent and complementary) by a seed development inhibitor at the 18th linkage group on the dominant locus was proposed [139, 140]. In addition, it has been reported that two SSR markers (VMC7f2 and p3_VvAGL11) are very close to the seed development inhibitor region and can be used in marker-based selection breeding studies [141, 142].

It has been reported that VvAGL11, one of these two markers, is located in a region between the promoter region and can be used successfully to identify seedless genotypes. Two markers selected in association with the Seed Development Inhibitor locus region were selected as candidate markers because of their low number of false positives [143, 144]. The VviAGL11 marker belongs to the D-lineage of the MADS-box genes controlling the identity of the grape ovules and stands out as the major functional candidate gene for seedless grape morphogenesis [145–147]. In addition, two SCAR markers (SCC8 and SCF27), which are related to seedlessness and could be used to identify seedless genotypes, have also been developed [140, 148]. Of these, the SCC8 marker was used to distinguish seedless from hybrid genotypes belonging to seeded × seedless combinations [129, 149]. Mejía and Hinrichsen [148], on the other hand, used both markers to determine seedless genotypes in Ruby seedless' X "Sultanina" combination and reported that SCF27 marker can be used with a much higher percentage to identify seedless ones in F1 hybrid genotypes. Studies on seedlessness trait at the molecular level have shown that there is a very important relationship between the efficacy of the markers used and the genetic background when evaluating the seedlessness property of different hybrid genotypes. In the absence of lignified seeds in seedless grapes, the p3_VvAGL11 marker can accurately identify seedlessness in approximately 85% of hybrid genotypes [150]. When some seedless hybrid populations with different genetic backgrounds were evaluated with some markers (VvIn16, p3—VvAGL11, SCF27 andVMC7f2), it was reported that the VMC7f2 and p3—VvAGL11 markers showed the most accurate allelic variability. In addition, researchers reported that each combination of parents should be evaluated specifically by markers related to seedlessness [144]. In **Table 1**, primers and their sequences used by different researchers to identify seedless hybrid genotypes are given.

2.6 Rootstock breeding

Grape rootstocks are used around the world, especially against phylloxera, but despite many difficulties in choosing a better rootstock, research is being carried out. The studies on rootstock breeding started after Phylloxera damage, especially in the vineyard areas in Europe towards the end of the nineteenth century [10]. The use of a very limited number of rootstocks in the viticulture industry is expected to change in the coming years. The large-scale application of microsatellite markers has become the preferred and most reliable tool for *Vitis* spp. identification, although data on rootstock genotyping are very limited. There are rootstock collections in different centers around the world and these are of great importance for rootstock and variety breeding studies. One of the richest of these rootstock collections is at the University of Milan in Italy. It was established to collect most of the genetic diversity of *Vitis* species useful for rootstock genetic breeding programs. The idea was to select the most suitable parents for new breeding programs to develop sustainable viticulture models [157]. In another grapevine rootstock breeding study carried out in the same institution, a study was conducted on drought, which is an important problem for many vineyard regions. Although the grapevine is not very sensitive to drought, irrigation is very important in terms of fruit quality, especially in areas where table grapes are

Primer	Forward primer Sequence 5′–3′	Reverse primer Sequence 5′–3′	Tm (°C)	Band size (bp)	References
SCC8	GGTGTCAAGTTGGAAGATGG	TATGCCAAAAACATCCCC	60	1018	Lahogue et al. [140]
SCF27	CAGGTGGGAGTAGTGGAATG	CAGGTGGGAGTAAGATTTGT	62	2000	Mejía and Hinrichsen [148]
P3_VvAGL11	CTCCCTTTCCCTCTCCCTCT	AAACGCGTATCCCAATGAAG	Touch down	198/188	Bergamini et al. [143]
VMC7F2	AAGAAAGTTTGCAGTTTATGGTG	AAGATGACAATAGCGAGAGAA	61	198	Adam-Blondon et al. [151]
GSLP1	CCAGTTCGCCCGTAAATG	—	32	569	Wang and Lamikanra [152]
ScORA7-760	GAAACGGGTGTGAGGCAAAGGTGG	GGCCATTAGGAAATCAACATTAC	56	760	Akkurt et al. [153]
OPB 15_{1274}	CGGCAGACCTTACTGAAAGGATGAGT	CAACCACCATCCAATGATGGCGGGCTTC	37	1274	Kim et al. [154]
S382-615	TGGGCGTCAA	—	36	615	Zhang et al. [155, 156]

Table 1.
Primers, sequences, and references are used by different researchers to identify seedless hybrid genotypes [56].

grown. For this reason, a study was carried out to develop more suitable rootstocks for areas with drought problems. Researchers assessed the drought tolerance of M-rootstocks physiological (gas exchange and stem water potential) and transcriptomic performances (genes involved in ABA synthesis and ABA-mediated responses to drought) were evaluated under well-watered and water-stressed conditions. In the study, drought-resistant genotypes were determined by using novel genotypes (M-rootstocks) developed by the University of Milan [158].

Due to the demand for new grapevine genotypes that are especially resistant to biotic and abiotic stress conditions, grape and rootstock breeding studies have increased in recent years and many studies have been carried out on gene regions that may be associated with these stress conditions, especially with the help of molecular methods. Due to their different advantages as rootstock, the most preferred species are *V. berlandieri*, *V. riparia* and *V. rupestris*. The rootstocks not only protect the grapevines from phylloxera pests but also provide the water and nutrients needed by the cultivar/genotype in different soil types. It has been reported that different rootstock species show differences in supply the nutritional needs of the grafted cultivars/genotypes. Gautier et al. [159] investigated the extent to which rootstocks with different genetic backgrounds modify the mineral composition of the petioles of the scion. In a study conducted by grafting vines of the Cabernet-Sauvignon cultivar on 13 different rootstock genotypes, it was reported that the mineral content of the petioles of the genotypes containing *Vitis riparia* in their genetic history can vary greatly. This is important proof that rootstocks can greatly affect mineral intake. An example of the effective use of rootstocks against different biotic stress factors is the study of Dalbó and Souza [160]. Researchers used rootstocks for grapevine decline and dieback, which is a big problem especially in Southern Brazil. This disease is characterized a set of symptoms that lead to the weakening and death of affected plants. In soils with insufficient aeration, root rot fungi can seriously damage the roots of vines and even cause death. The development of root rot resistant grape rootstocks is the one of the main goal in grapevine and rootstock breeding programs. In these breeding programs, first of all, resistant cultivars/genotypes should be determined. In one of the studies, some of the *Vitis caribaea* hybrids were resistant to root rot, while at the same time some undesirable characteristics such as overgrowth and absence of winter dormancy were detected. *Vitis palmata* and *Vitis shuttleworthii* species were also reported to have high resistance. But, the performance of the genotypes and their hybrids was very poor in dry soil conditions and further cross-breeding programs would be required to eliminate undesirable characteristics. Selected lines are evaluated for productivity and fruit quality in rootstock trials with different scion cultivars.

Resistance to phylloxera has been investigated for a long time at UC Davis, which has been working on grape rootstocks and breeding for many years. Recent studies here are investigating the evolution of possible phylloxera strains capable of aggressive nodosite feeding against some resistant rootstocks and foliar feeding strains once rare in the region. In these studies, the reproduction of phylloxera strains was investigated and molecular markers were developed for physical maps associated with resistance genes. As a result of the studies, rootstocks resistant to aggressive root knot nematodes have been developed [161].

Different studies are carried out on rootstock breeding, especially in the USA, Brazil, Europe, Australia, Iran, and China. In these studies, crosses are made in order to develop new rootstocks that are tolerant or resistant to different biotic and abiotic stress conditions, and then their compatibility with the varieties grafted on these rootstocks and their effects on fruit quality are examined. Since rootstock breeding studies are much more laborious and time-consuming than other cultivar breeding programs, they are preferred by a limited number of researchers. However, as the mechanisms that cause stress factors become more understandable at the physiological and genetic level, much more successful breeding programs with promising results will become more preferable all over the world. In addition, with interspecies breeding, there is an increasing number of studies in the world that these new interspecies hybrid genotypes can be used directly in growing instead of using rootstock. Because it is reported that new interspecies hybrids will be much more tolerant in terms of different stress conditions and therefore the need for rootstock usage may decrease.

3. Conclusion

Since grapes are among the most traded fruit types in the world, they have a significant impact on the agricultural economies of the countries. In particular, grapes are preferred by many people due to their different consumption patterns and important effects on human health. However, the demands of consumers can vary rapidly. In order to respond to these rapidly changing consumer demands, quality new grape cultivars should be developed through breeding programs. In these breeding programs, while mainly trying to develop new table and wine grape cultivars, rootstock breeding studies have started to increase in recent years. As a result of research studies carried out by many researchers on the grapevine genome, many characters related to gene regions have been determined and it has become possible to reach the targeted results in grapevine breeding studies in a much shorter time. In recent years, the number of new grape cultivars that are more resistant to biotic and abiotic stress conditions has been increasing rapidly in parallel with climate change and consumer preferences. It is predicted that in the coming years, hybrid cultivars between species will have a greater share in the market with their more resistance to different stress conditions and more friendly characteristics in terms of human and environmental health. As a result of the intensive studies carried out by many scientists with the grapevine genome, it is expected that transgenic grapevine plants, which are more accepted by many parts of the society, will take their place in the markets in a short time. With the integration of biotechnology into breeding programs in a way that will address the ethical concerns of consumers, breeding studies will gain momentum and it will be possible to feed the growing world population and make more sustainable viticulture.

Funding

This research did not receive any external funding.

Conflicts of interest

The author declares that there is no conflict of interest in this article.

Data availability statement

Not applicable here.

Author details

Arif Atak
Agriculture Faculty, Department of Horticulture, Uludag University, Bursa, Turkey

*Address all correspondence to: atakarif@gmail.com; arifatak@uludag.edu.tr

References

[1] FAOSTAT. Available from: https://www.fao.org/faostat/en/#data/QCL [Accessed: February 10, 2022]

[2] OEC. Available from: https://oec.world/en/profile/hs/grapes [Accessed: May 01, 2022]

[3] Alston JM, Sambucci O. Grapes in the world economy. In: Cantu D, Walker A, editors. The Grape Genome. Basel: Springer Nature; 2019. pp. 1-24. DOI: 10.1007/978-3-030-18601-2_1

[4] Dong X, Chen W, Liang Z, Li X, Nick P, Chen S, et al. VitisGDB: The multifunctional database for grapevine breeding and genetics. Molecular Plant. 2020;**13**(8):1098-1100. DOI: 10.1016/j.molp.2020.05.002

[5] Reisch BI, Owens CL, Cousins PS. In: Badenes ML, Byrne DH, editors. "Grape," in Fruit Breeding. New York, NY: Springer; 2012. pp. 225-262. DOI: 10.1007/978-1-4419-0763-9_7

[6] Atak A, Şen A. A grape breeding programme using different *Vitis* species. Plant Breeding. 2021;**140**:1136-1149. DOI: 10.1111/pbr.12970

[7] Paul HW. Science, Vine, and Wine in Modern France. Cambridge: Cambridge University Press; 1996

[8] Campbell C. The botanist and the vintner: How wine was saved for the world. Chapel Hill: Algonquin Books of Chapel Hill; 2005

[9] Cahoon GA. French hybrid grapes in North America. In: Ferree DC, editor. A History of Fruit Cultivars. Yakima, Washington: Good Fruit Grower Magazine; 1998. pp. 152-168

[10] Kambiranda D, Obuya J, Snowden J. Grapevine improvement through biotechnology. In: Genetic Transformation in Crops. London, UK: IntechOpen; 2020. DOI: 10.5772/intechopen.91851

[11] Ferreira R, Monteiro S, Picarra-Pereira M, Teixeira AR. Engineering grapevine for increased resistance to fungal pathogens without compromising wine stability. Trends in Biotechnology. 2004;**22**(4):168-173. DOI: 10.1016/j.tibtech.2004.02.001

[12] Naidu RA, Maree HJ, Burger JT. Grapevine leafroll disease and associated viruses: A unique pathosystem. Annual Review of Phytopathology. 2015;**53**:613-634. DOI: 10.1146/annurev-phyto-102313-045946

[13] Armijo G, Espinoza C, Loyola R, Restovic F, Santibáñez C, Schlechter R, et al. Grapevine biotechnology: Molecular approaches underlying abiotic and biotic stress responses. In: Morata A, Loira I, editors. Grape and Wine Biotechnology [Internet]. London: IntechOpen; pp. 1-40. Available from: https://www.intechopen.com/chapters/52240 DOI: 10.5772/64872

[14] Lund K. Grapevine breeding programs for the wine industry. Traditional and molecular techniques. Woodhead Publishing Series in Food Science, Technology and Nutrition. Oxford, UK: Woodhead Publishing; 2015. pp. 359-378. DOI: 10.1016/B978-1-78242-075-0.00015-6

[15] Montaigne E, Coelho A, Zadmehran SA. Comprehensive economic examination and prospects on innovation in new grapevine varieties dealing with global warming and fungal diseases.

Sustainability. 2021;**13**(23):13254. DOI: 10.3390/su132313254

[16] Kicherer A, Herzog K, Bendel N, Klück H-C, Backhaus A, Wieland M, et al. Phenoliner: A new field phenotyping platform for grapevine research. Sensors. 2017;**17**:1625. DOI: 10.3390/s17071625

[17] Rist F, Herzog K, Mack J, Richter R, Steinhage V, Töpfer R. High-precision phenotyping of grape bunch architecture using fast 3D sensor and automation. Sensors. 2018;**18**:763. DOI: 10.3390/s18030763

[18] Zhu NY, Liu X, Liu ZQ, Hu K, Wang YK, Tan JL, et al. Deep learning for smart agriculture: Concepts, tools, applications, and opportunities. International Journal of Agricultural and Biological Engineering. 2018;**11**:32-44. DOI: 10.25165/ijabe.v11i4.4475

[19] Grimm J, Herzog K, Rist F, Kicherer A, Töpfer R, Steinhage V. An adaptable approach to automated visual detection of plant organs with applications in grapevine breeding. Biosystems Engineering. 2019;**183**:170-183. DOI: 10.1016/j.biosystemseng.2019.04.018

[20] Martinson TE, Reisch B. Grapevine Breeding and Genetics, Grapes 101, Appellation Cornell. USA: Cornell University; 2020. DOI: hdl.handle.net/1813/103808

[21] Wang Q, Li P, Hanania U, Sahar N, Mawassi M, Gafny R, et al. Improvement of Agrobacterium mediated transformation efficiency and transgenic plant regeneration of *Vitis vinifera* L. by optimizing selection regimes and utilizing cryopreserved cell suspensions. Plant Science. 2005;**168**:565-571. DOI: 10.1016/j.plantsci.2004.09.033

[22] Kikkert JK, Ali GS, Wallace PG, Reisch B, Reustle GM. Expression of fungal chitinase in *Vitis vinifera* L. 'Merlot' and 'Chardonnay' plants produced by biolistic information. Acta Horticulturae. 2000;**528**:299-306. DOI: 10.17660/ActaHortic.2000.528.41

[23] Franks T, He DG, Thomas MR. Regeneration of transgenic *Vitis vinifera* L. Sultana plants: Genotypic and phenotypic analysis. Molecular Breeding. 1998;**4**:321-333. DOI: 10.1023/A:1009673619456

[24] Iocco P, Franks T, Thomas MR. Genetic transformation of major vine grape cultivars of *Vitis vinifera* L. Transgenic Research. 2001;**10**:105-112. DOI: 10.1023/a:1008989610340

[25] Colby SM, Juncosa AM, Meredith CP. Cellular differences in Agrobacterium susceptibility and regenerate capacity restrict the development of transgenic grapevines. Journal of the American Society for Horticultural Science. 1991;**166**:356-361. DOI: 10.21273/JASHS.116.2.356

[26] Berres R, Otten L, Tinland B, Malgarini-Clog E, Walter B. Transformation of *Vitis* tissue by different strains of Agrobacterium tumefaciens containing T-6b gene. Plant Cell Reports. 1992;**11**:192-195. DOI: 10.1007/bf00232531

[27] Sabbadini S, Capriotti L, Limera C, Navacchi O, Tempesta G, Mezzetti B. A plant regeneration platform to apply new breeding techniques for improving disease resistance in grapevine rootstocks and cultivars. BIO Web of Conferences. 2019a;**12**:01019. DOI: 10.1051/bioconf/20191201019

[28] Sabbadini S, Capriotti L, Molesini B, Pandolfini T, Navacchi O, Limera C, et al.

Comparison of regeneration capacity and Agrobacterium-mediated cell transformation efficiency of different cultivars and rootstocks of *Vitis* spp. via organogenesis. Scientific Reports. 2019;**9**:582. DOI: 10.1038/s41598-018-37335-7

[29] Maillot P, Deglène-Benbrahim L, Walter B. Efficient somatic embryogenesis from meristematic explants in grapevine (*Vitis vinifera* L.) cv. Chardonnay: An improved protocol, Trees. 2016;**30**:1377-1387. DOI: 10.1007/s00468-016-1374-9

[30] Zhang XM, Wu YF, Li Z, Song CB, Wang XP. Advancements in plant regeneration and genetic transformation of grapevine (*Vitis* spp.). Journal of Integrative Agriculture. 2021;**20**(6):1407-1434. DOI: 10.1016/S2095-3119(20)63586-9

[31] Badouin H, Velt A, Gindraud F, et al. The wild grape genome sequence provides insights into the transition from dioecy to hermaphroditism during grape domestication. Genome Biology. 2020;**21**:223. DOI: 10.1186/s13059-020-02131-y

[32] Mortazavi A, Williams BA, McCue K, Schaeffer L, Wold B. Mapping and quantifying mammalian transcriptomes by RNA-seq. Nature Methods. 2008;**5**(7):621-628. DOI: 10.1038/nmeth.1226

[33] Gambino G, Dal Molin A, Boccacci P, Minio A, Chitarra W, Avanzato CG, et al. Whole-genome sequencing and SNV genotyping of 'Nebbiolo' (*Vitis vinifera* L.) clones. Scientific Reports. 2017;**7**(1):17294. DOI: 10.1038/s41598-017-17405-y

[34] Minio A, Massonnet M, Figueroa-Balderas R, Vondras AM, Blanco-Ulate B, Cantu D. Iso-Seq allows genome-independent transcriptome profiling of grape berry development. G3 (Bethesda, MD). 1 March 2019;**9**(3):755-767. DOI: 10.1534/ g3.118.201008

[35] Cantu D, Walker A. In: Cantu D, Walker MA, editors. The Grape Genome. Cham: Springer International Publishing (Compendium of Plant Genomes); 2019:367. DOI: 10.1007/978-3-030-18601-2

[36] Cheng B, Furtado A, Henry RJ. Long-read sequencing of the coffee bean transcriptome reveals the diversity of full-length transcripts. Gigascience. 2017;**6**(11):1-13. DOI: 10.1093/gigascience/gix086

[37] Kim MA, Rhee JS, Kim TH, Lee JS, Choi AY, Choi BS, et al. Alternative splicing profile and sex-preferential gene expression in the female and male Pacific *Abalone Haliotis discus hannai*. Genes. 2017;**8**(3):99. DOI: 10.3390/genes8030099

[38] Li J, Harata-Lee Y, Denton MD, Feng Q, Rathjen JR, Qu Z, et al. Long read reference genome-free reconstruction of a full-length transcriptome from *Astragalus membranaceus* reveals transcript variants involved in bioactive compound biosynthesis. Cell Discovery. 2017a;**3**:17031. DOI: 10.1038/celldisc.2017.31

[39] Zhu FY, Chen MX, Ye NH, Shi L, Ma KL, et al. Proteogenomic analysis reveals alternative splicing and translation as part of the abscisic acid response in *Arabidopsis* seedlings. The Plant Journal. 2017;**91**(3):518-533. DOI: 10.1111/tpj.13571

[40] Zou C, Karn A, Reisch B, et al. Haplotyping the *Vitis c*ollinear core genome with rhAmpSeq improves

marker transferability in a diverse genus. Nature Communications. 2020;**11**:413. DOI: 10.1038/s41467-019-14280-1

[41] Zaman QU, Li C, Cheng H, Hu Q. Genome editing opens a new era of genetic improvement in polyploid crops. Crop Journal. 2019;7(2):141-150. DOI: 10.1016/j.cj.2018.07.004

[42] Li G, Jain R, Chern M, Pham NT, Martin JA, et al. The sequences of 1504 mutants in the model rice cultivar Kitaake facilitate rapid functional genomic studies. Plant Cell. 2017b;**29**(6):1218-1231. DOI: 10.1105/tpc.17.00154

[43] Bhat AI, Rao GP. Characterization of plant viruses. New York, NY: Humana; 2020. pp. 479-489. DOI: 10.1007/978-1-0716-0334-5_48

[44] Golino DA, Fuchs M, Sim S, Farrar K, Martelli G. Improvement of grapevine planting stock through sanitary selection and pathogen elimination. In: Meng B, Martelli G, Golino D, Fuchs M, editors. Grapevine Viruses: Molecular Biology, Diagnostics and Management. Cham: Springer; 2017. DOI: 10.1007/978-3-319-57706-7_27

[45] Julião SA, Ribeiro CDV, Lopes JML, Matos EMD, Reis AC, et al. Induction of synthetic polyploids and assessment of genomic stability in *Lippia alba*. Frontiers in Plant Science. 2020;**11**:292. DOI: 10.3389/fpls.2020.00292

[46] Longa YL, Qiaoc F, Jianga XF, Congc HQ, Suna ML, Xu ZJ. Screening and analysis on the differentially expression genes between diploid and autotetraploid watermelon by using of digital gene expression profile. Brazilian Journal of Biology. 2019;**79**(2):180-190. DOI: 10.1590/1519-6984.174475

[47] Sattler MC, Carvalho CR, Clarindo WR. The polyploidy and its key role in plant breeding. Planta. 2016;**243**:281-296. DOI: 10.1007/s00425-015-2450-x

[48] Touchell DH, Palmer IE, Ranney TG. In vitro ploidy manipulation for crop improvement. Frontiers in Plant Science. 2020;**11**:722. DOI: 10.3389/fpls.2020.00722

[49] Kuliyev VM. The study of polyploid mutant forms of grapes. Cytology & Histology International Journal. 2020;**4**(1):1-6

[50] Xie X, Agüero CB, Wang Y, Walker MA. In vitro induction of tetraploids in *Vitis* X *Muscadinia* hybrids. Plant Cell, Tissue and Organ Culture. 2015;**122**:675-683. DOI: 10.1007/s11240-015-0801-8

[51] Sinski I, Dal Bosco D, Pierozzi NI, Maia JDG, Ritschel PS, Quecini V. Improving in vitro induction of autopolyploidy in grapevine seedless cultivars. Euphytica. 2014;**196**:299-311. DOI: 10.1007/s10681-013-1034-8

[52] Acanda Y, Martínez Ó, González MV, Prado MJ, Rey M. Highly efficient in vitro tetraploid plant production via colchicine treatment using embryogenic suspension cultures in grapevine (*Vitis vinifera* cv. Mencía). Plant Cell, Tissue and Organ Culture. 2015;**123**:547-555. DOI: 10.1007/s11240-015-0859-3

[53] Zlenko VA, Likhovskoy VV, Volynkin VA, Khvatkov PA, Vasylyk IA, Dolgov SV. Induction of in vitro somatic embryogenesis in grapes (*Vitis vinifera* L.) of domestic and foreign breeding. Biotechnology. 2017;**33**(5):35-44. DOI: 10.21519/02342758-2017-33-5-35-44

[54] Klimenko V, Lushchay E, Zlenko V. BIO Web Conferences. 2021;**34**:03002. DOI: 10.1051/bioconf/20213403002

[55] Guo Y, Zhao Y, Li K, Liu Z, Lin H, Guo X, et al. Embryo rescue of crosses between diploid and tetraploid grape cultivars and production of triploid plants. African Journal of Biotechnology. 2011;**10**(82):19005-19010. DOI: 10.5897/AJB11.977

[56] Li S, Li Z, Zhao Y, Zhao J, Luo Q, Wang Y. New disease-resistant, seedless grapes are developed using embryo rescue and molecular markers. 3 Biotech. 2020;**10**:4. DOI: 10.1007/s13205-019-1993-0

[57] Giancaspro A, Mazzeo A, Carlomagno A, Gadaleta A, Somma S, Ferrara G. Optimization of an in vitro embryo rescue protocol for breeding seedless table grapes (*Vitis vinifera* L.) in Italy. Horticulturae. 2021;**8**:121. DOI: 10.3390/horticulturae8020121

[58] Sun L, Zhang G, Yan A, Xu H. The study of triploid progenies crossed between different ploidy grapes. African Journal of. Biotechnology. 2011;**10**(32):5967-5971. DOI: 10.5897/AJB10.1850

[59] Wakana A, Hiramatsu M, Park SM, Hanada N, Fukudome I, Yasukochi K. Seed abortion in crosses between diploid and tetraploid grapes and recovery of triploid plants through embryo culture. Journal of the Faculty of Agriculture, Kyushu University. 2003;**48**(1):39-50. DOI: 10.5109/4524

[60] Motosugi H, Naruo T. Production of triploid grape rootstocks by embryo culture and their growth characteristics. Journal of the Japanese Society for Horticultural Science. 2003;**72**(2):107-115

[61] Yang DL, Li W, Li S, Yang XL, Wu JL, Cao ZY. In vitro embryo rescue culture of F1 progenies from crosses between diploid and tetraploid grape cultivars. Plant Growth Regulation. 2007;**51**(1):63-71. DOI: 10.1007/s10725-006-9148-9

[62] Park SM, Wakana A, Kim JH, Jeong CS. Male and female fertility in triploid grapes (*Vitis* complex) with special reference to the production of aneuploid plants. Vitis. 2002;**41**(1):11-20

[63] Wang X, Cheng Z, Zhi S, Xu F. Breeding triploid plants: A review. Czech Journal of Genetics and Plant Breeding. 2016;**52**:41-54. DOI: 10.17221/151/2015-CJGPB

[64] Shengjian Z, Zijuan G, Shuyun Z, Xinzhong Z, Licun Z. The breeding of the new triploid grape cultivar 'Hongbiao seedless' with large berries and high quality. Chinese Agricultural Science Bulletin. 2005;**21**:230-232

[65] Wu JH, Ferguson AR, Murray BG, Jia Y, Datson PM, Zhang J. Induced polyploidy dramatically increases the size and alters the shape of fruit in *Actinidia chinensis*. Annals of Botany. 2012;**109**:169-179. DOI: 10.1093/aob/mcr256

[66] Ruiz M, Oustric J, Santini J, Morillon R. Synthetic polyploidy in grafted crops. Frontiers in Plant Science. 2020;**11**:540894. DOI: 10.3389/fpls.2020.540894

[67] Pieterse CM, Leon-Reyes A, Van der Ent S, Van Wees SC. Networking by small molecule hormones in plant immunity. Nature Chemical Biology. 2009;**5**(5):308-316. DOI: 10.1038/nchembio.164

[68] Qiu W, Feechan A, Dry I. Current understanding of grapevine defense mechanisms against the biotrophic fungus (*Erysiphe necator*), the causal

agent of powdery mildew disease. Horticulture Research. 2015;**2**:15020. DOI: 10.1038/hortres.2015.20

[69] Atak A. Determination of downy mildew and powdery mildew resistance of some grape cultivars. South African Journal of Enology and Viticulture. 2017;**38**(1):11-17. DOI: 10.21548/38-1-671

[70] Gadoury DM, Cadle-Davidson L, Wilcox WF, Dry IB, Seem RC, Milgroom MG. Grapevine powdery mildew (*Erysiphe necator*): A fascinating system for the study of the biology, ecology and epidemiology of an obligate biotroph. Molecular Plant Pathology. 2012;**13**(1):1-16. DOI: 10.1111/j.1364-3703.2011.00728.x

[71] Pauquet J, Bouquet A, This P, Adam-Blondon AF. Establishment of a local map of AFLP markers around the powdery mildew resistance gene Run1 in grapevine and assessment of their usefulness for marker assisted selection. Theoretical and Applied Genetics. 2001;**103**(8):1201-1210. DOI: 10.1007/s001220100664

[72] Coleman C, Copetti D, Cipriani G, Hoffmann S, Kozma P, Kovacs L, et al. The powdery mildew resistance gene REN1 cosegregates with an NBS-LRR gene cluster in two Central Asian grapevines. BMC Genetics. 2009;**10**:89. DOI: 10.1186/1471-2156-10-89

[73] Hoffmann S, Di Gaspero G, Kovacs L, Howard S, Kiss E, Galbacs Z, et al. Resistance to *Erysiphe necator* in the grapevine 'Kishmish vatkana' is controlled by a single locus through restriction of hyphal growth. Theoretical and Applied Genetics. 2008;**116**(3):427-438. DOI: 10.1007/s00122-007-0680-4

[74] Hausmann L, Maul E, Ganesch A, Töpfer R. Overview of genetic loci for traits in grapevine and their integration into the VIVC database. Acta Horticulturae. 2019;**1248**:221-226. DOI: 10.17660/ActaHortic.2019.1248.32

[75] Table of loci for traits in grapevine relevant for breeding and genetics. Update: April 30, 2021. Available from: https://www.vivc.de

[76] Kozma P, Kiss E, Hoffmann S, Galbacs Z, Dula T. Using the powdery mildew resistant *Muscadinia rotundifolia* and *Vitis vinifera* 'Kishmish vatkana' for breeding new cultivars. Acta Horticulturae. 2009;**827**:559-564. DOI: 10.17660/ActaHortic.2009.827.97

[77] Vezzulli S, Vecchione A, Stefanini M, Zulini L. Downy mildew resistance evaluation in 28 grapevine hybrids promising for breeding programs in Trentino region (Italy). European Journal of Plant Pathology. 2017;**150**:485-495. DOI: 10.1007/s10658-017-1298-2

[78] Vezzulli S, Malacarne G, Masuero D, Vecchione A, Dolzani C, Goremykin V, et al. The Rpv3-3 haplotype and stilbenoid induction mediate downy mildew resistance in a grapevine interspecific population. Frontiers in Plant Science. 2019;**10**:234. DOI: 10.3389/fpls.2019.00234

[79] Bove F, Savary S, Willocquet L, Rossi V. Designing a modelling structure for the grapevine downy mildew pathosystem. European Journal of Plant Pathology. 2020;**157**:251-268. DOI: 10.1007/s10658-020-01974-2

[80] Possamai T, Migliaro D, Gardiman M, Velasco R, De Nardi B. Rpv mediated defense responses in grapevine offspring resistant to *Plasmopara viticola*. Plants. 2020;**9**:781. DOI: 10.3390/plants9060781

[81] Eibach R, Zyprian E, Welter L, Töpfer R. The use of molecular markers for pyramiding resistance genes

in grapevine breeding. Vitis. 2007;**46**:120-124

[82] Zini E, Dolzani C, Stefanini M, Gratl V, Bettinelli P, Nicolini D, et al. R-loci arrangement versus downy and powdery mildew resistance level: A *Vitis* hybrid survey. International Journal of Molecular Sciences. 2019;**20**:3526. DOI: 10.3390/ijms2014

[83] Yıldırım Z, Atak A, Akkurt M. Determination of downy and powdery mildew resistance of some *Vitis* spp. Ciência e Técnica Vitivinícola. 2019;**34**(1):15-24. DOI: 10.1051/ctv/20193401015

[84] Shidfar M, Akkurt M, Atak A, Ergül A, Söylemezoğlu G. Evaluation of grapevine resistance to downy and powdery mildew in 'Regent' × 'Boğazkere' hybrid population segregating for resistance genes. Acta Scientiarum Polonorum Hortorum Cultus. 2019;**18**(1):181-188. DOI: 10.24326/asphc.2019.1.18

[85] Brown MV, Moore JN, Fenn P, McNew RW. Comparison of leaf disk, greenhouse, and field screening procedures for evaluation of grape seedlings for downy mildew resistance. HortScience. 1999;**34**:331-333. DOI: 10.21273/HORTSCI.34.2.331

[86] Boso S, Martinez MC, Unger S, Kassemeyer HH. Evaluation of foliar resistance to downy mildew in different cv. Albariño clones. Vitis. 2006;**4**:23-27

[87] Boso S, Alonso-Villaverde V, Santiago JL, Gago P, Dürrenberg M, Duggelin M, et al. Macro- and microscopic leaf characteristics of six grapevine genotypes (*Vitis* spp.) with different susceptibilities to grapevine downy mildew. Vitis. 2010;**49**:43-50

[88] Boso S, Alonso-Villaverde V, Gago P, Santiago JL, Martínez MC. Susceptibility to downy mildew (*Plasmopara viticola*) of different *Vitis* varieties. Crop Protection. 2014;**63**:26-35. DOI: 10.1016/j.cropro.2014.04.018

[89] Prajongjai T, Poolsawat O, Pornbungkerd P, Wongkaew S, Tantasawat PA. Evaluation of grapevines for resistance to downy mildew (*Plasmopara viticola*) under laboratory and field conditions. South African Journal of Enology and Viticulture. 2014;**35**:43-50. DOI: 10.21548/35-1-983

[90] Atak A, Akkurt M, Polat Z, Çelik H, Kahraman KA, Akgül DS, et al. Susceptibility to downy mildew (*Plasmopara viticola*) and Powdery mildew (*Erysiphe necator*) of different Vitis cultivars and genotypes. Ciência e Técnica Vitivinícola. 2017;**32**(1):23-32. DOI: 10.1051/ctv/20173201023

[91] Fischer BM, Salakhutdinov I, Akkurt M, Eibach R, Edwards KJ, Töpfer R, et al. Quantitative trait locus analysis of fungal disease resistance factors on a molecular map of grapevine. Theoretical and Applied Genetics. 2004;**108**:501-515. DOI: 10.1007/s00122-003-1445-3

[92] Welter LJ, Göktürk-Baydar N, Akkurt M, Maul E, Eibach R, Töpfer R, et al. Genetic mapping and localization of quantitative trait loci affecting fungal disease resistance and leaf morphology in grapevine (*Vitis vinifera* L). Molecular Breeding. 2007;**20**:359-374. DOI: 10.1007/s11032-007-9097-7

[93] Zendler D, Topfer R, Zyprian E. Confirmation and fine mapping of the resistance locus Ren9 from the grapevine cultivar 'Regent'. Plants. 2021;**10**(1):24. DOI: 10.3390/plants10010024

[94] Karn A, Zou C, Brooks S, Fresnedo-Ramírez J, Gabler F, Sun Q,

et al. Discovery of the REN11 locus from *Vitis aestivalis* for stable resistance to grapevine powdery mildew in a family segregating for several unstable and tissue-specific quantitative resistance loci. Frontiers in Plant Science. 2021;**12**:733899. DOI: 10.3389/fpls.2021.733899

[95] Dry I, Feechan A, Anderson C, Jermakow A, Bouquet A, Adam-Blondon A, et al. Molecular strategies to enhance the genetic resistance of grapevines to powdery mildew. Australian Journal of Grape and Wine Research. 2010;**16**:94-105. DOI: 10.1111/j.1755-0238.2009.00076.x

[96] Katula-Debreceni D, Lencsés AK, Szőke A, Veres A, Hoffmann S, Kozma PK, et al. Marker-assisted selection for two dominant powdery mildew resistance genes introgressed into a hybrid grape population. Scientia Horticulturae. 2010;**126**(4):448-453. DOI: 10.1016/j.scienta.2010.08.012

[97] Cadle-Davidson LE. A perspective on breeding and implementing durable powdery mildew resistance. Acta Horticulturae. 2019;**1248**:541-548. DOI: 10.17660/ActaHortic.2019.1248.72

[98] Cavaco AR, Figueiredo J, Laureano G, Sousa Silva M, Matos AR, Figueiredo A. Subtilisin-like proteins and lipid signalling events: The missing links in grapevine resistance to *Plasmopara viticola*. Acta Horticulturae. 2019;**1248**:567-574. DOI: 10.17660/ActaHortic.2019.1248.76

[99] Romero-Perez AI, Lamuela-Raventos MR, Andres-Lacueva C, de La Torre-Boronat MC. Method for the quantitative extraction of resveratrol and piceid isomers in grape berry skins. Effect of powdery mildew on the stilbene content. Journal of Agricultural and Food Chemistry. 2001;**49**:210-215

[100] Atak A, Göksel Z, Çelik H. Relations between downy/powdery mildew diseases and some phenolic compounds in *Vitis* spp. Turkish Journal of Agriculture and Forestry. 2017;**2017**(41):69-81. DOI: 10.3906/tar-1610-61

[101] Atak A, Göksel Z, Yılmaz Y. Changes in major phenolic compounds of seeds, skins, and pulps from various *Vitis* spp. and the effect of powdery and downy mildew diseases on their levels in grape leaves. Plants. 2021;**10**:2554. DOI: 10.3390/plants10122554

[102] Ricciardi V, Marcianò D, Sargolzaei M, Fassolo EM, et al. Dissecting the susceptibility/resistance mechanism of *Vitis vinifera* for the future control of downy mildew. BIO Web Conferences. 2022;**44**:4002. DOI: 10.1051/bioconf/20224404002

[103] Ciubotaru RM, Franceschi P, Zulini L, Stefanini M, Škrab D, et al. Mono-locus and pyramided resistant grapevine cultivars reveal early putative biomarkers upon artificial inoculation with *Plasmopara viticola*. Frontiers in Plant Science. 2021;**12**:693887. DOI: 10.3389/fpls.2021.693887

[104] Gray DJ, Li ZT, Dhekney SA. Precision breeding of grapevine (*Vitis vinifera* L.) for improved traits. Plant Science. 2014;**228**:3-10. DOI: 10.1016/j.plantsci.2014.03.023

[105] Nirala NK, Das DK, Srivastava PS, Sopory SK, Upadhyaya KC. Expression of a rice chitinase gene enhances antifungal potential in transgenic grapevine (*Vitis vinifera* L.). Vitis. 2010;**49**(4):181-187

[106] Villano C, Aversano R. Towards grapevine (*Vitis vinifera* L.) mildews resistance: Molecular defence mechanisms and new breeding

technologies. Italus Hortus. 2020;**27**:1-17. DOI: 10.26353/j.itahort/2020.3.0117

[107] Ruiz-García L, Gago P, Martínez-Mora C, Santiago JL, Fernádez-López DJ, Martínez MC, et al. Evaluation and pre-selection of new grapevine genotypes resistant to downy and powdery mildew, obtained by cross-breeding programs in Spain. Frontiers in Plant Science. 2021;**12**:674510. DOI: 10.3389/fpls.2021.674510

[108] Lukšić K, Zdunić G, Hančević K, Mihaljević MJ, Mucalo A, Maul E, et al. Identification of powdery mildew resistance in wild grapevine (*Vitis vinifera* subsp. *sylvestris* Gmel Hegi) from Croatia and Bosnia and Herzegovina. Scientific Reports. 2022;**12**:2128. DOI: 10.1038/s41598-022-06037-6

[109] Pirrello C, Malacarne G, Moretto M, Lenzi L, Perazzolli M, Zeilmaker T, et al. Grapevine DMR6-1 is a candidate gene for susceptibility to downy mildew. Biomolecules. 2022;**12**:182. DOI: 10.3390/biom12020182

[110] Ollat N, Cookson SJ, Destrac-Irvine A, Lauvergeat V, et al. Grapevine adaptation to abiotic stress: An overview. Acta Horticulturae. 2019;**1248**:497-512. DOI: 10.17660/ActaHortic.2019.1248.68

[111] Gambetta GA, Herrera JC, Dayer S, Feng Q, Hochberg U, Castellarin SD. The physiology of drought stress in grapevine: Towards an integrative definition of drought tolerance. Journal of Experimental Botany. 2020;**71**(16):4658-4676. DOI: 10.1093/jxb/eraa245

[112] Marguerit E, Boury C, Manicki A, Donnart M, Butterlin G, Némorin A, et al. Genetic dissection of sex determinism, inflorescence morphology and downy mildew resistance in grapevine. Theoretical and Applied Genetics. 2009;**118**(7):1261-1278. DOI: 10.1007/s00122-009-0979-4

[113] Simonneau T, Lebon E, Coupel-Ledru A, Marguerit E, Rossdeutsch L, Ollat N. Adapting plant material to face water stress in vineyards: Which physiological targets for an optimal control of plant water status? OENO One. 2017;**51**(2):167-179. DOI: 10.20870/oeno-one.2017.51.2.1870

[114] Padgett-Johnson M, Williams LE, Walker MA. Vine water relations, gas exchange, and vegetative growth of seventeen *Vitis* species grown under irrigated and nonirrigated conditions in California. Journal of the American Society for Horticultural Science. 2003;**128**(2):269-276. DOI: 10.21273/JASHS.128.2.0269

[115] Chaves MM, Zarrouk O, Francisco R, Costa JM, Santos T, Regalado AP, et al. Grapevine under deficit irrigation: Hints from physiological and molecular data. Annals of Botany. 2010;**105**(5):661-676. DOI: 10.1093/aob/mcq030

[116] Serra I, Strever A, Myburgh PA, Deloire A. Review: The interaction between rootstocks and cultivars (*Vitis vinifera* L.) to enhance drought tolerance in grapevine. Australian Journal of Grape and Wine Research. 2014;**20**(1):1-14. DOI: 10.1111/ajgw.12054

[117] Tomás M, Medrano H, Escalona JM, Martorell S, Pou A, Ribas-Carbó M, et al. Variability of water use efficiency in grapevines. Environmental and Experimental Botany. 2014;**103**:148-157. DOI: 10.1016/j.envexpbot.2013.09.003

[118] Luchaire N, Rienth M, Romieu C, Nehe A, Chatbanyong R, Houel C, et al. Microvine: A new model to study grapevine growth and developmental

patterns and their responses to elevated temperature. American Journal of Enology and Viticulture. 2017;**68**(3): 283-292. DOI: 10.5344/ajev.2017.16066

[119] Torregrosa L, Bigard A, Doligez A, Lecourieux D, Rienth M, Luchaire N, et al. Developmental, molecular and genetic studies on the grapevine response to temperature open breeding strategies for adaptation to warming. OENO One. 2017;**51**(2):155-165. DOI: 10.20870/oeno-one.2017.51.2.1587

[120] Pons A, Allamy L, Schüttler A, Rauhut D, Thibon C, Darriet P. What is the expected impact of climate change on wine aroma compounds and their precursors in grape? OENO One. 2017;**51**(2):141-146. DOI: 10.20870/oeno-one.2017.51.2.1868

[121] Duchêne E, Butterlin G, Claudel P, Jaegli N. Consequences of elevated temperatures during ripening on the biosynthesis of monoterpenols in grape berries. In: Sustainable Grape and Wine Production in the Context of Climate Change – Climwine2016; Bordeaux, France, 10-13 April 2016

[122] Ollat N, Peccoux A, Papura D, Esmenjaud D, Marguerit E, et al. Rootstocks as a component of adaptation to environment. In: Geros H, Chaves MM, Medrano H, Delrot S, editors. Grapevine in a Changing Environment: A Molecular and Ecophysiological Perspective. 1st ed. Chichester, UK: John Wiley & Sons; 2016. pp. 68-108

[123] Bert PF, Bordenave L, Donnart M, Hévin C, Ollat N, Decroocq S. Mapping genetic loci for tolerance to lime-induced iron deficiency chlorosis in grapevine rootstocks (*Vitis* sp.). Theoretical and Applied Genetics. 2013;**126**(2):451-473. DOI: 10.1007/s00122-012-1993-5

[124] Henderson SW, Dunlevy JD, Wu Y, Blackmore DH, Walker RR, Edwards EJ, et al. Functional differences in transport properties of natural HKT1;1 variants influence shoot Na+ exclusion in grapevine rootstocks. The New Phytologist. 2018;**217**(3):1113-1127. DOI: 10.1111/nph.14888

[125] Gautier A, Cookson SJ, Lagalle L, Ollat N, Marguerit E. Influence of the three main genetic backgrounds of grapevine rootstocks on petiolar nutrient concentrations of the scion, with a focus on phosphorus. OENO One. 2020;**54**:1-13. DOI: 10.20870/oeno-one.2020.54.1.2458

[126] Chenu K, Van Oosterom EJ, McLean G, Deifel KS, Fletcher A, Geetika G, et al. Integrating modelling and phenotyping approaches to identify and screen complex traits: Transpiration efficiency in cereals. Journal of Experimental Botany. 2018;**69**(13):3181-3194. DOI: 10.1093/jxb/ery059

[127] Tardieu F, Simonneau T, Muller B. The physiological basis of drought tolerance in crop plants: A scenario-dependent probabilistic approach. Annual Review of Plant Biology. 2018;**69**(1):733-759. DOI: 10.1146/annurev-arplant-042817-040218

[128] Varshney RK, Thudi M, Pandey MK, Tardieu F, Ojiewo C, et al. Accelerating genetic gains in legumes for the development of prosperous smallholder agriculture: Integrating genomics, phenotyping, systems modelling and agronomy. Journal of Experimental Botany. 2018;**69**(13):3293-3312. DOI: 10.1093/jxb/ery088

[129] Akkurt M, Tahmaz H, Veziroglu S. Recent developments in seedless grapevine breeding. South African Journal of Enology

and Viticulture. 2019;**40**:2. DOI: 10.21548/42-2-3342

[130] Ledbetter CA, Ramming DW. Seedlessness in grapes. Horticultural Reviews. 1989;**11**:159-184

[131] Spiegel-Roy P, Sahar N, Baron J, Lavi U. In vitro culture and plant formation from grape cultivars with abortive ovules and seeds. Journal of the American Society for Horticultural Science. 1985;**110**:109-112

[132] Tian L, Wang Y. Seedless grape breeding for disease resistance by using embryo rescue. Vitis. 2008;**47**:15-19

[133] Tian L, Wang Y, Niu L, Tang D. Breeding of disease-resistant seedless grapes using Chinese wild *Vitis* spp. I. In vitro embryo rescue and plant development. Scientia Horticulturae. 2008;**117**:136-141. DOI: 10.1016/j.scienta.2008.03.024

[134] Ji W, Wang Y. Breeding for seedless grapes using Chinese wild *Vitis* spp. II. In vitro embryo rescue and plant development. Journal of the Science of Food and Agriculture. 2013;**93**:3870-3875. DOI: 10.1002/jsfa.6342

[135] Ji W, Li ZQ, Zhou Q, Yao WK, Wang YJ. Breeding new seedless grape by means of in vitro embryo rescue. Genetics and Molecular Research. 2013;**12**:859-869. DOI: 10.4238/2013.March.26.1

[136] Li T, Li Z, Yin X, Guo Y, Wang Y, Xu Y. Improved in vitro *Vitis vinifera* L. embryo development of F1 progeny of 'delight' × 'ruby seedless' using putrescine and marker-assisted selection. In Vitro Cellular & Developmental Biology-Plant. 2018;**54**:291-301. DOI: 10.1007/s11627-018-9895-0

[137] Emershad RL, Ramming DW, Serpe MD. In ovulo embryo development and plant formation from stenospermic genotypes of *Vitis vinifera*. American Journal of Botany. 1989;**76**:397-402

[138] Pommer CV, Ramming DW, Emershad RL. Influence of grape genotype, ripening season, seed trace size, and culture date on in ovule embryo development and plant formation. Bragantia. 1995;**54**:237-249

[139] Bouquet A, Danglot Y. Inheritance of seedlessness in grapevine (*Vitis vinifera* L.). Vitis. 1996;**35**:35-42

[140] Lahogue F, This P, Bouquet A. Identification of a codominant scar marker linked to the seedlessness character in grapevine. Theoretical and Applied Genetics. 1998;**97**:950-959. DOI: 10.1007/s001220050976

[141] Pellerone FI, Edwards KJ, Thomas MR. Grapevine microsatellite repeats: Isolation, characterisation and use for genotyping of grape germplasm from Southern Italy. Vitis. 2001;**40**:179-186

[142] Mejía N, Soto B, Guerrero M, et al. Molecular, genetic and transcriptional evidence for a role of VvAGL11 in stenospermocarpic seedless in grapevine. BMC Plant Biology. 2011;**11**:57. DOI: 10.1186/1471-2229-11-57

[143] Bergamini C, Cardone MF, Anaclerio A, Perniola R, Pichierri, et al. Validation assay of p3_VvAGL11 marker in a wide range of genetic background for early selection of stenospermocarpy in *Vitis vinifera* L. Molecular Biotechnology. 2013;**54**(3):1021-1030. DOI: 10.1007/s12033-013-9654-8

[144] Bennici S, Guardo MD, Distefano G, Malfa SL, et al. Influence of the genetic background on the performance of molecular markers linked to seedlessness in table grapes. Scientia Horticulturae.

2019;**252**:316-323. DOI: 10.1016/j.scienta.2019.03.060

[145] Ocarez N, Mejía N. Suppression of the D-class MADS-box AGL11 gene triggers seedlessness in fleshy fruits. Plant Cell Reports. 2016;**35**:239-254. DOI: 10.1007/s00299-015-1882-x

[146] Malabarba J, Buffon V, Mariath JEA, Maraschin FS, Margis-Pinheiro M, Pasquali G, et al. Manipulation of VviAGL11 expression changes the seed content in grapevine (*Vitis vinifera* L.). Plant Science. 2018;**269**:126-135. DOI: 10.1016/j.plantsci.2018.01.013

[147] Royo C, Torres-Perez R, Mauri N, Diestro N, et al. The major origin of seedless grapes is associated with a missense mutation in the MADS-box gene VviAGL11. Plant Physiology. 2018;**177**:1234-1253. DOI: 10.1104/pp.18.00259

[148] Mejía N, Hinrichsen P. A new, highly assertive scar marker potentially useful to assist selection for seedlessness in table grape breeding. Acta Horticulturae. 2003;**603**:559-564

[149] Fatahi R, Zamani Z, Ebadi A, Mehlenbacher SA. The inheritance of seedless SCC8-SCAR and SSRs loci alleles in progeny of 'Muscat Hamburg' x 'Bidane Quermez' grapes. Leuven, Belgium: International Society for Horticultural Science (ISHS); 2004. pp. 329-335. DOI: 10.17660/actahortic.2004.652.42

[150] Conner PJ, Gunawan G, Clark JR. Characterization of the p3-VvAGL11 marker for stenospermocarpic seedlessness in *Euvitis* × *Muscadinia* grape hybrid progenies. Journal of the American Society for Horticultural Science. 2018;**143**:167-172. DOI: 10.21273/jashs 04366-18

[151] Adam-Blondon AF, Lahogue-Esnault F, Bouquet A, et al. Usefulness of two SCAR markers for marker-assisted selection of seedless grapevine cultivars. Vitis. 2001;**40**(3):147-156

[152] Wang YJ, Lamikanra O. Application and synthesis on the DNA probe for detecting seedless genes in grapevine. Journal of Northwest Sci-Tech University of Agriculture and Forestry. 2002;**30**:42-46

[153] Akkurt M, Welter L, Maul E, Töpfer R, Zyprian E. Development of SCAR markers linked to powdery mildew (Uncinula necator) resistance in grapevine (*Vitis vinifera* L. and *Vitis* sp.). Molecular Breeding. 2007;**19**:103-111. DOI: 10.1007/s1103 2-006-9047-9

[154] Kim G et al. Identification of AFLP and RAPD markers linked to anthracnose resistance in grapes and their conversion to SCAR markers. Plant Breeding. 2008;**127**:418-423. DOI: 10.1111/j.1439-0523.2008.01488.x

[155] Zhang Y, Zhang J, Wang Y. Screening the RAPD markers linked to the gene resistant to downy mildew in Chinese wild species of *Vitis*. Journal of Fruit Science. 2008;**25**:816-820

[156] Zhang J, Wang Y, Wang Y, Zhang Y. Marker-assisted selection for hybrids from cross between seedless and disease-resistant grape. Journal of Northeast Agricultural University. 2010;**41**:55-63

[157] Migliaro D, De Lorenzis G, Di Lorenzo GS, De Nardi B, Gardiman M, Failla O, et al. Extensive genotyping of a large collection of rootstocks, population structure analysis and core collection extrapolation for new breeding programs. Acta Horticulturae. 2019;**1248**:301-304. DOI: 10.17660/ActaHortic.2019.1248.43

[158] Bianchi D, Caramanico L, Grossi D, Brancadoro L, Lorenzis GD. How do novel M-rootstock (*Vitis* Spp.) genotypes cope with drought? Plants. 2020;**9**:1385. DOI: 10.3390/plants9101385

[159] Gautier A, Cookson SJ, Hevin C, Vivin P, Lauvergeat V, Mollier A. Phosphorus acquisition efficiency and phosphorus remobilization mediate genotype-specific differences in shoot phosphorus content in grapevine. Tree Physiology. 2018;**38**(11):1742-1751. DOI: 10.1093/treephys/tpy074

[160] Dalbó MA, Souza ALK. Rootstock breeding for resistance to grapevine decline and dieback in southern Brazil. Acta Horticulturae. 2019;**1248**:123-128. DOI: 10.17660/ActaHortic.2019.1248.18

[161] Walker MA, Lund K, Agüero C, Riaz S, Fort K, Heinitz C, et al. Breeding grape rootstocks for resistance to phylloxera and nematodes—It's not always easy. Acta Horticulturae. 2014;**1045**:89-97. DOI: 10.17660/ActaHortic.2014.1045.12

Chapter 14

An Update on Radish Breeding Strategies: An Overview

Raman Selvakumar

Abstract

In tropical, subtropical, and temperate climates, radish (*Raphanus sativus* L.) is a popular root vegetable. Radish diversity is intense from the eastern Mediterranean to the Caspian Sea. Many radish varieties have varied leaf morphology, root color, size, shape, flavor, vernalization requirements, and maturity times. Early radish variants were long and tapered rather than cylindrical, bulbous, elliptic, or spherical. For black Spanish radish, European-cultivated variety, and Asian-cultivated radish, three separate domestication processes occurred. The original radishes were black, followed by white in the 1500s then red and round in the 1700s. These are *R. sativus* L. var. *radicula* (sativus) or *R. sativus* L. var. *niger* radishes. Because of protogyny, self-incompatibility, open architecture, and biennial bolting, radish crosses readily. The fundamental methods for using heterotic breeding potential are SI, CMS, and doubled haploids (DH). This chapter discusses the various breeding strategies like inbred line development by the use of self-incompatibility, hybrid development by using male sterility system, population improvement, mutation breeding, haploid breeding, breeding strategies for biotic and abiotic stresses, QTL mapping, and genome wide and genomic tool in radish. Rapid developments in our understanding of advanced biotechnology technologies will increase our ability to identify cultivars and parental lines, check seed genetic purity, analyze phylogenetic links and genetic diversity, and add specific transgenic traits.

Keywords: radish, breeding, genetics, F_1 hybrid, *Alternaria*, *Fusarium*

1. Introduction

Radish (*Raphanus sativus* L.) is an annual herbaceous plant with two sets of nine chromosomes [1]. It is a member of the Cruciferae family and consumed raw as a salad component, garnish, and shredded radish [2]. Furthermore, radish has been used in cuisines all throughout the globe. Radish is often used in eastern Asian cuisines [3]. Priority is given to the creation of superior radish cultivars suitable for tropical and subtropical climates [4]. Furthermore, breeding research on a variety of agronomic qualities, such as disease resistance and suitability to human use, has been done. High yield, early maturity, late bolting, pungency, cold-hardiness, drought resistance, heat tolerance, and soil adaptation are all important features for radish breeding [2]. There are correlations between the consistency of the radish and its

IntechOpen

sugar concentration, pungency, cell complexity, water content, and pore size [5]. To develop radish varieties, mass selection or pedigree techniques are being used, with an emphasis on the red globe, oval red, and white forms [6]. The most difficult problem has been adapting radish cultivation to many growth seasons [7]. For a successful radish breeding procedure, significant genetic data on chromosomes and inheritance information for numerous genes relevant for agronomic, biochemical, and biotic and abiotic stressors must be collected [8, 9]. It is required to undertake research utilizing novel methods, such as chromosomal or gene modification [10].

Physical attractiveness, including length, form, size, and skin tone, has a significant influence on consumer desire and marketing judgment [11]. The skin is generally white, but it may take on pink, red, purple, yellow, and green tones. Red radishes, on the other hand, are around 40 cm long and have a mild flavor (not as spicy) [5]. The anthocyanin pelargonidin is responsible for red colors, whereas a cyanidin derivative is responsible for purple hues. Even while quality-related characteristics are highly heritable, cultivation practices often have a significant impact on them. Radishes' swelling taproots may take the shape of an oval, tapering, or cylindrical object [12]. Furthermore, cylindrical root variations are often collected mechanically [13]. Radish roots are high in antler velvet, generate beneficial phytochemicals, have cancer-preventive effects, and increase the flavor of Brassica products greatly [14]. Furthermore, radishes provide us with complex carbohydrates, dietary fiber, and organic nutrients and minerals [15].

Omics methodologies based on next-generation sequencing (NGS) techniques provide a significant amount of genomic research [16], which also allows for the distribution and acquisition of positional markers on the chromosomes as well as the identification of new genes and sequences [17]. Furthermore, genome-wide studies show the genetic foundation of some characteristics [18]. The ability to re-sequence genomes allows for genome-wide investigation of important markers and higher-throughput genotyping [19]. Less research has been generated that examines the most crucial historical events as well as current achievements in radish breeding. As a consequence, we have a wealth of information on all aspects of radish breeding and its countless accomplishments in this sector. We anticipate that vegetable breeders will benefit from this chapter in the future.

2. An overview of radish breeding

2.1 Origin and distribution

Radish is an annual vegetable in the Cruciferae family. From the Mediterranean to the Black Sea, the genus Raphanus was separated into the Raphanus DC. and Hesperidopsis Boiss sections. Each part has six species (*R. sativus, Raphanus raphanistrum, R. microcarpus, R. rostratus, R. landra,* and *Rumex maritimus*) and one species (*Raphanus aucheri*) [19]. Kitamura [19] also proposed that *R. sativus* may be grown in the Mediterranean region by natural or artificial hybridization of *R. landra* with *R. maritimus*. Banga [20] and Hida [21] proposed that four wild species (*Raphanistrum raphanistrum*, *R. maritimus, R. landra*, and *R. rostratus*) may have aided in the evolution of radish. Panetsos and Baker's [22] study on wild *R. sativus* and *R. raphanistrum* confirmed the species differentiation. The wild *R. sativus* has a white or partially purple flower on a white background as well as a delicate, fairly thick pod made up of spongy parenchyma. *R. raphanistrum*, on the other hand, bears yellow blooms and slender, robust pods. The ripe pods disintegrate into bits. The aforementioned two species flourished in California in intermediate forms that may have arisen from natural hybridization. The F_1 plants produced by artificially crossing the two species demonstrated intermediate features, including chromosomal configurations of 1IV + 7II at initial metaphase (MI) of pollen mother cells (PMCs) and fertility of 50% in both pollen production and seed setting. Based on these findings, it was proposed that the F_1 plants had reciprocal translocations in one pair of chromosomes. When *R. sativus* wild type was spontaneously backcrossed with cultivated radish, some progeny lacked quadrivalent chromosomes and had a high seed setting rate. It was therefore suggested that gene flow (or introgression) from *R. raphanistrum* into radish cultivars be encouraged. Eber [23, 24] found comparable findings in their study of hybrids between the wild *R. raphanistrum* and the cultivated *R. sativus* in France. In F_1 hybrids of *R. sativus* and *R. raphanistrum* L. ssp. *landra*, Kato and Fukuyama [25] revealed correct chromosomal organization of 9II during meiosis and robust seed setting. Based on these findings, it was hypothesized that *R. raphanistrum* might undergo chromosomal reconstruction. Harberd [26] proposed that the genus *Raphanus* be classified as a cytodeme based on chromosomal number, chromosome layout at MI in PMCs, and fertility studies. This finding was corroborated by Prakash [27]. Tsunoda [28, 29] thought that all wild radish species belonged to R. raphanistrum and evolved around the Mediterranean-Black Sea coast. *R. raphanistrum* was common in Russia and the New World, but it was not found in China, Japan, or India [30, 31]. *Raphanus* was recently separated into two species, *R. sativus* and *R. raphanistrum*, the latter of which contains additional wild species as *R. raphanistrum* subspecies [31]. *R. sativus* var. *hortensis* f. *raphanistroides* Makino [19], also known as Hama-daikon or *R. raphanistrum* ssp. *maritimus* [31], grew wild along the East Asian shoreline. Another kind of wild radish, Nora-daikon or No-daikon, thrived in areas far from the sea. It was thought that these wild radishes were the result of cultivated radishes escaping [19, 32–34] or the migration of weedy radishes tainted with cereals such as wheat and oat. Numerous research, however, support the first point of view. Germplasm resources for understanding the origins of farmed radish and improving the radish crop include Hama-daikon and Nora-daikon. Molecular studies of DNA and genomes, in addition to morphology, ecology, and cytogenetics, may provide insight into the origin, differentiation, and domestication of radish.

2.2 Botany of radish

Radish (*R. sativus* L.), an entomophilous flower, is an allogamous plant [35]. When it appears as three florets at the tip of each branch of the panicle during normal flowering, each flower is capable of producing a pod up to 1 to 3 inches long and holding one to six seeds [36]. The radish blossom's fresh corolla blooms in the morning and lasts till the following day [37]. The flower's pollen receptivity is present only for a brief duration each day, according to Kremer. Its clawed petals, four erected sepals, six stamens, and 1.5 to 2 cm broad, pink to purplish with purple veins, blooms in a 3 to 4 cm long style [38, 39]. A siliqua, sometimes called a seedpod, is a radish seed capsule that is 1.5 cm wide and 3 to 7 cm long. It bears a long, conical, and seedless beak and 6–12 seeds per pod [5]. The inflorescence of the radish is a typical Cruciferae raceme that is long, erect, and rectangular [40, 41]. Radchenko [42] studied the pollination of radish. When Crane and Mather [43] investigated how to cross-pollinate radish, they found that the "Icicle" and "Scarlet Globe" cvs were self-incompatible and pollinated by bees [44]. The research found that the number of honeybees visiting the radish blooms significantly affected the quantity of seeds produced [37]. Honeybees pollinate radish blossoms at a rate of 77 to 99 percent on average, according to Radchenko [42], which increases crop yield by 22% and enhances seed quality. Consequently, it is thought that radish is almost entirely insect-pollinated [45]. While the fruit is developing, the color of the seeds is somewhat yellow, and they eventually become reddish-brown [46, 47]. The lyrate, pinnately distinct mature radish leaves feature a larger terminal lobe and smaller lateral lobes. They are arranged in a rosette condition, and alternate form [48]. Longer root types include winter radishes, daikon or mooli, and oriental radishes, which may grow up to 60 cm in length and with leaves as large as 45 cm by 60 cm in width [48].

2.3 Characteristics traits

It is a self-incompatible allogamous species. The considerable genetic variety of radish landraces and wild radish populations is paralleled in cultivar DNA polymorphism. Self-incompatibility may be overcome by bud pollination or high-concentration CO2 treatment, permitting the development of self-compatible progeny [49]. However, they exhibit inbreeding depression, making it difficult to get inbred lines. S-receptor kinase (SRK) is the recognition molecule of the stigma, similar to the self-incompatibility of Brassica species, whereas SP11, also known as SCR, is the recognition molecule of pollen [50]. These recognition molecule genes, SRK and SP11/SCR, have several alleles and are passed down through generations as the S haplotype. There are numerous S haplotypes in *R. sativus* [51], and the nucleotide sequences of some S haplotypes in *R. sativus* are similar to those of *Brassica rapa*, indicating that S haplotypes possessed by an ancient species were inherited by species in both *Raphanus* and *Brassica* without significant nucleotide sequence modification [50]. The majority of radish cultivars are root vegetables. The size, shape, and color of radish roots are all important. There has been evidence that quantitative trait loci (QTLs) influence root structure and color [52, 53]. Other factors, such as blossoming, influence root thickness. The QTL with the highest LOD score corresponded to a QTL for bolting time in our QTL analysis of root thickness using offspring obtained by crossing "Aokubi" with a white thick root and a rat's tail radish cultivar. This might be natural since early blossoming is thought to reduce root thickness. The transcriptome of developing roots was studied, and genes involved in root thickening were found. The color of the radish root surface is caused by anthocyanins. Pelargonidin and cyanidin are the pigments responsible for the red and purple colors of radish varieties. The finding of purple roots in a hybrid of a red root line and a white root line shows that the red and white had knockout mutations in separate genes involved in cyanidin synthesis and that the functional alleles in the red and white functioned as complementing genes. In red root cultivars, alleles of the flavonoid 3′-hydroxylase (*F3′H*) gene exhibit *Ty3/gypsy* transposon or helitron insertions [54]. A dihydroflavonol reductase (RsDFR) and anthocyanidin synthase (RsANS) gene are expressed in the epidermal tissues of red-skinned cultivar roots but not in white-skinned cultivar roots [55]. The huge seed size of radish is a distinguishing feature among Brassicaceae species. Radish seeds weigh nearly five times as much as *B. rapa* seeds. Because of the large seed size, the cotyledons and hypocotyls of seedlings are larger than those of Brassica. The bigger seedling size allows for direct sowing in the field and produces sprouts that are larger than Brassica. The form of siliques is connected with the property of large seeds. Amphidiploid plants of intergeneric hybrids of *R. sativus* and *B. rapa* have intermediate siliques with a few seeds in both the beak and valvar regions. Although the amphidiploid plants are mostly sterile, a small number of seeds may be obtained. Furthermore, the seed size is about midway between *Raphanus* and Brassica. Isothiocyanates, which are responsible for the pungent flavor of radish, are produced when glucosinolates are digested. The flavor of grated fresh radish and radish salad is significantly influenced by glucosinolates. The major glucosinolate in radish roots is glucoraphasatin (also known as 4-methylthio-3-butenyl glucosinolate, dehydroerucin), and the glucosinolate composition changes slightly. The glucoraphasatin content in Japanese radish cultivars, on the other hand, varies greatly [56]. There have been reports of QTLs impacting glucosinolate concentration in the root [57], and the genes inferred are putatively involved. Ishida et al. (2015) identified a mutant with a high quantity of glucoerucin but no glucoraphasatin, and the gene responsible for this mutation was discovered [56]. Although most radishes are salt tolerant, *R. sativus* var. *raphanistroides* is especially so [58]. Although genetic research into the salt tolerance of R. sativus var. raphanistroides has not yet developed, radish salt tolerance genes will be significant in the evolution of Brassica crops. High-temperature stress has become a major concern in radish growing. High-temperature stress causes the center of a radish root to become reddish brown, resulting in unmarketable products. However, since sensitivity to high temperature stress varies, it should be possible to develop a resistant cultivar. When radishes bolt, their roots become fibrous and unsuitable for sale. As a result, the characteristic of late bolting is favored. On the other hand, cultivars for oil production or rat's tail radish are needed to bloom even in tropical settings. Although vernalization is required for floral induction, rat's tail radish may bloom without it. Radishes, like many other winter crops, have varying vernalization needs. A QTL with a significant LOD score in a region containing an FLC gene was found using progeny from a hybrid between "Aokubi" and rat's tail radish. Although *Plasmodiophora brassicae's* clubroot poses a serious danger to crop yield, there are techniques to control it [58].

Among Brassica crops, Japanese and South Korean radish cultivars are often resistant to clubroot. Radish may be used to breed resistance in Brassica plants [59]. A quantitative trait locus (QTL) promoting clubroot resistance has been found on LG1 [60], which correlates with *Rs5* [61]. *Fusarium* yellow, caused by *Fusarium oxysporum*, is one of the most serious diseases in radish production. This disease causes wilting of the leaves and browning of the vascular tissue in the root. Certain cultivars and landraces have a high degree of resistance to *Fusarium yellow*. Because of its low ability for plant regeneration, radish is difficult to produce in tissue and cell cultures. There have been few reports of successful protoplast, anther, or isolated microspore culture [62], and protoplast, anther, and isolated microspore cultures are unusual [63, 64]. The function of isolated genes in radish cannot be shown due to the intricacy of plant transformation. The development of an efficient plant transformation technique is critical for scientific and practical radish research. Because plant regeneration capability must be genetically diverse, the first step in developing *in vitro* culture techniques would be to find cultivars or lines with high regeneration potential.

2.4 Breeding goals

The strong nutritional value and chemical components of the Brassicaceae family have significantly improved human health and well-being [65]. Root length is highly valued by consumers, who constantly rate radishes in terms of root length, diameter, and color. These factors are all highly important throughout the purchasing experience, and therefore, visual signals like the label's color and location on the product are crucial. Too many gene-mapping studies have been devoted to the categorization of significant color-gene relationships in various plants [66]. The development of color genetics, which is essential to the success of radish crops, would be aided by the discovery and detection of key plant genes involved in radish color [67]. Numerous studies have been conducted thus far on the inheritance patterns of radish skin [68]. It was discovered that the most extensively studied varieties of radish, including red (30%), white (13%), and black (6%), had a total of 609 distinct chemical elements, distributed across 23 different groups [69]. The main plant sections from which the nutrients, anti-oxidants, and phytochemicals described in this study were derived were the roots, sprouts, and leaves [69]. Researchers have been interested in the natural red pigment that is abundant in red flesh radishes and is used widely in the food, wine, and cosmetics sectors [70]. Achieving uniformity in radish breeding in terms of color, size, and yield is becoming increasingly crucial [71]. Radishes' esthetic appeal and health advantages are significantly influenced by color [72]. However, there is little study on the detection, characterization, and quantification of flavonoids in multicolor radish. Despite this, Zhang [73] discovered that the anthocyanin molecules that gave red and purple radishes their color pigment were relatively similar. These substances included pelargonin, callistephin, and red cyanidin [73]. Purple ZJL contains cyanidin o-syringic acid and cyanin, but dark red TXH has more callistephin and pelargonin. The metabolites that give colorful radishes their distinctive colors are more often associated with the secondary plant chemical biosynthesis pathway than SZB genes, in contrast. This approach could be useful for creating new, high-quality varieties of radish [73].

2.5 Breeding methods

The form of the leaves, the color of the petiole and leaf blades, and the quantity of surface hair are the important factors in radish breeding. Roots that are beyond their prime, on the other hand, are picked based on their internal solidity and outer morphology. Watts [74] formulated an immersion method for the solidity test in which the roots are immersed in water; acceptable solid roots sink and are selected for planting, but unsuitable pithy roots float and are discarded. In the nineteenth and twentieth centuries, mass and repeating selections were utilized to boost productivity and uniformity. However, hybrid breeding began in the 1950s to examine the prospects of heterotic vigor. Twenty-first-century biochemical, molecular, and biotechnological tools facilitated and diversified breeding tactics for quality and stress tolerance. Population improvement techniques for radish, like those used for other crops, include mutation breeding, backcross breeding, hybrid breeding (synthetics, heterotic F_1 hybrids), molecular and transgenic methods, gamete selection, family selection, line breeding, recurrent selection, and mass-pedigree breeding [75, 76].

2.6 Radish breeding by using self-incompatibility system

Self-incompatibility (SI) is a mechanism that promotes stigmatization of self-pollen, prohibits self-fertilization and inbreeding, and demands outcrossing. Selfing may be prevented *via* embryo abortion; however, SI is pre-zygotic and precludes embryo development. The sporophytic type of incompatibility system causes pollen grains to fail to germinate and form pollen tubes on the surface of stigma epidermal cells (papilla) as well as the deposition of callose inside the papillae. Stout [77] discovered sporophytic SI in radish, and Bateman was the first to demonstrate how it was inherited. SI is caused by the pollen tube's inability to penetrate the papillae as well as a lack of adhesion, hydration, and pollen grain germination. Dickinson [78] reported that homomorphic SI is typically controlled by a single S-locus containing two multiallelic genes encoding the S-locus glycoprotein (SLG), S-locus receptor kinase (SRK), and S-locus cysteine rich protein/S-locus protein 11 (SCR/SP11), all of which are expressed on the stigma. So far, massive amounts of S-alleles have been discovered [79, 80]. A significant number of S-haplotypes in *Brassica oleracea*, *B. rapa*, and *R. sativus* have been discovered using a variety of techniques, including pollination tests, electrophoretic analysis of stigmatic proteins, DNA polymorphism in SLGs or SRKs, and determination of SLG, SRK, and SCR sequences [51, 81–83]. The SI technique has the benefit of enabling two parental lines to be homozygous for independent S alleles, allowing F_1 hybrid seed to be produced. Unlike cole crops, most radish genotypes have brittle and unstable SI systems. The majority of Indian radish genotypes tested at the IIVR in Varanasi, Uttar Pradesh, India, are self-compatible to mildly self-incompatible, with only a few genotypes, particularly red radish, exhibiting moderate self-incompatibility and a red genotype VRRAD-130 exhibiting severe self-incompatibility. *Raphanus* has been related to genetic variances in SI levels [84, 85]. However, it is less reliable since hybrid seeds always carry the danger of generating an unwanted number of siblings and because reproducing SI lines by bud-pollination (BP) is difficult. The SI system in Brassica, including radish, must be broken down by BP, CO_2 treatment, and NaCl treatment in order to maintain and propagate self-incompatible lines. In contrast to cole crops, most radish genotypes have rather weak and unstable SI systems; as a consequence, hybrid seeds including sib-seeds are always possible.

Because radish is a self-incompatibility crop with significant heterosis, the generation of F_1 hybrids based on self-incompatibility is desired to remove the time-consuming manual emasculation [86]. The basic purpose of a plant breeder is to identify S haplotype breeding lines. The plant breeder can keep the parental lines from crossing [87]. The S haplotypes of parental lines must be determined in order to achieve F_1 hybrid breeding because each parental line's S haplotype must indicate compatibility between parental lines [88, 89]. The abundance of S haplotype establishes a specific S haplotype using traditional procedures such as the test cross method, pollination, isoelectric focusing, immunoblot analysis, and pollen tube fluorescence analysis [85, 90]. The S alleles of the S haplotype are highly diverse [89]. In addition, Nikura and Matsuura found 37 alleles in Radish [91]. *Raphinus sativus* contains many S haplotypes, which are labeled S-1, S-2, S-3, and so on based on polymorphism in the SLG, SRK, and SCR/SP11 sequences [91]. Despite the fact that radish is not a part of the *Brassica* genus, Brassica SP11, SRK, and SLG alleles are interleaved in the evolutionary trees of these genes, indicating that the diversification of these alleles predates the speciation of these taxa [87]. Some S haplotypes in radish feature SP11, SRK, and SLG alleles that are very similar to some S haplotypes in *Brassica*, and one S haplotype in radish has been shown to have the same recognition specificity as one S haplotype in *B. rapa* [87]. Comparisons of the nucleotide sequences of the SP11 and SRK alleles, as well as recognition specificities across related S haplotypes of radish and *Brassica*, may aid in understanding the molecular structures of SP11 and SRK proteins. However, different studies number S haplotypes in radish; therefore, the nucleotide sequence data on S haplotypes are uncertain (Nishio and Sakamoto 2017). S haplotype in Raphanus and Brassica is also identified using the PCR-RFLP (polymerase chain reaction-restriction fragment length polymorphism) approach, which examines SLG and SRK [47, 50, 86, 92]. However, PCR-RFLP has two inherent limitations: First, it is difficult to design a universal primer that can amplify SLG and SRK alleles, and second, the presence of several homologous genes in Brassicaceae plants makes PCR amplification of specific SLG or SRK alleles more difficult [87]. To help in radish hybridization, the *Ogura* CMS approach created further advanced radish cultivars (cultivars with improved yield and quality) [13]. Because variety displays such as bulk selection, mixed mass pedigree selection, or bud pollination might take eight to twelve years to develop a new variety, new varieties must be created by different genetic processes [93].

2.7 Radish breeding by use of male sterility system

Male sterility (MS) is a condition in which plants are unable to produce viable pollen, which is required for efficient hybrid seed commercial production. It often manifests itself in floral development as an incompatibility of nuclear-mitochondrial interaction in alloplasmic lines created by spontaneous mutation. It may also occur in wide crosses (intraspecific, interspecific, and intergeneric). Staminal MS systems are common in radish. Male sterility systems that contain nuclear and/or mitochondrial genomes include genic male sterility (GMS), cytoplasmic male sterility (CMS), and cytoplasmic-genic male sterility (CGMS). Transgenic male sterility (TMS), a new kind of male sterility, was developed using biotechnological technologies. Based on the process of male sterility induction and fertility restoration, all TMS systems developed to date may be classified into five kinds [94].

Plants with cytoplasmic male sterility, which is inherited from their mothers, are unable to generate effective pollen. The CGMS hermaphrodite state is restored by a collection of nuclear genes known as restorers of fertility (Rf), which inhibit the CMS genes' activity. The Ogura CMS system has been applied in a number of contexts and is now commercially available. Ogura [95] found the CMS in a Japanese radish cultivar, giving rise to the term Ogura CMS. Ogura CMS is regulated by a recessive nuclear gene (*msms*) in association with sterile cytoplasm (*S*-cytoplasm).

The genotype of male sterile plants is *Smsms*, whereas the genotype of the maintenance line is *Nmsms*. Several CMS systems have also been identified in the Brassicaseae family. *Polima* [96, 97], *napus* [98, 99], *Ogura* [95], and *Anand* [100] are well-characterized CMS systems from the *Brassica* genus; however, the following systems, such as Ogura CMS [101] *Raphanus* and Brassica species, lack Rf genes, but all fertile The *Rf* genes are widely distributed in the Japanese wild radish, regardless of cytoplasm type (*R. raphanistrum*). The bulk of cultivated radishes in Japan and India lacks restorative genes in their populations, although European and Chinese variations do [102]. However, the *Rf* gene is essential for crops that require pollination and fertilization for economic growth, such as chili, tomato, eggplant, and melon. CMS lacking the *Rf* gene benefits from simple transfer in diverse backgrounds and is utilized in a variety of vegetables where the vegetative element is economically valuable, such as root crops, cole crops, tuber crops, and leafy vegetables.

Similar to CMS, the genetic emasculation approach in radish allows for the harnessing of heterotic vigor for yield, uniformity, adoption, and earliest maturity as well as the production of high-quality seeds. Despite being one of India's most important salad crops, the first CMS-based radish hybrids and Public Sector CMS lines were reported in 2018 from ICAR-IIVR, Varanasi, UP, (2018) by Singh and colleagues.

2.8 Population improvement

Cross-pollinated crops and Brassica vegetables have benefited from population improvement. In India, mass selection has been routinely utilized to enhance the genetics of radish. This approach is useful for genetic gain of simply inherited monogenic characteristics, but it is inefficient for qualities regulated by polygenes. Significant genetic variety for variables of relevance, as found by several studies in radish [103–108], is a precondition for crop development. Many cultivars have been produced across the globe with various goals in mind. Based on progeny appraisal, changes such as mass pedigree and family selection are superior to basic mass selection. The decision between these strategies is determined by the population's homogeneity. Recurrent selection is a preferable choice for improving quantitative traits, particularly those controlled by additive gene activity. This approach is helpful for improving leaf morphology, root form, size, color, yield, and other economic features. Kashi Lohit, a red radish cultivar, was developed at ICAR-IIVR in Varanasi, Uttar Pradesh, using a simple recurrent selection to target red color root, tapering root form, and yield. Kashi Lohit has around 30–125% more nutrients such as ascorbic acid, total phenolics, anthocyanins, and antioxidants than white-rooted commercial radish cultivars [109]. Inbred lines may be produced by breaking down SI barriers using either BP or chemical induction procedures (NaCl or CO2). Many cultivars have been released in India by public sector organizations including Kashi Hans, Kashi Sweta, Kashi Mooli-40, Kashi Lohit, and Kashi Aardra as some of the characters. VRRAD-150 (ICAR-IIVR, Varanasi); Pusa Desi, Pusa Reshmi, Pusa Chetki, and others. Pusa Himani, Pusa Mridula, Pusa Jamuni, and Pusa Gulabi [ICAR-Indian Agricultural Research Institute (IARI), New Delhi]; Hisar Sel-1 [Hisar Agricultural Research Institute]. Hisar University (HAU)]; Kalyanpur-1 [Chandra Shekhar Azad University of Technology] Kanpur's College of Science, Agriculture, and Technology (CSAUAT)]; Chaudhary, Palam Hriday Sarwan Kumar Himachal Pradesh Krishi Vishvavidyalaya, Palampur]; Punjab Safed and Punjab Pasand [Punjab Agricultural University (PAU)]; as well as Arka Nishant [ICAR-Indian Institute of Horticulture Research] (IIHR), Bengaluru].

2.9 Mutation breeding

Mutation breeding refers to the technique of developing and using genetic variation *via* induced mutagenesis. It is an effective method for improving complex traits, especially in crops with limited genetic bases, vegetative reproduction, and self-pollination. More than 3000 mutant cultivars of various crops have been published in more than 60 countries; of these, 776 mutants have been generated for various nutritional quality traits, including minerals [110]. When compared with wild-type radish, mutants displayed a 30% higher net photosynthetic rate and a 36% higher total chlorophyll content.

2.10 Breeding for biotic and abiotic stresses

Heat, rain, Alternaria blight, Fusarium wilt, white rust, aphids, and beetles are all major abiotic and biotic elements affecting radish cultivation. As a consequence, improving radish stress tolerance is an important breeding target. Radish takes a lot of breeding during the off-season. In India, breeding lines, cultivars, and hybrids resistant to heat stress (38–43°C), such as cvs, have been developed. Despite the North Indian plains' subtropical environment which has three seasons—summer, rainy, and winter—Pusa Chetki, Kashi Mooli-40, VRRAD-200, Chetki group, VRRADH-41, and VRRADH-42; and tolerance to high humidity make it feasible to produce radish commercially practically year-round. There are several sources of resistance to Fusarium wilt, according to Ashizawa [111], Peterson and Pound [112], and Soh [113]. Furthermore, after screening 260 accessions from 9 Asian and European countries, Jeon [114] discovered 54 radish accessions that were resistant to Fusarium wilt. Ghimire [115] also tested radish for *Alternaria* leaf spot.

2.11 Molecular markers to QTL breeding

Radish has been demonstrated to have a variety of economically important characteristics, such as yield, insect resistance, and disease resistance [2]. Yield is a complex trait governed by polygenic characteristics; it is difficult to discover these traits through standard breeding since they depend on phenotypic expression and interact with the environment and genotype. These challenges are addressed by the novel molecular breeding approach, which uses DNA markers for quantitative trait identification and linkage mapping [116]. Several DNA markers are utilized in breeding programs, including restriction fragment length polymorphism (RFLPs), random amplified polymorphism DNA (RAPD), simple sequence repeats (SSRs), and single-nucleotide polymorphism (SNPs) [117]. *Raphanus hortensis* var. *sativus* and var. *niger* were demonstrated to have distinct origins and to have descended from distinct progenitors owing to the application of molecular markers like as RAPD [118]. Several Asian varieties feature darker skin and flesh as well as variations in root size, length, and weight. It is therefore hardly unexpected that var. hortensis has genetic heterogeneity. Furthermore, Lee [119] performed phenotypic studies after genome-wide association analysis (GWAS) using genotyping-by-sequencing (GBS) to find FW resistance loci [119]. The GWAS study revealed 20 possible candidate genes and 44 single nucleotide polymorphisms (SNPs) that were significantly associated with FW resistance. Four QTLs were discovered in an F_2 population derived from an FW resistant line and a susceptible line, one of which was co-located with the SNPs on chromosome 7. These markers are newly accessible tools for molecular breeding efforts and marker-assisted selection to generate resistant *R. sativus* cultivars Lee [119]. Furthermore, Yu [120] produced a genetic linkage map on the F_2 population to detect the disease Fusarium wilt, and they identified a total of 8 loci conferring FW resistance that were spread across 4LGs, namely 2, 3, 6, and 7 of the *Raphanus* genome. Synteny analysis using the linked markers QTL found similarities to *A. thaliana* chromosome 3, which contains clusters of disease-resistance genes, showing that resistance genes are conserved between both. The sites of important QTLs discovered in the radish are Crr3, Crs1, and Crs2 [121]. Researchers uncovered a novel QTL named qRCD9 that modulates root CD by using markers SRAP, RAPD, SSR, ISSR, RAMP, and RGA to generate a genetic map of an F_2 population [116]. Resistance to cyst nematode (*Heterodera schachtii*) was discovered using RAPD, dpRAPD, AFLP, and SSR markers [122]. They identified 8 and 10 quantitative characteristics in radish for morphological aspects such as ovule number per silique, seed number per silique, plant shape, pubescence, whole plant weight (g), upper part weight (g), whole root weight (g), and main root weight using recombinant inbred lines (g) [53]. In the locations where QTLs were discovered, nine SNP markers were recently developed. The expression and nucleotide sequences of these genes suggested a possible function in the production of 4MTB-GSL in radish roots [57]. Fan [123] discovered that the *R2R3-MYB* transcription factor, which is responsible for creating the anthocyanin pigment 2, is located on chromosome 2 (PAP2). The *RsPAP2* gene, which encodes the amino acid sequence that gives radish its red skin color, was readily differentiated from previously identified RsMYB genes.

Biochemical- and DNA-based markers enable the identification and description of cultivars and parental lines of hybrids, assessing the genetic purity of seed, diversity in agricultural cultivars and their wild variants, phylogenetic analysis, and pinpointing the origin of the germplasm. The identification of isoenzymes was the first tool for genetic analysis [124]. This method was used by Tai-Young [125] to identify cultivars and validate the purity of radish seeds. DNA is now frequently examined by directly using PCR-based molecular approaches such as amplified fragment-length polymorphism (AFLP), random amplified polymorphic DNA (RAPD), and inter-simple sequence repeats (ISSR). Wang [126] classified 65 cultivated radish accessions from 21 European, Asian, and North African countries into four groups based on their origin (Europe, Middle East, South Asia, and East Asia) in a neighbor-joining tree. Along with RAPD and ISSR marker information, reliable descriptors based on isoenzymes were established for the different stages of radish development [127]. Furthermore, Nakatsuji [128] generated 417 radish SSR markers utilizing cDNA data and SSR-enriched genomic libraries, which may be used for genetic research in radish and related species. The genetic diversity of the collection was studied using 144 radish cultivars. An SSR-enriched collection was used to generate genomic SSR markers [129]. Furthermore, the genetic diversity of 126 radish F1 cultivars was assessed using 60 SSRs and 29 agronomic parameters [130].

The expression of the *orf138* gene causes Ogura CMS in Brassicaceae [131, 132]. The *orf138* gene contains at least nine known nucleotide sequence variants, including one by Yamagishi and Terachi [133] with a 39-nucleotide deletion (Kosena type). Additionally, a primer pair at the 3′ region of the atp6 gene (5′-cgcttggactatgctatgtatga-3′) and the 5′ area of the nad3 gene (5′-tcatagagaaatccaatcgtcaa-3′) produced a 2-kbp fragment that was unique to the NWB CMS type of male sterility and absent from other CMS kinds of radish. Through *de novo* transcriptome analysis, Nie [134] found crucial genes involved for bolting and blooming in radish.

2.12 Haploid breeding

In traditional breeding, the inbreds that are selfed and chosen during 6–8 generations of selfing constitute a major component of the commercial seed production of F_1 hybrids, which is detrimental to inbred vigor and inbreeding-depressive effects in Brassica crops, including radish. The microspore-produced doubled haploid (DH), which creates inbred lines with 100% homozygosity in only one generation, is an attractive tool because using DHs as parental lines may speed up breeding operations, create novel hybrids and varieties, conduct fundamental genetic research, and save time [135, 136]. Male gametophytic cells are cultured *in vitro* to form haploid plants, which are then treated with chromosome doubling procedures to produce haploid cells at the microspore or immature pollen developing stage. Since the initial report of effective isolation and culture of microspores in *Brassica napus*, microspore culture technology is being used in *Brassica* breeding [137]. Chun [138], Sugimoto [139], Takahata [62], and Tuncer [140] all successfully used microspore cultivation in radish.

2.13 Genomics and genomics tools

Today, genomics drive crop production, and the radish crop has been employed to investigate the underlying genotypic differences. The rapid rise of genomic data has spurred research into the genetic basis of plant characteristics such as better production, flowering, and disease resistance [141]. Several comprehensive studies on radish genome structure and chromosomal rearrangement during polyploidy events have been conducted [142], from which several genomic sequences have been generated [61]. Another research demonstrated that by combining the 454, Illumina, and PacBio sequencing technologies with bacterial artificial chromosome clones created by end sequencing, the whole genome of the Asian radish cultivar WK10039 was sequenced [143]. Numerous genetic investigations on the cultivated radish have been undertaken during the last 10 years [144]. A chromosome-scale genome assembly (*rs1.0*) of the Asian radish cultivar WK10039 was also generated, and the findings were compared to prior assemblies [145]. It provided more information than previously known due to increased genome coverage, contigs, and chromosomal anchoring [146]. However, radish mitochondrial genome sequences are now available in Radish Base, a genomic and genetic resource [147]. This resource now includes the mitochondrial genomes of two newly sequenced radish species, one from the normal cytoplasm and the other from the male-sterile cytoplasm of *Ogura* [148]. A recent study's bioinformatics analysis of the radish genome discovered 54,357 coding genes and 20 COL transcription factors [149]. Each COL gene in the "Aok I daikon" cultivar had a match with its corresponding COL gene in the "kaz sa" cultivar. Furthermore, the cultivar "WK10039" was screened for a total of 20 radish COL genes [149]. Furthermore, BLASTP analysis of the radish genome revealed 35 different *RsOFP*s and five *RsOFP*-like genes (with no/partial OVATE domains), with the majority of genes being intron-less and containing the bulk of the genome's coding sequences [149]. The HiSeq2000 technology was also used to generate whole-genome shotgun sequences on the *R. sativus* inbred line XYB36–2, a 119.75 GB dataset [150]. Based on 17-mer analysis, the estimated genome size was 530 MB. A 387.73 MB was assembled into 44,820 high-quality scaffolds using SOAP *denovo* and SSPACE [151, 152]. This study's assembly produced outstanding results using fosmid clones (98.86% coverage). The assembly was much greater in quality than the previously released entire genome of *R. raphanistrum* (254 Mb contigs) and two assemblies of *R. sativus* 'Aok I (116.0 and 179.8 Mb). The "Okute-Sakurajima" genome was reconstructed from scratch, yielding an estimated haploid genome size of 498.5 MB. The *de novo* assembly showed a largely heterozygous genome [153]. Further, long-read sequencing produced 36.0 GB of data from 2.3 million reads with an N50 length of 29.1 kB (60.7 coverage of the predicted genome size). The long-read assembly of 504.5 MB primary contigs, including 1437 sequences with an N50 length of 1.2 MB, and 263.5 Mb alternative contigs, including 2373 sequences with an N50 length of 154.6 kB, contains the other haplotypes with different alleles, also known as haploid sequences, after two rounds of data polishing [153]. Following polyploidy, research on the radish genome has revealed vital insights about the radish genome's origin and evolution, providing deep knowledge on radish genetics and breeding [66]. The detailed information and genomic approaches obtained as a result of these studies help to a better understand the radish triplicated genome structure. Furthermore, these strategies improve radish breeding by increasing the use of marker-assisted collection, comparative genomic study, and the distribution of knowledge from reference data to new radish accessions [154]. As a consequence, a gateway with a large volume of genomic data and many linkages to specific genome analysis methodologies is very useful for radish research and breeding.

2.14 Genetic engineering

Genetic engineering is significant in agriculture since it improves agricultural characteristics and meets the needs of undernourished nations. The improvement of metabolic engineering methods and gene technology has sped up the development of usable germplasms [155]. Plant approaches advance by enhancing attributes; scientists have successfully developed transgenic radishes with a variety of agronomic qualities [156]. According to Tzfira and Citovsky [157] and Lacroix and Citovsky [158], some radish types contain beneficial features, such as better yield, that are passed on to the host plant. Gene transfer is carried out with the assistance of the pathogen *Agrobacterium*, which is widely employed as a strategy for plant hairy root lines, which seem to develop better than other types of root systems [159]. Herbaceous hairy roots are sought after because of their robustness, quick development, and ability to promote root-up growth in plants [160]. *Agrobacterium*, which grows in nutrient solution and has unique capabilities such as biochemically and biotransforming different metabolites, produces hairy roots. *Agrobacterium* is the greatest choice for producing secondary metabolites since it aids growth regulators [161]. New sources of natural chemicals may be identified by focusing on the hairy roots [162]. In a cultivated situation, chromosome disruption or amplification may potentially affect a plant's fertility. Herbicides, antibiotics, metabolic mimics, and non-toxic substances all aid in the survival of changed cells. Radish regeneration is inhibited by kanamycin and hygromycin B. Floral dipping is a process that might be used to genetically edit radish, according to current plant biotechnology breakthroughs. In this strategy, the photoperiodic gene GIGANTEA in radish is co-suppressed, which helps the plant postpone bolting and flowering. It might be used to increase the medicinal potential of a crop [163]. It is addressed how to improve transformation efficiency and choose new characteristics to produce late-flowering radish [118]. Transgenic radish (*R. sativus* L. *longipinnatus* Bailey) was created in 2001 using plants that had been dipped into an *Agrobacterium* solution containing both the beta-glucuronidase (GUSA) gene and the herbicide resistance gene (bar) between the flanking T-DNA border sequences [156]. Finally, Southern blotting data demonstrated that the GUSA and bar genes had been incorporated into altered plant genomes and were segregating as dominant Mendelian features [156]. The radish RHA2b gene encodes a transcription factor implicated in the abscisic acid (ABA) signal transduction process as well as preharvest sprouting and seed dormancy, according to one research [164]. The *RsRHA2b* gene was cloned and introduced into Zhengmai 9023 by *Agrobacterium*-mediated stem apex transformation, according to the researchers [164]. *Agrobacterium*-mediated transformation was found to be a superior method for genetic modification [165]. Transgenic radish (*Raphanus sativa* L., cv. Jin Ju Dae Pyong) grown on Murashige and Skoog medium was used to study the use of adventitious shoot development on hypocotyl explants for Agrobacterium-mediated radish genetic transformation [64]. Furthermore, Northern blot findings revealed that the GUS gene transcript was found in a few regenerated plants, indicating genetic alteration [64]. In his study, Curtis also investigates strategies for delivering therapeutic proteins into radish for on-site administration of consumable proteins [163]. Concerns have been raised about pollen-mediated gene transfer after the introduction of transgenic radish into the wild. Potential risks and the field planting of transgenic radish are sometimes raised in talks concerning transgenic crops [156, 166, 167]. Plant regeneration from hypocotyl explants and somatic embryogenesis from hypocotyls was used to produce branches in radish. By adding aminoethoxyvinylglycine (AVG), an inhibitor of ethylene synthesis, and $AgNO_3$, an inhibitor of ethylene action, to the regeneration medium, cultured radish hypocotyl explants were able to regenerate shoots at a rate of 40% [168].

3. Conclusions and future perspectives

A few of the crucial radish breeding features are higher yield, early maturity, late bolting, pungency, cold hardiness, drought resistance, heat tolerance, and soil adaptation. Self-incompatibility alleles found in the radish genome make it possible to produce F_1 hybrids without the time-consuming and labor-intensive manual emasculation necessary for radish. To prevent hand emasculation, it is essential to know the S haplotypes of the parental lines when creating F_1 combinations. Inter- and intra-specific hybridizations are essential for the effective generation of radish yield because they allow the introduction of favorable agronomic features into the population. It is crucial to get comprehensive genetic data on chromosomes as well as the knowledge of inheritance. Researchers must comprehend the regulatory variables that synchronize at different developmental stages for each of the above-mentioned features in order to better understand and predict resistance, yield characteristics, and fruit quality. It is still essential to create a reliable and long-lasting plan for plant disease resistance, which is now being thought about. This is due to the ability of diseases to produce new bacterial strains, which may evade resistance. Scientists will now have accurate knowledge on disease resistance genes for a range of diseases as well as genes encoding essential biochemical features of the plant thanks to the completion of large-scale sequencing of the radish genomes earlier this year. One such method is speed breeding; as the cost of genome sequencing decreases, RAD-sequencing and DNA microarrays will be used more often, allowing for quicker genome mapping and tagging of novel quantitative trait loci. In order to enhance the number of resistant radish genotypes, these quantitative trait loci (QTLs) may introduce resistance into high-yielding radish genotypes and combine them with important resistance genes. Additionally, to increase radish crop output and quality, GWAS (genome-wide association studies) may map traits to particular candidate genes on a genome-wide scale. Using trait-specific genetic resources, heterotic potential, hundreds of molecular markers, highly saturated genetic maps, and effective contemporary technologies will all contribute to the development of prospective radish varieties and hybrids with improved quality and stress tolerance. Significant genetic and metabolic variety has been found, opening the door to breeding for genetic improvement and controlled harvest variability in agriculture. Finally, the use of trait-specific genetic resources, as well as the availability of thousands of molecular markers, highly saturated genetic maps, and efficient modern tools, will undoubtedly aid in the development of potential radish varieties and hybrids with improved quality and stress tolerance. Future radish breeding strategies that are crucial for boosting output and productivity as well as the effective use of input resources include targeted breeding strategies to create model crop ideotypes, improve nutritional quality, increase sustainability of production, increase adaptability to various climatic conditions, and increase tolerance to insects and diseases and efforts to pyramid two- or multi-tiered breeding approaches to widen the genetic base.

Author details

Raman Selvakumar
Centre for Protected Cultivation Technology, ICAR-Indian Agricultural Research Institute, New Delhi, India

*Address all correspondence to: selvakumarsingai@gmail.com

References

[1] Richharia RH. Cytological investigation of *Raphanus sativus*, Brassica oleracea, and their F 1 and F2 hybrids. Journal of Genetics. 1937;**1**:19-44

[2] Kaneko Y, Matsuzawa Y. Radish: Raphanus sativus L. In: Genetic Improvement of Vegetable Crops. London, UK: Elsevier; 1993. pp. 487-510

[3] Patra JK, Das G, Paramithiotis S, Shin HS. Kimchi and other widely consumed traditional fermented foods of Korea: A review. Frontiers in Microbiology. 2016;**7**:1493

[4] Ebert AW. Ex situ conservation of plant genetic resources of major vegetables. In: Conservation of Trop. Plant Speci. New York: Springer; 2013. pp. 373-417

[5] Singh A, Sharma S. Radish. In: Antioxidants in Vegetables and Nuts-Properties and Health Benefits. Singapore: Springer Nature; 2020. pp. 209-235

[6] Kaneko Y, Kimizuka-Takagi C, Bang SW, Matsuzawa Y. Radish. In: Kole C, editor. Vegetables. Genome Mapping and Molecular Breeding in Plants. Heidelberg, Berlin: Springer; 2007;**5**:141-160

[7] Simard MJ, Légère A. Synchrony of flowering between canola and wild radish (*Raphanus raphanistrum*). Weed Science. 2004;**6**:905-912

[8] Kumar R, Solankey SS, Singh M. Breeding for drought tolerance in vegetables. Vegetable Science. 2012;**39**:1-15

[9] Dixon GR. Vegetable Brassicas and Related Crucifers. Walllingford, Oxfordshire: CABI; 2007;**14**:327

[10] Porteus MH, Carroll D. Gene targeting using zinc finger nucleases. Nature Biotechnology. 2005;**8**:967-973

[11] Petropoulos SA, Sampaio SL, Di Gioia F, Tzortzakis N, Rouphael Y, Kyriacou MC, et al. Grown to be blue—Antioxidant properties and health effects of colored vegetables. Part I: Root vegetables. Antioxidants. 2019;**12**:617

[12] Rubatzky VE, Yamaguchi M. Cole crops, other brassica, and crucifer vegetables. In: World Vegetables. Boston, MA: Springer; 1997. pp. 371-417

[13] Bhardwaj RK, Kumari R, Vikram A. Efficient methods for the improvement of temperate root vegetables. Accel Plant Breed. 2020;**2**:155-196

[14] Cartea ME, Velasco P, Obregón S, Padilla G, de Haro A. Seasonal variation in glucosinolate content in Brassica oleracea crops grown in northwestern Spain. Phytochemistry. 2008;**2**:403-410

[15] El-Ramady HR, Domokos-Szabolcsy É, Abdalla NA, Taha HS, Fári M. Postharvest management of fruits and vegetables storage. In: Sustainable Agriculture Review. Cham: Springer; 2015;**15**:65-152

[16] Wang Y, Xu L, Chen Y, Shen H, Gong Y, Limera C, et al. Transcriptome profiling of radish (*Raphanus sativus* L.) root and identification of genes involved in response to lead (*Pb*) stress with next generation sequencing. PLoS One. 2013;**6**:66539

[17] Kim J, Manivannan A, Kim DS, Lee ES, Lee HE. Transcriptome sequencing assisted discovery and computational analysis of novel SNPs

associated with flowering in *Raphanus sativus* in-bred lines for marker-assisted backcross breeding. Horticulture Research. 2019;**1**:1-12

[18] Kobayashi H, Shirasawa K, Fukino N, Hirakawa H, Akanuma T, Kitashiba H. Identification of genome-wide single-nucleotide polymorphisms among geographically diverse radish accessions. DNA Research. 2020;**27**:dsaa001

[19] Kitamura S. Varieties and transitions of radish. In: Nishiyama I, editor. Japanese Radish. Japan: Jpn Soc From Sci Tokyo; 1958. pp. 1-19 (in Japanese)

[20] Banga O. Radish, Raphanus sativus (Cruciferae). In: Simmonds NW, editor. Evolution of Crop Plants. London: Longman; 1976. pp. 60-62

[21] Hida K. Root crops: Radish. In: Matsuo T, editor. Collected Data of Plant Genetic Resources. Vol. 2. Tokyo: Koudansha; 1990. pp. 823-834. (in Japanese)

[22] Panetsos CA, Baker HG. The origin of variation in wild Raphanus sativus (Cruciferae) in California. Genetica. 1967;**38**:243-274

[23] Eber F, Chevre AM, Baranger A, Vallee P, Tanguy X, Renard M. Spontaneous hybridization between a malesterile oilseed rape and two weeds. Theoretical and Applied Genetics. 1994;**88**:362-368

[24] Eber F, Boucherie R, Broucqsault LM, Bouchet Y, Chevre AM. Spontaneous hybridization between vegetable crops and weeds. 1. Garden radish (Raphanus sativus L.) and wild mustard (Sinapis arvensis L.). Agronomie. 1998;**18**:489-497

[25] Kato M, Tokumasu S. Anelectrophoretic study of esterase and peroxidase isozyme in Brassicoraphanus. Euphytica. 1979;**28**:339-349

[26] Harberd DJ. Cytotaxonomic studies of brassica and relatedgenera. In: Vaughan JG, Mac Leod AJ, Jones BMG, editors. The Biology and Chemistry of the Cruciferae. London: Academic; 1976. pp. 47-68

[27] Prakash S, Takahata Y, Kirti PB, Chopra VL. Cytogenetics. In: Gomez-Campo C, editor. Biologyof Brassica Coenospecies. Amsterdam: Elsevier; 1999. pp. 59-106

[28] Tsunoda S (1979) Ecology of wild species of the tribe Brassiceae and origin of cultivation. In: Recent Advances in Breeding. Jpn Soc Breed, 20: 41-45. (in Japanese)

[29] Tsunoda S. The history of cultivated plants in the tribe Brassiceae. (1) cultivated plants and wild allies in the tribe Brassiceae. Agriculture & Horticulture. 1991;**66**:233-239. (in Japanese)

[30] Warwick SI. Guide to the wild germplasm of brassica and allied crops. IV. Wild species in the tribe Brassiceae (Cruciferae) as sources of agronomic traits. Technology and Bulletin. 1993;**17E**:1-19

[31] Hinata K. Radish Asahi World Plant, Encyclopedia, Japan. 1995;**67**:206-207. (in Japanese)

[32] Furusato K, Miyazawa A. Varieties of Japanese radish viewed from horticulture. In: Nishiyama I, editor. Japanese Radish. Tokyo: Jpn Soc From Sci; 1958. pp. 138-161. (in Japanese)

[33] Aoba T. Geographical distribution of domestic varieties in radish. In: Vegetables. Tokyo: Publication Department, Housei University; 1981. pp. 232-251. (in Japanese)

[34] Aoba. Pedigree and variation of wild radish in Japan. Technology & Agriculture. 1988;**12**:94-114 (in Japanese)

[35] Bateman AJ. Self-incompatibility systems in angiosperms. 3. Cruciferae (No. RESEARCH). Heredity. 1955;**9**:53-68

[36] Bailey LH. Manual of Cultivated Plants. Rev. ed. USA: Macmillan Company, Cornella University; 1949

[37] Kremer JC. Influence of honeybee habits on radish seed yield. Q. Bull. Mich. Agric. Exp. Stn. 1945;**27**:413-420

[38] Nishio T. Economic and academic importance of radish. In: The Radish Genome. Compendium of Plant Genomes. Cham: Springer; 2017. pp. 1-10

[39] Singh A, Sharma S, Dolly. Radish. In: Nayik GA, Gull A, editors. Antioxidants in Vegetables and Nuts-Properties and Health Benefits. Singapore: Springer; 2020. pp.209-235

[40] Kercher S, Conner JK. Patterns of genetic variability within and among populations of wild radish, Raphanus raphanistrum (Brassicaceae). American Journal of Botony. 1996;**11**:1416-1421

[41] Gopalakrishnan TP. Vegetable Crops. India: New India Publication; 2007. pp. 244-247

[42] Radchenko TG. Role of Honeybees as Pollinators in Increasing the Seed Crop from Cabbage and Radish. Beijing, China: Bdzhilnistvo; 1966;**2**:72-74

[43] Crane M, Mather K. The natural cross-pollination of crop plants with particular reference to the radish. Annals of Applied Biology. 1943;**4**:301-308

[44] Young HJ, Stanton ML. Influences of floral variation on pollen removal and seed production in wild radish. Ecology. 1990;**2**:536-547

[45] Jones HA, Rosa JT. Truck Crop Plants. London: Macmillan & Co., Ltd.; 1928. p. 538

[46] Kalia P. Root vegetable crops. Journal ofNew Seeds. 2005;**2-3**:247-275

[47] Lim S, Lee EJ, Kim J. Decreased sulforaphene concentration and reduced myrosinase activity of radish (*Raphanus sativus* L.) root during cold storage. Postharvast Biology and Technology. 2015;**100**:219-225

[48] Brickell C. The Royal Horticultural Society Encyclopedia of Gardening (Print). London: Dorling Kindersley; 1992. pp. 356-357

[49] Niikura S, Matsuura S. Genetic analysis of the reaction level of self-incompatibility to a 4% CO2 gas treatment in the radish (Raphanus sativus L.). Theoretical and Applied Genetics. 2000;**101**:1189-1193

[50] Okamoto S, Sato Y, Sakamoto K, Nishio T. Distribution of similar self-incompatibility (S) haplotypes in different genera, Raphanus and brassica. Sexual Plant Reproduction. 2004;**17**:33-39

[51] Sakamoto K, Kusaba M, Nishio T. Polymorphism of the S-locus glycoprotein gene (SLG) and the S-locus related gene (SLR1) in Raphanus sativus L. and self-incompatible ornamental plants in the Brassicaceae. Molecular & General Genetics. 1998;**258**:397-403

[52] Tsuro M, Suwabe K, Kubo N, Matsumoto S, Hirai M. Mapping of QTLs controlling root shape and red pigmentation in radish, Raphanus sativus L. Breeding Science. 2008;**58**:55-61

[53] Hashida T, Nakatsuji R, Budahn H, Schrader O, Peterka H, Fujimura T, et al. Construction of a chromosome-assigned, sequence-tagged linkage map for the radish, Raphanus sativus L. and QTL analysis of morphological traits. Breeding Science. 2013;**63**:218-226

[54] Ozeki Y. Study on the relationship between the anthocyanin molecular species and color phenotype of roots in the inbred lines of red radish (Raphanus sativus L.). Japan Food Chem Res Found Res Rep. 2010;**16**:33-39 (in Japanese with English summary)

[55] Park NI, Xu H, Li X, Jang IH, Park S, Ahn GH, et al. Anthocyanin accumulation and expression of anthocyanin biosynthetic genes in radish (Raphanus sativus). Journal of Agricultural and Food Chemistry. 2011;**59**:6034-6039

[56] Ishida M, Kakizaki T, Morimitsu Y, Ohara T, Hatakeyama K, Yoshiaki H, et al. Novel glucosinolate composition lacking 4-methylthio-3-butenyl glucosinolate in Japanese white radish (Raphanus sativus L.). Thoer Appl Genet. 2015;**128**:2037-2046

[57] Zou Z, Ishida M, Li F, Kakizaki T, Suzuki S, Kitashiba H, et al. QTL analysis using SNP markers developed by next-generation sequencing for identification of candidate genes controlling 4-methylthio-3-butenyl glucosinolate contents in roots of radish, Raphanus sativus L. PLoS One. 2013;**8**:e53541

[58] Nasu S, Kitashiba H, Nishio T. "Na-no-hana project" for recovery from the tsunami disaster by producing salinity-tolerant oilseed rape lines: Selection of salinity-tolerant lines of brassica crops. Journal of Integrated Field Science. 2012;**9**:33-37

[59] Akaba M, Kaneko Y, Hatakeyama K, Ishida M, Bang SW, Matsuzawa Y. Identification and evaluation of clubroot resistance of radish chromosome using a Brassica napus–Raphanus sativus monosomic addition line. Breeding Science. 2009;**59**:203-206

[60] Kamei A, Tsuro M, Kubo N, Hayashi T, Wang N, Fujimura T, et al. QTL mapping of clubroot resistance in radish (Raphanus sativus L.). Theoretical and Applied Genetics. 2010;**120**:1021-1027

[61] Kitashiba H, Li F, Hirakawa H, Kawanabe T, Zou Z, Hasegawa Y, et al. Draft sequences of the radish (Raphanus sativus L.) genome. DNA Research. 2014;**21**:481-490

[62] Takahata Y, Komatsu H, Kaizuma N. Microspore culture of radish (Raphanus sativus L.): Influence of genotype and culture conditions on embryogenesis. Plant Cell Reports. 1996;**16**:163-166

[63] Park B-J, Liu Z, Kanno A, Kameya T. Transformation of radish (Raphanus sativus L.) via sonication and vacuum infiltration of germinated seeds with agrobacterium harboring a group 3 LEA gene from B. napus. Plant Cell Reports. 2005;**24**:494-500

[64] Cho MA, Min SR, Ko SM, Liu JR, Choi PS. Agrobacterium-mediated genetic transformation of radish (Raphanus sativus L.). Plant Biotechnol. 2008;**25**:205-208

[65] Al-Shehaz A. Brassicaceae (mustard family). Encyclopedia of life Sciences. Chichester: Wiley; 2001. pp. 1-5

[66] Yu D, Gu X, Zhang S, Dong S, Miao H, Gebretsadik K, et al. Molecular basis of heterosis and related breeding strategies reveal its importance in vegetable breeding. Horticulture Research. 2021;**8**:1-17

[67] Schreiner M, Peters H-K, Schonhof KA, Widell S. Seasonal climate effects on root colour and compounds of red radish. Journal ofthe Science of Food and Agriculture. 2002;**11**:1325-1333

[68] Yarnell SH. Cytogenetics of the vegetable crops. II. Crucifers. The Botonical Review. 1956;**2**:81-166

[69] Gamba M, Asllanaj E, Raguindin PF, Glisic M, Franco OH, Minder B, et al. Nutritional and phytochemical characterization of radish (Raphanus sativus): A systematic review. Trends in Food Science & Technology. 2021;**113**:205-218

[70] Jing P, Zhao SJ, Ruan SY, Xie ZH, Dong Y, Yu LL. Anthocyanin and glucosinolate occurrences in the roots of Chinese red radish (Raphanus sativus L.), and their stability to heat and pH. Food Chemistry. 2012;**4**:1569-1576

[71] Elsayed AYA, Hamdino MI, Ahmed A, Shabana AI. Expected genetic gain in radish (Raphanus sativus var. red radicula) submitted to different procedures of selection. Egypt Journal of Plant Breed. 2016, 2016;**20**:313-328

[72] Zhang J, Zhao J, Tan Q, Qiu X, Mei S. Comparative transcriptome analysis reveals key genes associated with pigmentation in radish (Raphanus sativus L.) skin and flesh. Scientific Reports. 2021;**1**:1-11

[73] Zhang J, Qiu X, Tan Q, Xiao Q, Mei S. A comparative metabolomics study of flavonoids in radish with different skin and flesh colors (Raphanus sativus L.). Journal of Agriculture and Food Chemistry. 2020;**49**:14463-14470

[74] Watts LE. The use of a new technique in breeding for solidity in radish. Journal ofHorticulture Science. 1960;**35**:221-226

[75] Ram HH. Vegetable Breeding – Principles and Practices. New Delhi: Kalyani Publishers; 2005. pp. 324-328

[76] Singh BK, Singh B, Singh PM. Breeding cauliflower: A review. International Journal of Vegetable Science. 2018;**24**(1):58-84

[77] Stout AB. Further experimental studies on self-incompatibility in hermaphroditic plants. Journal of Genetics. 1920;**9**:85-129

[78] Dickinson HG. Pollen-stigma interactions: So near yet so far. Trends in Genetics. 2000;**16**:373-376

[79] Karron JD, Marshall DL, Olilveras DM. Numbers of sporophytic self-incompatibility alleles in populations of wild radish. Theoretical and Applied Genetics. 1990;**79**:457-460

[80] Okazaki K, Hinata K. Analysis of S-alleles and S-glycoproteins in F1 hybrid varieties of Japanese radish (Raphanus sativus L.). Japan Journal of Breed. 1984;**34**:237-245

[81] Haseyama Y, Kitashiba H, Okamoto S, et al. Nucleotide sequence analysis of S-locus genes tounify S haplotype nomenclature in radish. Molecular Breeding. 2018;**38**:116

[82] Oikawa E, Takuno S, Izumita A, et al. Simple and efficient methods for S genotyping and S screening in genus brassica by dot-blot analysis. Molecular Breeding. 2011;**28**:1-12

[83] Wang Q, Zheng P, Zhang L. Identification and classification of S-haplotypes in radish (Raphanus sativus). Plant Breeding. 2019;**138**:121-130

[84] Haruta T. Studies on the genetics of self-and cross-incompatibility in

cruciferous vegetables. Res Bull Takii Plant Breed Exp Stn. 1962;**2**:1-169

[85] Ruffio-Chable V, Herve Y, Dumas C, Gaude T. Distribution of S-haplotypes and its relationship with self-incompatibility in Brassica oleracea part 1- inbred lines of cauliflower (B. oleracea var. botrytis). Theoretical and Applied Genetics. 1997;**94**:338-346

[86] Nishio T, Sakamoto K. Polymorphism of self-incompatibility genes. In: The Radish Genome. Cham, Switzerland: Springer International Publishing; 2017. pp. 177-188

[87] Wang R, Mei Y, Xu L, Zhu X, Wang Y, Guo J, et al. Genome-wide characterization of differentially expressed genes provides insights into regulatory network of heat stress response in radish (*Raphanus sativus* L.). Functional and Integrative Genomics. 2018;**18**:225-239

[88] Nou S, Watanabe M, Isogai A, Hinata K. Comparison of S-alleles and S-glycoproteins between two wild populations of Brassica campestris in Turkey and Japan. Sex Plant Reproduction. 1993;**6**:79-86

[89] Ockendon DJ. The S-allele collection of Brassica oleracea. In: III International Symposium on Brassicas and XII Crucifer Genetics Workshop. Wellesbourne, United Kingdom: Acta Hortic. 2000;**539**:25-30

[90] Georgia AVS, Ockendon J, Gabrielson R, Ma ire. Self-incompatibility alleles in broccoli. Horticulture Science. 1982;**17**:748-749

[91] Niikura S, Matsura S. Genetic variation of the self-incompatibility alleles (S-alleles) in the cultivated radish (*Raphanus sativus*) the R-R method. Acta Horticulture. 2001;**546**:359-366

[92] Nishio T, Kusaba M, Sakamoto K, Ockendon DJ. Polymorphism of the kinase domain of the S-locus receptor kinase gene (SRK) in Brassica oleracea L. Theory and Applied Genetics. 1997;**95**:335-342

[93] Cox K, Beaton C. Fruit and Vegetables for Scotland: What to Grow and How to Grow it. Scotland: Birlinn Ltd.; 2020

[94] Kumar S. Male sterility in vegetables. In: Singh KP, Bahadur A, editors. Olericulture-Fundamental of Vegetable Production, I. Ludhiana: Kalyani Publication; 2014. pp. 431-439

[95] Ogura H. Studies of male sterility in Japanese radish with special reference to the utilization of this sterility towards the practical raising of hybrid seeds. Meun Fac Agric Kagoshima Univ. 1968;**6**:39-78

[96] Handa H, Gualberto JM, Grienenberger JM. Characterization of the mitochondrial orfB gene and its derivative, orf224, a chimeric open reading frame specific to one mitochondrial genome of the "Polima" male-sterile cytoplasm in rapeseed (Brassica napus L.). Current Genetics. 1995;**28**(6):546-552

[97] Jean M, Brown GG, Landry BS. Genetic mapping of fertility restorer genes for the Polima cytoplasmic male sterility in canola (Brassica napus L.) using DNA markers. Theoretical and Applied Genetics. 1997;**95**:321-328

[98] Homme YL, Brown GG. Organizational differences between cytoplasmic male sterile and male fertile brassica mitochondrial genomes are confined to a single transposed locus. Nucleic Acids Research. 1993;**21**(8):1903-1909

[99] Homme YL, Stahl RJ, Li X, et al. Male sterility induced by Brassica napus

cytoplasm is correlated with expression of a chimeric gene homologous to the pol CMS associated orf224 gene. Current Genetics. 1997;**31**:325-335

[100] Cardi T, Earle ED. Production of new CMS Brassica oleracea by transfer of 'Anand' cytoplasm from B. rapa through protoplast fusion. Theoretical and Applied Genetics. 1997;**94**:204-212

[101] Bonnet A. Behaviour of cytoplasmic male sterility of Japanese origin in European radish varieties. Versailles: Eucarpia Cruciferae Newsletter. 1970. pp. 83-88

[102] Yamagishi H. Distribution and allelism of restorer genes for Ogura cytoplasmic male sterility in wild and cultivated radishes. Genes & Genetic Systems. 1998;**73**:79-83

[103] Jatoi SA, Javaid A, Iqbal M, et al. Genetic diversity in radish germplasm for morphological traits and seed storage proteins. Pakistan Journal of Botany. 2011;**43**(5):2507-2512

[104] Park KW, Lee JM, Munchen TU, Fritz D. A study of correlation between pithiness in petilole and root of radish. Korean Society of Horticulture and Science. 1989;**7**(1):36-37

[105] Rabbani MA, Murakami Y, Kuginuki Y, Takayanagi K. Genetic variation in radish germplasm from Pakistan using morphological traits and RAPDs. Genetic Resources and Crop Evolution. 1998;**45**:307-331

[106] Singh BK, Koley TK, Karmakar P, et al. Pigmented radish (Raphanus sativus L.): Genetic variability, heritability and inter-relationships of total phenolics, anthocyanins and antioxidant activity. Indian Journal of Agricultural Sciences. 2017;**87**(12):1600-1606

[107] Sudan SK, Samnotra RK, Gupta AK. A short note on field screening of radish varieties under sub-tropical conditions. Himachal Journal Agriculture Research. 1991;**17**(1-2):104

[108] Yamane K, Lü N, Ohnishi O. Multiple origins and high genetic diversity of cultivated radish inferred from polymorphism in chloroplast simple sequence repeats. Breeding Science. 2009;**59**:55-65

[109] Singh BK, Singh J. Lal muli karegi tanaw dur, aankhon ko roshani degi bharpur. Dainik Jagran. 2019;**15**:14

[110] Jain SM, Suprasanna P. Induced mutations for enhancing nutrition and food production. Geneconserve. 2011;**40**:201-215

[111] Ashizawa M, Hida K, Yoshikawa H. Studies on the breeding of Fusarium resistance in radish. I. Screening of radish varieties for Fusarium resistance. Bull Veg Ornam Crops Res Sta. 1979;**A6**:39-70

[112] Peterson J, Pound G. Studies on resistance in radish to Fusarium oxysporum f. sp. conglutinance race 2. Phytopathology. 1960;**50**:807-816

[113] Soh JW, Han KS, Lee SC, Lee JS. Evaluation of resistance of Fusarium wilt in radish genetic resources. Korean Journal of Horticulture Science and Technology. 2013;**31**(S2):92

[114] Jeon YA, Sung JS, Rhee JH, et al. Evaluation of resistance to Fusarium oxysporum f. sp. raphani in radish germplasm. Korean Journal Breed Science. 2017;**49**(4):344-350

[115] Ghimire MS, Khana P, Pokhrel A, et al. Response of different cultivars of radish (Raphanus Sativus L.) to Alternaria leaf spot on seed production during winter at Rupandehi Nepal.

International Journal of Applied Sciences Biotechnology. 2016;**4**(3):318-324

[116] Xu L, Wang L, Gong Y, Dai W, Wang Y, Zhu X, et al. Genetic linkage map construction and QTL mapping of cadmium accumulation in radish (Raphanus sativus L.). Theory and Applied Genetics. 2012;**125**:659-670

[117] Muminović J, Merz A, Melchinger AE, Lübberstedt T. Genetic structure and diversity among radish varieties as inferred from AFLP and ISSR analyses. J. of the a. Society for Horticulture Science. 2005;**130**:79-87

[118] Curtis IS. Genetic engineering of radish: Current achievements and future goals. Plant Cell Reports. 2011;**30**:733-744

[119] Lee O, Koo H, Yu JW, Park HY. Genotyping-by-sequencing-based genome-wide association studies of Fusarium wilt resistance in radishes (Raphanus sativus L.). GEN. 2021;**12**:858

[120] Yu X, Choi SR, Ramchiary N, Miao X, Lee SH, Sun HJ, et al. Comparative mapping of Raphanus sativus genome using brassica markers and quantitative trait loci analysis for the Fusarium wilt resistance trait. Theory and Applied Genetics. 2013;**126**:2553-2562

[121] Kamei A, Tsuro M, Kubo N, Hayashi T, Wang N, Fujimura T, et al. QTL mapping of clubroot resistance in radish (Raphanus sativus L.). Theory and Applied Genetics. 2010;**120**:1021-1027

[122] Budahn H, Peterka H, Mousa M, Ding Y, Zhang S, Li J. Molecular mapping in oil radish (Raphanus sativus L.) and QTL analysis of resistance against beet cyst nematode (Heterodera schachtii). Theory and Applied Genetics. 2009;**118**:775-782

[123] Fan L, Wang Y, Xu L, Tang M, Zhang X, Ying J, et al. A genome-wide association study uncovers a critical role of the RsPAP2 gene in red-skinned Raphanus sativus L. Horticulture Research. 2020;**7**:1-13

[124] Byrne DH, Littleton TG. Electrophoretic characterization of diploid plums of southeastern United States. Journal of American Society of Horticulture Sciences. 1988;**113**(6):918-924

[125] Tai-Young P, Jinkug K, Kyu-Yong S, Wah-Mo Y. Cultivar identification and seed purity test by electrophoresis of isozymes in radish and Chinese cabbage. Journal of Korean Society for Horticulture and Sciences. 1995;**36**:805-811

[126] Wang N, Kitamoto N, Ohsawa R, Fujimura T. Genetic diversity of radish (Raphanus sativus) germplasms and relationships among worldwide accessions analyzed with AFLP markers. Breeding Science. 2008;**58**(2):107-112

[127] Cruz SM, Nery MC, Pinho EVRV, de Laia ML. Molecular characterisation of radish cultivars. Rev Ciên Agron. 2014;**45**(4):815-822

[128] Nakatsuji R, Hashida T, Matsumoto N, et al. Development of genomic and EST-SSR markers in radish (Raphanus sativus L.). Breeding Science. 2011;**61**:413-419

[129] Bae KM, Sim SC, Hong JH, et al. Development of genomic SSR markers and genetic diversity analysis in cultivated radish (Raphanus sativus L.). Horticulture, Environment and Biotechnology. 2015;**56**:216-224

[130] Lee ON, Park HY. Assessment of genetic diversity in cultivated radishes (Raphanus sativus) by agronomic traits and SSR markers. Scientia Horticulturae. 2017;**223**:19-30

[131] Duroc Y, Gaillard C, Hiard S, et al. Nuclear expression of a cytoplasmic male sterility gene modifies mitochondrial morphology in yeast and plant cells. Plant Science. 2006;**170**:755-767

[132] Grelon M, Budar F, Bonhomme S, Pelletier G. Ogura cytoplasmic male-sterility (CMS)-associated orf138 is translated into a mitochondrial membrane polypeptide in male sterile *brassica* cybrids. Molecular & General Genetics. 1994;**243**:540-547

[133] Yamagishi H, Terachi T. Intra- and inter-specific variations in the mitochondrial gene orf138 of Ogura-type male sterile cytoplasm from Raphanus sativus and Raphanus raphanistrum. Theoretical and Applied Genetics. 2001;**103**:725-732

[134] Nie S, Li C, Xu L, et al. De novo transcriptome analysis in radish (Raphanus sativus L.) and identification of critical genes involved in bolting and flowering. BMC Genomics. 2016;**17**:389-405

[135] Forster BP, Heberle-Bors E, Kasha KJ, Touraev A. The resurgence of haploids in higher plants. Trends in Plant Science. 2007;**12**:368-375

[136] Gu HH, Zhao ZQ, Sheng XG, et al. Efficient doubled haploid production in microspore culture of loose-curd cauliflower. Euphytica. 2014;**195**:467-475

[137] Lichter R. Induction of haploid plants from isolated pollen of Brassica napus L. ZPflanzenphysiology. 1982;**105**:427-434

[138] Chun C, Park H, Na H. Microspore-derived embryo formation in radish *R. sativus* L. according to nutritional and environmental conditions. Horticulture, Environment and Biotechnology. 2011;**52**(5):530-535

[139] Sugimoto K. F1 hybrid production using doubled haploid of microspore-derived 'Moriguchi' radish (Raphanus sativus L.). Journal of Japan Socirty for Horticulture and Sciences. 2002;**71**(2):187-191

[140] Tuncer B. Callus formation from isolated microspore culture in radish (Raphanus sativus L.). Journal of Animal and Plant Sciences. 2017;**27**(1):277-282

[141] Tuberosa R, Salvi S. Genomics-based approaches to improve drought tolerance of crops. Trends in Plant Sciences. 2006;**11**:405-412

[142] Li Z, Zhou M, Zhang Z, Ren L, Du L, Zhang B, et al. Expression of a radish defensin in transgenic wheat confers increased resistance to Fusarium graminearum and Rhizoctonia cerealis. Functional and Integrated Genome. 2011;**11**:63-70

[143] Jeong YM, Chung WH, Choi AY, Mun JH, Kim N, Yu HJ. The complete mitochondrial genome of cultivated radish WK10039 (Raphanus sativus L.). Mitochondrial DNA Part A. 2016;**27**:941-942

[144] Jeong YM, Chung WH, Mun JH, Kim N, Yu HJ. De novo assembly and characterization of the complete chloroplast genome of radish (Raphanus sativus L.). Gene. 2014;**551**:39-48

[145] Mitsui Y, Michihiko S, Kenji K, Nobukazu N, Mari S, Misaki I, et al. The radish genome and comprehensive gene expression profile of tuberous root

formation and development. Scientific Reports. 2015;**5**:1-14

[146] Lee YJ, Mun JH, Jeong YM, Joo SH, Yu HJ. Assembly of a radish core collection for evaluation and preservation of genetic diversity. Horticulture, Environment and Biotechnology. 2018;**59**:711-721

[147] Shen D, Sun H, Huang M, Zheng Y, Li X, Fei Z. Radish Base: A database for genomics and genetics of radish. Plant and Cell Physi. 2013;**54**:33

[148] Tanaka Y, Tsuda M, Yasumoto K, Yamagishi H, Terachi T. A complete mitochondrial genome sequence of Ogura-type male-sterile cytoplasm and its comparative analysis with that of normal cytoplasm in radish (Raphanus sativus L.). BMC Genomics. 2012;**13**:1-12

[149] Hu T, Wei Q, Wang W, Hu H, Mao W, Zhu Q, et al. Genome-wide identification and characterization of CONSTANS-like gene family in radish (Raphanus sativus). PLoS One. 2018;**13**:e0204137

[150] Xiaohui Z, Zhen Y, Shiyong M, Yang Q, Xinhua Y, Xiaohua C, et al. A de novo genome of a Chinese radish cultivar. Horticulture Journal. 2015;**1**:155-164

[151] Li F, Hasegawa Y, Saito M, Shirasawa S, Fukushima A, Ito T, et al. Extensive chromosome homoeology among Brassiceae species were revealed by comparative genetic mapping with high-density EST-based SNP markers in radish (Raphanus sativus L.). DNA Research. 2011;**18**:401-411

[152] Boetzer M, Henkel CV, Jansen HJ, Butler D, Pirovano W. Scaffolding pre-assembled contigs using SSPACE. Bioinformatics. 2011;**27**:578-579

[153] Shirasawa K, Hirakawa H, Fukino N, Kitashiba H, Isobe S. Genome sequence and analysis of a Japanese radish (Raphanus sativus) cultivar named 'Sakurajima Daikon'possessin giant root. DNA Research. 2020;**27**:10

[154] Varshney RK, Bohra A, Yu J, Graner A, Zhang Q, Sorrells ME. Designing future crops: Genomics-assisted breeding comes of age. Trends in Plant Science. 2021;**26**:631-649

[155] Wang J, Qiu Y, Wang X, Yue Z, Yang X, Chen X, et al. Insights into the species-specific metabolic engineering of glucosinolates in radish (Raphanus sativus L.) based on comparative genomic analysis. Scientific Reports. 2017;**1**:1-9

[156] Curtis IS, Nam HG. Transgenic radish (Raphanus sativus L. longipinnatus Bailey) by floral-dip method–plant development and surfactant are important in optimizing transformation efficiency. Transgenic Research. 2001;**10**:363-371

[157] Tzfira T, Citovsky V. Agrobacterium: From Biology to Biotechnology. New York, NY: Springer; 2008

[158] Lacroix B, Citovsky V. Pathways of DNA transfer to plants from agrobacterium tumefaciens and related bacterial species. Annual Review of Phytopathology. 2019;**25**:231-251

[159] Ali MA, Azeem F, Abbas A, Joyia FA, Li H, Dababat AA. Transgenic strategies for enhancement of nematode resistance in plants. Frontiers in Plant Sciences. 2017;**8**:750

[160] Gelvin SB. Agrobacterium in the genomics age. Plant Physiology. 2009;**150**:1665-1676

[161] Giri A, Narasu ML. Transgenic hairy roots: Recent trends and applications. Biotechnology Advances. 2000;**18**:1-22

[162] Berkov S, Pavlov A, Kovatcheva P, Stanimirova P, Philipov S. Alkaloid spectrum in diploid and tetraploid hairy root cultures of Datura stramonium. Z Natur f. 2003;**58**:42-46

[163] Curtis IS. The noble radish: Past, present and future. Trends in Plant Sciences. 2003;**8**:305-307

[164] Li S, Zhan N, i J, Xian J. A new male sterile line of "Duane-13" radish (Raphanus sativus L.) produced by ethyl methanesulfonate mutagenesis. Genetics Research and Crop Evolution. 2019;**66**:981-987

[165] Hayta S, Smedley MA, Demir SU, Blundell R, Hinchliffe A, Atkinson N, et al. An efficient and reproducible agrobacterium-mediated transformation method for hexaploid wheat (Triticum aestivum L.). Plant Methodology. 2019;**15**:1-15

[166] Rissler J, Mellon M, Mellon MG. The Ecological Risks of Engineered Crops. USA: MIT Press; 1996

[167] Snow AA, Palma PM. Commercialization of transgenic plants: Potential ecological risks. Biological Sciences. 1997;**47**:86-96

[168] Pua EC, Sim GE, Chi GL, Kong LF. Synergistic effect of ethylene inhibitors and putrescine on shoot regeneration from hypocotyl explants of Chinese radish (Raphanus sativus L. var. longipinnatus Bailey) in vitro. Plant Cell Reports. 1996;**15**:685-690

Chapter 15

Pollination Biology and Environmental Water Pollution Indicator of Onion (*Allium cepa* L.)

Anup Kumar Sarkar

Abstract

Numerous genes in flowering plants, including onion (*Allium cepa* L.) govern morphological character differences in structure, shape, orientation, weight and number, etc. arising from their assortment and recombination due to pollination. Pollination in onion flower occurs mainly by insects (91%) and wind (9%), with gravity also contributing to the pollination process. The hybrid vigour seeds through cross pollination as an essential input in enhancing crop productivity. The present study reveals that an onion plant generally takes around 63 days to attain flowering position, and complete flowering condition in 70–72 days, which include 15–18 days for sprouting of the green hollow fleshy shoots along with about 45 days to complete the peduncle formation on the top small part of the bud. *A. cepa* L. takes around 160 days to produce mature seeds with life cycle completion. The onion bulb roots are used for the last 50 years to study chromosomal behaviour as an indicator of environmental water pollution. The presence of different impurities and heavy metals in the polluted water causes reduction in reproductive capacity of cells due to the occurrence of peculiarity from the normal mitotic cell division in onion . Cytotoxicity influences all morphological characters, including root growth retardation, mitotic index, chromosomal aberration, etc. Thus, the present investigation explores the effect of pollutant water on pollination biology, cytotoxicity, root apical meristem cells in onion. We report a significant ($p < 0.05$) in the mitotic index in polluted water as compared to normal water.

Keywords: cross pollination, heavy metals, chromosomal aberration, cytotoxicity and environmental indicator, water pollution, pollination biology

1. Introduction

The *Allium* (family:Amaryllidaceae) is a large genus of onion or garlic fragrant perennial bulbous herbs and globally represents about 700 species, but only seven species are cultivated. Members of the family are the world's oldest cultivated plants after potatoes, yams, and tomatoes, comprising an important group of vegetables, except the tropics, New Zealand and Australia. *Allium* is a native to southern western Asia with cultivation throughout the world, predominantly in the temperate regions [1].

There are different species of *Allium* which are grown in the field for the next generation in the form of flowers instead of bulbous structures. Generally, the seedlings are used for bulb production as vegetable. Two development phases occur to complete the life cycle of onions. One phase is entirely responsible for the production of seeds only whereas the other herbaceous annual phase is for the production of bulb from the seedling. Pollination in onion takes by an array of insects [2, 3]. The hermaphrodite onion flowers cannot fertilise themselves since the anthers exhibit protandry, releasing sticky and wet pollens before the stigma becomes receptive [4]. Thus, a cross-pollination between two flowers of the same plant (Geitonogamy) or two flowers from different plants (Allogamy) accomplishes seed production in the onion. The cross-pollinated seeds are used for the production of the onion bulb. FAO [5] reported that the production of onion in India is 12.5 ton/ha only whereas it is much lower than the production 41.12 ton/ha in the USA. Different sources of Municipal, Industrial, Agricultural, and advanced technological waste components ingredients can increase significant amounts of impurities in surface water and later on slowly deposited at a lower base as a consequence water pollution contributed a serious problem for the health of the biological organism along with human, those interact with this aquatic ecosystem in developed and developing countries. Those waste materials released from the different sources are mainly toxic metals and metalloids which are not converted into harmless nontoxic forms by the biological process but engaged in the environmental system which react the suppression activity of metabolism and translocation of reserve food materials into low concentration and impose to damage of the growing regions cells of the living organism [6]. Living organisms encounter nasty toxic heavy materials responsible for damage/modification of the genetic materials happening in the cell cycle. Meristematic root tips of *A. cepa* L. are used all over the world for testing the level of environmental water pollution [7–10]. To evaluate water quality being used for cytological studies of *A. cepa* L. root tips squash technique offers one of the best and quick methods, which also provides a reliable estimate for the genotoxic effect of heavy metal and chloride interaction on the environment. The water samples from three locations, Preonagar, Mathpara and Harishnagar 24 Parganas (N), West Bengal, India have been assessed on the basis of mitotic cell activity (Mitotic Index) and different chromosomal abnormalities (CAs) in the meristematic root tips cells of *A. cepa* L. (**Table 1**). The effect was compared with distilled water. The wastewater from the above locations enters, in different ways, either the agriculture fields for irrigation or river flow which makes a hazard to the ecosystem of that environment. With this background, the hazardous elements in the water samples assumingly react with the chromosome of the meristematic part of root tip cells of *A. cepa* L. that acts as an indicator for natural water pollution.

2. Materials and methods

2.1 Selection of land

To avoid harbouring of root rot or wilt pathogen, the selection of healthy edaphic conditions is the first priority for a normal luxuriantly growing plant. Neutral pH with loam clay loam soil was selected for the experiment at the Village-Belu, P.O. Madhappur, Amdanga, North 24 Parganas (Latitude 22°11′6″ N to 23°15′2″ N and Longitude 88°20′ to 89°5′ E) West Bengal, India.

Sl. No.	Heavy metals, Chloride and pH	Limit as per IS 10500: 2012		Different elements presence in the three locations waste water samples		
		Minimum	Maximum	Preonagar	Mathpara	Harishnagar
1	Mn (mg/l)	0.01	0.30	0.321	0.275	0.101
2	Cr (mg/l)	0.001	0.05	< 0.001	< 0.001	< 0.001
3	Cu (mg/l)	0.001	1.5	< 0.001	< 0.001	0.006
4	Cd (mg/l)	0.001	0.003	< 0.001	< 0.001	<0.001
5	Fe (mg/l)	0.001	0.30	0.340	0.501	0.520
6	Pb (mg/l)	0.001	0.01	< 0.001	< 0.001	< 0.001
7	Zn (mg/l)	0.001	15.0	0.450	< 0.001	0.002
8	Ni (mg/l)	0.001	0.02	0.006	< 0.001	0.002
9	Cl (mg/l)	N/A	1000	89.19	79.55	269.98
10	pH	-	-	5.30	5.15	5.10
11	Colour	-	-	Bluish	Blackish	Blackish

Table 1.
Heavy metal, Chloride and pH analysis of three Experimental locations water samples (Preonagar, Mathpara and Harishnagar).

2.2 Bulb selection and environmental condition

Onion bulbs around 5–6 cm in diameter were selected for sowing of the above field. The good flowerings were obtained by cool weather after bulb planting. It was also noticed that the good sunshine at the time of full blooming stages helped in attracting the beneficial pollinators for the higher rate of hybrid vigour seeds through cross-pollination. One more important parameter was kept in mind that relative humidity (RH) would be in the lower range during the time of seed development. As for the farmer's concern, the experiment was started in the month of the middle of October for sowing the bulb in the field of the above area.

2.3 Experiential laying out

The buds were planted in beds of size 1×3 m with 20–25 cm spacing during October. The experiments were conducted in Randomised Complete Block Design (RCBD) with 3 replications.

2.4 Floral biology

Morphology, physiology and phenology of plant flowering play a vital role in the reproductive capability of individual plant species. A flower's attractiveness to the visiting fauna and efficiency of pollen transfer to those visitors depend on the morphological characters, i. e. shape, colour and flower architecture [11]. The cross-pollinated onion plants are completely dependent on the different types of insects for their pollination [12]. Complete field observation was conducted to invent the onion floral characteristics, flower (bud) initiation and duration, number of days for maximum flowering, number of peduncles per bulb. The number of florets per inflorescence, life of single floret, colour of the florets and odour were observed in the field materials.

Phenology

Plant phenology is concerned with the timing of recurring events such as

- leaf flushing
- flowering and
- fruiting
- The timing, intensity and duration of the flowering among plants dictate effective successful insect visits and cross-pollination, resulting in the success of the plant reproductive cycle (**Table 2**).

Sl. No.	Floral characters	Remarks
1.	Sprouting of the green fleshy shoots	15.67± 2.45 days
2.	Days from showing to peduncle formation	45.03± 0.98 days
3.	No. of days for maximum plants in flowering	62.21± 0.65 days
4.	Bud breaking and Flower initiation	70.05± 1.09 days
5.	Number of Peduncle per bulb	4.25 ± 1.00 cm
6.	Length of Peduncle	68.45± 0.58 cm
7.	Type of inflorescence	Umbel
8.	Number of Florets per inflorescence	306.78 ± 29.0
9.	Life of single floret	7.01 ± 0.52
10.	Colour of Inflorescence	Dull white in colour
11.	Type of overy	Superior
12.	Length of pistil	4.50 ± 0.58 mm
13.	Length of style	3.01± 0.67 mm
14.	Number of anthers per floret	6
15.	Type of anther	Bilocular
16.	Length of Stamen	0.71 ± 0.19 cm
17.	Length of Filament	0.52 ± 0.13 cm
18.	Length of Anther	0.22 ± 0.10 cm
19.	Mode of anther dehiscence	Longitudinal
20.	Time of Anthesis	Early in the morning
21.	Anther dehiscence	Whole day
22.	Mode of pollination	Mostly done by insects
23.	Type of Pollination	Cross pollination

Table 2.
Different morphological characters of the Allium cepa L.

Volume of nectar

- Insulin-pushing syringes were used for the collection of nectar from 15 number of florets from each inflorescence of the 20 number peduncle of each replication of the experiment.

- The portable instrument Refractometer was used to quantify the total soluble solids (TSS) of nectar.

Pollinators

- In 15 number of randomly selected inflorescence were used for counting the pollinators in 10 minutes of an hour of the day.

- Visitors were observed under the dissection microscope for identification of the pollinators through trapping and killing by the rectified spirit.

2.5 Collection of effluents sample

The coloured polluted sewage water was collected from the municipality channel of the Preonagar, Mathpara and Harishnagar, North 24 Parganas (22.61680 N, 88.40290 E), West Bengal, India in the winter season from three locations at the depth of 5–7 inches from five random points within the municipality drains of each location. The dirty water was filtered by muslin cloth so many times to remove all visible muddy materials present in the water sample and later on storage having been done in a clean container for conducting physiochemical analysis followed by use in cytological studies of the experimental onion crop.

2.6 Physiochemical parameters

The heavy metals (Manganese, Nickel, Copper, Zinc, Cadmium, Chromium, Iron and Lead) and chloride responsible for different chromosomal mutations were estimated for their concentration (mg/l) in the water sample of the different parts of North 24 Parganas (N), West Bengal (**Table 1**).

Sample assay.

Onion (*A. cepa* L.)

Onion (*A. cepa* L.) bulbs of about the same size (42 mm) weighing about 33 gm and 10 months old were selected by removing the loose outer scale, older roots without damaging meristematic tissues and scrapped the bottom part with the help of a blade so that root primordia were formed in wastewater samples of different locations (treatments) and distilled water (control condition). Keep in mind that the bottom part of each bulb slightly emerged out of the water. The incubated time of the sample in each treatment was 48 hours at the temperature of 22°C in dark conditions.

2.7 Chromosome preparation

The squash preparation demonstrated by Sharma and Sharma [13] was used for the chromosome preparation of the treated onion roots. The following methods were used to investigate the root meristematic tissue exhibiting mitotic cell division:

Pre-treatment

- The developing root tips 1 cm in length were cut and pre-treated with super-saturated aqueous solution of pDB-Aesculine initially kept in an ice chamber (0°C) of a refrigerator for 10 minutes in the form of adequate fixation with good preserved chromosome structure.

- After that, the refrigerator temperature was changed to 10°C for 3 hours.

- After washing in distilled water, the onion roots were fixed in Carnoy's fluid-I (Glacial acetic acid: Dehydrated alcohol = 1:3) for overnight at 20°C (about 24 hours).

Staining

- Before staining, the treated root tips were kept in 45% acetic acid for 10 minutes at room temperature.

- The staining of all the roots were done in a mixture of 2% Aceto-Orcein (Sigma-aldrich) acid and 1 N HCl (9:1) for 45 minutes after slightly warming at 60°C by a sprit lamp.

- The deeply stained tip portion of onion roots (1 mm size) were cut and placed in a drop of 45% acetic acid which has a very remarkable penetrating property even higher than alcohol due to its smaller ions.

- Squash preparation was made using thumb pressure on a clean grease-free slide.

Observation

- The slides were prepared from random samples each of five root tips for each treatment along with control.

- The five random microscopic fields were scored for each slide.

- The mean mitotic index for each sample and each treatment were compared with those corresponding to control and "t" test was applied to find out the significant difference, if any.

- The slides were temporarily sealed with paraffin wax for observation.

- The optical microscope used in the investigation was Olympus with the Prog-Res Capture Pro 2.1 photosystem.

- *Mitotic index (MI)* was computed using following formulae:

$$\text{Active Mitotic Index } (AMI)\,\% = \frac{\text{Total number of dividing cells}}{\text{Total number of cells observe}} \times 100$$

- *Total Abnormality Percentage (TAP) %* was computed using following formulae:

$$\text{Total Abnormality Percentage } (TAP)\ \% = \frac{\text{Number of abnormal cells}}{\text{Total number of cells scored}} \times 100$$

2.8 Statistical analysis

The experiments were conducted according to randomised complete block design (RCBD) with three replications. Data were expressed as mean ± standard error (SEM) [14].

3. Result

3.1 Phenological and growth variable

The onion (*A. cepa* L.) is a biennial crop that completes the life cycle in 2 years i.e., 1 year after plantation, using the nutrients from the soil, we get the onion bulb which is the vegetable. Second year that particular size of vegetable bulbs is used for the production of fleshy peduncles and seeds. In the study after sowing the onion bulbs the peduncle started flowering within 45.03 ± 0.98 days (**Figure 1b** and **Table 2**). The flowers are raised on the top of single very compressed hollow internode as a peduncle. Each plant contained around 4.25 ± 1.00 number of peduncles with many other chlorophyll less fleshy leaves and each hollow peduncle length was about 68.45 ± 0.58 cm. Each peduncle held on the top of an umbel inflorescence consisted of 306.78 ± 29.0 florets (**Figure 1f**). Initially, small size inflorescence of *A. cepa* L. covered by 2 layers of white colour membrane called spathe. These spathes protect the juvenile buds of the umbel inflorescence. Due to the increase in the internal pressure of the florets, the spathe splits open which took 8–10 days. It was noticed that around 6 O'clock early in the morning anthesis takes place but the temperature is the crucial factor for the initiation of the anthesis. Fertility of the pollen along with stigma receptivity was observed the highest on the day of anthesis. In two whorls arrangement hold both the perianth and six number stamens. The stamen was found to be 0.71 ± 0.19 cm in length and consists of a bilocular anther 0.22 ± 0.10 cm long (**Figure 1j**). The anther splits longitudinally to release the pollen grains and takes the entire day for its dehiscence. Pistil length (with superior ovary) was 4.50 ± 0.58 mm (**Figure 1i, j**). In the normal condition, only one pedicel was carried by a single floret but in this experiment, it was observed that there were two florets attached to the single pedicel (**Figure 1g, h**). Different species of the genus *Allium* show the variable colour combination among the inflorescences (**Figure 1c**). The onion flower releases a very strong odour which attracts various insects for cross-pollination. The presence of nectar in flowering plants attracts insects which makes interspecific relationship for cross-pollination in the plants for hybrid vigour seeds production. Morphologically bowl-shaped florets of the onion inflorescence produce nectar with hidden nectarines (**Figure 1o, p**) thus act as a good food reward and attract a wide range of vectors among which few of them are pollinators. These vectors belong to different families. One of the family Apidae including honey bees such as *Apis cerana* and *Apis florea* were found in the field most of the time (**Figure 1l**). Another insect was noticed with elongated siphon which might help to suck the nectar from inside of the floret

(**Figure 1 k**, **m**, **n**). Very good relationships between pollinators and temperature were observed as the pollinator movement was maximum in first half of the day and slowly decreased when temperature increased.

3.2 Different elements present in the samples

Table 1 enlists the different types of toxic metals along with chloride in different water samples. Water samples from three locations attained a higher range of iron quantities as per the standard limits of iron concentration (0.001–0.30 mg/l). Heavy metals like Cd, Cr, Cu and Pb were found in all water samples, lesser than the permissible limits (0.001 mg/l). Nevertheless, manganese (0.32 mh/l) was found in higher concentration in Preonagar water sample than Mathpara and Harishnagar samples as compared to standard limits (**Figure 2**).

3.3 Root growth inhibition

The onion root growth inhibition effect was clearly visible in plants receiving wastewater from the three locations compared with the healthy, elongated roots in distilled water (**Figure 3**).

3.4 Effect on mitotic phase frequency

A study of the mitotic phase frequency revealed that the total number of cells in wastewater of different locations was less than the cells present in distilled water treatment. It exhibited that dividing cells present in distilled water was higher than the remaining three sources of wastewater. In the anaphase stages, it was noticed in the sample of Mathpara to have a higher number of metaphase stages as compared to Preonagar, Harishnagar and distilled water samples (**Figure 4**).

Figure 1.
Sprouting shoots along with floral parts of the A. cepa L. *(a) Sprouting shoots of the onion (15 days old) (b) Cylindrical fleshy leaf with Peduncle (45 days old) (c) Very good No. of Peduncle with Inflorescence (70 days Old) (d) Immature florets (e) Pedicel of a about equal length all arising from the apex of the Peduncle with single open floret (f) Onion inflorescence at mature bud and bloom stage 98% (100 days old) (g) Two florets are attach with single pedicel (h) Distinctly showing the two parts attached at lower part of pedicel (i) Pistil of the floret (j) Petals, Pistil and Stamens of the floret (k, l, m, n) Vector/pollinator (arrow) observed on the individual flower (o, p) Floral nectar (arrows).*

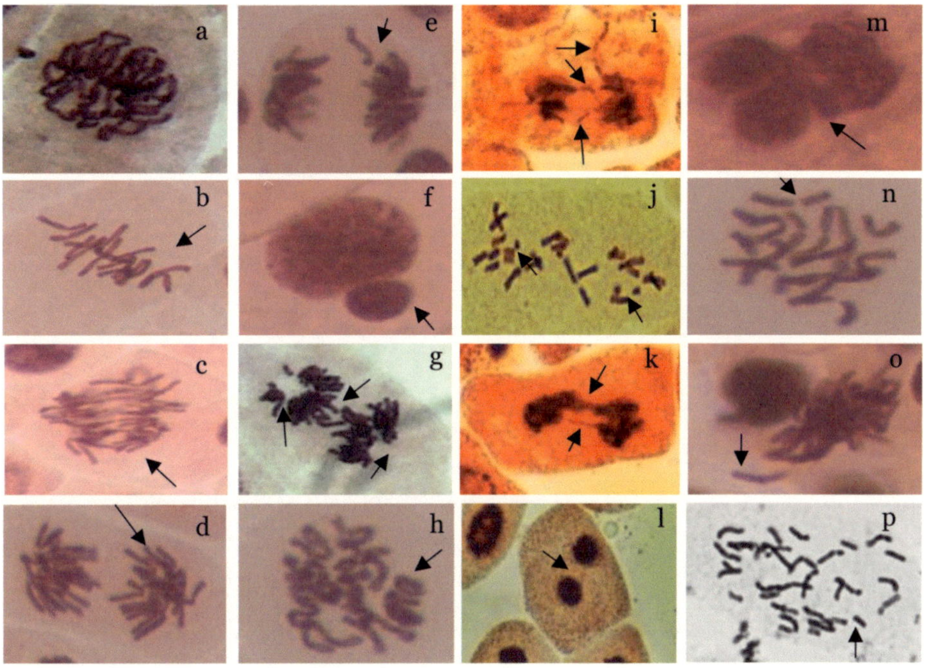

Figure 2.
Normal division and Chromosomal abnormalities observed in A. cepa L. (2n=16) meristematic cells exposed to three different polluted waters with control (distilled water) (a) Normal Late Prophase (b) Normal Metaphase (c) Normal Early Anaphase (d) Normal Telophase respectively treated with distilled water (e) Telophase with chromosome break (f) Micronucleus of large size in Interphase (g) Multipolar telophase with chromosome bridge (h) Tripolar anaphase respectively treated with Preonagar sample (i) Anaphase bridge with laggard chromosome (j) Fragmented chromosome (k) Anaphase with multi chromosome bridges (l) Bi-nucleus cell, respectively treated with Mathpara sample and (m) Multi nucleus cell (n) Fragmented chromosome (o) Metaphase with chromosome adherence and chromosome break (p) C-metaphase chromosome respectively treated with Harishnagar sample.

3.5 Chromosomal abnormalities (CA)

Meristematic cells present in root tips of *A. cepa* L. were exposed to 72 hours of the three different locations water sample and exhibited various chromosomal aberrations induced by chemical agents in the form of changes either in chromosomal structure or in the total numbers of chromosomes compared with distilled water as control. The different types of observed abnormalities included telophase with chromosome break, micronucleus, multipolar telophase with chromosome bridge, tripolar anaphase, anaphase bridge with laggard chromosome, fragmented chromosome, anaphase with multi chromosome bridges, bi-nucleus cell, multi nucleus cell, metaphase with chromosome adherence and chromosome break and c-metaphase chromosome (**Figure 2**). Studies indicated that all the toxic elements present in the different locations sample water caused a significant increase in the total aberrant cells except distilled water (**Figure 3**). The toxic elements in wastewater samples induced the highest aberrant cells frequency of about 18% which was observed to be in Mathpara, whereas medium in Preonagar (11%) and the lowest frequency in Harishnagar (8%) water samples (**Figure 3**). The analysis of chromosomal abnormalities was made mainly on the anaphase and telophase stages of the cell cycle division.

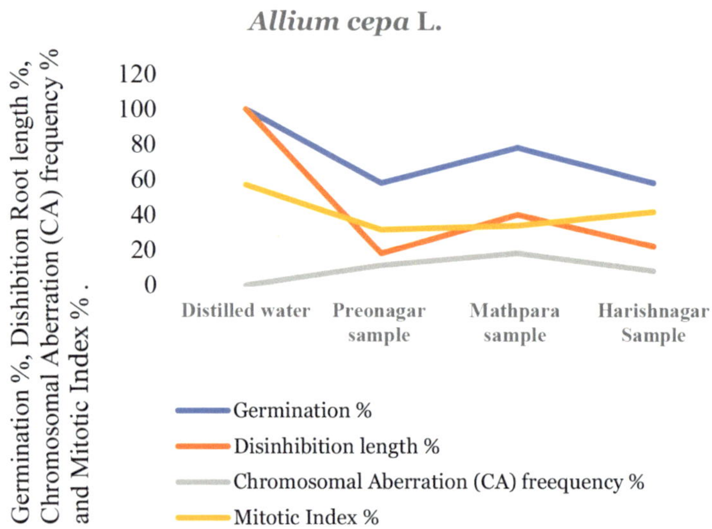

Figure 3.
Effect of Distilled water, Preonagar, Mathpara, and Harishnagar water samples on Germination %, Disinhibition Root length %, Chromosomal Aberration (CA) frequency % and Mitotic Index (MI) % in A. cepa L.

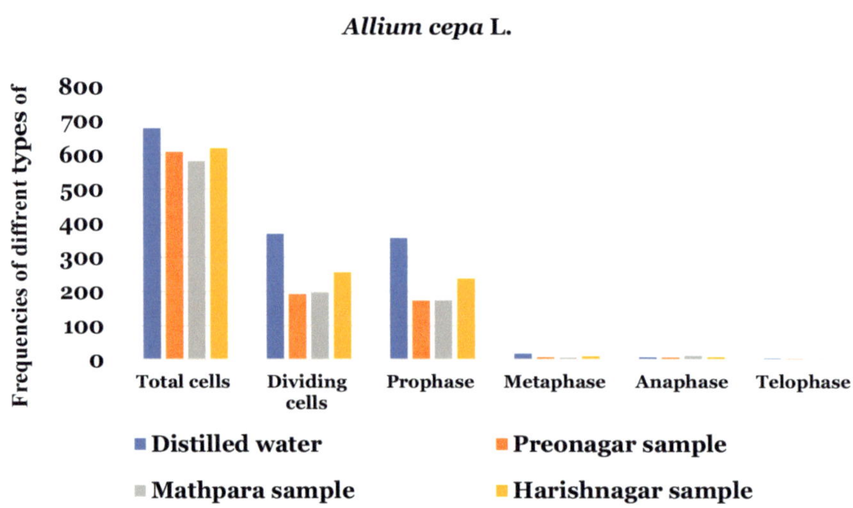

Figure 4.
Frequencies of different types of cells after treatment with different water samples (Distilled water, Preonagar, Mathpara and Harishnagar water samples).

The upward direction of Chromosomal aberrations (CA) of the *A. cepa* root tips meristematic cells reflected a significant decrease in the Mitotic Index (MI) due to taxological elements existing in the wastewater samples (**Figure 3**).

4. Discussion

Environment and its interaction with the genome help in the expression of physio-morphological characters of all organisms. The inflorescences *A. cepa* L. is of dull white colour whereas *Allium aflatunense* bears violet colour then slowly turned into purple colour [15]. *Allium giganteum* inflorescences produce red-coloured florets from the purple colour [16]. Nevertheless, the onion inflorescence displays a very strong flavour and odour due to the presence of chemical alteration of a volatile secondary metabolite like S-methyl cysteine sulphoxide compound [17]. Temperature strongly influences the reproductive structure of the onion [18]. Interspecies relation may be a cause of the very good method for cross-pollination in the different types of organisms [19]. It is no wonder that nectar offers as a food reward to the insect vectors pollinating the plant to their pollen vectors [20]. In the present investigation, different species of the Genus *Apis* viz., Apis dorsata, *A. cerana,* Apis mellifera and *A. florea* visit the onion inflorescence florets as a gatherer of pollens and nectar (**Figure 4**) [21].

The present study establishes a steady relationship between the wastewater heavy metal toxicity and abnormal cellular behaviour of onions having been indicated by frequencies of the different mitotic phases, MI and the types of structural chromosomal abnormalities. The cytological studies of the meristematic cells of the *A. cepa* L. roots assay provide one of the most reliable and useful protocols for the investigation of environmental pollution, biological monitoring and determination of the toxicity of the different elements present in the different sources.

The lower level of the MI in the *A. cepa* L. meristematic cells can be indicated as a reliable process to determine the presence of cytotoxic agents such as heavy metals in the environment and considered as a real test to evaluate the pollution level in the natural water bodies [22]. Our observation corroborates earlier investigations using the Mitotic Index evaluation as a tool for the detection of cytotoxicity mediators and pollution agents present in the environment [10, 23, 24].

Fiskesjo [25] has established drinking water contamination by copper using the *A. cepa* L. test. Subsequently, the author has successfully extended the same test for others to detect different toxic metals such as Hg, Ni, Cu, Cd, Ne, Al, Mn and Li, establishing chromosomal abnormalities in the form of C-metaphase to be associated with heavy metal Ni [26].

A. cepa L. meristematic root cells assay efficiently evaluates different aqueous concentrations of copper mine waste causing cytogenetic effects such as chromosomal abnormalities. The 100 percent concentration of copper mine waste i.e., raw sample presented the highest toxicity and exhibit a relationship with the inhibition of MI along with chromosome breaks, delay, bridges and adherence were the most frequent Chromosomal aberration (CA) observed [27].

Borboa and Torre [28] have assessed heavy metals Zinc and Cadmium genotoxicity linked with chromosomal abnormalities in the *A. cepa* L. test system with cadmium inflicting a greater genotoxic effect. Seth et al. [29], in evaluation studies of the genotoxicity effects of cadmium by the *A. cepa* L test, have revealed inhibition of CA with MN induction.

Contamination by tannery effluents found in river water causes significant frequencies of chromosomal abnormalities and micronucleus in *A. cepa* L. meristematic cells noticed [30].

5. Conclusion

A. cepa L. test in the present investigation emerges as the best model and gentle assay to detect the presence of environmentally hazardous elements, which causes genotoxicity and mutagenesis. Thus, this test provides an important tool for screening environmental contaminants and their result can be used as a warning signal for human health.

Conflicts of interest

The author declares no conflict of interest.

Author details

Anup Kumar Sarkar
Department of Botany, West Bengal State University, Kolkata, India

*Address all correspondence to: ak94sarkar@yahoo.com

References

[1] Thompson CH, Kelly CW. Vegetable Crops. USA: McGrawHill Book Co., Inc.; 1957

[2] Benedek P, Gaal E. The effect of insect pollination in seed onion, with observations on the behaviour of honeybees on the crop. Journal of Apicultural Research. 1972;**11**:175-180

[3] Free JB. Insect Pollination of Crops. London: Academic Press; 1970

[4] Delaplane KS, Mayer DF. Crop Pollination by Bees. New York, USA: CABI Publishing; 2000

[5] FAO. Production Year Book. 2002;**56**

[6] Ayansina AS, Olubukola BO. A new strategy for heavy metal polluted environments: A review of microbial biosorbents. International Journal of Environmental Research and Public Health. 2017;**14**(94):1-16

[7] Fiskesjo G. *Allium* test for screening chemicals; evaluation of cytological parameters. In: Plants for Environmental Studies. Boca Raton, New York: Lewis Publishers; 1997. pp. 307-333

[8] Fiskesjo G. The *allium* test as a standard in environmental monitoring. Hereditas. 1985;**102**:99-112

[9] Ivanova E, Staikova TA, Velcheva I. Cytogenetic testing of heavy metal and cyanide contaminated river water in a mining region of Southwest Bulgaria. Journal of Cell and Molecular Biology. 2005;**4**:999-106

[10] Sarkar AK, Saha R, Halder R. Chromosomes damage by sewage water studies in the *Allium cepa* L. and *Zea mays* L. Caryologia. 2022;**75**(1):55-63

[11] Muchhala N. Exploring the boundary between pollination syndromes: Bats and hummingbirds as pollinators of *Burmeistera cyclostigmata* and *B. tenuiflora (Campanulaceae)*. Oecologia. 2003;**134**:373-380

[12] Baswana KS. Role of insect-pollinators on seed production in coriander and fennel. South Indian Horticulture. 1984;**32**:117-118

[13] Sharma AK, Sharma A. Chromosome Techniques, Theory and Practice. 3rd Edn., London: Butterworth & Co. (Publisher) Ltd.; 1980

[14] Gomez KA, Gomez AA. Statistical Procedures for Agricultural Research. New York: Willey; 1984

[15] Żuraw B, Weryszko-Chmielewska E, Laskowska H, Pogroszewska E. The structure of septalnectaries and nectar presentation in the flowers of *Allium aflatunense* B Fedtsch. Acta Agrobotanica. 2009;**62**(2):31-41

[16] Żuraw B, Weryszko-Chmielewska E, Laskowska H, Pogroszewska E. The location of nectaries and nectar secretion in the flowers of *Allium giganteum* REGEL. Acta Agrobotanica. 2010;**63**(2):33-40

[17] Jones MG, Hughes J, Tregova A, Milne J, Tomsett AB, Collin HA. Biosynthesis of the flavour precursors of onion and garlic. Journal of Experimental Botany. 2004;**55**(404):1903-1918

[18] Branca F, Ruggeri A. Reproductive response of onion seed plants to photo thermal conditions. Acta Horticulturae. 1994;**362**:25-33

[19] Bronstein JL, Alarcón R, Geber M. The evolution of plant-insect

mutualisms. The New Phytologist. 2006;**172**:412-428

[20] Simpson BB, Neff JL. Floral rewards-alternatives to pollen and nectar. Annals of the Missouri Botanical Garden. 1981;**68**:301-322

[21] Devi S, Gulati R, Tehri K, Poonia A. The pollination biology of onion (*Allium cepa* L.)-a review. Agricultural Reviews. 2015;**313**

[22] Smaka-Kincl V, Stegnar P, Lovka M, Toman MJ. The evaluation of waste, surface and ground water quality using the *Allium* test procedure. Mutation Research. 1996;**368**:171-179

[23] Leme DM, Marin-Morales MA. Chromosome aberration and micronucleus frequencies in *Allium cepa* cells exposed to petroleum polluted water—A case study. Mutation Research. 2008;**650**:80-86

[24] Fernandes TCC, Mazzeo DEC, Marin-Morales MA. Mechanism of micronuclei formation in polyploidizated cells of *Allium cepa* exposed to trifluralin herbicide. Pesticide Biochemistry and Physiology. 2007;**88**:252-259

[25] Fiskesjo G. *Allium* test on copper in drinking water. Vatten. 1981;**37**:232-240

[26] Fiskesjo G. The *allium* test—An alternative in environmental studies: The relative toxicity of metal ions. Mutation Research. 1988;**197**:243-260

[27] Inceer H, Beyazoglu O, Ergul HA. Cytogenetic effects of wastes of copper mine on root tip cells of *Allium cepa* L. Pakistan Journal of Biological Sciences. 2000;**3**:376-377

[28] Borboa L, Torre CDL. The genotoxicity of Zii (II) and Cd (II) in *Allium cepa* root meristematic cells. The New Phytologist. 1996;**134**:481-486

[29] Seth CS, Misra V, Chauhan LKS, Singh RR. Genotoxicity of cadmium on root meristem cells of *Allium cepa*: Cytogenetic and comet assay approach. Ecotoxicology and Environmental Safety. 2008;**1**:711-716. DOI: 10.1016/jecoenv.2008.02.003

[30] Matsumoto ST, Mantovani MS, Malaguttii MIA, Dias AL, Fonseca IC, Marin-Morales MA. Genotoxicity and mutagenicity of water contaminated with tannery effluents, as evaluated by the micronucleus test and comet assay using the fish *Oreochromis niloticus* and chromosome aberrations in onion root-tips. Genetics and Molecular Biology. 2006;**29**:148-158

Chapter 16

Genetic Improvement of Stevia: A Natural Non-Calorie Sweetener

Rahul Dev Gautam, Ravi Kumar, Ujala Kashyap, Pawan Kumar, Satbeer Singh, Sanatsujat Singh and Ashok Kumar

Abstract

Stevia rebaudiana, a native of South America, is a perennial herb of the Asteraceae family, also known as a natural sweetener due to the presence of steviol glycosides (SGs) in the leaves. China is the largest producer and exporter of stevia, while Japan is the primary consumer. The increasing demand for natural low-calorie sweeteners in the medicine and food industry has increased the pressure over stevia cultivation. Still, its cultivation and region-specific agrotechnologies need to be developed. The major bottleneck in stevia production are the lack of region specific cultivation technologies, non-availability of quality planting material, and uncharacterized and not properly conserved plant genetic resources. All these constraints have limited the stevia production to some specific regions of the world. Development of high-yielding cultivars with enhanced SGs content using modern breeding techniques is of prime importance to meet its increasing demand. Among the glycosides present in the leaves, rebaudioside-A is the most desirable glycosides having 250–300 times sweeter than sucrose, while, after bitter taste is due to the presence of stevioside and dulcoside. Therefore, the development of varieties with high rebaudioside-A and low stevioside content is highly desirable. This chapter focused on the improvement of propagation methods, characterization and conservation of genetic resource in stevia and its utilization in crop improvement programs.

Keywords: natural sweetener, rebaudioside-A, stevia, stevia rebaudiana, steviol glycosides

1. Introduction

Stevia rebaudiana Bertoni, also known as 'candy leaf', 'honey leaf' and 'sweet herb', is a zero-calorie source alternative to sugar or artificial sweeteners as a natural sweetener [1]. It is primarily grown in forests, mountainous areas, dry valleys and on the banks of rivers [2, 3]. There are ~230 species of the genus Stevia have been identified throughout the world based on their growth behavior and chemical compounds. Out of 230 species, *S. rebaudiana* and *S. phlebophylla* are two species that have been identified with higher steviol glycosides and a sugary taste [4–8].

SG (Steviol glycosides), including stevioside, Rebaudioside-A, B, C, D, M, and dulcoside, are secondary metabolites (diterpene glycosides), which are extracted from the leaves of the *Stevia rebaudiana* plant that are approximately 250–300 times more sweet than sucrose or cane sugar [9, 10]. According to a WHO survey, approximately 500 million people worldwide will be diabetic by 2030 [11, 12]. The use of stevia extract in high concentrations results in a licorice aftertaste [13]. Most sugar consumers prefer low-calorie, natural sweeteners in their food to reduce the risk of cardiovascular disease, obesity, diabetes, and tooth decay [14, 15]. Because of its pH stability, stevia has no influence on blood glucose and insulin levels [16]. Nonetheless, no harmful effects have been documented with its usage [17, 18].

Steviol Glycosides (SGs) have been reported as safer to use as a sweetener in Japan after the demonstration of about 40,000 studies. Extract of stevia has also been reported to be antioxidant, reduce hypertension and reduce blood pressure [7]. Stevia was formerly prohibited in the United States/Nations for commercial usage as a food ingredient in food items or food industry, but in 1995, it was approved as a dietary supplement by Food and Drug Administration (FDA) throughout the world [19]. SGs have wide use in herbal medicine to make sugar-free tonics for diabetic patients, cosmetic industries for making face creams, mouth-wash, toothpaste, and food industries for making ketchup, drinks, fruit juices sauces, desserts, and energy drinks [13]. Due to the high demand for natural sweeteners, farmers are growing *Stevia rebaudiana* on a large scale as commercial cultivation in different parts of the world from Asia to America [20]. In Europe, undefined varieties of stevia are propagated through traditional method, and they show high genetic diversity, which does not assure the production of good quality steviol glycosides.

Most of the steviol glycosides content is found just before the plant transitions from the vegetative phase to the reproductive stage and initiation of flower buds [21, 22]. Some efforts have been undertaken to alter the flowering time in stevia and the long day—short day relationship, leading to an increase in steviol glycoside content [21, 23]. According to certain research, the temporary production of SGs is also controlled by nutrient availability, temperature, and the plant's requirement for their GAs/steviol glycoside with inter/intra genotypic variation [21, 23].

In the late 1990s, the Stevia plant was first imports to India by the University of Agricultural Sciences in Bangalore, where research on plant adaptation began. After that, the Institute of Himalayan Bioresource Technology (CSIR), introduced two accessions of *Stevia rebaudiana* for cultivation and domestication in Himachal Pradesh. The production of enhanced stevia varieties/cultivars is vital for boosting steviol glycoside chemicals (Rebaudioside-A, D, & M) with the breeding process to meet the industry's continuously expanding demand. Selection of possible genotypes based on vegetative development and adequate characteristic for maximum accumulation of steviol glycoside with an increase in ontogenetic period provide for high metabolic flux. Although incredible approaches are taken in the field of agronomic practices [24–26], purification and extraction of steviol glycoside [27], seed germination, self-incompatibility, cross-pollination, and a lack of wild germplasm to access inhibit breeding approaches to achieve genetic changes in [28].

1.1 Botanical description and systemic classification

S. rebaudiana is a 60–80 cm tall perennial herb that is part of the Asteraceae family. However, under certain climatic conditions, it behaves as both an annual and a perennial. It is typically grown in subtropical and tropical climates. It has a brittle stem, elliptical leaves with an alternate pattern of leaf arrangement on the stem [29], and a broad root system; yet, are more vigorous when compared to wild plants [2, 30].

Stevia plant bears small sessile leaves that are 1.2 cm wide and 3–8 cm long, serrated, lanceolate to oblong, irregularly curled upwards and elevated in the center [31]. In addition to growing procedures, the quality of stevia leaves is affected by environmental factors such as air purity, sunshine intensity, irrigation method, cleanliness, soil type during the cultivation and processing, and storage of dry leaves after harvesting. For the best growth of stevia plants, the soil pH ranges from 6.5 to 7.5 [32]. Stevia leaves have sweetness with a pleasant flavor that lasts for hours, but they also have an aftertaste of bitterness due to the presence of bitter constituents in leaf veins [33].

Stevia has a white inflorescence with pale purple corollas that are small in size and placed in an unequal sympodial cyme, loose paniculate head on the opposite side of the bract [31, 34]. "Stevia flowers have both reproductive organs are present in corymbs and coated with small 2–6 white florets" with "Stevia flowers have both reproductive organs present in corymbs coated with small 2–6 white florets" [35]. Because of the several stages of flower development, the plant takes more than 30 days to achieve its full blooming [36, 37]. The stevia flower has five little anthers and carries extremely allergenic pollen with a viability of 65 percent [38], while, the stigma is bi-lobed/bifurcated in the middle part and style is covered with anthers. Due to self-incompatibility, stevia is a cross-pollinated (insect-pollinated) crop [39, 40]. For optimum seed production in the stevia crop, three to four hives per hectare with a high density of bees are recommended [41]. Stevia is a short-day plant whose blooming is affected by the photoperiod. As a result, a photoperiod of at least 12 hours is thought to be best for stevia flowering.

In the northern hemisphere, the months of September to December are favorable for blossoming whereas, in the southern hemisphere, it begins in January and lasts through March. Flowering takes 54–104 days following seedling transplanting in the field, depends on the cultivar's day-length sensitivity, which ranges from 8 to 14 hours under short-day circumstances [39, 40]. A photoperiod of 8 hours is more favorable for flowering but does not allow for full vegetative growth due to the long dark period [42]. Stevia seeds are roughly 3 mm in size, with very little endosperm and 20 hairy pappus bristles that aid in seed distribution through wind [35]. Stevia seeds germinate poorly [43, 44] viable seeds are usually dark colored, whereas pale yellow/transparent seeds are sterile [35, 45, 46].

Stevia rebaudiana is one of the most important plants in Asteraceae family [44, 47–49]. In 1888, Moises Santiago Bertoni discovered *S. rebaudiana* for the first time in Paraguay [50]. In the honor of Paraguayan chemist Rebaudi, this plant was known as *Eupatorium rebaudianum* Bertoni. The name was later changed to *S. rebaudiana* and also recommended in pharmaceutical as well as in food industries [51].

A systemic classification of *Stevia rebaudiana* according to hierarchy is given below [52].

Kingdom—Plantae
Division—Magnoliophyta
Class—Magnoliopsida
Group—Monochlamydae
Order—Asterales
Family—Asteraceae
Tribe—Eupatorieae
Genus—*Stevia*
Species—*rebaudiana*
Botanical name: *Stevia rebaudiana* Berto

1.2 Origin and distribution

Stevia is a South American native plant, primarily found in Paraguay, Brazil, and Argentina [53]. It is grown commercially in Canada, India, China, Brazil, Japan, The United Kingdom, the United States, Spain, Australia, Belgium, South Korea, Taiwan, Israel, and Thailand [54, 55]. Japan and China are the world's leading producers and exporters of stevia [53].

For the first time, Paraguayans and Brazilians employed the leaves of *S. rebaudiana* as a sweetener [56]. Dr. Rebaudi reported various sweetening agents such as stevioside and rebaudioside (white crystalline substance) in 1905 [57]. England attempted to grow stevia as a commercial crop in 1942, but they were unsuccessful. However, Paraguay became the first country to cultivate it commercially in 1964 [58, 59]. Following that, Japan made significant efforts to establish stevia and its cultivation as a commercial crop, as well as to conduct various studies to assess stevia's potential [60]. Nowadays, Japan is the major market for the consumption of stevia all over the world [37] and China is the largest producer as well as exporter of stevia (approx. 2–3 billion/year) in the market [17]. Furthermore, it has become well-known as a crop in several nations, including the United States, India, Korea, Canada, Indonesia, Brazil, Mexico, and Tanzania [10, 61].

2. Propagation

2.1 Propagation through seeds

Stevia is mainly reproduced through seeds in its wild habitat. Stevia seeds are small in size, and approximately 40-gram seed is required for a one-hectare stevia plantation. Furthermore, 1 ha may produce approximately 8.1 kg seed, which is sufficient for a plantation of 200 ha area [44]. Nursery plants are typically cultivated from January to March, although in polyhouse conditions, they may be raised all year. Seedlings with 5–7 leaves that are two months old are put in the field. However, in temperate locations, seeds are frequently ineffectual in germination [62]. In the short growing season of the northern region, farmers use glasshouses or greenhouses for better establishment of stevia crop. The seeds should be sown up to a depth of 1–2 cm in soil and Irrigation should be necessary at regular intervals. Germination studies have revealed the presence of two types of seeds i.e. tan and black colored. Black seeds show more viability and germination in light as compared to tan seeds [38].

2.2 Vegetative propagation

Stem cuttings are used for vegetative propagation, which is an excellent strategy for multiplying stevia plants [30]. Because some plants do not produce viable seeds for germination, vegetative propagation is sometimes the only method of replication. During February, a hardy branch cutting from a fresh stem or shoot with three to four nodes is ideal for planting in the soil [30, 63–65]. A one-year plant's leaf axials exhibit 98–100 percent rooting [64] however cutting from other areas of the plant affects its roots and growth. In February, axial stem cutting with two pairs of leaves produces the best rooting results, while three pairs of leaves cutting produce outstanding results in April [66]. In comparison to other seasons, late winter is the greatest period for stevia rooting [67]. Growth regulators can sometimes encourage roots and sprouting. Several growth regulators raise the content of stevioside in leaves [68, 69].

2.3 Micro-propagation

Propagation through tissue is a technique in which a single tissue (or explant), such as seeds, auxiliary shoots, leaves, sprouts, inter-nodal explant, and shoot primordia, is used for successful development of a new plant [39, 70–75]. Micro-propagation via tissue culture is a quick procedure of multiplying disease-free plants of a selected or bred clone, and it has also been recorded in the literature [39, 43, 74, 76–78]. Maximum shoot development and biomass growth occurred in Murashige and Skoog (MS) media improved with 0.25 mg/l kinetin and 0.5 mg/l BAP (6-benzylamino purine) solution, while roots developed well in MS (Murashige and Skoog) media supplied with 1.0 mg/l IBA (Indole-3-butyric acid) solution [79]. The cultures were always incubated at 24 ± 2°C with 60–80 percent relative humidity and photoperiods of 16 hours with a light intensity of 3000 lux [80].

2.4 Conservation of plant genetic resources

The plant genetic resources include genetic stocks, active collection, general germplasm, pre-breeding material, base collection, breeder's collection, interspecific derivatives, etc. Synthetic seed technology with added osmotic agents was used for germplasm conservation of *Stevia rebaudiana* Bert [81]. Synthetic seed production in *Stevia rebaudiana* with micropropagation may solve many problems [82]. In vitro propagation protocol of *S. rebaudiana* has been also established to meet the demand [83, 84]. Shoot tips obtained from in vitro shoot cultures of *Stevia rebaudiana* Bertoni encapsulated in 4% calcium alginate used as synthetic seeds. Synthetic seeds capsulated with 0.05 M mannitol after 6 weeks are the most suitable for conversion [85].

Due to the presence of self-incompatibility, seeds produced from the individual plant would be representative of half-sib progeny [10]. Lack of homozygous populations due to self-incompatibility can be overcome by double haploid production which can be used for the further breeding program. The crucial restraint in stevia cultivation is the lack of evergreen plant cultivars. Germplasm for delayed flowering to attain a prolonged vegetative phase has been developed through mutation breeding and can be transferred to high Rab-A consisting varieties or any other desirable genotype. Diverse lines from various breeding institutions can be shared for hybrid production. Recovery of phytochemicals through processing technology precisely, green technology should be boosted. The development of tetraploid and triploid germplasm can be used for hybridization and commercial exploitation.

2.5 Characterization and evaluation

The increasing rate of obesity and diabetes, people are getting admired to natural sweeteners as compared to sucrose. The sweet glycoside which is present in *Stevia rebaudiana* Bertoni is known as stevioside. Stevioside is a diterpenoid glycoside due to the presence of an aglycone with three molecules of glucose. Along with this, several other sweet compounds are also present which include stevioside, rebaudioside A, B, C, D, E, dulcoside and steviolbioside [13] while, another study defined steviol, stevioside, β-carotene, riboflavin, nilacin, austroinullin, rebaudi oxides, dulcoside and thiamine in stevia [86]. Previously, chlorogenic acid, caffeic acid, trans-ferulic acid and rutin presence in stevia leaves has also been reported by the process of softening [87]. Phytochemicals like, tannins, alkaloids, glycosides, flavonoids, saponins, triterpenes and quinone have also been reported from stevia leaf extract [88, 89].

Ethanolic leaf extract's chief constituents were the glycosides, followed by tannins. Though, alkaloids were also in greater quantities but significantly smaller compared to glycosides. The constituents like glycosides and tannins followed by alkaloids and flavonoids were extracted through leaf extracts in ethyl acetate, however they were present in lesser quantity than glycosides. The presence of triterponoids and flavonoids was observed in moderate quantity while quinine was in lowest quantity. Some studies have revealed that dehydrated extract from stevia leaves consist of xanthophylls, flavonoids, water-soluble chlorophylls, neutral water-soluble oligosaccharides, sweet diterpene glycosides, hydroxynnamic acid (Caffeic, and chlorogenic, among others), free sugars, amino acids, essential oils and lipids [90, 91].

Stevioside and rebaudioside-A, which are chief SG in stevia leaves, are stable molecules at wide range of temperatures and pH in aqueous solution. Stevioside is highly thermostable due to which the commercialization of stevia has nurtured worldwide [90]. The antioxidant activity of extracts is due to the presence of phenols [92]. Gas-chromatography the leaf oil validated the existence of linolenic, stearic, oleic, palmitic, and palmitoleic acids. The nutrient analysis of leaves through atomic absorption spectrophotometry had shown the amount of phosphorus, potassium, magnesium, calcium, sulfur and sodium. Minerals like cobalt, iron, manganese, molybdenum, selenium, copper, and zinc were found to be in trace amounts [93]. The GC-MS of leaves has shown their presence of phytol, β-amyrin, γ-sitosterol, heptatriacotanol, agatholic acid, dihydroxanthin, lupenone, 1-duvatrienediol and fatty acids. From the leaves of stevia many phenylethanoid glycosides like steviophethanoside, cuchiloside, icariside D, salidroside and tyrosol have also been separated [94]. Some phenolic compounds like caffeic acid, 4-O-caffeoylquinic acid, 3,4-O-dicaffeoylquinic acid, 3-O-caffeoylquinic acid, quercetin and quercetin-3-O-rhamnoside were extracted from stevia leaf residue [95]. The medicinal properties can be justified by the occurrence of composites of phenolic and flavonoid group and can be applied in food/nutraceutical and pharmaceutical industries. Rebaudioside A (2–4% total dry weight), dulcoside A (0.4–0.7%), Stevioside (5–10%) and rebaudioside C (1–2%) are the principal components present in stevia leaves [96]. The sweetness fold of the glycosides related to sugar are 250–450 in rebaudioside D, 300–350 in rebaudioside B, 250–450 in rebaudioside A, 150–300 in rebaudioside E, 100–125 in steviolbioside, 300-fold in stevioside, 50–120 in dulcoside A and 50–120 in rebaudioside C [60]. Stevioside is hydrolyzed into glucose and steviol in gastrointestinal tract by the bacterial activity [97]. Apart from being sweet, stevioside is also having a bitter aftertaste [98] which can be decreased through alteration of enzymes of stevioside by b-galactosidase, isomaltase, pullanase [99] or dextrin saccharase [100]. Stevioside and stevia extract had been used even as a routine medicine by South Americans [47].

2.6 Development/identification of gene pools and core collections

Fifteen genes in Stevia [*Stevia rebuadiana* (Bertoni); family: Asteraceae] had been identified to produce diterpenoid steviol glycosides (SGs), which are ~300 times sweeter than sugar. Several genes of the pathway, including SrDXS, SrDXR, SrCPPS, SrKS, SrKO, and three glucosyltransferases, SrUGT85C2, SrUGT74G1, and SrUGT76G1, have been identified in stevia. Seven more complete cDNA sequences were cloned, including SrGGDPS, SrMDS, SrMCT, SrCMK, SrHDR, SrHDS, and SrIDI. Except for SrDXR and SrKO, gene expression was highest in the first nodal leaf and lowest in fifth node's leaf. The expression of SrKO was highest in the leaf at the third node, whereas SrDXR expression increased up to the third leaf and then decreased.

The highest concentrations of SGs were found in a sequence of leaf stem and root tissue, with a similar expression pattern of all 15 genes. The genes reacted to terpenoids biosynthesis modulators. Treatment with gibberellin (GA3) increased the expression of SrMCT, SrCMK, SrMDS, and SrUGT74 G1, whereas treatment with methyl jasmonate and kinetin decreased the expression of all fifteen genes in the pathway [101].

Genetic divergence plays an important role in any breeding program or selection of parents for target traits. For the flowering phase, stevia has shown significant genetic diversity. For a better understanding of the genotypic control of SGs, difference in their composition may be useful prior to breeding for it, some of the preliminary studies on SGs structure and their genotypic changeability in a given population have identified three clusters: (1) plants with primarily glucose (glc)-type glycosides (Stev and Reb-A); (2) plants with primarily rhamnose (rhm)-type glycosides (Reb-C, Dulc A); and (3) plants with almost equal amounts of glc- and rhm-type glycosides. Because of variable glycosylation, each SG has individual organoleptic and biological features. It is well known that a sugar unit or a carboxyl group in the C19 position, as well as a sugar with a hydroxyl group in the C13 position, are required for it sweet taste. Rhamnosylation, on the other hand, reduces the organoleptic qualities, and the resulting sweetness and taste quality of rhamnosylated SGs (such as Dulc A and Reb-C) is poorer to that of their glycosylated equivalents.

Leaf yield (h^2 = 62.1), leaf-to-stem ratio (h^2 = 78.8), and SGs content (h^2 = 76.6) of stevia are economically important breeding traits with high variability within populations and heritability. Because of their high heredity, they can be adjusted through selection [18]. The genetic regulation of the quantities of Reb-A and Reb-C was investigated, and it was discovered that they were both made by the same enzyme and differed only in the composition of one sugar unit [102]. Barbet-Massin and colleagues evaluated genotypic inconsistency for SG content and structure in 96 stevia genotypes in multiple trials. At the INP-EI Purpan, five genotypes were transplanted. High variability was observed in SG content and composition, with high amount of Reb-A than Steviol content and a high proportion of minor SGs in some genotypes. Among the different environmental conditions, it was found that SG composition remained stable while SG content varied.

In the temperate European climatic conditions, six stevia genotypes were studied in comparison with Gawi [103]. Fifteen stevia clones were assessed for genetic divergence in order to choose genitors in a hybridization procedure based on their total SGs performance and significant Reb-A to Stev ratio. Genetic variability was observed among the clones for fresh and dry matter, plant height and SG concentration and Reb-A/Stevioside ratio. Four clones were found to have considerable mean genetic divergence in comparison to the entire genotypic pool investigated. So the generation of the segregated population with high genetic potential can be produced from these four clones which could provide for superior individuals' selection [104].

About 90 varieties have been developed throughout the world [29, 37, 105]. Criolla and Morita II are well studied and known varieties. Criolla is an original stevia variety native to Paraguay and Morita II is selected for high Reb-A content. Eirete is developed for intensive cultivation in Paraguay. A variety Katupyry, characterized for greater sweetening power was selected recently for cultivation in arid soils. Morita II was further improved and Morita III was obtained which is known for its low water requirement. Some more varieties like SW 107, AKH L4, AKH L1 and SW 201 have been released with improved traits. The selection of parents for a wider variety of different traits is determined by genetic divergence. For developing a breeding program, the genotypic and phenotypic diversity should be studied [106].

Natural variations already existing within the species are used by breeders and need to observe variations in its expression to measure any character. This variation reflects genetic variation. Genetic variation, environmental variation and their interaction were found to be the source of variations. Breeders are supposed to understand the extent and nature of the genetic and environmental control to modify the quantitative and qualitative properties of stevia. Genotype selection is the major interest of breeders [107]. Program for successful Stevia breeding is dependent on the plant's selection for desirable features in order to anticipate the genotypic value of the selected plants [108]. Growing conditions are the major factor affecting some characters. SGs accumulation and composition in stevia depends upon phenological stages and growth conditions such as irradiance, photoperiod, available nutrients and temperature [22, 37, 108–110].

2.7 Molecular characterization

Due to the availability of specific molecular markers the molecular evaluation of stevia is reported limitedly. The simple sequence repeats can be developed using Expressed sequence tags (ESTs). 5548 stevia EST sequence was studied by Kaur and co-workers (2015) from the leaf tissues. A non-redundant set of sequences was observed after clustering and assembly of ESTs in which 168 SSRs, 471 contigs and 3845 singletons were identified. 82.2% of EST SSRs can be used for putative function. 61.11% polymorphism from the 18 primers was found which were synthesized from SSR containing 18 singletons by using Primer3 software. As the EST-SSRs exhibit cross-species transferability so they can be used for the molecular work in stevia which would make the work simple and cost-effective.

Genetic diversity among 16 collections was assessed for efficiency comparison of two marker systems against one marker using RAPD and ISSR markers [111]. Sixty six scorable bands were observed in 22 selected ISSR primers, 54 (81.8%) of which were polymorphic. Forty nine bands were amplified in 23 ISSR primers, 44 (89.8%) of which were polymorphic. On analyzing pooled data of ISSR and RAPD using UPGMA revealed 0.365 to 0.887 variations among the accession for genetic similarity. A contrast for levels of rebaudiana-A and stevioside in genotypes A&B collected from Solan and singled out by dendrograms generated from different techniques. Both techniques could be used for evaluating genetic diversity, though ISSR results in more polymorphism.

Germplasm is a very significant material for crop development and notably in developing nations, the introducing accession to new areas is still a vital breeding strategy [109]. It is commonly utilized in breeding programs as a source of more genes and to increase genetic diversity among parental groups. Introductions are commercial cultivars that can be used right away. Exotic germplasm adaptation, on the other hand, is a long-term endeavor. Intermating occurs over numerous generations, with incremental selection pressure exerted to desired gene pairings. Institutions worldwide that have conducted stevia research and/or evaluation experiment have acquired planting material from Paraguay wild, where stevia has adapted and become a hot spot for its diversity. The goal of seed collecting is to preserve genes rather than genotypes, because no genotype in stevia is real breeding owing to heterozygosity. The Institute of Himalayan Bioresource Technology maintains and multiplies many *Stevia rebaudiana* introductions that are morphologically varied in terms of growth habit and sweetness. Additional choices for desired plant types are being made by separating progenies of particular selections.

3. Major constraints in crop production

Zaman and co-workers found that different soils types have significantly affected the growth and leaf yield. Maximum plant height, leaf area, fresh and dry weight, branch and leaf number were observed in the plants grown on non-calcareous soil and were similar to the plants grown on acidic soil. Leaf biomass yield was found to be maximum in non-calcareous soil. Díaz and co-workers [112] studied the effect of altitude and fertilizers on biomass production of stevia and found that the organo-mineral fertilizer decreased the differences in nutrient uptake between different altitudinal gradients than the mineral fertilizers and increased biomass production by 49%. The highest stevioside yield of 30 g/m^2 was gained before flowering during the starting of September with an increased yield of leaf biomass when the crop was raised during the optimal planting season of onset of spring [113]. Micro-propagated shoots were exposed to drought stress by using polyethylene glycol (PEG) 6000 for four weeks and it was observed that 4% PEG 6000 concentration enhanced the growth dynamics and pharmaceutical compounds as a defensive response against reactive oxygen species produced in water deficit conditions [114]. Weeds are a common problem during agricultural operations which hinders crop production and requires a large number of synthetic chemicals for controlling the weed population.

Taak and co-workers [115] reported that the use of herbicides controlled the weed growth but the use of mulches like rice straw and eucalyptus leaves increased the plant growth characteristics as well. However, mulching not only control the weed proliferation in stevia but also have a significant effect on the dry weight and leaf biomass. The effect of different concentrations of titanium dioxide nanoparticles on physiological and phytochemical properties was investigated by Rezaizad and co-workers [116] and found that 400 mg/l significantly affected the rebaudioside-A and B content while 80 mg/l had maximum effect on rebaudioside-C and F content. Nitrogen content is negatively correlated with steviol glycoside in leaves while the increased steviol glycoside is correlated with a decreased ratio of rebaudioside A over stevioside [117]. Stevia cultivation in temperate climatic conditions depends mainly on the genotype's ability to withstand the overwinter and is found that there is yield loss if crop is harvested in the first year where first-year harvest modality impacts the SG yield for three years [118]. Light favors germination and at least 20°C temperature are required. Germination can be accelerated at increased temperature for 24 h but it reduces the total germination [119].

3.1 Studies to overcome production constraints

About 2–25% yield loss is reported due to weeds, to overcome this problem stable transgenic stevia plants were produced through Agrobacterium-mediated genetic transformation of nodal explants derived in vitro using herbicide resistance gene. The presence, expression, stability and copy number of the bar gene in putative trans-formants by various molecular techniques like- PCR, RT-PCR, qRT-PCR and southern blot hybridization have been confirmed. This procedure can be used for the inter-kingdom transfer of genes into stevia genome [120]. Guleria and Yadav adopted the gene silencing approach for understanding the genetic regulation of steviol glycoside biosynthesis and found SrKA13H and SrUGT85C2 as carbon flux influencing regulatory genes between steviol glycoside and gibberellin biosynthesis [3].

Taak and co-workers studied the use of different herbicides like pendimethalin, atrazine, paraquat, and 2, 4-D against common weeds, *Erigron sumatrensis*, *Parthenium hysterophorus*, and *Solanum nigrum* found in the stevia field and recommended 2,4-D as best herbicide for controlling weeds in stevia [121]. Better adaptation in different environmental conditions is due to the role of SG as the highest amount of SG and phenolic compounds were found in the plants that showed the highest value of PS-II converted from the energy fraction photochemically [122]. To overcome the poor seed germination rate, an experiment was set up by Gorzi and co-workers under drought stress conditions by application of various seed priming techniques [123]. They found that the use of salicylic acid, zinc and iron or their integrated use at suitable concentrations can promote germination and seedling growth due to increased antioxidant capacity under drought conditions. Ameliorative treatment of stevia with sodium nitroprusside and putrescine or their combination decreased the negative effects of drought stress [124]. Melatonin is not only reported to increase the seed germination in salinity conditions but also in enhancing the production of SG in stevia plants, where the highest amount of stevioside and rebaudioside A were obtained with 5 and 20 μM melatonin [125]. Global warming and climate change are the biggest issues in the current century that affecting agriculture at a greater pace.

Tursun and co-workers studied the effects of high carbon dioxide and temperature effects on stevia and found non-significant changes in aromatic compounds [126]. Generally, aldehyde, ketone and alcohol concentration decreased on the other hand terpene concentration increased with increased carbon dioxide and temperature concentration. Vascular wilt caused by a soil-borne pathogen, *Fusarium oxysporum* is an emerging pathogen in the crop. UDEAGIEM-H01 strain of *Trichoderma asperellum* was found to be a preventive agent with high ability to control *Fusarium oxysporum* in stevia plants [112]. By controlling the quality of light, seed germination and the quality of plantlets produced can be improved. Blue LED light promoted the development of roots and leaves, increased the number and opening of stomata whereas stem and root length increased under influence of red light while chlorophyll and carotenoid synthesis was least affected under red light [125]. Seeds germinate better in red light (660 nm) than white light (400–700 nm) [127]. Different media used for seed germination showed different results and it was found that soil and combination of soil and rice husk is better for seed germination while minimum germination was found in vermiculite [128]. Better germplasm can be maintained using synthetic seed technology. Nower encapsulated shoot tips of in vitro cultures in 4% calcium alginate and the highest multiple shoots from nodal segments were found in MS medium supplemented with 1.0 mg l^{-1} benzyl adenine [85]. A novel approach was developed using bacterium *Bacillus safensis* STJP extracted from rhizospheric soil of stevia for the formation of Paneer-whey based bio-formulation which increased the fresh, dry weight and stevioside content through nutrient(s) linked mechanism [129]. RITA, BIT and SETIS temporary immersion systems were studied for scaling up micropropagation for biomass evaluation in the field using different concentrations of calcium pantothenate, sucrose and gibberellic acid and it was found that temporary immersion systems can boost plant production [130].

The use of potential varieties will boost the yield in stress conditions as stevia is resistant to moderate stresses [131]. Seed viability can be maintained for up to 3 years if stored in darkness with low humidity [132]. Two types of seed colors are produced with 76.7% viability of black seeds against 8.3% of tan-colored seeds [133]. Jain and co-workers used Moringa leaf extracts as a foliar spray and it was observed that the Jaffna variety of Moringa significantly improved the growth and physiological parameters of stevia [134].

3.2 Breeding options

The success of stevia breeding is determined by the selection of parents, the creation of crosses, the development of a sufficient population, and further selections. Most breeding programs are based on crossbreeding and selection in stevia. Stevia is self-incompatible species and cross-pollination is brought about by insects [10]. Ability to biosynthesize steviol glycoside is the most characteristic trait so the breeding program is mainly focused on content and composition modification. As the content of steviol glycoside is highest in leaves so the biomass and leaf yield is another modified trait.

Recurrent selection could be used for the improvement of quantitative traits in cross-pollinated crops like stevia which involves the selection followed by crossing between the desired recombinants. RSIT 94-1306 and RSIT 751 lines were produced with a crossing and selection approach [135]. AC Black Bird and PTA-444 are the synthetic and composite cultivars, respectively with altered glycoside content. PTA-444 could be reproduced with seeds [133, 136]. The method for obtaining seed was suggested by Sun (2001). Wang (2006) patented the method for breeding stevia hybrids which used the technique for vegetative hybrid production [135].

Through mutagenesis (potential tool to create and isolate the new variability for anticipated commercial characters) variability can be created and isolated in a shorter time period compared convention breeding. Genetic diversity can be obtained at a faster rate by the use of physical and chemical mutagens. Several mutagenic agents, such as X-rays, g-rays, fast neutrons, thermal neutrons and chemicals such as EMS, DES, MNUA, ENUA, MNU, ENU, can be used to produce useful mutations. The development of plants with desired traits can also be achieved through mutation induction. Induced mutations may be used for the improvement of traits with low variability within the population.

The induction of polyploidy to improve agronomic yields has been used successful in many commercial crop plants. It improves the adaptability of individuals and vigorness by increased organ and cell sizes (associated with polyploidy). Higher content of rebaudioside can also be linked to triploidy. Stevia triploid plants can be obtained either by placing stevia seeds in colchicine solution or by breeding tetraploid females with diploid males. Tetraploid plants have bigger leaves which can increase the biomass yield, however, nonfunctional pollens are also found in all the Polyploids [137, 138]. Haploid plants can be obtained using the anther culture technique in which immature anthers are used for in vitro cultures which can be used for the formation of a double haploid plant or a population that is completely homozygotic. Plant from this homozygotic population can be used for hybridization. Anthers of stevia were cultured in vitro in Murashige and Skoog's liquid medium supplemented with 0.1 mg/L (−1) and 1 mg/L (−1) BAP and plants were regenerated. The diploid number of chromosomes was observed through the cytological studies of root tips [135]. A gain in steviol glycoside content was noticed by manipulating the photoperiod and flowering time [22, 139, 140]. Controlled mutagenesis can be used for altering the flowering time through the identification of floral integrator genes while CRISPR-Cas and VIGS techniques can be used for gene silencing [141].

4. Present status of use or incorporation of desired traits

Him Stevia is a registered variety at CSIR- Institute of Himalayan Bioresource Technology which was produced through the crossing of two parents CSIR-IHBT-ST-2009-2 × CSIR-IHBT-ST-2009-4. This variety is characterized by large club-shaped, dark green leaves having serrated margins on the upper half and rugose leaf surface. It is rich in stevioside 5.87% and rebaudioside A 7.34% content. A mutant variety with accession number IHBT-ST-02 is produced through gamma irradiation method in which leaves are medium ovate with dark green color and serrate margin producing stevioside 7.02% and rebaudioside A 2.30% content.

A tetraploid stevia plant with accession number C-7-3-4 was formed through colchicine treatment which is having large ovate leaves with bluntly rounded leaf apex and is dark green in color with a serrulate margin. It is characterized by 8.40% stevioside and 4.33% rebaudioside content with a low Reb-A/stevioside ratio of 0.52. Colchicine treatment was used for the formation of polyploid plants having accession number C-8-3-4 with the same leaf characters as that of the tetraploid plant but differ in the percentage of stevioside (8.47%) and rebaudioside (2.19%) with lowest Reb-A/stevioside ratio of 0.26.

5. Looking forward or future perspective

The increasing demand for natural low-calorie sweeteners in the personal hygiene and food industry has increased the pressure over stevia cultivation. Bitter taste due to the presence of stevioside, a compound in leaf veins that need to be eradicated or the development of varieties with low stevioside content is highly desirable. Through the use of mutagenesis, molecular and biotechnological approaches the quantity of Rebaudioside-A should be increased as it is the most desirable glycoside compound in the leaves. The development of such varieties with improved tolerance to biotic and abiotic stress is a major need. Getting over the seed germination problems will require the improvement of methods for the germination and field stand of the crop. Identification of plants with best Rebaudioside-A: Stevioside ratio for generating optimum planting material through plant tissue culture techniques.

Acknowledgements

We are grateful to the Director, CSIR-IHBT, Palampur, for his encouragement and providing the necessary facilities for the study. The authors are also grateful to Council of Scientific and Industrial Research, New Delhi, India for providing financially support for the study. CSIR-IHBT Publication No. is.

Conflict of interest

None.

Author details

Rahul Dev Gautam, Ravi Kumar, Ujala Kashyap, Pawan Kumar, Satbeer Singh, Sanatsujat Singh and Ashok Kumar*
Agrotechnology Division, Council of Scientific and Industrial Research—Institute of Himalayan Bioresource Technology (CSIR—IHBT), Palampur, HP, India

*Address all correspondence to: ashok@ihbt.res.in

References

[1] Penner R, Shanks K, Timcke J, Krigbaum J, Uno J. Stevia from Paraguay: Market Perspectives, Development and Potential; Use and International Regulations. Asunción: Paraguay Vende; 2004

[2] Frederico AP, Ruas PM, Marin-Morales MA, Fuas CF, Nakajima JN. Chromosome studies in some Stevia Cav. (Compositae) species from southern Brazil. Brazilian Journal of Genetics. 1996;**19**:605-609

[3] Guleria P, Yadav SK. Agrobacterium mediated transient gene silencing (AMTS) in Stevia rebaudiana: Insights into Steviol glycoside biosynthesis pathway. PLoS One. 2013;**8**(9):e74731

[4] Savita SM, Sheela K, Sharan S, Shankar AG, Ramakrishna P, Sakey S. Health implications of Stevia Rebaudiana. Journal of Human Ecology. 2004;**15**:191-194

[5] Brandle JE, Telmer PG. Steviol glycoside biosynthesis. Phytochemistry. 2007;**68**:1855-1863

[6] Prakash DG, Clos J, Wilkens K, Fosdick L. Development of rebaudiana, a natural, non-caloric sweetener. Food Chemistry. Toxicology. 2008;**46**:575-582

[7] Gupta E, Purwar S, Sundaram S, Rai GK. Nutritional and therapeutic values of Stevia rebaudiana: A review. Journal of Medicinal Plant Research. 2013;**7**(46):3343-3353

[8] Shivanna N, Naika M, Khanum F, Kaul VK. Antioxidant, anti-diabetic and renal protective properties of Stevia Rebaudiana. Journal of Diabetes and its Complications. 2013;**27**(2):103-113

[9] Chatsudthipong V, Muanprasat C. Stevioside and related compounds: Therapeutic benefits beyond sweetness. Pharmacology Therapeutics. 2009;**121**:41-54

[10] Yadav AK, Singh S, Dhyani D, Ahuja PS. A review on the improvement of stevia [Stevia rebaudiana (Bertoni)]. Canadian Journal of Plant Science. 2011;**91**(1):1-27

[11] Danaei G, Finucane MM, Lu Y, Singh GM, Cowan MJ, Paciorek CJ, et al. National, regional, and global trends in fasting plasma glucose and diabetes prevalence since 1980: Systematic analysis of health examination surveys and epidemiological studies with 370 country-years and 2· 7 million participants. The lancet. 2011;**378**(9785):31-40

[12] Vaz JA, Patnaik A. Diabetes mellitus: Exploring the challenges in the drug development process. Perspectives in Clinical Research. 2012;**3**(3):109-112

[13] Goyal SK, Samsher GRK. Stevia (*Stevia rebaudiana*) a bio-sweetener: A review. International Journal of Food Sciences and Nutrition. 2010;**61**(1):1-10

[14] Jeppesen PB, Gregersen S, Rolfsen SE, Jepsen M, Colombo M, Agger A, et al. Antihyperglycemic and blood pressure-reducing effects of stevioside in the diabetic Goto-Kakizaki rat. Metabolism. 2003;**52**(3):372-378

[15] Fujita H, Edahira F. Safety utilization of stevia sweetener. Shikukim Kagyo. 1979;**82**:65-72

[16] Gao J, Brennan MA, Mason SL, Brennan CS. Effect of sugar replacement with stevianna and inulin on the texture

and predictive glycaemic response of muffins. International Journal of Food Science and Technology. 2016;**51**(9):1979-1987

[17] Kinghorn AD, Soejarto DD. Current status of stevioside as a sweetening agent forhuman use. In: Wagner H, Hikino H, Farnsworth NR, editors. Economic and Medical Plant Research. London, UK: Academic Press; 1985. pp. 1-52

[18] Brandle JE, Rosa N. Heritability for yield, leaf: Stem ratio and stevioside content estimated from landrace cultivar of Stevia rebaudiana. Canadian Journal of Plant Science. 1992;**72**:1263-1266

[19] Bespalhok-Filho JC, Hattori K. Embryogenic callus formation and histological studies from Stevia rebaudiana (Bert.) Bertoni floret explants. Revista Brasileira de Fisiologia Vegetal. 1997;**9**:185-188

[20] Šic Žlabur J, Voca S, Dobricevic N, Ježek D, Bosiljkov T, Brncic M. Stevia rebaudiana Bertoni—A review of nutritional and biochemical properties of natural sweetener. Agriculturae Conspectus Scientificus. 2013;**78**:25-30

[21] Gantait S, Das A, Banerjee J. Geographical distribution, botanical description and self-incompatibility mechanism of genus Stevia. Sugar Tech. 2018;**20**(1):1-10

[22] Ceunen S, Compernolle F, Mai AH, Geuns JM. Diterpene glycosides from Stevia phlebophylla A. Carbohydrate Research. 2013;**379**:1-6

[23] Singh SK, Singh AK, Dwivedi P. Modulating effect of salicyclic acid in tomato plants in response to waterlogging stress. Journal of Agriculture Environment & Biotechnology. 2017;**10**(1):1-8

[24] Pal PK, Mahajan M. Tillage system and organic mulch influence leaf biomass, steviol glycoside yield and soil health under sub-temperate conditions. Industrial Crops and Products. 2017;**104**:33-44

[25] Munz S, Präger A, Merkt N, Claupein W, Graeff-Hönninger S. Leaf area index, light interception, growth and steviol glycoside formation of Stevia rebaudiana Bertoni under field conditions in southwestern Germany. Industrial Crops and Products. 2018;**111**:520-528

[26] Shahverdi MA, Omidi H, Tabatabaei SJ. Plant growth and SG as affected by foliar application of selenium, boron and iron under NaCl stress in Stevia rebaudiana Bertoni. Industrial Crops and Products. 2018;**125**:408-415

[27] Pol J, Hohnová B, Hyötyläinen T. Characterisation of Stevia Rebaudiana by comprehensive two-dimensional liquid chromatography time-of-flight mass spectrometry. Journal of Chromatography. A. 2007;**1150**(1):85-92

[28] Mathur S, Bulchandan N, Parihar S, Shekhawat GS. Critical review on SG: Pharmacological, toxicological and therapeutic aspects of high potency zero caloric sweetener. International Journal of Pharmacology. 2017;**13**(7):916-928

[29] Singh S, Rao G. Stevia: The herbal sugar of 21st century. Sugar Tech. 2005;**7**(1):17-24

[30] Shock CC. Rebaudi's stevia: natural non-caloric sweeteners. California Agriculture. 1982;**36**:4-5

[31] Dwivedi RS. Unnurtured and untapped super sweet nonsacchariferous plant species in India. Current Science. 1999;**76**:1454-1461

[32] Chalapathi MV, Thimmegowda S, Kumar ND, Chandraprakash J, Rao GGE. Vegetative propagation of stevia (Stevia rebaudiana) under field conditions. Crop Research. 1999;**18**:319-320

[33] Maiti RK, Purohit SS. Stevia: A Miracle Plant for Human Health. Agrobios (India) Jodhpur, India: Agrobios Publisher; 2008

[34] Marsolais AA, Brandle J, Sys EA. Stevia plant named 'RSIT 94-751' United States Patent PP10564. 1998

[35] Goettemoeller J, Ching A. Seed germination. In: Janick J, editor. Perspectives on New Crops and New Uses. Alexandria, VA: ASHS Press; 1999. pp. 510-511

[36] Taiariol DR. Characterization of the Stevia rebaudiana Bert. 2004. Available from: http://www.monografias.com/trabajos13/stevia/stevia.html

[37] Ramesh K, Singh V, Megeji NW. Cultivation of stevia [*Stevia rebaudiana* (Bert.) Bertoni]: A comprehensive review. Advances in Agronomy. 2006;**89**:137-177

[38] Monteiro R. Taxonomia e biologia da reproducao da Stevia rebaudiana Bert. Brazil: Universidade Estadual de Campinas; 1980

[39] Miyagawa H, Fujikowa N, Kohda H, Yamasaki K, Taniguchi K, Tanaka R. Studies on the tissue culture of Stevia rebaudiana and its components: (II). Induction of shoot primordia. Planta Medica. 1986;**4**:321-324

[40] Chalapathi MV, Thimmegowda S, Sridhara S, Ramakrishna Parama VR, Prasad TG. Natural non-calorie sweetener stevia (*Stevia rebaudiana* Bertoni)—A future crop of India. Crop Research. 1997;**14**:347-350

[41] Oddone B. How to Grow Stevia. Pawcatuck, Connecticut: Guarani Botanicals, Inc; 1999. pp. 1-30

[42] Valio IFM, Rocha RF. Effect of photoperiod and growth regulators on growth and flowering of Stevia rebaudiana Bertoni. Japanese Journal of Crop Science. 1977;**46**:243-248

[43] Carneiro JWP, Muniz AS, Guedes TA. Greenhouse bedding plant production of Stevia rebaudiana (Bert) Bertoni. Canadian Journal of Plant Science. 1997;**77**:473-474

[44] Lester T. Stevia rebaudiana (sweet honey leaf) Australia. New Crops News Letters. 1999;**11**:120

[45] Felippe GM. Stevia rebaudiana—A review. Journal of Chromatography. 1978;**161**:403-405

[46] Oddone B. How to Grow Stevia. Pawtucket, CT: Guarani Botanicals; 1997

[47] Soejarto D, Compadre CM, Medon PJ, Kamath SK, Kinghorn AD. Potential sweetening agents of plant origin. II. Field search for sweet-tasting Stevia species. Economic Botany. 1983;**37**:71-79

[48] Madan S, Sayeed A, Singh GN, Kohli K, Singh YKR, Garg M. Stevia rebaudiana Bertoni: A review. Indian Journal of Natural Products and Resources. 2010;**1**(3):267-287

[49] Barroso M, Barros L, Angelo Rodrigues M, Joao Sousa M, Santos-Buelga C, Ferreira I. Stevia rebaudiana Bertoni cultivated in Portugal: A prospective study of its antioxidant potential in different conservation conditions. Industrial Crops and Products. 2016;**90**:49-55

[50] Abbas Momtazi-Borojeni A, Esmaeili SA, Abdollahi E, Sahebkar A. A review on the pharmacology and toxicology of steviol glycosides extracted from Stevia rebaudiana. Current Pharmaceutical Design. 2017;**23**(11):1616-1622

[51] Bertoni M. Le Kaà He-é Sa nature et ses proprietes. Anales Cientificos Paraguayos. 1905;**5**:1-14

[52] Grashoff JL. A Systematic Study of the North and Central American Species of stevia. Austin, TX: University of Texas; 1972. p. 624

[53] Shinde MR, Winnier J. Health benefits and application of Stevia rebaudiana bertoni in dentistry. Journal of Drug Delivery and Therapeutics. 2020;**10**(4-s):271-274

[54] Ranjan R, Jaiswal J, Jena J. Stevia as a natural sweetener. International Journal of Research in Pharmacy and Chemistry. 2011;**4**(1):1199-1202

[55] Soufi S, D'Urso G, Pizza P, Rezgui S, Bettaieb T, Montoro P. SG targeted analysis in leaves of Stevia rebaudiana (Bertoni) from plants cultivated under chilling stress conditions. Food Chemistry. 2016;**190**:572-580

[56] Soejarto D. Botany of Stevia and Stevia rebaudiana. In: Kinghorn AD, editor. Stevia: The Genus Stevia. London: Taylor and Francis; 2002. pp. 18-39

[57] Puri M, Sharma D. Antibacterial activity of stevioside towards food-borne pathogenic bacteria. Engineering in Life Sciences. 2011;**11**(3):326-329

[58] Katayama O, Sumida T, Hayashi H, Mitsuhashi H. The Practical Application of Stevia and Research and Development Data. Japan: I.S.U Company; 1976. p. 747

[59] Lewis WH. Early uses of Stevia rebaudiana (Asteraceae) leaves as a sweetener in Paraguay. Economic Botany. 1992;**46**:336-337

[60] Crammer B, Ikan R. Sweet glycosides from the Stevia plant. Chemische Berichte. 1986;**22**:915-916

[61] Fors A. A new character in the sweetener scenario. Sugar Journal. 1995;**58**:30

[62] Shaffert EE, Chebotar AA. Structure, topography and ontogeny of Stevia rebaudiana. Botanicheskii Zhurnal. 1994;**79**:38-48

[63] Lee JI, Kang KK, Lee EU. Studies on new sweetening resource plant Stevia (Stevia rebaudiana Bert.) in Korea. I. Effects of transplanting date shifting by cutting and seeding dates on agronomic characteristics and dry leaf yields of Stevia. Research Reports. 1979;**21**:171-179

[64] Gvasaliya VP, Kovalenko NV, Garguliya MC. Studies on the possibility of growing honey grass in Abkhazia conditions. Subtropicheskie Kul Tury. 1990;**5**:149-156

[65] Nishiyama P, Kusumoto IT, Costa SC, Alvarez M, Vieira LGE. Correlation between the contents of total carbohydrates and steviosides in leaves of Stevia rebaudiana. Arquivos De Biologia E Tecnologia. 1991;**34**:425-434

[66] Zubenko VF, Rogovskii SV, Chudnovskii BD. Effect of the leafiness of cuttings and of day length on the rooting and transplant growth of Stevia rebaudiana. Fiziologiya i Biokhimiya Kul'Turnykh Rastenii. 1991;**23**:407-411

[67] Carvalho MAM, Zaidan LBP. Propagation of Stevia rebaudiana from stem cuttings. Pesquisa Agropecuária Brasileira. 1995;**30**:201-206

[68] Chen SY, Li QR. Effect of growth substances on the stevioside content of Stevia rebaudiana. Plant Physiology Communications. 1993;**29**:265-267

[69] Acuna I, Nepovim A, Valicek P. Micropropagation of plants Stevia rebaudiana Bertoni in vitro and content of stevioside in leaves after application of growth regulators under field conditions. Agricultura Tropica et Subtropica (Czech Republic). 1997;**30**:51-60

[70] Handro W, Hell KG, Kerbauy GB. Tissue culture of Stevia rebaudiana, a sweetening plant. Cargo Cult Science. 1977;**29**:1240-1248

[71] Yang YW, Chang CW. In vitro plant regeneration from leaf explants of Stevia rebaudiana Bertoni. Zeitschrift für Pflanzenphysiologie. 1979;**93**:337-343

[72] Ferreira CM, Handro W. Micropropagation of Stevia rebaudiana through leaf explants from adult plants. Planta Medica. 1987;**54**:157-160

[73] Akita M, Shigeoka T, Koizumi Y, Kawamura M. Mass propagation of shoots of Stevia rebaudiana using large scale bioreactor. Plant Cell Reports. 1994;**13**:180-183

[74] Kornilova OV, Kalashnikova EA. Clonal micropropagation of stevia (Stevia rebaudiana Bertoni). Izvestiya Timiryzevskoi-Sel Skokhozyaistvennoi Adademi Russia. 1996;**1**:99-104

[75] Constantinovici D, Cachita CD. Aspects of in vitro multiplication in Stevia rebaudiana Bert. Cercetari Agronomic in Moldova. 1997;**30**:80-86

[76] Ferreira CM, Handro W. Production, maintenance and plant regeneration from cell suspension cultures of Stevia rebaudiana (Bert.) Bertoni. Plant Cell Reports. 1988;**7**:123-126

[77] Bondarev NI, Nosov AM, Kornienko AV. Effects of exogenous growth regulators on callusogenesis and growth of cultured cells of Stevia Rebaudiana Bertoni. Russian Journal of Plant Physiology. 1998;**45**:770-774

[78] Nepovim AVT. In vitro propagation of Stevia rebaudiana plants using multiple shoot culture. Planta Medica. 1998;**64**:775-776

[79] Razak UN, Ong CB, Yu TS, Lau LK. In vitro micropropagation of Stevia rebaudiana Bertoni in Malaysia. Brazilian Archives of Biology and Technology. 2014;**57**(1):23-28

[80] Tadhani MB, Jadeja RP, Rema S. Micropropagation of stevia rebaudiana using multipleshoot culture. Journal of Cell and Tissue Research. 2006;**6**(1):545-548

[81] Wang YH, ElSohly MA, Khan IA, Lata H, Chandra S. In vitro germplasm conservation of elite Stevia rebaudiana Bertoni. International Symposium on Plant Cryopreservation. 2013;**1039**:303-308

[82] Ali A, Gull I, Majid A, Saleem A, Naz S, Naveed NH. In vitro conservation and production of vigorous and desiccate tolerant synthetic seeds in Stevia rebaudiana. Journal of Medicinal Plant Research. 2012;**6**(7):1327-1333

[83] Thiyagarajan M, Venkatachalam P. Large scale in vitro propagation of Stevia rebaudiana (bert) for commercial application: Pharmaceutically important and antidiabetic medicinal herb. Industrial Crops and Products. 2012;**37**(1):111-117

[84] Zayova E, Nedev T, Dimitrova L. In vitro storage of Stevia rebaudiana Bertoni under slow growth conditions and

mass multiplication after storage. The Biological Bulletin. 2017;**3**(1):30-38

[85] Nower AA. In vitro propagation and synthetic seeds production: An efficient method for Stevia rebaudiana Bertoni. Sugar Tech. 2014;**16**(1):100-108

[86] Jayaraman S, Manoharan MS, Illanchezian S. In-vitro antimicrobial and antitumor activities of Stevia rebaudiana (Asteraceae) leaf extracts. Tropical Journal of Pharmaceutical Research. 2008;**7**(4):1143-1149

[87] Lemus-Mondaca R, Vega-Gálvez A, Zura-Bravo L, Ah-Hen K. Stevia rebaudiana Bertoni, source of a high-potency natural sweetener: A comprehensive review on the biochemical, nutritional and functional aspects. Food Chemistry. 2012;**132**(3):1121-1132

[88] Milani PG, Formigoni M, Dacome AS, Benossi L, Costa CE. New seminal variety of Stevia rebaudiana: Obtaining fractions with high antioxidant potential of leaves. Anais da Academia Brasileira de Ciências. 2017;**89**(3):1841-1850

[89] Sheeja RR, Lawrence B. Phytochemical screening of the leaves of Stevia rebaudiana Bertoni. International Journal of Current Microbiology and Applied Sciences. 2015;**4**(3):344-347

[90] Abou-Arab AE, Abou-Arab AA, Abu-Salem MF. Physico-chemical assessment of natural sweeteners steviosides produced from Stevia rebaudiana Bertoni plant. African Journal of Food Science. 2010;**4**(5):269-281

[91] Komissarenko NF, Derkach AI, Kovalyov IP, Bublik NP. Diterpene glycosides and phenylpropanoids of Stevia rebaudiana Bertoni. Rast Research. 1994;**1**(2):53-64

[92] Kim IS, Yang M, Lee OH, Kang SN. The antioxidant activity and the bioactive compound content of Stevia rebaudiana water extracts. LWT- Food Science and Technology. 2011;**44**(5):1328-1332

[93] Tadhani MB, Subhash R. In vitro antimicrobial activity of Stevia rebaudiana Bertoni leaves. Tropical Journal of Pharmaceutical Research. 2006;**5**(1):557-560

[94] He J, Zhu NL, Kong J, Peng P, Li LF, Wei XL, Jiang YY, Zhang YL, Bian, BL, She GM Shi RB (2019) A newly discovered phenylethanoid glycoside from Stevia rebaudiana Bertoni affects insulin secretion in rat INS-1 islet β cells. Molecules, 24(22):4178.

[95] Zhao L, Yang H, Xu M, Wang X, Wang C, Lian Y, et al. Stevia residue extract ameliorates oxidative stress in d-galactose-induced aging mice via Akt/Nrf2/HO-1 pathway. Journal of Functional Foods. 2019;**52**:587-595

[96] Wood JHB, Allerton R, Diehl HW, Fletcher JHG. Stevioside- I. the structure of the glucose moieties. The Journal of Organic Chemistry. 1995;**20**(7):875-883

[97] Koyama E, Kitazawa K, Ohori Y, Izawa O, Kakegawa K, Fujino A, et al. In vitro metabolism of the glycosidic sweeteners, stevia mixture and enzymatically modified stevia in human intestinal microflora. Food and Chemical Toxicology. 2003;**41**(3):359-374

[98] Jakinovich JW, Moon C, Choi YH, Kinghorn AD. Evaluation of plant extracts for sweetness using the Mongolian gerbil. Journal of Natural Products. 1990;**53**(1):190-195

[99] Lobov SV, Kasai R, Ohtani K, Tanaka O, Yamasaki K. Enzymic production of sweet stevioside

derivatives: Transglucosylation by glucosidases. Agricultural and Biological Chemistry. 1991;**55**(12):2959-2965

[100] Yamamoto K, Yoshikawa K, Okada S. Effective production of glycosyl-steviosides by α-1, 6 transglucosylation of dextrin dextranase. Bioscience, Biotechnology, and Biochemistry. 1994;**58**(9):1657-1661

[101] Kumar H, Kaul K, Bajpai-Gupta S, Kaul VK, Kumar S. A comprehensive analysis of fifteen genes of steviol glycosides biosynthesis pathway in Stevia rebaudiana (Bertoni). Gene. 2012;**492**(1):276-284

[102] Brandle J. Genetic control of rebaudioside A and C concentration in leaves of the sweet herb, Stevia rebaudiana. Canadian Journal of Plant Science. 1999;**79**(1):85-91

[103] Lankes C, Zabala UM, Müller V. Performance of Stevia rebaudiana bertoni genotypes under European temperate zone conditions. In: XXVIII International Horticultural Congress on Science and Horticulture for People IHC2010. 2010

[104] Anami ET, Poletine JP, Goncalves-Vidigal MC, Vidigal F, Lacanallo GF, Kvitschal MV, et al. Characterization and genetic divergence in Stevia rebaudiana (Bert.) Bertoni clones based in agronomical and morphological characteristics. Journal of Food, Agriculture and Environment. 2010;**8**:463-469

[105] Ibrahim IA, Nasr MI, Mohammedm BR, El-Zefzafi MM. Nutrient factors affecting in vitro cultivation of Stevia rebaudiana. Sugar Tech. 2008;**3**:248-253

[106] Othman HM, Osman M, Zainuddin Z. Morphological assessment of Stevia rebaudiana Bertoni. A accession in IIUM's germplasm as initial material forstevia breeding. Australian journal of basic and applied. Science. 2015;**9**(25):1-9

[107] Caligari PD. Plant Breeding and Crop Improvement. Wiley Online Library. 2001

[108] Lorenzana RE, Bernardo R. Accuracy of genotypic value predictions for marker-based selection in biparental plant populations. Theoretical and Applied Genetics. 2009;**120**(1):151-161

[109] Tavarini S, Passera B, Angelini LG. Crop and Steviol glycoside improvement instevia by breeding. In: Steviol Glycosides: Cultivation, Processing, Analysis and Application in Food. Royal Society of Chemistry. 2018. pp. 1-31

[110] Brandle JE, Starratt AN, Gijzen M. Stevia rebaudiana: Its agricultural, biological, and chemical properties. Canadian Journal of Plant Science. 1998;**78**:527536

[111] Sharma N, Kaur R, Era V. Potential of RAPD and ISSR markers for assessing genetic diversity among *Stevia rebaudiana* Bertoni accessions. 2016

[112] Díaz-Gutiérrez C, Arroyave C, Llugany M, Poschenrieder C, Martos S, Peláez C. Trichoderma asperellum as a preventive and curative agent to control fusarium wilt in Stevia rebaudiana. Biological Control. 2021;**2011**:10453

[113] Serfaty M, Ibdah M, Fischer R, Chaimovitsh D, Saranga Y, Dudai N. Dynamics of yield components and stevioside production in Stevia rebaudiana grown under different planting times, plant stands and harvest regime. Industrial Crops and Products. 2013;**50**:731-736

[114] Ahmad MA, Javed R, Adeel M, Rizwan M, Yang Y. PEG 6000-stimulated

drought stress improves the attributes of In-vitro growth, SG production, and antioxidant activities in Stevia rebaudiana Bertoni. Plants. 2020;**9**:1552

[115] Taak P, Koul B, Chopra M, Sharma K. Comparative assessment of mulching and herbicide treatments for weed management in *Stevia rebaudiana* (Bertoni) cultivation. South African Journal of Botany. 2020a:303-311

[116] Rezaizad MS, Abbaspour H, Hashemi Moghaddam H, Gerami M, Ramezani M. Photocatalytic activity of titanium dioxide nanoparticles (TiO_2) on the physiological and phytochemical properties of Stevia (Stevia rebaudiana Bertoni). Journal of Medicinal plants and By-product. 2020;**2**:169-177

[117] Barbet-Massin C, Giuliano S, Alletto L, Daydé J, Berger M. Nitrogen limitation alters biomass production but enhances Steviol glycoside concentration in Stevia rebaudiana Bertoni. PLoS One. 2015;**10**(7):e0133067

[118] Le Bihan Z, Hastoy C, Cosson P, Boutié P, Rolin D, Schurdi-Levraud V. To cut or not to cut? That is the question in first year harvest of Stevia rebaudiana Bertoni production. Industrial Crops and Products. 2021;**162**:113209

[119] Tanaka O. Application of C-nuclear magnetic resonance spectrometry to structural studies on glycosides: Saponins of Panax spp. and natural sweet glycosides Yakugaku Zasshi.105323351. 1985

[120] Taak P, Ahmad MS, Koul B. Comparative assessment of the effect of different herbicides on weeds reported in cultivation of natural sweetener plant: Stevia rebaudiana bertoni. Plant Archives. 2020b;**20**(2):2594-2599

[121] Taak P, Tiwari S, Koul B. Optimization of regeneration and Agrobacterium-mediated transformation of Stevia (Stevia rebaudiana Bertoni): A commercially important natural sweetener plant. Science Reports. 2020c;**10**:16224

[122] Libik-Konieczny M, Capecka E, Kąkol E, Dziurka M, Grabowska-Joachimiak A, Sliwinska E, et al. Growth, development and SG content in the relation to the photosynthetic activity of several Stevia rebaudiana Bertoni strains cultivated under temperate climate conditions. Scientia Horticulturae. 2018;**234**:10-18

[123] Gorzi A, Omidi H, Bostani AB. Morpho-physiological responses of Stevia (Stevia rebaudiana Bertoni) to various priming treatments under drought stress. Applied Ecology and Environmental Research. 2018;**16**:4753-4771

[124] Pradhan N, Singh P, Dwivedi P, Pandey DK. Evaluation of sodium nitroprusside and putrescine on polyethylene glycol induced drought stress in Stevia rebaudiana Bertoni under in vitro condition. Industrial Crops and Products. 2020;**154**:112754

[125] Simlat M, Szewczyk A, Ptak A. Melatonin promotes seed germination under salinity and enhances the biosynthesis of steviol glycosides in Stevia rebaudiana Bertoni leaves. PLoS One. 2020;**15**(3):e0230755

[126] Tursun AÖ, Jabran K, Gürkan H, Telci İ. Effects of elevated temperature and carbon dioxide concentrations on aromatic compounds of *Stevia rebaudiana*. Sugar Technology. 2021;**2021**:1-8

[127] Abdullateef RA, Osman M. Effects of visible light wavelengths on seed germinability in *Stevia rebaudiana* Bertoni. International Journal of Biology. 2011;**3**(4):83

[128] Kumar R. Seed germination of Stevia rebaudiana influenced by various potting media. Octa Journal of Biosciences. 2013;**1**(2):143-146

[129] Prakash J, Arora NK. Development of bacillus safensis -based liquid bioformulation to augment growth, stevioside content, and nutrient uptake in *Stevia rebaudiana*. World Journal of Microbiology and Biotechnology. 2020;**36**(1):1-13

[130] Rosales C, Brenes J, Salas K, Arce-Solano S, Abdelnour-Esquivel A. Micropropagation of Stevia rebaudiana in temporary immersion systems as an alternative horticultural production method. Revista Chapingo Serie Horticultura. 2018;**24**(1):69-84

[131] Debnath M, Ashwath N, Midmore DJ. Physiological and morphological responses to abiotic stresses in two cultivars of Stevia rebaudiana (Bert.) Bertoni. South African Journal of Botany. 2019;**123**:124-132

[132] Kawatani T, Kaneki Y, Tanabe T. The cultivation of kaa he-e (Stevia rebaudiana). II seed germination with special reference to the optimum temperature and light sensitivity Japanese. Journal of Tropical Agriculture. 1977;**1977**:20137142

[133] Brandle J (2001) Stevia rebaudiana with altered steviol glycoside composition [on-line]. US Patent no. US6255557 B1

[134] Jain P, Farooq B, Lamba S, Koul B. Foliar spray of Moringa oleifera lam. Leaf extracts (MLE) enhances the stevioside, zeatin and mineral contents in Stevia rebaudiana Betoni. South African Journal of Botany. 2020;**132**:249-257

[135] Luwańska A, Perz A, Mańkowska G, Wielgus K. Application of in vitro stevia (Stevia rebaudiana Bertoni) cultures in obtaining steviol glycoside rich material. Herba Polonica. 2015;**61**(1):50-63

[136] Morita T, Bu Y. Variety of Stevia rebaudiana Bertoni [on-line]. US Patent no. US 6031157. 2000

[137] Shuichi H, Tsuneo Y, Satoshi F. Breeding of tyriploid plants of stevia (Stevia rebaudiana Bertoni) with highrebaudioside A content. Japanese Journal of Tropical Agriculture. 2001;**45**(4):281-289

[138] Vde O, Forni-Martins ER, Magalhães PM, Alves MN. Chromosomal and morphological studies of diploid and polyploid cytotypes of Stevia rebaudiana (Bertoni) Bertoni (Eupatorieae, Asteraceae). Genetics and Molecular Biology. 2004;**27**(2):215-222

[139] Yoneda Y, Nakashima H, Miyasaka J, Ohdoi K, Shimizu H. Impact of blue, red, and far-red light treatments on gene expression and steviol glycoside accumulation Stevia rebaudiana. Phytochemistry. 2017a;**137**:57-65

[140] Yoneda Y, Shimizu H, Nakashima H, Miyasaka J, Ohdoi K. Effects of light intensity and photoperiod on improving SG content in Stevia rebaudiana. 2017

[141] Singh G, Pal P, Masand M, Seth R, Kumar A, Singh S, et al. Comparative transcriptome analysis revealed gamma-irradiation mediated disruption of floral integrator gene (s) leading to prolonged vegetative phase in *Stevia rebaudiana* Bertoni. Plant Physiology and Biochemistry. 2020;**148**:90-102

Chapter 17

Monitoring the Aroma Compounds of *Vicia faba* L var. Major and var. Minor

Imene Rajhi, Bechir Baccouri, Fatma Rajhi, Moez Amri, Guido Flamini and Haythem Mhadhbi

Abstract

The volatile compounds of *Vicia faba*. L var. minor and var. major seeds were evaluated by headspace-solid phase micro-extraction (HS-SPME) coupled to gas chromatography-mass spectrometry (GC-MS). The total identification percentages of the extracted volatiles were 95.5% and 98.3%, respectively. The number of aroma compounds detected was 28. Among them, 15 compounds were determined in the emission of whole legume seeds of minor cultivar and 22 from the major one. The volatiles were classified into five chemical classes, i.e., monoterpene hydrocarbons, oxygenated monoterpenes, sesquiterpene hydrocarbons, apocarotenes, and non-terpene derivatives. Aldehydes and alkanes were considered as the most abundant constituents in non-terpene derivatives, followed by esters, alcohols, phenols, phenones, and hydrocarbons. A wide difference in term of volatiles was observed between major and minor faba bean cultivars. This study can provide useful information about the specific volatile characteristics for each cultivar and its possible use in the conception of legume-based ingredients and for pertinent breeding programs.

Keywords: HS-SPME-GC-MS, aroma, faba bean minor, faba bean major, chemical classes, specific compounds

1. Introduction

Vicia faba, also known in the culinary sense as the broad bean, fava bean, or faba bean, horse bean, field bean, bell bean, Windsor or tic bean is a flowering plant in the pea and bean family Fabaceae native to North Africa and south-western Asia, and extensively cultivated elsewhere [1, 2]. It is grown as a winter annual in warm temperate and subtropical areas. Hardier cultivars grown in the Mediterranean region can tolerate winter temperatures of −10°C without serious injury, whereas the hardiest European cultivars can tolerate up to −15°C [3]. Although usually classified in the same genus Vicia as the vetches, some botanists treat it in a separate monotypic genus Faba. Four subspecies including *V. faba* ssp. minor, *V. faba* ssp. equina, *V. faba* ssp. major, and *V. faba* ssp. Paucijuga were reported by Cubero [4]. However, in [5],

Muehlbauer and Tullu classified faba beans based on seed size. Botanically, faba beans were divided into two subspecies: paucijuga and faba. Faba subspecies was, also, subdivised into *V. faba* var. minor with small, and rounded seeds (1 cm long), *V. faba* var. equina with medium-sized seeds (1.5 cm) and *V. faba* var. major with large broad flat seeds (2.5 cm).

Faba bean seeds are very important crops due to their high protein content ranging from 20% to 41%, depending on variety [6]. Using legumes in human diet could potentially prolong life and maintain well-being [7]. They also contain adequate proportions of carbohydrates and oil which increases their food value [8]. Recently, the consumption of legumes was increased over the world and that due to their high nutritional value and beneficial health effects. Neverthelss, the pulses off flavors, in general and in faba bean seeds in special, limit their use [9]. The undesirable aroma of legumes is due to their emission of volatiles developping from the fatty acid oxidation catalyzed by lipoxygense. Aldehydes, alcohols, ketones, acids, and pyrazines categories were responsible for the musty, beany, grassy, earthy, and leafy unwanted odors in faba bean seeds [10].

To enhance the use of pulses worldwide, it is necessary to define methods to ameliorate the flavor of legumes. In fact, an agreeable flavor is an important aspect for any successful product [11]. Technologies and methods such as control of oxidation, germination, dehulling, enzymatic treatment, heat processing, fermentation, milling, and cultivar selection has been described to prevent the off-flavors development and accumulation in pulses [12–14]. The off-flavors in faba bean seeds were released during harvest, processing, and storage time [15]. To enhance the suitablity of legumes odors, it is relevant to start with raw pulses that emitted low desirable odors to guarantee that the development of off-flavors keeps at the lowest level during processing conditions [10, 16]. Because of the limited information available which dealing with the unwanted odors in faba beans, the purpose of this study was the monitoring of the volatiles emitted by faba bean seeds var. major and var. minor using the headspace solid-phase microextraction coupled to gas chromatography with mass spectrometry HS-SPME-GC-MS (**Figure 1**).

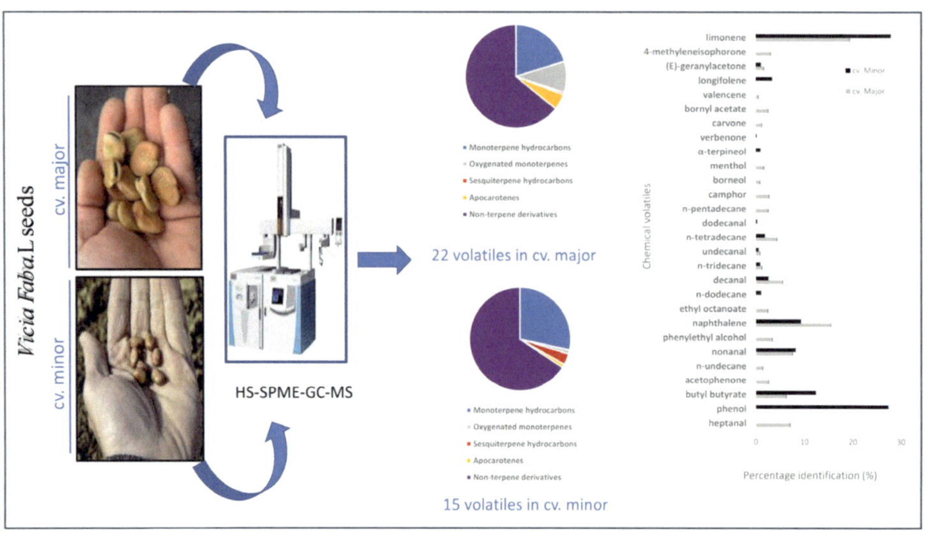

Figure 1.
Graphical abstract of the presented study.

2. Materiels and methods

2.1 Chemicals and reagents

A pure reference compounds including heptanal, benzaldehyde, octanal, *(E)*-2-octenal, decanal, dodecanal, all *n*-hydrocarbons, butyl butyrate, 6-methyl-5-hepten-2-one, naphthalene, acetophenone, phenol, β-pinene, *p*-cymene, limonene, γ-terpinene, 1,8-cineole, α-terpineol, verbenone, bornyl acetate, longifolene, geranylacetone (*E* + *Z* isomers), 1-octanol, 1-decanol, ethyl benzoate, 2-undecanone, α-thujone + β-thujone mixture, and pulegone were purchased from Sigma, Aldrich, Supelco and Merck and used to compare retention times and mass spectra.

2.2 Sample preparation

Major and minor faba bean cultivars were used in this study, named Mamdouh and Najeh, respectively. The bean seeds were offred by the Field Crops Laboratory of the National Institute of Agricultural Research (INRAT). 5 g of similar size seeds, without any physical damages, were choosen. The seeds were stored at 4°C in an opaque aluminum bag until analysis.

2.3 Headspace solid-phase micro extraction (HS-SPME)

The headspace spontaneous volatile emissions of the two vicia faba cultivars were sampled by means of HS-SPME. Three replications were considered for each legume. The sample was kept at room temperature for 30 min. A Solid Phase Micro-Extraction (SPME) device (Supelco, Bellefonte, PA, USA) covered with poly-dimethyl-siloxane (PDMS, 100 μm) was used, adapted according to the fabricator instructions. After that, the fiber was removed and transmitted to the injection port of the GC-MS system. All the desorption conditions were the same for all the seeds. Furtheremore, blanks were conducted before starting the first SPME extraction. Quantitative comparisons of relative peaks areas were assessed among the same chemicals in the different samples.

2.4 Gas chromatography coupled with mass spectrometry (GC-MS)

Gas chromatography-electron impact mass spectrometry (GC-EIMS) analyses were assessed using an Agilent 7890B gas chromatograph (Agilent Technologies Inc., Santa Clara, CA, USA) equipped with an Agilent HP-5MS (Agilent Technologies Inc., Santa Clara, CA, USA) capillary of 12 column (30 m × 0.25 mm; coating thickness 0.25 μm) and an Agilent 5977B single quadruple mass detector (Agilent Technologies Inc., Santa Clara, CA, USA). The conditions of sample analysis were as following: injector and transfer line temperatures 220 and 240°C, respectively; oven temperature programmed from 60 to 240°C at 3°C/min; carrier gas helium at 1 mL/min; split ratio 1:25. The acquisition parameters were as follows: full scan; scan range: 30–300 m/z; scan time: 1.0 s.

2.5 Compounds identification

The determination of the volatile compounds was established by comparability of the retention times with those of the authentic samples, comparing their linear retention indices (LRI) relative to the series of *n*-hydrocarbons. Computer matching was also used against commercial and laboratory-developed mass spectra library built up from pure substances and components of known mixtures and MS literature data [17].

3. Results and discussion

In total, 28 volatiles were detected by HS-SPME-GC-MS. Among them, 15 were identified in var. minor and 22 in var. major; accounting for 98.3% and 95.4% of total emission, respectively. The flavor constituents were divided into five chemical classes, including monoterpenes hydrocarbons (MH), sesquiterpenes hydrocarbons (STH), apocarotenes (AP), non-terpenes derivatives (NTD), and oxygenated monoterpenes (OM). Faba bean var. major emitted 61.4% of NTD, 19.4% of MH, 9.3% of OM, 4.8% of AP, and 0.6% of STH volatiles. On the other hand, 64.7% of NTD, 27.9% of MH, 3.4% of STH, 1.2% of OM, and 1.1% of AP were detected in the bouquet of the minor cultivar (**Table 1**). This data is in accordance with the studies of *Khrisanapant* [18] and Rajhi et al. [10]. Indeed, these authors demonstrated that NTD group was the major chemical class in legumes.

3.1 Non terpenes derivatives constituents (NTD)

The individual volatile profiles of both cultivars are summarized in **Figure 2**. NTD volatiles identified in this study were: 5 aldehydes (heptanal, nonanal, decanal, undecanal, and dodecanal), 5 alkanes (*n*-dodecane, *n*-undecane, *n*-tridecane, *n*-tetradecane, and *n*-pentadecane), esters (butyl butyrate and ethyl octanoate), one phenol (phenol), 1 phenone (acetophenone), alcohol (phenylethyl alcohol), and hydrocarbon (naphthalene). The most abundant chemical group was aldehyde; including nonanal, which has a fat, citrus, and green flavor, decanal which is soap, orange peel and tallow like aroma, and undecanal, which has a waxy type of flavor. These chemicals were emitted by var. minor (8.2%, 2.6%, and 0.6%, respectively) and var. major (7.6%, 5.6%, and 0.9%, respectively). The heptanal aldehyde, which has a strong fruity odor [19], was registered only in the individual profile of var. major (7.1%). However, the dodecanal aldehyde, which has a soapy type of odor [19], was only found in the volatile bouquet of var. minor (0.3%). These results are in good agreement with a previous investigation of volatile

Constituents	*Vicia faba* major	*V. faba* minor
Monoterpene hydrocarbons	19.4	27.9
Oxygenated monoterpenes	9.3	1.2
Sesquiterpene hydrocarbons	0.6	3.4
Phenylpropanoids	0	0
Apocarotenes	4.8	1.1
Non-terpene derivatives	61.4	64.7
Total identified (%)	95.5	98.3

Table 1.
Chemical classes percentage of volatiles compounds in major and minor faba bean seeds.

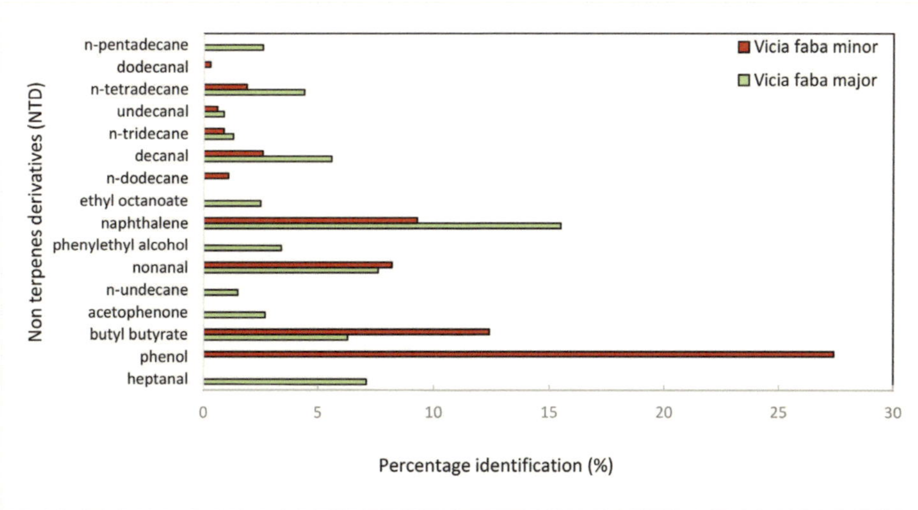

Figure 2.
NTD volatile compounds identified in faba bean var. minor and var. major.

compounds of other legumes, in particular soybeans, mung beans, cowpeas, peas, chickpeas, orange lentils and adzuki beans [18, 20, 21].

Alkanes were the second dominant class in the NTD volatiles. Among the detected alkanes, two constituents were emitted by both cultivars: the *n*-tridecane and *n*-tetradecane (**Figure 2**). *n*-undecane and *n*-pentadecane were registered in the emission profile of var. major (1.6% and 2.6%, respectively). However, the alkane *n*-dodecane was characterized in the individual profile of var. minor. The high content in alkanes, was also previously reported by Oomah et al. [20] for different types of *P. vulgaris*. Their abundance legume seeds may be explained by the occurrence of lipid peroxidation, which causes the formation of the characteristic aroma of dry pulses, since alkanes are mainly obtained from oxidative reaction of lipids [11, 20].

Two esters were identified using the HS-SPME-GC-MS: the butyl butyrate, which has a fruity odor [19], is emitted by major and minor cultivars (6.3% and 12.4%, respectively), and the ethyl octanoate, presented a strong fruity and flowers odor [19], was found only in the bouquet of the latter one (2.5%). The NTD phenol was emitted by var. minor with a percentage identification of 27.4%.

This result agrees with that of Mebazaa et al. [22] and Ramadan et al. [23]. In their study on black, dark red kidney and pinto beans, Oomah et al. [20] demonstrated that the volatile profiles of studied *P. vulgaris* pulses were rich in aldehydes, alkanes, esters, and ketones.

3.2 Oxygenated monoterpenes

Seven constituents belong to the oxygenated monoterpenes were detected in both cultivars using HS-SPME-GC-MS. Among them five such as camphor, borneol, menthol, carvone, and bornyl acetate were appeared in the emission profile of var. major (2.8%, 0.9%, 1.7%, 1.3%, and 2.6%, respectively). However, the minor cultivar emitted much more α-terpineol (1%) and verbenone (0.2%) (**Figure 3**).

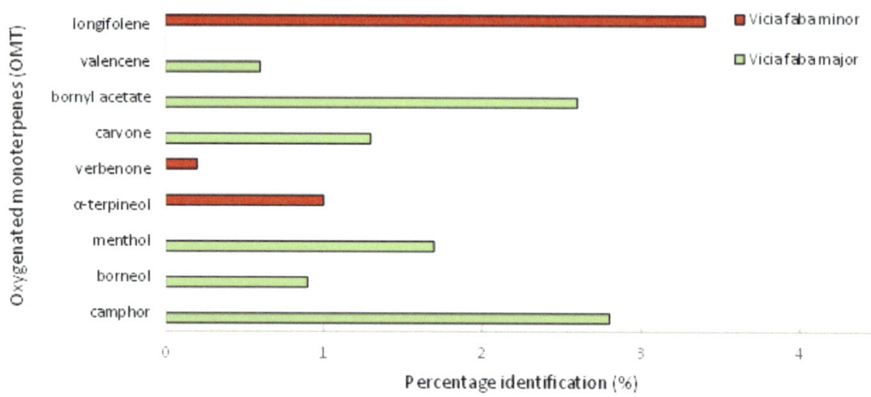

Figure 3.
OMT volatile compounds identified in faba bean var. minor and var. major.

3.3 Sesquesternes hydrocarbons, apocarotenes, and monoterpenes hydrocarbons constituents

The individual profiles of emitted aroma revealed that two sesquiterpene hydrocarbons were detected; the valencene and longifolene. The major compound emitted by each cultivar was the monoterpene hydrocarbon limonene (19.4 for var. major and 27.9%) (**Figure 4**).

3.4 Beany flavors

Hexanal, (*E,E*)-2,4-decadienal, (*E,E*)-2,4-nonadienal, nonanal, 2-pentylfuran, and 1-hexanol have been screened and considered as beany flavor markers in pulses [24, 25]. These are generated from the oxidation of unsaturated fatty acids (linoleic and linolenic) by an enzymatic pathway [15, 26, 27]. In this study, only one beany flavor marker was identified: nonanal, which emitted in a comparable amount by both cultivars (8.2 and 7.6% for var. minor and var. major, respectively). These beany flavors are responsible for the unwanted and unpleasant flavor of faba beans.

3.5 Volatiles and faba bean breeding

Unfortunately, pulses including faba bean seeds are not widely used in the world to their highest potential due to their inappreciated flavors, the poor digestibility of their proteins and their long cooking time [9]. One of the most significant sensory elements of food is its aroma, which is the result of the interaction between taste and smell perceptions. Unprocessed legumes emit volatile compounds as a result of the oxidation of fatty acids, which is mediated by lipoxygenase, during the harvesting, processing, and storage processes, giving them the musty, grassy, and bean-like flavors [9, 28]. Aldehydes, alcohols, ketones, acids, and pyrazines are frequently linked to the components of off-flavors in legumes. It is crucial to discuss technology that enhance the flavor of pulses in order to promote the consumption of legumes worldwide. A successful product must, in fact, have a flavor that consumers like. To prevent the development of off odors in legumes, germination, oxidation control,

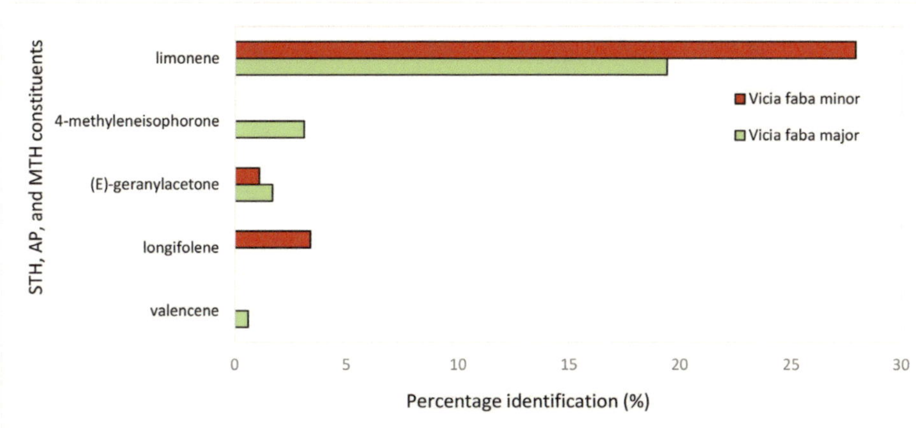

Figure 4.
STH, AP, and MTH volatile compounds identified in faba bean var. minor and var. major.

heat processing, enzymatic treatment, and cultivar selection have all been described [9]. To ensure that the formation of unwanted scent stays at the lowest possible level during processing and storage, it is crucial to start with raw materials that exhibit few off flavors. A selection of raw material, in term of volatiles of different cultivars of legumes should be performed. Thus, this type of study is very important to select a legume cultivars which had the lowest emission of off-flavors and then these latter cultivars can be selected as important breeding lines based on their organoleptic proprieties.

4. Conclusions

In this study, we performed the monitoring of the volatiles of faba bean var. minor and var. major using HS-SPME-GC-MS. The data revealed a difference between both profiles; that can be explained by the varietal effect. This study can provide useful information about the specific volatile characteristics for each cultivar and its possible use in the conception of legume-based ingredients and for pertinent breeding programs.

Author details

Imene Rajhi[1*], Bechir Baccouri[2], Fatma Rajhi[3], Moez Amri[4,5], Guido Flamini[6,7] and Haythem Mhadhbi[1]

1 Laboratory of Legumes and Sustainable Agrosystems, Biotechnology Center of Borj Cedria, Hammam-Lif, Tunisia

2 Laboratory of Olive Biotechnology, Biotechnology Center of Borj Cedria, Hammam-Lif, Tunisia

3 Higher Institute of Water Sciences and Techniques of Gabes, University of Gabes, Gabes, Tunisia

4 Agro-BioSciences (AgBS), University Mohammed VI Polytechnics (UM6P), Benguerir, Morocco

5 Crop Laboratory, National Institute of Agricultural Research of Tunisia (INRAT), Ariana, Tunisia

6 Dipartimento di Farmacia, Pisa, Italy

7 Centro Interdipartimentale di Ricerca "Nutraceutica e Alimentazione per la Salute" Nutrafood, Università di Pisa, Pisa, Italy

*Address all correspondence to: imenrajhi@yahoo.fr

References

[1] Neji I, Rajhi I, Baccouri B, Barhoumi F, Amri M, Mhadhbi H. Leaf photosynthetic and biomass parameters related to the tolerance of *Vicia faba* L. cultivars to salinity stress. Euro-Mediterranean Journal for Environmental Integration. 2021;**6**:22

[2] Rajhi I, Moussa S, Neji I, Baccouri B, Chikha M, Mhadhbi H. Photosynthetic and physiological responses of small seeded faba bean genotypes (*Vicia faba* L.) to salinity stress: Identification of a contrasting pair towards salinity. Photosynthetica. 2020;**58**(1):174-185

[3] Robertson LD, Singh KB, Erskine W, Abd El Moneim AM. Useful genetic diversity in germplasm collections of food and forage legumes from West Asia and North Africa. Genetic Resources and Crop Evolution. 1996;**43**:447-446

[4] Cubero JI. Evolutionary trends in *Vicia faba*. Theoretical and Applied Genetics. 1973;**43**:59-65

[5] Muehlbauer F, Tullu A. Cent. New Crops Plants Production. NewCrop Fctsfeet: Vicia Faba L. Purdue Univ; 1997

[6] VidalValverde C, Frias J, Sotomayor C, DiazPollan C, Fernandez M, Urbano G. Nutrients and antinutritional factors in faba beans as affected by processing. Zeitschrift für Lebensmittel-Untersuchung und -Forschung. 1998;**207**:140-145

[7] Patterson CA, Maskus A, Dupasquier C. Pulse crops for health. Cereal Foods World. 2009;**54**:108-112

[8] Rodino AP, Nistal JH, Hermida M, Santalla M, De Ron AM. Sources of variation for sustainable field pea breeding. Euphytica. 2009;**166**:95-107

[9] Roland WSU, Pouvreau L, Curran J, van de Velde F, de Kok PMT. Flavor aspects of pulse ingredients. Cereal Chemistry. 2017;**94**:58-65

[10] Rajhi I, Baccouri B, Rajhi F, Mhadhbi H, Flamini G. Monitoring the volatile compounds status of whole seeds and flours of legume cultivars. Food Bioscience. 2021;**41**:101105

[11] Akkad R, Kharraza E, Hanb J, Housec JD, Curtisa JM. Characterization of the volatile flavor compounds in low and high tannin faba beans (*Vicia faba* var. minor) grown in Alberta, Canada. Food Research International. 2019;**120**:285-294

[12] Rajhi I, Boulaaba M, Baccouri B, Rajhi F, Flamini G, Mhadhbi H. Assessment of dehulling effect on volatiles, phenolic compounds and antioxidant activities of faba bean seeds and flours. South African Journal of Botany. 2022;**147**:741-753

[13] Rajhi I, Baccouri B, Rajhi F, Mhadhbi H, Flamini G. HS-SPME-GC-MS characterization of volatile chemicals released from microwaving and conventional processing methods of fenugreek seeds and flours. Industrial Crops Production. 2022:182. [ahead of print]

[14] Rajhi I, Baccouri B, Rajhi F, Hammami J, Souibgui M, Amri M, et al. Evaluation of germination effect on volatile compounds of different faba bean cultivars using HS-SPME/GC-MS. Journal of Food Composition and Analysis. 2022;**112**:104692

[15] Akkad R, Kharraz E, Han J, House J, Curtis JM. The effect of short-term storage temperature on the key

headspace volatile compounds observed in Canadian faba bean flour. Food Science and Technology International. 2021:1-9

[16] Azarnia S, Boye JI. Flavour compounds in legumes: Chemicaand sensory aspects. In: Azarnia S, Boye JI, editors. Progress in Food Science and Technology. New York, NJ, USA: Nova Science Publishers, Inc; 2011

[17] Davies NW. Gas chromatographic retention indices of monoterpenes and sesquiterpenes on methyl silicon and Carbowax 20M phases. Journal of Chromatography A. 1990;**503**:1-24

[18] Khrisanapant P, Kebede B, Ying Leong S, Indrawati O. A comprehensive characterisation of volatile and fatty acid profiles of legume seeds. Food. 2019;**8**:651

[19] Acree T, Arn H. Flavornet and human odor space. 2004. Available from: http://www.flavornet.org/flavornet.html

[20] Oomah BD, Liang LSY, Balasubramanian P. Volatile compounds of *Phaseolus vulgaris* L. Plant Foods and Human Nutrition. 2007;**62**:177

[21] del Rosario R, de Lumen BO, Habu T, Flath RA, Mon TR, Teranishi R. Comparison of headspace volatiles from winged beans and soybeans. Journal of Agricultural and Food Chemistry. 1984;**32**:1011-1016

[22] Mebazaa R, Rega B, Camel V. Analysis of chemical composition and antioxidant activity of human male armpit sweat after fenugreek ingestion: Characterization of odor active compounds by gas chromatography coupled to mass spectrometry and olfactometry. Food Chemistry. 2011;**128**:227-235

[23] Ramadan MM, Yehia HA, Shaheen MS, Abed EL-Fattah MS. Aroma volatiles, antibacterial, antifungal and antioxidant properties of essential oils obtained from some spices widely consumed in Egypt. American-Eurasian Journal of Agricultural & Environmental Sciences. 2014;**14**(6):486-494

[24] Jiang ZQ, Pulkkinen M, Wang YJ, Lampi AM, Stoddard FL, Salovaara H. Faba bean flavour and technological property improvement by thermal pre-treatments. LWT – Food Science and Technology. 2016;**68**:295-305

[25] Xu M, Jin Z, Lan Y, Rao J, Chen B. HS-SPME/GCMS/olfactometry combined with chemometrics to assess the impact of germination on flavor attributes of chickpea, lentil, and yellow pea flours. Food Chemistry. 2019;**280**:83-95

[26] Hall C, Hillen C, Robinson JG. Composition, nutritional value, and health benefits of pulses. Cereal Chemistry. 2017;**94**(1):11-31

[27] Jelen H, Wasowicz E. Lipid-derived flavor compounds. In: Jelen H, editor. Food Flavors: Chemical, Sensory and Technological Properties. 1st ed. Boca Raton: CRC Press; 2012. pp. 65-93

[28] Rackis JJ, Sessa DJ, Honig DH. Flavor problems of vegetable food proteins. Journal of the American Oil Chemists' Society. 1979;**56**:262-271

Chapter 18

Application of Tissue Culture Techniques to Improve the Productivity of Medicinal Secondary Products from Medicinal Plants

Ahmed M. Hassanein

Abstract

The plant kingdom is considered the most important source of medicinal chemicals. *In vitro* culture techniques are being considered a promising alternative to traditional agricultural processes to improve medicinal plants multiplication and their production of pharmaceutical compounds. In this chapter, several *in vitro* culture strategies are discussed to improve secondary metabolites production, including (1) plant kingdom as a source of medicinal chemicals, (2) *in vitro* culture of medicinal plants, (3) culture media optimization, (4) application of suspension cell culture for production of secondary metabolites, (5) elicitation to enhance the productivity of the culture, (6) precursor intermediates feeding, (7) selection of high-yielding cell lines, (8) overexpression of genes that control the production of bioactive compounds, and (9) scale-up production. Also, challenges that hinder the *in vitro* culture of medicinal plants using different techniques and the use of those techniques to produce pharmaceutical compounds are discussed in this chapter, including (a) secondary metabolites toxicity, (b) low growth rate, (c) culture browning, (d) limitation in the application of transformation, (e) somaclonal variation, and (f) vitrification. Therefore, the principal objective of the current chapter was to shed light on the studies on some medicinal plants and the used protocols to overcome some difficulties in terms of *in vitro* propagation that maximize their economic values.

Keywords: *In vitro* plant culture, medicinal plants, pharmaceutical compounds, tissue culture, medicinal chemicals

1. Introduction

Unlike animals, plants do not have the ability to move, making them vulnerable to attack by pests and sometimes animals. To overcome this problem, plant tissues synthesize enormous compounds, such as terpenes, polyphenols, cardenolides, steroids, alkaloids, and glycosides, and use them as defense strategies [1]. These defense compounds are called secondary metabolites and are not necessary for essential plant functions, such as growth, photosynthesis, and reproduction. These compounds are accumulated in the plant body to use by man as pharmaceutical, agrochemicals, aromatics, and food additives [1, 2]. Despite the progress in synthetic chemistry, plants are considered the most successful sources of drugs due to their bioactive compounds produced through secondary metabolism pathways [2].

In industrialized and developing countries, raw plant materials and plant-derived pharmaceuticals have naturally an essential component of present-day human healthcare systems. A known fact is that over 80% of the human beans use herbal medicines for healthy living [3]. In this respect, at present, more than 40% of the used pharmaceuticals by Western countries are derivatives of natural resources [4]. Worldwide, man uses about 35000–70000 plant species to prevent and cure diseases, most of them are reported in China (10,000–11,250), India (7500), Mexico (2237), and others [5]. Quality assurance and standardization of herbal medicines during the collection, handling, processing, and production of herbal medicine are essential prerequisites to ensure safety for the global herbal market. Wild plant materials are collected from gardens, open pasture, or forest land. In some cases, medicinal plants grow like weeds on agricultural land. While the bulk of the medicinal plant materials is still wild-harvested, a very small number of plant species are cultivated commercially [6]. However, increase populations and urban growth were associated with an over-exploitation of natural resources. Unfortunately, several medicinal plant species are disappeared due to the expansion of land for the purpose of growing crops, urban expansion, uncontrolled deforestation, and intensive collection [7]. Now, the increase in demand for these compounds encouraged the cultivation of large areas of medicinal plants and the application of new technologies, such as plant tissue culture (PTC) to preserve them from extinction and improve their productivity in quality and quantity.

Manufacturing of medicinal products from soil-grown plants faces some challenges, such as: (1) The wild-targeted plant does not exist in sufficient abundance in the local environment or is rare in general, (2) Cultivation of the target plant may need certain conditions, (3) Production of the target substance may require to grow plants for a long time, (4) The target substance may present at low concentration in cultivated or harvested plants, (5) Variations in environmental conditions may result in the production of bioactive compounds at a non-homogeneous quantity or quality, (6) Collection of plants for pharmaceuticals may be unsafe, (7) Harvest of propagated medicinal plants for drug industries is time- and money-consuming [8]. To overcome all the obstacles, PTC techniques express the great potential for bioproduction of phytoconstituents of high therapeutic value. By application of artificial techniques, regulation of the biosynthetic pathway of the certain plant to enhance the production of valuable compounds or avoidance of production of an unwanted substance become possible.

With the aid of gene technology and molecular techniques, *in vitro* culture procedures, such as cell, organ or tissue culture, somatic embryogenesis, somatic hybridization, genetic transformation, hairy roots, and induction of somaclonal variation, and others can be applied to the improvement of bioactive compounds yields. For example, recombinant DNA technology can be used to direct metabolic pathways and produce pharmaceuticals, such as antibodies and hormones. These *in vitro* culture techniques are better than others where they are carried out under precisely controlled physical and chemical conditions. PTC techniques are a resolution for the propagation of seedless medicinal plants and others with small or unviable seeds that not be able to germinate in soil [9, 10]. In addition, PTC techniques hold significant promise for true to type, disease-free, rapid and mass multiplication, and plant development [11].

Application of PTC technologies in the medicinal plant does not free from problems but avoiding their problems can be precisely controlled, which makes *in vitro* cultivation an ideal alternative to produce medicinal compounds from plants [12]. One of the obstacles is that the prices of the products resulting from biotechnology are higher than other products resulting from cultivated or wild plants. In this concern, the application of large-scale PTC techniques have been found to be an attractive alternative tools to the traditional plantations, where they offer a controlled supply of secondary metabolites independent of plant availability and a more consistent product in quantity and quality [13]. In the last decade, to meet pharmaceutical industry demand and conserve natural sources, researchers concentrated their efforts on optimizing culture conditions for maximizing the obtained yield of targeted secondary metabolites by application of several artificial-developed techniques [14].

Through PTC techniques, a whole plant can be regenerated from an organ, small tissue, or a plant cell but it should carry out on a suitable culture medium and under a controlled environment [15]. Under these conditions, the obtained plantlets are true to type and show characteristics identical to the mother plant. On the other hand, the culture conditions can be controlled to stimulate genetic variation for plant improvement, but it requires the construction of a selection procedure to select an elite mutant. For several decades, *in vitro* culture techniques are being used increasingly as a supplement to traditional breeding tools for the modification and improvement of plants. For example, *Coryodalis yanhusuo*, an important medicinal plant was improved through the application of the somatic embryogenesis technique to produce disease-free lines [16]. While PTC can be established from any part of a plant, meristematic tissues, such as shoot tip or nodal segments, are usually recommended [15, 17, 18]. In addition, the physiological state of the donor plant affects strongly on regeneration ability of the cultured plant materials [9, 18].

The application of PTC techniques in the medicinal and other plant species becomes an essential prerequisite for plant propagation and improvement [15, 17]. The application of plant tissue culture has several advantages: (1) It results in the production of thousands of plantlets in a short period from a small segment of the tested plant. (2) It is a main procedure to obtain pathogen-free plants. (3) It can be used to culture plants round the year, irrespective of weather or season. (4) It needs little space for the propagation of the southlands of plants. (5) It can be used as the main procedure to produce a new cultivar of a certain plant. (6) It can be used to understand the effect of a specific biotic or abiotic factor on a tested plant beyond the interaction of other factors. (7) It helps to understand the molecular biology of plant differentiation. (8) It is an essential prerequisite during the production of genetically engineered plants. (9) It is an effective procedure for the production of pharmaceutical compounds. (10) It is an essential procedure for the preservation of endangered plant species, genetic assets, and gene banks.

2. Plant kingdom as a source of medicinal chemicals

Phytotherapy becomes a complementary and important part of pharmacotherapy and modern medicine. It is a type of treatment based on natural medicinal resources (drugs) and herbal remedies for the purposes of prevention and treatment of illness.

Herbal drugs mean using the whole plant or part of it, fresh or dry, to treat or prevent human disease. Any plant part (flower, leaf, root, bark, fruit, and seed), resins, balsams, rubber, plant exudates, algae, fungi, or lichen can be used as herbal drugs for its medicinal properties. Herbal drugs or herbal remedies contain active ingredients of herbal medicinal products. The aerial plant parts, such as leaves, seeds, and flowers, are often able to synthesize and accumulate secondary metabolites more than those obtained by underground parts, such as roots or rhizomes [19]. For example, in *Scrophularia kakudensis*, the total phenol and flavonoid, as well as free radical scavenging compounds, were higher in shoot than root extract [20]. The variable contents of bioactive compounds in different plant tissues may be due to the specialized ability of each tissue to synthesize the bioactive ingredients or their ability to store them considering the physiological condition and endogenous hormone levels [19].

Based on their biosynthetic origins, reports classify the bioactive secondary metabolites of the plant into major groups, including phenolic compounds, terpenoids, nitrogen-containing alkaloids, and sulfur-containing compounds [21]. Phenolic compounds were the most important group where they are largely used to enhance human health and they naturally occur in fruits, vegetables, cereals, and beverages. Phenols are classified into different groups, including phenolic acids, flavonoids, stilbenes, and lignans, and they include apigenin, diosmin, quercetin, kaempferol, eriodictyol, naringenin, hesperetin, baicalein, chrysin, catechin, morin, genistein, curcumin, colchicine, resveratrol, and emodin. For the production and extraction of hundreds of these secondary products, plant cell, tissue, or organ cultures were used [21].

As a part of complementary and alternative medicine, medicinal plant extracts are widely used in chronic diseases like diabetes, hypertension, cancer, etc. Melatonin and serotonin, as antioxidants, were detected in the field and greenhouse-grown *Ocimum sanctum* L. plants [22]. Extract of *in vitro* cultures of *Hovenia dulcis* has antitumor effects [23]. *Aegle marmelos* can be used as antibacterial, antifungal, antidiabetic, and antioxidant [24]; it is also useful to treat several symptoms, such as stomachalgia, diarrhea, dysentery, malaria, and fever [25]. *In vitro* propagated *Artemisia japonica* was used to obtain antioxidant, insecticidal, antimalarial, antisporulant, antimicrobial, cytotoxic, and osteoinductive activities [26]. Acacetin (5,7-dihydroxy-4-methoxyflavone) has several therapeutic effects, it is found in more than 200 plant species belonging to 60 plant families especially *Asteraceae and Lamiaceae* families [27]. Acacetin is used for antiplasmodial, anticancerous, antidiabetic, antiperoxidative, antipyretic, anti-inflammatory, and antiproliferative activities [27]. Several compounds with anti-uveal melanoma activity were extracted from *Acacia nilotica,* including gallocatechin 5-O-gallate, methyl gallate, gallic acid, catechin 5-O-gallate, catechin, 1-O-galloyl-β-D-glucose, digallic acid, and 1,6-di-O-galloyl-β-D-glucose [28]. Biotechnological systems can be used to obtain vaccines from many plant species to provide immune protection against diseases [29]. Production of plant-based edible vaccines is mainly manipulated by the integration of the transgene into *in vitro* cultured plant cells to produce the antigen protein for specific diseases [30].

Screening of 346 methanol extracts of 281 native and cultivated plant species in Egypt indicated that *Agave americana, A. lophantha, Furcraea selloa, Calotropis procera, Pergularia tomentosa, Asclepias sinaica, Alkanna orientalis, Khaya grandifoliola, Swietenia mahogani, Pimenta racemosa, Pinus canariensis, Verbascum sinaiticum, Solanum elaeagnifolium, S. nigrum, and Brachychiton rupestris* have strong antischistosomal activity [31]. In addition, the antioxidant activity of the extract of 90 plants was determined by 2, 2 diphenyl-1-picrylhydrazyl (DPPH) assay [32], and extracts of some plant species expressed high antioxidant and cytotoxic activities that inhibited the growth of cancer cells [33]. Leaves of *A. marmelos* contain several medicinal compounds including π-sitosterol, lupeol, aegelin, rutin, flavone, glycoside, marmesinine, oisopentenyl halfordiol, phenylethyl cinnamides, and marmeline [24].

3. Application of *in vitro* culture techniques on medicinal plants

Plant tissue culture is the most promising savior of medicinal plants that face problems of low yield and susceptibility to biotic or abiotic stress. Also, PTC can be used for *in situ* and *ex situ* conservation, propagation, polyploidy or aneuploidy induction, plant engineering, and bioreactor applications. *In vitro* multiplication was established in many threatened and endemic medicinal plants, such as *Bacopa monnieri* [34], *Paedaria foetida* [35], *Picrorhiza kuroa* [36], *Salvadora persica* [37], *Potentilla fulgens* [38], *Eryngium foetidum* [39], and *H. dulcis* [40].

High multiplication using seedling tissues or shoot meristems was achieved in several plant species, such as *Citrullus colocynthis* [41], *Zephyranthes bulbous* [42], *Plectranthus vetiveroids* [43], *Glossocardia bosvallea* [44], *Cannabis sativa* [45], *O. sanctum* L. [22], *Caralluma retrospeciens* [46], *Solanum nigrum* [15], *Moringa oliefera* [47], *Pulicaria incisa* [48], *Rosa damascena* [49], *A. marmelos* [50], *Artemisia judaica* [51], and *Hyoscyamus muticus* [52].

For long-term storage of medicinal plant materials, cryopreservation is recommended where it is carried out in liquid nitrogen (−196°C). Different plant organs or parts, including seeds, corms, bulbs, rhizomes, roots, tubers, buds, and cuttings, can be stored for conservation purposes [11], especially in medicinal plants with recalcitrant seeds. The main applied techniques of cryopreservation of medicinal plants are vitrification, desiccation, and encapsulation–dehydration. Vitrification-cryopreservation of shoot tips of *Dioscorea floribunda* medicinal plant indicated that the genome of cryopreserved shoot tips was stable upon application of molecular, morphological, and biochemical procedures [53]. Vitrification–encapsulation–dehydration techniques of *Dioscorea deltoidei* medicinal plant shoot tips proved that the secondary metabolites of cryopreserved shoot tips were like control plants [54].

PTC is more efficient than naturally grown plant materials to assess the effect of different experimental conditions on the production of secondary metabolites of medicinal plants [55]. PTC opens the way for the production of engineered molecules and produces new forms of plant secondary metabolites [56]. These new forms of compounds may have a valuable effect on biological control, food, pharmaceutical, and other strategies. Transformation techniques are widely dependent on PTC for enhancing the *in vitro* production of valuable plant secondary metabolites [57].

Different types of PTC techniques are successfully exploited for *in vitro* propagation as well as synthesis and extraction of secondary metabolites [12]. Sometimes root culture is recommended because it provides valuable biomass in a short time and stable metabolite productivity. In addition, root cultures express genetic stability for long-term culture compared to other forms of *in vitro* cultures, such as cell aggregates and rhizoids. Roots are fully organized plant organ, ensures biochemical stability, and usually express the full biosynthetic capacity as same as soil-grown plant root. *In vitro* root cultures could be a better alternative for the accumulation of elevated contents of secondary metabolites. For example, root cultures of *Hemidesmus indicus* were used as a tool for *in vitro* production of 2-hydroxy 4-methoxy benzaldehyde [11, 58].

In vitro-produced hairy roots are formed without connection with any other plant organs. Then, the synthesized metabolites are not transported to other plant parts and are accumulated where they are synthesized. The produced secondary metabolites may be present in minor, undetectable quantities *in vivo* but they are present in higher levels in hairy roots due to the optimized culture conditions (14). Consequently, mass production of secondary compounds in the bioreactor was established using hairy root cultures [59].

4. Application of *in vitro* culture techniques for the production of pharmaceuticals

The synthetic capacity of secondary metabolites of the dedifferentiated tissue often differs substantially from that of differentiated one, both quantitatively and qualitatively. The differences in synthetic capacities are a direct response to differences in enzyme patterns between differentiated and undifferentiated tissues, they are mirrors for gene expression of these tissues. The culture of differentiated plant materials often shows biochemical and genetic stability, it offers a high-productivity system that does not need wide-ranging optimization. For example, the major alkaloid (vindoline) is scarcely produced by *Catharanthus roseus* suspension cultures but shoot cultures produce it in high quantity [60]. In addition, while the callus culture of *Taraxacum officinale* synthesizes and accumulates α and γ-amyrins, differentiated tissue synthesizes and accumulates taraxasterol and lupeol [61]. The previous studies indicate that different classes of secondary metabolites need different phases of cell or tissue differentiation.

Generally, *in vitro* conditions can be easily modulated to enhance the synthesis of secondary metabolites through modulation of the pathway of primary metabolism in plants. The *in vitro* obtained compounds are important as dyes, drugs, cosmetics, flavors, food additives, perfumes, agrochemicals, etc. Some of these compounds, such as flavors, fragrances, and colorants, cannot be produced by microbial cells or chemically synthesized but they can be synthesized by plant cell culture systems [62]. Several reports indicated that *in vitro* cultures were found to be more efficient than whole plants for the formation of bioactive secondary metabolites such as ajmalicine, ajmaline, anthraquinones, benzylisoquinoline alkaloids, berberine, bisoclaurine, coniferin, diosgenin, ginseng, ginsenoside, glutathione, nicotine, rosmarinic acid, raucaffricine, shikonin, taxol, terpentine, tripdiolide, and ubiquinone-10 [1, 2, 14, 62].

Under aseptic conditions, cultured plant materials can be used to generate bioactive or secondary metabolites, including flavonoids, alkaloids and other phenolics, terpenoids, saponins, steroids, tannins, glycosides, colorants, fragrances, and volatile oils [14]. Production of high-value active secondary metabolites at industrial levels, such as shikonin, berberine, and sanguinarine, was fulfilled from cell cultures of *Lithospermum erythrorhizon*, *Coptis japonica,* and *Papaver somniferum*, respectively [63]. Secondary bioactive metabolites in *in vitro* cultured *Swertia chirayita* were higher than *in vivo* plants [2]. The more antimicrobial property of the *in vitro* regenerated plant products was related to more bioactive metabolites. In addition, Manivannan et al. [20] reported that since the contents of phytochemicals in seed and *in vitro* derived plants were similar, the *in vitro* plantlets can be used as alternate for the seed grown plants for the production of bioactive metabolites. Also, acacetin (an individual flavonoid) was slightly increased in *in vitro* grown plantlets than that of *in vivo* grown plants due to the artificial conditions of the *in vitro* culture and modulation of endogenous hormone [20].

Pharmaceutical compounds that are obtained from *in vitro* cultured plant materials may be more easily extracted and purified due to the absence of significant amounts of pigments, thus resulting in lower manufacturing expenses [64]. Control of the production of secondary metabolites can be carried out using *in vitro* culture techniques. For example, low biomass and hypericin production of *Hypericum perforatum* shoots was improved by prolonging the time of culture for more than 30 days [65].

5. Strategies are used to improve secondary metabolites production

The biosynthesis of secondary metabolites using unorganized cultured cells or organized organs, such as roots, can be enhanced by altering the environmental conditions or selecting an elite variant clone [66]. There are many procedures that can be controlled to increase the productivity of *in vitro* cultured medicinal plants from the active substances with medicinal effects, and this is what will be discussed in this chapter.

5.1 Culture media optimization

To understand factors that control the biosynthesis of pharmaceutical compounds by cultured plant materials, studies on gene expression, enzyme activity, and signal transductions were carried out [12, 14]. The establishment of desired productivity of the PTC needs optimization of overall culture conditions to enhance both culture biomass and metabolites productivity. For example, while sulfate and ammonium nitrate ions increased the colchicine content of *Gloriosa superba* callus, a higher concentration of phosphate and calcium decreased alkaloid biosynthesis [67]. The differences in the composition of various PTC media formulations affect on water potential of the cultural environment [68]. Then, different media exerted different values of water potential. *In vitro* culture of certain medicinal plant materials on different media expresses different values of biomass and secondary metabolites. Medium selection is a major step in optimizing the culture conditions to produce an abundance of plant matter capable of producing an abundance of biological compounds [69]. Through media optimization of the *in vitro* cultured medicinal plants, the chemical composition is changed, the content of toxic compounds is reduced and novel chemical compounds may be formed [70]. In general, media optimization is an essential prerequisite to enhancing the production of antioxidants and other valuable secondary metabolites, it means that plant growth regulators and specific additives should be modulated to enhance *in vitro* production of biomass and secondary metabolites [71].

When nodal segments of *Ocimum basilicum* were cultured under the influence of different culture media including MS medium in different strengths and different combinations of PGRs, they expressed different values of methyl eugenol, linalool, and 1,8-cineole fractions [71]. Nodal segments are of *Cunila menthoides* medicinal plant cultured on MS medium containing different concentrations of PGRs resulting in biosynthesis of phenols, alkaloids, and terpenes in regenerated plants [71]. Media containing different types and concentrations of PGRs express different differentiation pathways and biomass values [72, 73], and it was associated with the expression of different types and concentrations of pharmaceuticals in cultured plant materials [74]. Contents of bioactive compounds in embryogenic callus and regenerated shoots of *Rosa rugosa* petal explants were influenced by PGRs type, concentration, and the nitrogen source [75]. Also, in *Chonemorpha fragrance*, the amount of synthesized camptothecin was influenced by the PGRs type and concentrations [76].

The effect of carbon source concentration and type on culture biomass and metabolites productivity should be investigated. To enhance the biomass and biosynthesize of secondary metabolites, sucrose is widely used as a carbon source and it was better than maltose, glucose, and others [77]. During *in vitro* propagation, the optimal concentration of sucrose depends on plant species [15, 47, 68, 73]. For example, feeding the culture medium with 60 mM nitrogen and rise sucrose concentration from 3% sucrose to 5% increase the biomass production and camptothecin accumulation by 2.4-fold in the cell suspension cultures of *Nicotiana nimmoniana* [77]. In *Panax vietnamensis*, 2–5% sucrose enhanced the biomass and ginsenoside production in the cell suspension but 6–7% sucrose inhibited ginsenoside accumulation [78]. Geraniol production in transgenic tobacco cell suspension cultures was influenced by several culture conditions including carbon source light, and inoculums size [14].

Physical culture conditions can also affect the ability of *in vitro* cultured plants for the production of secondary metabolites. For example, light as one of these physical conditions can affect strongly on the production of secondary metabolites in *Abelmoschus esculentus* [79]. Culture media pH is an essential factor for the production of valuable plant material mass and its content of secondary metabolites. The optimum pH for normal plant tissue cultures is 5.8 but it should be changed if the purpose of the culture is to produce bioactive compounds. In a comparative study by Hagendoom et al. [80], on different plant species, they detected a positive correlation between acidification of the cytoplasm and the accumulation of different secondary metabolites including coniferin and lignin. In cell suspension of *C. roseus,* an increase in the pH of culture media between 4.3 and 9.0 was associated with a sharp increase in alkaloid production [81]. In general, low and high pH of the medium retard biomass and withanolide production in *Withania somnifera* cell culture [82], but the optimal pH was 4.5 for enhancing biomass production and Bacoside A formation [83].

In a scale-up production system, modulation the composition of the culture media is an essential prerequisite to enhance the production efficiency of a selected cell line, but long-term cultivation may lead to the reduction of the yield [64] due to an increase in somaclonal variation [84]. Consequently, genetic stability of the cultured plant materials should be established using determined indicators, such as molecular markers, stability of growth parameter index over extended subculture cycles, and metabolite production.

5.2 Suspension and callus cultures

Callus culture is an undifferentiated-unorganized mass obtained by cell division on cultured plant material on an agar medium. Then, calli are subcultured either for *in vitro* propagation through organogenesis or embryogenesis or used to establish suspension culture [85]. When callus in suitable texture is obtained on solid or semisolid agar medium and suspended in a specific liquid growth medium, the cells disperse and divide more and more producing cell suspensions. Then, cells can have faster and uniform growth rates associated with secondary metabolite production. Suspension cultures are the most widely employed PTC techniques in the production of secondary metabolites. When cells are grown in aqueous media to produce cell suspensions, some cells do not disperse in the medium and form tissue clumps, which disrupts growth and weakens the production of targeted secondary compounds.

Suspension cultures are also amenable for growth in small and giant fermenters, but these cultures may show genetic and biochemical variation. Under selected conditions, exploitation of cell cultures capable of producing medicinal compounds at a level similar or superior to that of intact plants.

Callus culture itself is exploited to produce and study secondary compounds in many medicinal plant species [66]. For induction of callus formation, specific culture conditions should be established, which means that cultured cells divide and proliferate rapidly as long as the cultural environment has sufficient nutrients and suitable growth regulators. Conditions for callus induction and proliferation are not favorable for the production of secondary metabolites. For induction of secondary metabolites, calli culture conditions should be changed or transferred to a new medium with a different composition [11]. High yields of proteolytic enzymes from the callus tissue culture of *Allium sativum* L. on MS medium containing NAA and BAP were obtained [86].

The advantages of the application of suspension-cell cultures are obvious including: (1) The biomass production is usually more rapid than that of other *in vitro* culture types as well as a whole plant, (2) Chemical and physical conditions can be easily controlled allowing the production of certain pharmaceuticals throughout the year if necessary, (3) Producers can provide their products in a sustainable manner that does not depend on large areas and leave the arable land areas to grow other crops, (4) The size and quality of the product can be controlled according to the market demand, (5) Producers can select plant cell line that ensures or improve product quality, and (6) Producers can combine more than one method, which leads to the development of new products.

Application of specific cell lines and selective culture of that cell lines lead to the production of secondary compounds more than those obtained from original tissues and normal culture conditions [87]. The addition of plant growth regulators enhances the production of target secondary metabolites in several medicinal plant species [88]. Cell immobilization [89] and genetic makeup [90] can be optimized to enhance the synthesis of secondary compounds under *in vitro* culture conditions. Cultured cells can be immobilized to form aggregates to enhance secondary metabolite production [91]. Cell immobilization is achieved through growing cells as aggregates or using substances such as alginate or polyurethane foam cubes [92].

Two-phase cell suspension cultures establish a growth medium for maximizing cell biomass and production of naphthoquinone pigment in the first phase, but the second phase was established at the dark condition and room temperature with alkaline pH. These two phases system enhanced biomass production six-fold and optimized metabolite production in *Arnebia* sp. [93]. In suspension cultures of *C. roseus*, cultures produced up to 20 g DW L − 1 of biomass. In addition, two phases culture technique increased active cell biomass with 10 times higher indole alkaloids production in comparison to that of the one-phase culture [94]. Under dark conditions at 25°C for 40 days, the two phases of co-culture of *Panax ginseng* and *Echiancea purpurea* adventitious root in bioreactors containing MS medium supplemented with IBA (25 μM), sucrose (50 g L − 1), and methyl jasmonate (200 μM) as elicitor for 30 days enhanced the production of ginsenosides and caffeic acid derivatives [95].

5.3 Elicitation as an effective strategy to enhance the productivity of *in vitro* cultures

In vitro or *in vivo* cultured plants show physiological and morphological responses to physical, chemical, or microbial agents which are called elicitors. Therefore, elicitation describes any processes that induce or enhance the synthesis of secondary metabolites to ensure plant survival and competitiveness [96, 97]. During *in vivo* growth, plant secondary metabolites are elicited in plant cells in response to environmental stresses as a defensive strategy against the abiotic agent or invading pathogen [2]. Elicitation effectiveness depends on several parameters, some of them are related to elicitor agents themselves, and others are related to the elicited *in vitro* cultured plant materials. The elicitor-related effects include elicitor type, concentration, and exposure duration. Cultures' age, cultivated line, medium composition, type, and concentration of growth regulators are essential parameters during the application of elicitation strategies. Hence, the application of factors, such as biotic or abiotic agents, that trigger the defense response in *in vitro*-cultivated plant materials enhanced the productivity of bioactive compounds [98].

Most of the used biotic elicitors are either exogenous or endogenous microbial agents but abiotic is a wide range of materials, mainly heavy metals [14, 99]. Methyl jasmonate, salicylic acid, yeast extract, chitosan, inorganic salts, UV radiation, or others can be used as elicitors to improve secondary metabolites production of the cultured plant materials [97, 100]. Citric acid, L-ascorbic acid, and casein hydrolysate were also used as elicitors to enhance the total phenolic content in the callus of *Rosa damascene* [49].

In the suspension culture of *Mentha pulegium*, when media were supplemented with yeast extract and salicylic acid, a significant increase of limonene, menthone, menthol, and α-pinene was detected [101]. Fifty different substances were detected in an *in vitro* cultured *Anemia tomentosa* upon jasmonic acid application, whereas 20 substances were only detected in wild-type plants [102]. Secondary metabolites production in callus, cell suspension, or hairy roots of *Ammi majus* L. were elicited by autoclaved lysate of cell suspension of *Enterobacter sakazaki* bacteria [103]. Anthraquinone production in *Rubia akane* cell culture was elicited by chitosan [104]. Genetically stable *in vitro* regenerated plants of *Capparis spinosa* were confirmed by RAPD analysis with a two-fold increase in flavonoid content than those of the wild plants when plants were regenerated under the influence of methyl jasmonate elicitor [105]. Elicitation of *Ambrosia artemisiifolia* hairy root cultures to produce thiorubrine A was dependent on cultures' age as well as elicitor concentration and exposure time. Maximum of eight-fold thiorubrine A production was achieved when 16-day-old cultures were elicited with 50 mg l^{-1} vanadyl sulfate elicitor for 72 h [106].

Abiotic stresses for a given period can be used as an elicitor. Temperature, light parameters (intensity, photoperiod, and wavelength), and water potential of the medium influence the fresh and dry biomass [15] as well as the concentration of active metabolites [107]. Any factor that affects the water stress of the media should affect growth and bioactive compound synthesis. The profound change in the culture water potential due to the addition of NaCl, mannitol, or polyethylene glycol can elicit the production of secondary metabolites [107]. The relationship between abiotic-nutritional deficiency stress and enhancement of the production of secondary metabolites was reported [108]. Deficiencies of nitrogen, phosphate, potassium, sulfur, or magnesium increase the production of phenolic compound accumulation in different plant species [109], which may be due to oxidative stress and modulation of the expression of some genes [110]. The combination between target gene overexpression and elicitors increased the yield of secondary metabolites. Across studied plant species, elicitors promoted the yield of secondary metabolites from 1.0 to a maximum of 2230-fold [100]. Abiotic elicitors were applied to enhance growth and ginseng saponin biosynthesis in *P. ginseng* hairy roots [111].

Specific microorganisms can be used for elicitor purposes [112]. It takes place through the co-cultivation of plant cells with microorganisms. Compared to non-elicited control tissues, coculture of *Aspergillus flavus* with *C. roseus* resulted in increases in vinblastine (7.88%) and vincristine (15.5%) concentrations [112]. Cocultivation between microorganisms and cultured plant tissue should avoid conditions that stimulate microorganism toxic components [12].

5.4 Precursor feeding

Under perfect and controlled conditions, *in vitro* cultured plants not only have a higher metabolic rate than differentiated or soil-grown plants but also compressed biosynthesis cycles in shorter periods of time. In addition, the addition of precursors and elicitors plays an important role in promoting the secondary metabolism of cells and tissues grown under well-controlled industrial conditions (PTC). Precursor feeding is a strategy that is based on the assumption that if intermediates of bioactive molecules are added at the beginning or during, the *in vitro* culture period, they can serve as a substrate to improve the production of secondary metabolites in cultured plant materials. Precursors refer to any compounds that can be converted by the *in vitro* cultured plant materials into secondary metabolites through biosynthetic pathways [14, 113], and they depend on the type and concentration of precursor, and addition timing [114]. According to the World Health Organization definition any plant that contains a substance that can be used for medicinal use or as a precursor to synthesize new or semi-synthetic pharmaceuticals as a medicinal plant. The addition of alanine precursor was used to stimulate the biosynthesis of plumbagin in *Plumbago indick* when it was added to the root cultures on the 14th day of cultivation along with sequential addition of Diaion HP-20 36 h after it was fed, this increased the target output 14 times [115]. Phenylalanine precursor was needed for the biosynthesis of silymarin in hairy roots of *Silymarin marianum* [116] or the biosynthesis of podophyllotoxin in the cell suspension cultures of *Podophyllum hexandrium* [117]. Combining elicitation with chitosan and precursor feeding with squalene was used to produce 27.49 mg/g DW withanolides [118].

Feeding the culture medium with organic compounds, such as vitamins or amino acids enhanced *in vitro* production of many secondary compounds. In callus and cell suspension cultures of *Centella asiatica,* amino acid feeding enhanced the production of triterpenes and asiaticoside [96]. Also, valine, threonine, and isoleucine enhanced adhyperforin production in shoot cultures of *Hyraceum perforatum* [119]. Feeding the suspension cultures of *C. roseus* with L- tryptophane or L-glutamine resulted in the production of the highest value of cell mass and indole alkaloids production [120]. Feeding the culture medium of *Spilanthes acmella* with casein hydrolysate and L-phenylalanine promoted biomass and scopoletin production [121]. Feeding squalene into culture medium of *C. asiatica* calli promoted production of madecassoside and asiaticoside [96]. In *Solanum lyratum* cell cultures, feeding with sterols such as cholesterol, stigmasterol or mixed sterols promoted the biosynthesis of solasodine, solasonidine, and solanine without effect on culture biomass [122].

The yield of salidroside was improved by feeding Rhodiola genus plants with an appropriate concentration of precursors and elicitors such as precursors, phenylalanine, tyrosol, and tyrosine [123]. Tyrosol feeding (0.5 mM) expressed the most obvious effect on salidroside content in the cell suspension cultures of *R. sachalinensis* [124]. When feeding the culture medium with precursors promoted the production of secondary metabolites without biomass accumulation, it needs a combination between precursors and elicitors to overcome the obstacle. This strategy was used to enhance the biosynthesis of sennoside A and B in callus cultures of *Cassia augustifolia* [125].

5.5 High-yielding cell lines selection

Genetic diversity within medicinal plants has great importance and can be used for plant improvement and the selection of an elite line. The selection of high biomass and metabolite(s) producing cell lines plays an important role in optimizing the productivity of *in vitro* cultivated plant materials. The yield of biomass and active metabolites may vary within varieties, genotypes, or populations of plant species [See 14]. The genotype has direct effects on the ability of the plant to produce valuable biomass and pharmaceutical compounds. To avoid high coast, the genotype with high yield and secondary metabolites contents should be carefully selected. For example, wright selection of *Pilocarpus microphyllus* resulted in the production of pilocarpine content ranging from 16.3 to 235.9 μg g^{-1} in dry weight [126], it was 15 times higher than the content found in wild plants.

To get a high yield of metabolites, Briskin [127] described the biotechnological methods for the selection of high-yielding cell lines in medicinal plants by addressing several topics, including media components, elicitation, immobilization, physical stress, and transformation. This means that the identification and establishment of high producing and fast-growing *in vitro* cultures are essential prerequisites, especially when the target secondary metabolite content of the selected cell line should be high. Selecting the higher-yielding cell lines was the essential step for optimizing the production of the anticancer drugs camptothecin [128].

Qualitative and quantitative estimation of active metabolites may show variability depending on the spatial and temporal changes that may happen during the process. Variation in secondary metabolites yield may be due to their repression or losses before or during the extraction processes. Consequently, the determined secondary metabolite value may not exactly indicate the actual content of secondary metabolite in a given tissue or plant species. Nevertheless, quantitative and qualitative methods can be applied to select high-yielding cell lines [14]. Selection of the high-yielding lines can be established by exposing the population of plant materials to toxic inhibitors, biosynthetic precursors, or stressful environments and followed by selecting cells that show higher production of targeted components [2]. Selection can be carried out using callus, cell suspension, or through any other *in vitro* culture procedure. In this regard, the answers to the following questions must be quite clear: Does diversity occur naturally or by using chemical, physical or biological substances that help in mutation to produce genetic diversity from which it can be selected? What are the methods used to identify and isolate the most qualitatively and quantitatively productive line?

5.6 Overexpression of genes that control the production of bioactive compounds

The production of secondary metabolites is a metabolic process that is influenced by several physicochemical factors. These factors can be controlled and optimized in large-scale production. Traditional mutagenesis programs have been used by the pharmaceutical industry for yield improvement of medicinal plants. Recently, the development of recombinant DNA technology has provided new and effective tools to obtain elite strains with high content of secondary metabolites through overexpression of specific enzymes involved in their biosynthetic pathways aiming to increase the production levels and speed the metabolic processes [67, 96]. Consequently, plant genetics, recombinant DNA technologies, and PTC have developed to improve the ability of several medicinal plants to biosynthesize secondary metabolites efficiently.

To control the synthesis of certain natural products, the enzymes involved in the synthesis of these reactions and how they are influenced by *in vitro* culture conditions should be carefully determined. Niggeweg et al. [129] identified the enzymes that control the pathway of synthesis of an important bioactive compound through controlling these pathways. This control can be investigated on a gene expression and genome level [1] but it is not enough because it does not always give clear and specific information on the nature of the encoded enzyme that controls the intended reaction. Consequently, genomic studies have been used in combination with physiological and biochemical aspects to understand the biosynthetic pathways of specific secondary metabolites [1]. In this concern, metabolic engineering strategies concentrate on the stimulation of certain pathways over others by overexpressing certain genes.

Using PTC, key gene overexpression that involved in the biosynthetic of valuable biologically active compounds can be controlled leading to produce compounds in high quantity and quantity. For example, the overexpression of geranyl diphosphate synthase and geraniol synthase genes in *C. roseus* led to a significant improvement in plant production from monoterpene indole alkaloids of vinblastine and vincristine [130]. In periwinkle cell lines, overexpression of the strictosidine synthase (Str) gene resulted in tenfold activity than wild type leading to the accumulation of high content of ajmalicine, strictosidine, serpentine, tabersonine, and catharanthine [131]. Overexpressing tryptophan decarboxylase (Tdc) gene resulted in accumulation of TIAs (serpentine, catharanthine, strictosidine) more than wild type in transgenic cell suspension culture of periwinkle [132]. In addition, overexpression of H6H (hyoscyamine 6β-hydroxylase) from *Hyoscyamus niger* in *Atropa belladonna* hairy roots enhanced scopolamine production [133]. In addition, suppression of the rosmarinic acid synthase gene led to an increase in the plant content of 3,4-dihydroxyphenyllactic acid which led to improving the quality of rosmarinic acid in *Salvia miltiorrhiza* [134].

Bioactive secondary metabolites are under coordinated control of the biosynthetic genes, and transcription factors (TFs) play an important role in this regulation [135]. Transcriptional regulation means the change in gene expression levels by modulation of transcription rates. Studies on the regulation of the production of secondary metabolite pathways are focused on the regulation of structural genes through TFs [135]. For example, the expression of genes involved in TIAs (terpenoid indole alkaloids, such as vincristine and vinblastine) metabolic pathway is elicited by jasmonates, it is regulated biosynthesis of terpenoid indole alkaloid (TIAs) and artemisinin [135]. Jasmonate was demonstrated as a regulator of deacetylvindoline 4-O-acetyltransferase (DAT) expression [136]. Expressed DAT is involved in the biosynthesis of TIAs member-vindoline through transferring an acetyl group to deacetylvindoline for vindoline production. It was clear that most of the genes codded for TIA pathway enzymes are tightly regulated by specific TFs under the regulation of JAs but it is carried out in coordination with developmental growth stage and environmental factors [135].

TFs of TIA genes respond to JAs and/or other elicitors. In *C. roseus* a few TFs (CrORCA2, CrORCA3, CrBPF1, CrWRKY1, CrMYC1, and CrMYC2) have been characterized, two of them (ORCA2 and ORCA3) are positively influenced by JAs [137].

ORCA2 plays a critical role in the regulation of TIA metabolism where it regulates gene expression of both feeder pathways as well as STR and SGD, genes that codded for enzymes catalyzing the first two steps in biosynthesis of TIA [138]. In addition, ORCA3 overexpression resulted in the increase of some genes such as TDC, STR, and desacetoxyvindoline- 4-hydroxylase (D4H) leading to the accumulation of vinblastine and other metabolites in the TIA pathway [139]. Other TF such WRKY family that is induced by JAs is involved in TIA biosynthesis [140]. In *Catharanthus* hairy roots, overexpression of CrWRKY1 results in up-regulation of TIA pathway genes, especially the TDC gene. TF-CrWRKY1 binds the TDC promoter resulting in and transactivation of the TDC promoter in *Catharanthus* cells [141]. Preferential expression of CrWRKY1 and its interaction with other TFs (including CrORCAs and CrMYCs) play an essential role in the accumulation of vinblastine in *C. roseus* [135].

5.7 Transformation

The genetic transformation was used as a powerful tool to improve the productivity of secondary metabolites. In general, *Agrobacterium rhizogenes* was used to transfer genes in several dicotyledonous plants where roots are formed at the site of infection; what is called "hairy roots." Agrobacterium-mediated transformation technology may be better than direct gene transfer techniques including particle bombardment and electroporation [129]. Transformed hairy roots mimic the biochemical machinery of normal roots and are used to produce secondary metabolites where they are stable and have high productivity under growth regulators free culture [88]. Hairy roots transformed systems have great potential for commercial production of viable secondary metabolites and become a good alternative for raw plant materials.

Gene transfer using *Agrobacterium* can possibly be used to transfer DNA fragments that contain the genes of interest at higher efficiencies and lower cost. In *Raphanus sativus* L., a medicinal plant, plants formed hairy roots using *A. rhizogenes*, it was associated with the production of higher content of phenolic flavonoid and quercetin content compared to non-transformed plants [142]. Hairy roots were used for the production of phenolic acid, flavonoid, and wedelolactone from *Sphagneticola calendulacea* [143], tropane alkaloids of hyoscyamine, anisodamine, and scopolamine from *Scopolia lurida* [144].

Bacopa monnieri was transformed using *A. tumefaciens* with tryptophan decarboxylase and strictosidine synthase genes, which were obtained from *C. roseus.* Transformed tissues showed an increase in the terpenoid indole alkaloid pathway which led to an increase of 25-fold in tryptophan content in comparison with non-transformants [145]. Sharma et al. [146] used *A. tumefaciens* to transfer tryptophan decarboxylase and strictosidine synthase genes to *C. roseus*, it increased the content of terpenoid indole alkaloid metabolite due to the transient overexpression of these genes. In addition, several medicinal plants were subjected to genetic transformation including *Iphigenia indica* [88], *Artemisia annua* [57], *Aconitum heterophyllum* [100], *P. somniferum* L. and *Eschscholzia californica* [147]. *Solanum aviculare* [148], *Pueraria phaseoloides* [149], *Crataeva nurvala* [150], *Gymnema sylvestre* [151] and *Holostemma ada-kodien* [152] and *Araujia sericirfera* and *Ceropegia spp* [153].

5.8 Scale-up production

The application of PTC in medicinal plants can be scaled up using "bioreactors," which allow atomization and production of a high yield of medicinal secondary products [154]. Therefore, scale-up production is a bioreactor application for the cultivation of plant cells on large-scale aiming for the mass production of valuable bioactive compounds. Also, bioreactor-based micropropagation was found to increase shoot multiplication for the commercial propagation of *B. monnieri* plants and maximize the content of bacosides in shoot biomass using an airlift bioreactor system [154]. Production of secondary metabolites using *in vitro* culture techniques is recommended strategy, especially when studying morphological and physiological processes associated with metabolites biosynthesis is necessary [155].

Cell suspension offers the wright combination of physical and chemical environments that must be used in the large-scale production of secondary metabolites in the bioreactor process [156]. Consequently, scale-up production in the bioreactor was used to expand the production of secondary metabolites from research to the industrial level. Systems of various sizes and features of bioreactors were created and applied for the mass production of secondary metabolites [157]. The application of plant tissue culture techniques in bioreactors for scale-up production facilitates obtaining some expensive pharmaceuticals that are synthesized in low quantity during *in vitro* or *in vivo* cultures. Since scale-up production of skikonin substance was achieved using bioreactors by Tabata and Fujita [158], other successful scale-up productions were obtained such as ginseng [159] and taxol [160].

Bioreactor operating system should provide efficient oxygen and nutrient supply, homogenous distribution of cultivated plant materials, and other factors that ensure optimal biomass and metabolite production [161]. While most of these bioreactors rely on cell suspension cultures, few of which are rely on differentiated tissues such as somatic embryos and hairy roots [162]. Application of suspension culture facilitates metabolites isolation [157].

For scale-up production, automation becomes an essential prerequisite, where it controls the pH of the culture area, culture viscosity, osmolarity, temperature, redox potential, oxygen supply, production of carbon dioxide, nutrients, weight, and liquid levels, and follows the rate of cell density. This automation needs sensors and monitoring systems that ensure mass production of pharmaceuticals and monitoring of physical, chemical, and biological parameters [163].

Perfusion cultivation is a system where continuous feeding of fresh media into a bioreactor system and removal of cells-free media were carried out in a modified bioreactor. The aim of this type of bioreactor and perfusion cultivation is to scaling-up the production of pharmaceutical compounds using plant cell, tissue, and organ cultures. The perfusion system offers a great advantage where it overcomes nutrient depletion and accumulation of growth inhibitors within the cultivated system, and it resulted in the promotion of biomass and pharmaceutical compounds. Semi-continuous perfusion was established in *Anchusa officinalis* where it was carried out in the shake flasks with a manual exchange of media. It resulted in the promotion of more than two-fold cell density and rosmarinic acid production in comparison to batch cultures [164].

Advances in immobilization and scale-up production techniques increase the applications of plant cell cultures for the purpose of producing high added value secondary compounds such as compounds with chemotherapeutic or antioxidant properties. For example, cell cultures of *Plumbago rosea* were immobilized using an MS medium containing 10 mM $CaCl_2$ and calcium alginate for the production of important medicinal compounds, such as plumbagin [165]. Their studies indicated the impact of immobilization on the increased accumulation of plumbagin where immobilization in calcium alginate resulted in enhancement of plumbagin production up to three folds compared with that of control [156].

6. limitations over secondary metabolite production *in vitro*

In general, there are many factors that may hinder the application of PTC for various purposes in the field of medicinal and other plant species. The production of medicinal compounds using PTC has two important aspects—the amount of plant materials should be sufficient for the production of the target substance, as well as the quantity and quality of the produced substance. Hence, it is necessary to identify and avoid the conditions and phenomena that may negatively affect the growth efficiency of the *in vitro* cultured plant tissue.

6.1 Avoidance of secondary metabolites toxicity

Obstacles facing the production of medicinal compounds from wild or cultivated plants can be avoided by using cell and tissue cultures, but these compounds may be toxic to the *in vitro* cultured cells or tissues and result in retardation of plant material growth and metabolite yield. Consequently, the toxicity of any secondary metabolite should be assessed and culture conditions should be modulated to avoid the production obstacles. On the other side, the toxic effect of a secondary metabolite can be beneficially used for the treatment of some illnesses, for example, cancer [166].

Long-term culture can be used for the accumulation of desirable metabolite(s), but it can be a problematic and limiting factor that should be avoided by the application of certain techniques, such as medium enrichment or substitution in bioreactors [167]. These strategies include accumulation of metabolites in vacuoles, and other subcellular compartments or the exudation of metabolites into the culture medium [168]. The last strategy needs the application of additional techniques to decrease the concentration of the accumulated metabolite leading to further biosynthesis. It is accomplished by changing the medium of the culture manually or mechanically. In this regard, hairy roots were recommended, but not all secondary compounds are synthesized and accumulated in the roots [66].

6.2 Avoidance of low growth rate of cultured plant materials

While successful production of a wide range of valuable secondary metabolites can be obtained using unorganized callus or suspension cultures, the differentiated organ can be used but each of them may face some problems. The most important problems are the slow growth rate and somaclonal variation [84]. Consequently, the production of secondary compounds through the application of PTC techniques becomes unstable at a specific period. Generally, the problems facing the production of secondary metabolites using PTC can be easily solved by changing the culture conditions to avoid growth retardation and somaclonal variation [11]. Also, the application of PTC techniques in combination with other approaches could be used to avoid growth retardation and genetic variation [11].

The appropriate conditions for increasing the growth of the cultured plant materials may be different from the conditions for increasing the concentration of the active substance. To overcome these dilemmas, a two-step protocol is used, one of which provides optimal conditions for growth and the other provides optimal conditions to produce the active substance [12]. For example, while growth stimulators should be used during the growth phase, elicitors should be used to stimulate the biosynthesis of active compounds [169].

Accumulation of secondary metabolites is obtained under the influence of biotic or abiotic stress, but it retards the biological mass. To ensure a high yield of secondary metabolites, producers hope to conserve conditions to stimulate high biomass and biosynthesis of the targeted metabolite. Consequently, optimization of culture conditions to increase growth parameters or application of elicitors become an essential prerequisite [169].

6.3 Avoidance of problems constrain the application of transformation in the production of active compounds

Despite *Agrobacterium* is an essential tool for gene transformation; sometimes some technical problems retard its application in some plants, it depends on genotype and/or transformation technique [170]. On the other hand, many factors can affect the efficiency of *Agrobacterium*-mediated transformation such as *Agrobacterium*'s optical density [171], antibiotic [118] or acetosyringone concentrations, and inoculation time [172]. All these difficulties should be avoided for the successful application of transformation techniques in the field of secondary metabolites production.

6.4 Avoidance culture browning

In vitro cultured explants release phenol compounds, which are oxidized by polyphenol oxidase and turned the media brown [173]. In woody plants, phenolic exudation appears early during the excision of plants causing browning of the cultured medium [173]. Browning closes the base of explants and retards the movement of nutrients from the medium into the cultured plant materials leading to retardation of plant growth. To overcome tissue browning, antioxidants or phenol absorbents, such as ascorbic acid, glutathione, activated charcoal, and polyvinylpyrrolidone were used. Also, transferring explants into new culture media at regular intervals can control the negative effects of the browning phenomenon [173]. To overcome the browning effect in the culture media in *Glycyrrhiza inflata* cell cultures, cultures were optimized in a bioreactor containing maximum cell concentration [174]. Dark conditions help to reduce the browning problem may be due to the reduction of the activity of the enzymes concerned with phenols synthesis and oxidation [175, 176].

6.5 Avoidance of somaclonal variation of the cultured plant materials

Production of the secondary metabolites using the cell culture technique is low during the early stage of growth where high carbon utilization exists and is associated with enhancement of primary metabolism. On the other hand, the production of secondary metabolites is high at the late stage when carbon is less needed for the production of primary metabolism [14]. Prolonged the age of the cultured plant materials is necessary but it may be associated with genetic variation [47, 84]. Therefore, the enhancement of growth criteria of the cultured plant materials is not sufficient to confirm the optimization of *in vitro* culture techniques for the production of secondary metabolites, but also genetic stability at the DNA level of the cultured plant materials is an essential parameter. For example, regenerates with high genetic fidelity and improved chemical profile of endangered *C. spinosa* L were reported, where the two-fold increase in flavonoids content than that of wild plants was obtained using methyl jasmonate and BAP [105]. Plant material with genetic fidelity after propagated *in vitro* culture was detected and used for the isolation of 20-hydroxyecdysone and polypodine B [177]. Production of true to type regenerants in *Artemisia absinthium* is very important in the commercial production of secondary metabolites [178].

Somaclonal variation results from chromosomal changes in number or structure, transposable elements, or possibly pre-existing genetic changes in the donor plant. To detect somaclonal variation, several molecular techniques such as Random Amplified Polymorphic DNA (RAPD), Inter Simple Sequence Repeat (ISSR), and Simple Sequence Repeat (SSR) were recommended [18, 47].

6.6 Avoidance of vitrification

Sometimes, the production of secondary metabolites through some techniques such as cell suspension is not always an adequate procedure. Then, other techniques such as organ culture can be used as a supernumerary method for the production of secondary metabolites [85]. Shoot cultures as same as hairy root cultures are recommended for production of pharmaceuticals where they are genetically stable [179]. Shooty teratomas were produced for the production of secondary metabolites, such as vincristine in *C. roseus* [180] and naphthoquinone in *Drosera capensis* var. alba [181]. In some plant species, shoot culture showed vitrification problems, such as in moringa [47].

Generally, tissue culture plant materials were incubated in vessels to prevent microbial contamination and retard culture desiccation but these conditions may cause restriction of gases exchange between cultures and their surrounds. Under insufficient ventilation stress, the growth of the cultured plant materials was retarded due to retardation of photosynthesis, transpiration, and uptake of water and nutrients leading to the accumulation of ethylene and the appearance of vitrification or hyperhydricity [182]. The symptoms of vitrification are slowing growth rate, necrosis of shoot tips, loss of apical dominance, disorganized cell wall, fragile leaves, reduction of shoot multiplication, poor acclimatization, impaired stomatal function, reduction of some metabolites, alteration of ion composition, inhibition of H_2O_2 detoxification enzymes [183, 184].

Vitrification in medicinal and other plant species can be avoided by reducing the relative humidity and improving the aeration within culture vessels [183, 184], decreasing the concentration of free water by increasing the concentration of agar [185], and using anti-ethylene compounds including CoCl2, AgNO3 or salicylic acid [47, 183]. To confirm which anti-ethylene compounds can be used to conserve the genetic fidelity of *in vitro* cultured moringa shoots, fingerprinting profiles of the long-term culture (14 subcultures) were assessed using RAPD, SSR, and ISSR. While the application of silver nitrate improved plant multiplication and reduced vitrification but it resulted in higher somaclonal variation in comparison to salicylic acid [47].

7. Conclusion

An increase in the world's population imposed an important matter, which is the inevitability of leaving arable land for food production. Where modern agricultural techniques can be used to produce secondary metabolites and preserve the genetic assets of these plants, the most notable technique is PTC. In addition, different PTC techniques are used to propagate rare and endangered plant species. Changes in the physical and chemical conditions of *in vitro* culture are easy and under control in a way that cannot be provided at all under field conditions. The ease of controlling the conditions of PTC conditions made it possible to use certain conditions to obtain true-to-type clones and their products, but other conditions are used to establish somaclonal variation for noval line selection.

The use of plant tissue techniques has become dependent on it to produce pharmaceutical materials after laboratory and applied experiments have proven that *in vitro* cultured plant materials are able to produce pharmaceuticals with the same amount and quality that can be obtained from soil cultivated plants. Moreover, the application of elite physical and chemical conditions of *in vitro* cultured plant materials made their production of secondary metabolites superior in quantity and quality to that of wild or cultivated plants. Therefore, to produce pharmaceutical compounds in large quantities to suit the increase in the population and increase their demand for safe medical products, tissue and cell culture techniques have been improved under several names including culture media optimization, the establishment of suspension and callus cultures, elicitation to enhance the productivity of *in vitro* cultures, application of precursor feeding as a substrate to improve the production of secondary metabolites, high yielding cell lines selection, enhance the overexpression of genes that control the production of bioactive compounds, application of genetic transformation using *A. rhizogenes* and application of "bioreactors" for scale-up production.

The use of PTC techniques to produce pharmaceutical compounds depends on the availability of production of sufficient-viable plant biomass to produce pharmaceutical substances with the requested quality and quantity. Therefore, it is necessary to understand all the factors that limit the production of targeted mass to avoid them such as the toxicity of secondary metabolites, low growth rate of cultured plant materials, and problems that constrain the application of transformation on a wide spectrum of plant species, somaclonal variation during cell or tissue cloning and verification of the cultured plant organs.

Author details

Ahmed M. Hassanein
Central Laboratory of Genetic Engineering, Faculty of Science, Sohag University, Sohag, Egypt

*Address all correspondence to: ahmed.hassanein@science.sohag.edu.eg

References

[1] Oksman-Caldentey KM, Inzé D. Plant cell factories in the postgenomic era: New ways to produce designer secondary metabolites. Trends in Plant Science. 2004;**9**:433-440

[2] Rao SR, Ravishankar GA. Plant cell cultures: Chemical factories of secondary metabolites. Biotechnology Advances. 2002;**20**(2):101-153

[3] Folashade OK, Omoregie EH, Peter OA. Standardization of herbal medicines—A review. International Journal of Biodiversity and Conservation. 2012;**4**(3):101-112

[4] Gantait S, Debnath S, Ali MN. Genomic profile of the plants with pharmaceutical value. 3 Biotech. 2014;**4**:563-578

[5] Toledo VM. New paradigms for a new ethnobotany: Reflections on the case of Mexico. In: Schultes RE, Von Reis S, editors. Ethnobotany: Evolution of a Discipline. London, UK: Chapman and Hall; 1995. pp. 75-88

[6] FAO. Trade in Medicinal Plants. Raw Materials, Tropical and Horticultural Products Service Commodities and Trade Division, Economic and Social Department, Food and Agriculture Organization of the United Nations. 2004. Rome. Available from: http://www.fao.org/3/af285e/af285e00.pdf

[7] Nalawade SM, Tsay H. *In vitro* propagation of some important Chinese medicinal plants and their sustainable usage. In vitro cellular & developmental biology. Plant. 2004;**40**:143-154

[8] An C. *In Vitro* propagation of commonly used medicinal trees in Korea. Journal of Environmental Sciences. 2019;**35**:272-280

[9] Hassanein AM, Mazen AMA. Adventitious bud formation in *Alhagi graecorum*. Cell, Tissue and Organ Culture. 2001;**65**:31-35

[10] Hassanein AM, Galal DM, Soltan DM, Saad GK. Effect of medium strength and activated charcoal on *in vitro* shoot multiplication and growth of jojoba. Journal of Environmental Studies. 2015;**14**:81-90

[11] Thangavel K, Ebbie MG, Ravichandran P. Biotechnology and *In Vitro* conservation of Medicinal Plants. Annals of Plant Sciences. 2014;**3**(06)

[12] Cardoso JC, de Oliveira ME, de Cl Cardoso F. Advances and challenges on the *in vitro* production of secondary metabolites from medicinal plants. Horticultura Brasileira. 2019;**37**:124-132

[13] Sajc L, Grubisic D, Vunjak-Novakovic G. Bioreactors for plant engineering: An outlook for further research. Biochemical Engineering Journal. 2000;**4**:89-99

[14] Isah T, Umar S, Mujib A, Sharma MP, Rajasekharan PE, Zafar N, et al. Secondary metabolism of pharmaceuticals in the plant *in vitro* cultures: Strategies, approaches, and limitations to achieving higher yield. Plant Cell, Tissue and Organ Culture. 2018;**132**:239-265

[15] Hassanein AM, Soltan DM. *Solanum nigrum* is a model plant in tissue and protoplast culture. Biologia Plantarum. 2000;**43**:501-509

[16] Sagare AP, Lee YL, Lin TC, Chen CC, Tsay HS. Cytokinin-induced somatic embryogenesis and plant regeneration in *Corydalis yanhusuo* (Fumariaceae)—A

medicinal plant. Plant Science. 2000;**160**(1):139-147

[17] Hassanein AM, Faheed FA, Salem JM, El-nagish AA. Some important aspects in moringa micropropagation. Acta Agriculturae Slovenica. 2019;**113**(1):13-27. DOI: 10.14720/aas.2019.113.1.02

[18] Hassanein AM, Salem JM, Hassan RA, Soltan DM. Morpho-molecular characterization and differential totipotency of three *Solanum* taxa. Phyton (Horn, Austria). 2020;**60**:145-159

[19] Amoo SO, Aremu AO, van Staden J. *In vitro* plant regeneration, secondary metabolite production and antioxidant activity of micropropagated *Aloe arborescens* Mill. Plant Cell, Tissue and Organ Culture. 2012;**111**(3):345-358

[20] Manivannan A, Soundararajan P, Park YG, Jeong PR. *In vitro* propagation, phytochemical analysis, and evaluation of free radical scavenging property of *Scrophularia kakudensis* Franch tissue extracts. BioMed Research International. 2015;**2015**:11

[21] Kabera JN, Semana E, Mussa AR, He X. Plant secondary metabolites: Biosynthesis, classification, function and pharmacological properties. The Journal of Pharmacy and Pharmacology. 2014;**2**:377-392

[22] Shukla MR, Kibler A, Turi CE, Erland LAE, Sullivan JA, Murch SJ, et al. Selection and micropropagation of an Elite Melatonin Rich Tulsi (*Ocimum sanctum* L.) Germplasm Line. Agronomy. 2021;**11**:207

[23] Castro TC, Pelliccione VLB, Figueiredo MR, Soares RODA, Bozza MT, Viana VRC, et al. Atividade antineoplsica e tripanocida de Hovenia dulcis Thunb. Cultivada *in vivo* e *in vitro*. Brazilian Journal of Pharmacogenetics. 2002;**12**:96-99

[24] Mandal S, Parsai A, Tiwari PK, Nataraj M. The effect of additional additives on the axillary shoot micropropagation of medicinal plant *Aegle marmelos* (L.) Corrêa. World News of Natural Sciences. 2021;**34**:54-71

[25] Kothari S, Mishra V, Bharat S, Tonpay SD. Antimicrobial activity and phytochemical screening of serial extracts from leaves of *Aegle marmelos* (Linn.). Acta Polymer Pharmaceutical Drug Research. 2011;**68**(5):687-692

[26] Shinde S, Pooja R, Katewal PR, Shanbhag DD, Joseph KS, Murthy HN. *In Vitro* Propagation of *Artemisia japonica*. Journal of Herbs, Species & Medicinal Plants. 2017;**23**(1):36-43. DOI: 10.1080/10496475.2016.1256010

[27] Yao L, Zhu S, Liu W, Manzoor Z, Nisar M F, Li M. A comprehensive review of pharmacological and Analytical Aspects of Acacetin-Natural Resources for Human Health. Open access article—CC BY-NC-ND license 2021. Available from: http://creativecommons.org/licenses/by-nc-nd/4.0/

[28] Salem MM, Davidorf FH, Abdel-Rahman MH. *In vitro* anti-uveal melanoma activity of phenolic compounds from the Egyptian medicinal plant *Acacia nilotica*. Fitoterapia. 2011;**82**:1279-1284

[29] Sander VA, Corigliano MG, Clemente M. Promising plant-derived adjuvants in the development of coccidial vaccines. Frontier in Veterinary Science. 2019;**6**:20. DOI: 10.3389/fvets.2019.00020

[30] Gunasekaran B, Gothandam KM. A review on edible vaccines and their

prospects. Brazilian Journal of Medical and Biological Research. 2020;**53**(2). DOI: 10.1590/1414-431X20198749

[31] Yousif F, Hifnawy MS, Soliman G, Boulos L, Labib T, Mahmoud S, et al. Large-scale *in Vitro.* Screening of Egyptian Native and Cultivated Plants for Schistosomicidal Activity. Pharmaceutical Biology. 2007;**45**(6):501-510. DOI: 10.1080/13880200701389425

[32] El-Hallouty S, Fayad W, Meky N, Hasabo AA, editors. *In vitro* anticancer activity of some Egyptian plant extracts against different human cancer cell lines. International Journal of Pharm Tech Research. 2015, 2015;**8**(2):267-327

[33] Aboul-Enein AM, Abu El-Ela F, Shalaby EA, El-Shemy HA. Traditional medicinal plants research in Egypt: Studies of antioxidant and anticancer activities. Journal of Medicinal Plant Research. 2012;**6**(5):689-703

[34] Mathur S, Sharma S, Gupta MM, Kumar S. Evalution of an Indian germplasm collection of the medicinal plant *Bacopa monnieri* (L.) Pennell by use of multivariate approaches. Euphytica. 2003;**133**(3):255-265

[35] Srivastava SK, Srivastava N. *In Vitro* multiplication of *Paedaria foetida* L—A rare medicinal plant. Journal of Plant Biochemistry and Biotechnology. 2004;**13**:89-91

[36] Sood H, Chauhan HS. Development of a low cost micropropagation technology for an endangered medicinal herb (*Picorhiza kurroa*) of North-Western Himalayas. Journal of Plant Science. 2009;**4**(2):21-31

[37] Mathur S, Shekhawat GS, Batra A. Micropropagation of *Salvadora persica* Linn. via cotyledonary nodes. Indian Journal of Biotechnology. 2002;**1**(2):197-200

[38] Sambyal M, Dogra A, Koul S, Ahuja A. Rapid *in vitro* propagation of *Potentilla fulgens* Wall—A Himalayan alpine herb of medicinal value. Journal of Plant Biochemistry and Biotechnology. 2006;**15**:143-145

[39] Arockiasamy S, Prakash S, Ignacimuthu S. Direct organogenesis from mature leaves and petioles of *Eryngium foetidum* L. Biologia Plantarum. 2002;**45**:129-132

[40] Saavedra AM, Castrol T, Cordeiro L, Athayde T, Albarello N, Simoes C. *In vitro* propagation and cryopreservation of the medicinal species *Hovenia dulcis* Thunb. (Rhamnaceae). Plant Cell, Tissue and Organ Culture. 2021;**144**:577-591

[41] Meena MC, Patni V. *In vitro* clonal propagation of *Citrullus colocynthis* (Linn.) Scharad: Threatened medicinal plant. Plant Cell Biotechnology and Molecular Biology. 2007;**8**(3&4):147-152

[42] Gayathri D, Ramagopal G. *In vitro* propagation of *Amaryllis* and *Zephyranthes bulbous* species using household tissue culture medium. Plant Cell Biotechnology and Molecular Biology. 2007;**8**(3&4):131-136

[43] Sivasubramanian S, Vallinayagam S, Patric RD, Manickam VS. Microproagation of *Plectranthus vetiveroides* (Jacob) Sing & Sharma—A medicinal plant. Phytomorphology. 2002;**52**:55-59

[44] Geetha R, Gopal GV. Direct regeneration in *Glossocardia bosvallea*: A medicinal plant. Plant Cell Biotechnology and Molecular Biology. 2007;**8**:119-124

[45] Monthony AS, Page SR, Hesami M, Jones AMP. The past, present and future

of *Cannabis sativa* tissue culture. Planta. 2021;**10**:185. DOI: 10.3390/plants

[46] Shawky R, Mussa EA. *In vitro* propagation of the rare *Caralluma retrospeciens* plant in Egypt. International Journal of Advanced Research. 2020;**8**(8):812-818

[47] Hassanein AM, Salem JM, Faheed FA, El-nagish AA. Effect of anti-ethylene compounds on isoenzyme patterns and genome stability during long term culture of Moringa oliefera. Plant Cell, Tissue and Organ Culture. 2018;**132**:201-212

[48] Ghareb HE. *In vitro* propagation of the rare *Pulicaria incisa*. Journal of Basic Applied Science and Research. 2018;**8**(7):26-33

[49] Darwish HY, Ahmed SM. Elicitors enhancing phenolics content and related gene expression variation in petal—Derived calli of *Rosa damascena* Mill. Egyptian Journal of Bottomline. 2020;**60**(1):71-79

[50] Puhan P, Rath SP. *In vitro* propagation of *Aegle marmelos* (L.) corr., a medicinal plant through axillary bud multiplication. Advances in Bioscience and Biotechnology. 2012;**3**:121-125

[51] Liu CZ, Murch SJ, El-Demerdash M, Saxena PK. Regeneration of the Egyptian medicinal plant Artemisia judaica L. Plant Cell Reports. 2003;**21**:525-530

[52] Abed Elmaksood WM, Ebad FA, Bosila HA. *In vitro* propagation of the endangered medicinal plant *Hyoscyamus muticus* L. (Egyptian Henbane). Journal of Applied Environmental Biological Science. 2016;**6**(4):25-34

[53] Ahuja S, Mandal BB, Dixit S, Srivastava PS. Molecular, phenotypic and biosynthetic stability in *Dioscorea floribunda* plants derived from cryopreserved shoot tips. Plant Science. 2002;**163**:971-977

[54] Dixit-Sharma S. Cryopreservation of somatic embryos. Indian Journal of Biotechnology. 2005;**4**:47-55

[55] Lalaleo L, Testillano P, Risueňo MC, Cusidó RM, Palazon J, Alcazar R, et al. Effect of *in vitro* morphogenesis on the production of podophyllotoxin derivatives in callus cultures of *Linum album*. Journal of Plant Physiology. 2018;**228**:47-58

[56] Espinosa-Leal CA, Puente-Garza A, García-Lara CA. *In vitro* plant tissue culture: Means for production of biological active compounds. Planta. 2018;**248**:1-18

[57] Cai G, Li G, Ye H, Li G. Hairy root culture of *Artemisia annua* L. by Ri plasmid transformation and biosynthesis of artemisinin. Chinese Journal of Biotechnology. 1995;**11**:227-235

[58] Sreekumar S, Seeni S, Pushpangadan P. Production of 2 – hydroxy 4- methoxy benzaldehyde using root cultures of *Hemidesmus indicus*. Biotechnology Letters. 1998;**20**(7):631

[59] Vanisree M, Tsay HS. Studies on the production of some important secondary metabolites from medicinal plants by plant tissue cultures. Botanical Bulletin of Academia Science. 2004;**45**:1-22

[60] Hirata K, Yamanaka A, Kurano N, Miyamota K, Miura Y. Production of indole alkaloids in multiple shoot culture of (*Catharanthus roseus* 9L.) G. Don. Agricultural and Biological Chemistry. 1987;**5**:11311-11317

[61] Akashi T, Furuno T, Takahashi T, Ayabe S. Biosynthesis of triterpenoids in cultured cells regenerated and wild

plant organs of *Taraxacum officinale*. Phytochemistry. 1994;**36**:303-308

[62] Alamgir ANM. Therapeutic medicinal plants and their extracts. Pharmacognosy. 2017;**73**:403-426

[63] Sathuluri RR, Gokare R. Plant cell cultures: Chemical factories of secondary metabolites. Biotechnology Advances. 2002;**20**(2):101-153. DOI: 10.1016/S0734-9750(02)00007-1

[64] Chang WD, Huang WW, Chen CC, Chang YS, Tsay HS. The production of secondary metabolites from Chinese medicinal herbs by suspension cell and tissue culture. In: Proc. 7th Int. Congr of SABRAO and WASS. Taipei, Taiwan; 1994. pp. 535-540

[65] Figueiro ADA, Correa CM, Astarita LV, Santarem ER. Long-term maintenance of the *in vitro* cultures affects growth & secondary metabolism of St. John's Wort. Ciencia Rural. 2010;**40**(10):2115-2121

[66] Kim W, Pamela J. Weathers Secondary metabolism of hairy root cultures in bioreactors. Plant. 2002;**38**:1-10

[67] Bharathi P, Philomina D. Effect of nutritional factors and precursors on the formation of colchicine in *Gloriosa superba in vitro*. Research Biotechnology. 2010;**1**:29-37

[68] Hassanein AM, Galal A, Saad G, Soltan DM. Impact of slight changes in water potential of culture media on *in vitro* shoot multiplication, esterase, and protein patterns of *Simmondsia chinensis* L. Egyptian Journal of Botany. 2022;**62**(1):97-111. DOI: 10.21608/ejbo.2021.46723.1571

[69] Trejo-Tapia G, Arias-Castro C, Rodrguez-Mendiola M. Influence of the culture medium constituents & inoculum size on the accumulation of blue pigment & cell growth of Lavandula spica. Plant Cell, Tissue and Organ Culture. 2003;**72**:7-12

[70] Gandhi SG, Mahajan V, Bedi YS. Changing trends in biotechnology of secondary metabolism in medicinal and aromatic plants. Planta. 2015;**241**:303-317

[71] Monfort LEF, Bertolucci SKV, Lima AF, de Carvalho AA, Mohammed A, Blank AF, et al. Effects of plant growth regulators, different culture media and strength MS on production of volatile fraction composition in shoot cultures of *Ocimum basilicum*. Industrial Crops and Products. 2018;**116**:231-239

[72] Oliveira JPS, Hakimi O, Murgu M, Koblitz MGB, Ferreira MSL, Cameron LC, et al. Tissue culture and metabolome investigation of a wild endangered medicinal plant using high definition mass spectrometry. Plant Cell Tissue Organ Culture. 2018;**134**:153-162

[73] Hassanein AM. Hormonal requirements trigger different regeneration pathways in *Alhagi Graecorum*. Journal of Plant Biotech. 2004;**6**:171-179

[74] Raj D, Kokotkiewicz A, Drys A, Luczkiewicz M. Effect of plant growth regulators on the accumulation of indolizidine alkaloids in *Securinega suffruticosa* callus cultures. Plant Cell, Tissue and Organ Culture. 2015;**123**(1):39-45. DOI: 10.1007/s11240-015-0811-6

[75] Jang HR, Lee HJ, Park BJ, et al. Establishment of embryogenic cultures and determination of their bioactive properties in *Rosa rugosa*. Horticulture, Environment and Biotechnology. 2016;**57**(3):291-298

[76] Isah T, Umar S. Influencing *in vitro* clonal propagation of *Chonemorpha fragrans* (moon) Alston by culture media strength, plant growth regulators, carbon source and photoperiodic incubation. Journal of Forest Research. 2020;**31**(1):27-43

[77] Karwasara VS, Dixit VK. Culture medium optimization for camptothecin production in cell suspension cultures of *Nothapodytes nimmoniana* (J. Grah.) Mabberley. Plant Biotechnology Reports. 2013;7(3):357-369

[78] Thanh NT, Van KN, Paek KY. Effecting of medium composition on biomass and ginsenoside production in cell suspension culture of *Panax vietnamensis* Haet Grushv. VNUJ Science Natural Technology. 2007;**23**:269-274

[79] Irshad M, Debnath B, Mitra S, Arafat Y, Li M, Sun Y, et al. Accumulation of anthocyanin in callus cultures of red-pod okra (*Abelmoschus esculentus* (L) Hongjiao) in response to light and nitrogen levels. Plant Cell Tissue and Organ Culture. 2018;**134**:29-39

[80] Hagendoom JM, Wagner AM, Segers C, van der Plas LHW, André Oostdam A, van Walraven HS. Cytoplasmic acidification and secondary metabolite production in different plant cell suspensions. Plant Physiology. 1994;**106**:723-730

[81] Majerus F, Pareilleus A. Production of indole alkaloids by gelentrapped cells of *Catharanthus roseus* cells in a continuous flow reactor. Biotechnology Letters. 1986;**8**:863-866. DOI: 10.1007/BF01078646

[82] Nagella P, Murthy HN. Establishment of cell suspension cultures of *Withania somnifera* for production of withanolide A. Bioresource Technology. 2010;**101**(17):6735-6739. DOI: 10.1016/j

[83] Naik PM, Manohar SH, Praveen N, Murthy HN. Effects of sucrose & pH levels on the *in vitro* shoot regeneration from the leaf explants of *Bacopa monnieri* and accumulation of bacoside A in regenerated shoots. Plant Cell, Tissue and Organ Culture. 2010;**100**(2):235-239. DOI: 10.1007/s11240-009-9639-2

[84] Hassanein AM, Ahmed AM, Soltan DM. Study of somaclonal variation and gene expression as influenced by long term culture in sorghum. Current Opinion in Biotechnology. 2008;**4**:13-20

[85] Fazili MF, Bashir I, Ahmad M, Yaqoob U, Geelani SN. *In vitro* strategies for the enhancement of secondary metabolite production in plants: A review. Bulletin of the National Research Centre. 2022;**46**:35

[86] Pan ZG, Liu CZ, Murch SJ, El-Demerdash M, Saxena PK. Plant regeneration from mesophyll protoplasts of the Egyptian medicinal plants, *Artemisia judiica* L. and *Echinops spinosissimum* (Turra). Plant Science. 2003;**165**:681-687

[87] Kurz WGW, Constabel F. Production of secondary metabolites. In: Altman A, editor. Agricultural Biotechnology. New York: Marcel Dekker Inc; 1998. pp. 183-224

[88] Mukhopadhyay MJ, Mukhopadhyay S. Biotechnological Approach for enhancement of Colchicine Accumulation in *Iphigenia indica* Kunth. Journal of Plant Biochemistry and Biotechnology. 2008;**17**(2):185-188

[89] Feher A, Pasternak T, Otvos K, Miskolczi P, Dudits D. Induction of embryogenic competence in somatic plant cells: A review. Biologia (Bratislava). 2002;**57**:5-12

[90] Raghavan V, Srivastava PS. Embryo culture. In: Jhori JM, editor.

Experimental Embryology of Vascular Plants. Berlin, Heidelberg, New York: Springer Verlag; 1982. pp. 195-230

[91] Roberts SC, Shule ML. Large-scale plant cell culture. Current Opinion in Biotechnology. 1997;**8**(2):154-159. DOI: 10.1016/s0958-1669(97)80094-8

[92] Gontier E, Sangwan BS, Barbotin JN. Effects of calcium, alginate, and calcium-alginate immobilization on growth and tropane alkaloid levels of a stable suspension cell line of *Datura innoxia* Mill. Plant Cell Reports. 1994;**13**:533-536

[93] Gupta K, Garg S, Singh J, Kumar M. Enhanced production of naphthoquinone metabolite (shikonin) from the cell suspension culture of *Arnebia* sp. & its up-scaling through bioreactor. Biotech. 2014;**4**(3):263-273. DOI: 10.1007/s13205-013-0149-x

[94] Tom R, Jardin B, Chavarie C, Rho D, Archambault J. Effect of culture process on alkaloid production by *Catharanthus roseus* cells: II. Immobilized cultures. Journal of Biotechnology. 1991;**21**(1-2):21-42

[95] Wu CH, Murthy HN, Hahn EJ, Paek KY. Establishment of adventitious root co-culture of Ginseng & *Echinacea* for the production of secondary metabolites. Acta Physiologiae Plantarum. 2008;**30**(6):891-896. DOI: 10.1007/s11738-008-0181-3

[96] Kiong AL, Mahmood M, Fodzillan NM, Daud SK. Effects of precursor supplementation on the production of triterpenes by *Centella asiatica* callus culture. Pakistan Journal of Biological Sciences. 2005;**8**:1160-1169. DOI: 10.3923/pjbs.2005.1160.1169

[97] Giri CC, Zaheer M. Chemical elicitors versus secondary metabolite production *in vitro* using plant cell, tissue and organ cultures: Recent trends and a sky eye view appraisal. Plant Cell, Tissue and Organ Culture (PCTOC). 2016;**126**:1-18

[98] Zhao J, Liu W, Wang J. Recent advances regarding constituents and bioactivities of plants from the genus *Hypericum*. Chemistry & Biodiversity. 2015;**12**:309-349

[99] Garcia-Perez I, Posma JM, Chambers ES, Mathers JC, Draper J, Beckmann M, et al. Dietary metabotype modelling predicts individual responses to dietary interventions. Nature Food. 2020;**1**:355-364. DOI: 10.1038/s43016-020-0092-z

[100] Giri A, Banerjee S, Ahuja PS, Giri CC. Production of hairy roots in *Aconitum heterophyllum* Wall. using *Agrobacterium rhizogenes*. *In vitro* Cell Development Biology. 1997;**33**:293-306

[101] Darvishi E, Kahrizi D, Bahraminejad S, Mansouri M. *In vitro* induction of α-pinene, pulegone, menthol, menthone and limonene in cell suspension culture of pennyroyal (*Mentha pulegium*). Cellular and Molecular Biology. 2016;**62**:7-9

[102] Castilho CVV, Neto JFF, Leitão SG, Barreto CS, Pinto SC, da Silva NCB. *Anemia tomentosa* var. anthriscifolia *in vitro* culture: Sporophyte development and volatile compound profile of an aromatic fern. Plant Cell Tissue and Organ Culture. 2018;**133**:311-323

[103] Staniszewska I, Królicka A, Maliński E, Łojkowska E, Szafranek J. Elicitation of secondary metabolites in *in vitro* cultures of *Ammi majus* L. Enzyme and Microbial Technology. 2003;**33**:565-568

[104] Jin JH, Shin JH, Kim JH, Chung IS, Lee HJ. Effect of chitosan elicitation and media components on the

production of anthraquinone colorants in madder (*Rubia akane* Nakai) cell culture. Biotechnology and Bioprocess Engineering. 1999;**4**:300-304

[105] Goda SM, Ahmed SA, El Sherif F, Hassanean HA, Ibrahim AK. Genetically stable plants with boosted flavonoids content after *in vitro* regeneration of the endangered Capparis spinosa L. Global Drugs Therapy. 2017;**2**(4):1-7

[106] Bhagwath SG, Hjortso MA. Statistical analysis of elicitation strategies for thiarubrine A production in hairy root cultures of *Ambrosia artemisiifolia*. Journal of Biotechnology. 2000;**80**(2):159-167. DOI: 10.1016/S0168-1656(00)00256-X

[107] Isah T. Stress and defense responses in plant secondary metabolites production. Biological Research. 2019;**52**:39

[108] García-Pérez P, Lozano-Milo E, Landín M, Gallego PP. Combining medicinal plant *in vitro* culture with machine learning technologies for maximizing the production of phenolic compounds. Antioxidants. 2020;**9**:210

[109] Blasco B, Navarro-León E, Ruiz JM. Oxidative Stress in Relation with Micronutrient Deficiency or Toxicity. Amsterdam, The Netherlands: Elsevier Inc; 2018

[110] Sharma A, Shahzad B, Rehman A, Bhardwaj R, Landi M, Zheng B. Response of phenylpropanoid pathway and the role of polyphenols in plants under abiotic stress. Molecules. 2019;**24**:2452

[111] Jeong GA, Park DH. Enhanced secondary metabolite biosynthesis by elicitation in transformed plant root system: Effect of abiotic elicitors. Applied Biochemistry and Biotechnology. 2006;**129**:436-446

[112] Tonk D, Mujib A, Maqsood M, Ali M, Zafar N. *Aspergillus flavus* fungus elicitation improves vincristine and vinblastine yield by augmenting callus biomass growth in *Catharanthus roseus*. Plant Cell, Tissue and Organ Culture. 2016;**126**:291-303

[113] Kowalczyk T, Merecz-Sadowska A, Picot L, Brčič Karaċonji I, Wieczfinska J, Sliwiński T, et al. Genetic manipulation and bioreactor culture of plants as a tool for industry and its applications. Molecules. 2022;**27**:795. DOI: 10.3390/molecules27030795

[114] Jackson P, Attalla MI. *N*-Nitrosopiperazines form at high pH in post-combustion capture solutions containing piperazine: A low-energy collisional behavior study. Rapid Communications in Mass Spectrometry. 2010;**24**(24):3567-3577. DOI: 10.1002/rcm

[115] Jaisi A, Panichayupakaranant P. Enhanced plumbagin production in *Plumbago indica* root cultures by l-alanine feeding and in situ adsorption. Plant Cell, Tissue and Organ Culture. 2016;**38**(2):351-355. DOI: 10.1007/s11240-016-1155-6

[116] Rahimi S, Hasanloo T, Najafi F, Khavari-Nejad RA. Enhancement of silymarin accumulation using precursor feeding in *Silybum marianum* hairy root cultures. Plant Omics Journal. 2011;**4**(1):34-39

[117] Tumova L, Rimakova J, Tuma J, Dusck J. *Silybum marianum in vitro*-flavolignan production. Plant, Cell & Environment. 2006;**52**:454-458

[118] Sivanandhan G, Selvaraj N, Ganapathi A, Manickavasagam M. Enhanced biosynthesis of withanolides by elicitation and precursor feeding in cell suspension culture of *Withania somnifera* (L.) Dunal in Shake-Flask

Culture and Bioreactor. PLoS One. 2014;**9**:e104005

[119] Karppinen K, Hokkanen J, Tolonen A, Maltila S, Hohtola A. Biosynthesis of hyperforin & adhyperforin from amino acid precursors in the shoot cultures of *Hypericum perforatum*. Phytochemistry. 2007;**68**:1038-1045. DOI: 10.1016/j.phytochem.2007.01.001

[120] Taha H, El-Bahr MK, Seif-Nasr MM. *In vitro* studies on Egyptian *Catharanthus roseus* (L.) G.Don. IV: Manipulation of Some Amino Acids as Precursors for Enhanced of Indole Alkaloids Production in Suspension Cultures. Australian Journal of Basic and Applied Sciences. 2009;**3**(4):3137-3144

[121] Abyari M, Nasr N, Soorni J, Sadhu D. Enhanced accumulation of scopoletin in cell suspension culture of *Spilanthes acmella* Murr. using precursor feeding. Brazilian Archives of Biology and Technology. 2016;**59**:e16150533

[122] Lee MH, Cheng JJ, Lin CY, Chen YJ, Lu MK. Precursor-feeding strategy for production of solanine, solanidine, & solasodine by a cell culture of *Solanum lyratum*. Process Biochemistry. 2007;**42**(5):899-903

[123] Liu Y, Wang J, Wang L, Zhang C, Bai Q, Lyu X, et al. Biosynthesis and biotechnological production of salidroside from *Rhodiola* genus plants. Phytochemistry Reviews. 2022. DOI: 10.1007/s11101-021-09800-1

[124] Xu JF, Liu CB, Han AM, Feng PS, Su ZG. Strategies for the improvement of salidroside production in cell suspension cultures of *Rhodiola sachalinensis*. Plant Cell Reports. 1998;**17**:288-293. DOI: 10.1007/s002990050394

[125] Chetri SK, Kapoor H, Agrawal V. Marked enhancement of sennoside bioactive compounds through precursor feeding in *Cassia angustifolia* Vahl & cloning of isochorismate synthase gene involved in its biosynthesis. Plant Cell, Tissue and Organ Culture. 2016;**124**(2):431-446. DOI: 10.1007/s11240-015-0905-1

[126] Sandhu SS, Abreu IN, Colombo CA, Mazaferra P. Pilocarpine content and molecular diversity in jaborandi. Scientia Agrária. 2006;**63**:478-482

[127] Briskin DP. Biotechnological methods for selection of high-yielding cell lines and production of secondary metabolites in medicinal plants. In: Kayser O, Quax WJ, editors. Medicinal Plant Biotechnology. Weinheim: Wiley; 2007. pp. 187-201

[128] Isah T. Production of camptothecin in the elicited callus cultures of *Nothapodytes nimmoniana* (J.Graham) mabberly. Chemical Papers. 2017;**71**(6):1091-1106. DOI: 10.1007/s11696-016-0056-9

[129] Niggeweg R, Michael AJ, Martin C. Engineering plants with increased levels of the antioxidant chlorogenic acid. Nature Biotechnology. 2004;**22**(6):746-754

[130] Kumar S, Singh A, Kumar B, Singh B, Bahadur L, Lal M. Simultaneous quantitative determination of bioactive terpene indole alkaloids in ethanolic extracts of *Catharanthus roseus* (L.) G. Don by ultrahigh performance liquid chromatography-tandem mass spectrometry. Journal of Pharmaceutical and Biomedical Analysis. 2018;**151**:32-41

[131] Canel C, Lopes-Cardoso MI, Whitmer S, van der Fits L, Pasquali G, van der Heijden R, et al. Effects of over-expression of strictosidine synthase and tryptophan decarboxylase on alkaloid production by cell cultures

of *Catharanthus roseus*. Planta. 1998;**205**:414-419

[132] Whitmer S, van der Heijden R, Verpoorte R. Effect of precursor feeding on alkaloid accumulation by a tryptophan decarboxylase over-expressing transgenic cell line T22 of *Catharanthus roseus*. Journal of Biotechnology. 2002;**96**:193-203

[133] Hashimoto T, Yun DJ, Yamada Y. Production of tropane alkaloids in genetically engineered root cultures. Phytochemistry. 1993;**32**:713-718

[134] Zhou Z, Tan H, Li Q, Chen J, Gao S, Wang Y, et al. CRISPR/Cas9-mediated efficient targeted mutagenesis of RAS in *Salvia miltiorrhiza*. Phytochemistry. 2018;**148**:63-70

[135] Afrin S, Huang J, Luo Z. JA-mediated transcriptional regulation of secondary metabolism in medicinal plants. Scientific Bulletin. 2015;**60**(12):1062-1072

[136] Goldhaber-Pasillas GD, Mustafa NR, Verpoorte R, et al. Jasmonic acid effect on the fatty acid and terpenoid indole alkaloid accumulation in cell suspension cultures of *Catharanthus roseus*. Molecules. 2014;**19**:10242-10260

[137] Tang KX, Liu DH, Wang YL, et al. Overexpression of transcriptional factor ORCA3 increases the accumulation of catharanthine and vindoline in *Catharanthus roseus* hairy roots. Russian Journal of Plant Physiology. 2011;**58**:415-422

[138] Li CY, Leopold AL, Sander GW, Shanks JV, Zhao L, Gibson S. The ORCA2 transcription factor plays a key role in regulation of the terpenoid indole alkaloid pathway. BMC Plant Biology. 2013;**13**:155

[139] Pan Q, Wang Q, Yuan F, et al. Overexpression of ORCA3 and G10H in *Catharanthus roseus* plants regulated alkaloid biosynthesis and metabolism revealed by NMR-metabolomics. PLoS One. 2012;**7**:e43038

[140] Suttipanta N, Pattanaik S, Kulshrestha M, et al. The transcription factor CrWRKY1 positively regulates the terpenoid indole alkaloid biosynthesis in *Catharanthus roseus*. Plant Physiology. 2011;**157**:2081-2093

[141] Schluttenhofer C, Pattanaik S, Patra B, et al. Analyses of *Catharanthus roseus* and *Arabidopsis thaliana* WRKY transcription factors reveal involvement in jasmonate signaling. BMC Genomics. 2014;**15**:502-522

[142] Balasubramanian M, Anbumegala M, Surendran R, Arun M, Shanmugam G. Elite hairy roots of *Raphanus sativus* (L.) as a source of antioxidants and flavonoids. 3 Biotech. 2018;**8**:128

[143] Kundu S, Salma U, Ali MN, Hazra AK, Mandal N. Development of transgenic hairy roots and augmentation of secondary metabolites by precursor feeding in *Sphagneticola calendulacea* (L.) Pruski. Industrial Crops and Products. 2018;**121**:206-215

[144] Lan X, Zeng J, Liu K, Zhang F, Bai G, Chen M, et al. Comparison of two hyoscyamine 6β-hydroxylases in engineering scopolamine biosynthesis in root cultures of *Scopolia lurida*. Biochemical and Biophysical Research Communications. 2018;**497**:25-31

[145] Sharma A, Verma N, Verma P, Verma RK, Mathur A, Mathur AK. Optimization of a *Bacopa monnieri*-based genetic transformation model for testing the expression efficiency of pathway gene constructs of medicinal crops. *In*

Vitro Cell Development and Biological Plant. 2017;**53**:22-32

[146] Sharma A, Verma P, Mathur A, Mathur AK. Genetic engineering approach using early Vinca alkaloid biosynthesis genes led to increased tryptamine and terpenoid indole alkaloids biosynthesis in differentiating cultures of *Catharanthus roseus*. Protoplasma. 2018;**255**:425-435

[147] Park SU, Facchini PJ. *Agrobacterium rhizogenes* mediated transformation of opium poppy, *Papaver somniferum* and California poppy, *Eschscholzia californica* cham root cultures. Journal of Experimental Botany. 2000;**51**:1005-1016

[148] Argolo AC, Charlwood BV, Pletsch M. The regulation of solasodine production by *Agrobacterium rhizogenes* transformed roots of *Solanum aviculare*. Planta Medica. 2000;**66**:448-451

[149] Shi HP, Kintzios S. Genetic transformation of *Pueraria phaseoloides* with *Agrobacterium rhizogenes* and puerarin production in hairy roots. Plant Cell Reports. 2003;**21**:1103-1107

[150] Inamdar JA, Natraj M, Mohan JSS, Subramanian RB. Somatic embryogenesis from callus cultures of *Crataeva nurvala* Buch. Ham. Phytomorphology. 1990;**40**(3):280-284

[151] Kumar HGA, Murthy HN, Paek KY. Somatic embryogenesis and plant regeneration in *Gymnema sylvestre*. Plan Cell Tissues Organs. 2002;**71**:85-88

[152] Martin KP. Plant regeneration through somatic embryogenesis on *Holostemma ada-Kodian*, a rare medicinal plant. Plant Cell Tissue Organs. 2003;**72**:79-82

[153] Patil MV. Micropropagation studies in *Ceropegia spp*. *In Vitro* Cell Development. 1998;**34**:240-243

[154] Saha PS, Sarkar S, Jeyasri R, Muthuramalingam P, Ramesh M, Jha S. *In Vitro* Propagation, Phytochemical and Neuropharmacological Profiles of *Bacopa monnieri* (L.) Wettst.: A review. Plants. 2020;**9**:411. DOI: 10.3390/plants9040411

[155] Tejavathi DH, Rajanna MD, Sowmya R, Gayathramma K. Induction of somatic embryos from cultures of *Agave Veracruz* Mill. In Vitro Cellular & Developmental Biology. 2007;**43**:423-428. DOI: 10.1007/s11627-007-9088-8

[156] Hussain MS, Fareed S, Ansari S, Rahman A, Ahmad IZ, Saeed M. Current approaches toward production of secondary plant metabolites. Journal of Pharmacy & Bioallied Sciences. 2012;**4**(1):10-20. DOI: 10.4103/0975-7406.92725

[157] Park SY, Paek KY. Bioreactor culture of shoots and somatic embryos of medicinal plants for production of bioactive compounds. In: Paek KY, Murthy HN, Zhong JJ, editors. Production of Biomass and Bioactive Compounds Using Bioreactor Technology. New York: Springer; 2014. pp. 337-368

[158] Tabata M, Fujita Y. Production of shikonin by the plant cell cultures. In: Zatlin M, Day P, Hollaender A, editors. Biotechnology in the Plant Science. Cambridge: Academic Press; 1985. pp. 207-218

[159] Hibino K, Ushiyama K. Commercial production of ginseng by the plant tissue culture technology. In: Fu TJ, Singh G, Curtis WR, editors. Plant Cell and Tissue Culture for Production of Food Ingredients. New York: Springer; 1999. pp. 215-224

[160] Phyton Biotech. Capacity, reliability and quality in taxane API supply. 2014. Available from: https://phytonbiotech.com/apis/docetaxel/2014

[161] Steingroewer J, Bley T, Georgiev V, et al. Bioprocessing of differentiated plant *in vitro* systems. Engineering in Life Sciences. 2013;**13**(1):26-38. DOI: 10.1002/elsc.201100226

[162] Ferrie AM. Protocols for *in vitro* cultures and secondary metabolite analysis of aromatic and medicinal plants. Annals of Botany. 2010;**105**(4):vii-viii

[163] Pohlscheidt M, Charaniya S, Bork C, Jenzsch M, Noetzel TL, Luebbert A. Bioprocess and Fermentation Monitoring. Hoboken: John Wiley & Sons; 2013

[164] Su WW, Humphrey AE. Production of rosmarinic acid in highdensity perfusion cultures of *Anchusa officinalis* using a high sugar medium. Biotechnology Letters. 1990;**12**(11):793-798. DOI: 10.1007/BF01022597

[165] Vanisree M, Chen YL, Shu-Fung L, Satish MN, Chien YL, HsinSheng T. Studies on the production of some important secondary metabolites from medicinal plants by plant tissue cultures. Botanical Bulletin of Academia Sinica. 2004;**45**:1-22

[166] Madariaga-Mazón A, Hernández-Alvarado RB, Noriega-Colima KO, Osnaya-Hernández A, Martinez-Mayorga K. Toxicity of secondary metabolites. De Gruyter. 2019. DOI: 10.1515/psr-2018-0116

[167] Sharma AK, Sharma MK. Plants as bioreactors: Recent developments and emerging opportunities. Biotechnology Advances. 2009;**27**(6):811-832

[168] Shitan N. Secondary metabolites in plants: Transport and self-tolerance mechanisms. Bioscience, Biotechnology, and Biochemistry. 2016;**80**(7):1-11. DOI: 10.1080/09168451.2016.1151344

[169] Siddiqui MS, Thodey K, Trenchard I, Smolke CD. Advancing secondary metabolite biosynthesis in yeast with synthetic biology tools. FEMS Yeast Research. 2012;**12**(2):144-170. DOI: 10.1111/j.1567-1364.2011.00774.x

[170] Niazian M, Sadat-Noori SA, Galuszka P, Mortazavian M. Tissue culture-based Agrobacterium-mediated and in plant transformation methods. Czech Journal of Genetics and Plant Breeding. 2017;**53**(4):133-143. DOI: 10.17221/177/2016-CJGPB

[171] Pandey S, Mishra A, Patel MK, Jha B. An Efficient method for *Agrobacterium*-mediated genetic transformation and plant regeneration in cumin (*Cuminum cyminum* L.). Applied Biochemistry and Biotechnology. 2013;**171**(1):1-9. DOI: 10.1007/s12010-013-0349-1

[172] Fernando WGD, Zhang X, Amarasinghe CC. Detection of *Leptosphaeria maculans* and *Leptosphaeria biglobosa* causing blackleg disease in canola from *Canadian canola* seed lots and dockage. Plants. 2016;**5**:12. DOI: 10.3390/plants5010012

[173] Rout GR, Samantaray S, Das P. *In vitro* manipulation and propagation of medicinal plants. Biotechnology Advances. 2000;**18**:91-120

[174] Wang G, Qi NM. Perfusion culture of *Glycyrrhiza inflata* suspension cells in a stirtank bioreactor. Australian Journal of Botany. 2009;**57**(3):240-246. DOI: 10.1071/BT08187

[175] Hassanein AM, Ahmed AM, Abdel-Hafez AII, Soltan DM. Phenol oxidizing isoenzymes and malate dehyderogenases patterns, and organogenesis of *Solanum nigrum* L. as affected by light treatments. Acta Agronomica Hungarica. 1999;**47**:127-136

[176] Jones AMP, Saxena PK. Inhibition of Phenylpropanoid Biosynthesis in *Artemisia annua* L.: A Novel Approach to Reduce Oxidative Browning in Plant Tissue Culture. PLoS One. 2013;**8**(10):e76802. DOI: 10.1371/journal.pone.0076802

[177] Malinski MP, Budzianowsk J, Kikowska M, Derda M, Jaworska MM, Mlynarczyk DT, et al. Two ecdysteroids isolated from micropropagated *Lychnis floscuculi* and the biological activity of plant material. Molecules. 2021;**26**:904. DOI: 10.3390/molecules26040904

[178] Kour B, Kour G, Kaul S, Dhar MK. *In vitro* mass multiplication and assessment of genetic stability of *in vitro* raised *Artemisia absinthium* L. plants using ISSR and SSAP molecular markers. Advances in Botany. 2014;**2014**:7

[179] Massot B, Milesi S, Gontier E, Bourgaud F, Guckert A. Optimized culture conditions for the production of furanocoumarins by micropropagated shoots of *Ruta graveolens*. Plant Cell, Tissue and Organ Culture. 2000;**62**:11-19

[180] Begum F. Augmented production of vincristine in induced tetraploids of Agrobacterium transformed shooty teratomas of *Catharanthus roseus*. Medicinal Plants. 2011;**3**:59-64

[181] Krolicka A, Szpitter A, Stawujak K, Baranski R, Gwizdek-Wisniewska A, Skrzypczak A, et al. Teratomas of *Drosera capensis* var. alba as a source of naphthoquinone: Ramentaceone. Plant Cell, Tissue and Organ Culture. 2010;**103**:285-292

[182] Chen J, Ziv M. The effect of ancymidol on hyperhydricity, regeneration, starch and antioxidant enzymatic activities in liquid-cultured Narcissus. Plant Cell Reports. 2001;**20**:22-27

[183] Isah T, Mujib A. *In vitro* propagation & camptothecin production in *Nothapodytes nimmoniana*. Plant Cell, Tissue and Organ Culture. 2015;**121**:1-10. DOI: 10.1007/s11240-014-0683-1

[184] Salem JM. Effects of anti-ethylene compounds on vitrification and genome fidelity of *Stevia rebaudiana* Bertoni. Egyptian Journal of Botany. 2020;**60**:519-535

[185] Debergh PC, Harbaoui Y, Lemeur R. Mass propagation of globe artichoke (*Cynara scolymus*): Evaluation of different hypotheses to overcome vitrification with special reference to water potential. Physiologia Plantarum. 1981;**53**:181-187

Section 4

The Perspective for Plant Breeding

Chapter 19

Genetically Modified Crops and Their Impact on New Era of Agriculture

Palanisamy Dhamotharan, Sakthivel Surya Krishna, Sampathkumar Sruthi and Subburaj Monica

Abstract

Genetically engineered crops are different from genetically modified crops. Changes in genetic make-up of crops by any conventional or any means technology fall under genetically modified crops category. In general, among different traits, herbicide and pest tolerance were more exploited in which herbicide tolerant crops occupy more than 90% of the total cultivated area of genetically engineered crops as the mode of actions of herbicides are well known and tolerant gene sources are readily available in wild weedy forms and various microorganisms. By knowing the pathway of mode of action of herbicides and pest tolerance, herbicide and pest tolerant crops were transformed by altering the structure and functions of rate limiting enzymes without affecting the normal functionalities of plants. Other than these two economically important characters, several characters were engineered in different crop plants such as disease resistant, increased yield and biomass production, male sterility and quality characters like anti-allergy factors, modified composition of fatty acid, protein, increased micronutrients and enzyme contents, reduced contents of anti-nutritional factors and toxic substances. Through genetic engineering, transformed plants are used for establishing pharmaceutical products. In terms of area coverage, soya-bean followed maize, cotton, canola, alfalfa. Apart from genetically engineered crops, genome edited crops are developed for nullifying the negative effects and up-regulating the target traits having positive effects thus ultimately assisting in achieving food and nutritional security.

Keywords: agriculture, GE/GM crops, RNAi, ZFN, TALEN, CRISPR/Cas9

1. Introduction

Plant genome engineering aims to modify crops by incorporating agronomically desirable traits, which could not be achieved through conventional plant breeding methods. Genetically modified crops vary considerably from genetically engineered crops in that the former recruits the modification through both natural and artificial means, whereas the latter is distinguished only by artificial means that would not occur naturally. Genetically modified (GM) crops have proven to be great complement to conventional crops in meeting global demands for increased yields, increased food security, decreased pesticide use, and higher nutritional quality. The modified composition of fatty acid, protein and increased micronutrients, resistance to pests and disease, male sterility, quality characters like anti-allergy factors, reduced contents of anti-nutritional factors and toxic substances and medical benefits among which herbicide and pest tolerance are more remarkably exploited. The genetically modified herbicide tolerance crops occupy more than 90% of the total cultivated area of genetically engineered crops as the mode actions of herbicides are familiar and tolerant gene sources are more obtainable in wild weedy forms and microorganisms, which paved way for transformation by altering the structure and functions of rate-limiting enzymes without affecting the normal functionalities of plants. In terms of global acreage, genetically modified soya beans are widely cultivated followed by maize, cotton, canola, and alfalfa. However, apart from genetically modified crops, genome-edited crops are generated by nullifying the negative effects of traits of interest while up-regulating the positive effects of the target traits. Genetically modified crops benefit humanity by increasing the availability and quality of food and medical care, as well as contributing to a cleaner environment and alleviating hunger and disease around the world. In this chapter, a detailed discussion addressing the global need for resistance to insect pests, disease, herbicide tolerance, stress tolerance, quality improvement, male sterility and yield improvement through genetic modification in crop plants have been made.

2. Insect resistance

One of the most difficult issues in plant crop cultivation is dealing with insect pests. Insect pests are primarily managed using insecticides, but the rising occurrence of insect resistance genes in many organisms could be harnessed and introduced to crop plants through the effective use of transgenic technology. The cloning of genes codes for insecticidal δ-endotoxins dates back to the early 1980s [1]. Transgenic tomato and tobacco produce modified toxin genes which provide insect resistance res the first examples of genetically modified plants [2–4]. As insect-resistance genes transferred into plants predominantly act on the digestive system of insects, researchers are currently identifying genes with distinct modes of action to combat the development of resistance in the target insects, and to enhance potency. Few noteworthy insect control proteins (ICPs) such as protease inhibitors, different enzymes, ribosomal inactivating proteins, and lectins derived from various genus and domains that have an antimetabolic or toxic effect on insect digestion are being viewed as an alternative to control insect infestation or confer resistance to plants.

2.1 Source of transgenes

2.1.1 Resistance gene from microorganisms

2.1.1.1 Cry gene from Bacillus thuringiensis

The most important and successful example of a transgene derived from the bacterium *B. thuringiensis* is Bt gene toxin [5]. It is a significant soil borne spore-forming bacteria that produces insecticidal crystal (Cry) proteins encoded by *cry* genes (*cry*-represents gene; Cry-represents protein) during sporulation [5, 6]. Proteinases in the insect gut proteolytically cleave the inactive protoxins to produce the active 60–65Kd toxin made up of high homology regions interspersed with (hyper)variable regions. Sequence analysis [7, 8] and X-ray crystallography were used to infer the structure and functional roles of the toxin's three domains [9] that bind to glycoprotein receptors in the brush border membrane of the midgut epithelium of susceptible insects. The nature of the receptors that explicitly play a significant role in establishing susceptibility/resistance to a specific Bt toxin, is under intensive investigation, with a number of midgut integral membrane glycoproteins, including aminopeptidase and a cadherin-like protein, being identified [10–13]. Following binding, the toxin rapidly and irreversibly inserts into the cell membrane resulting in the formation of a pore supposedly by Cry proteins leading to disruption of the electrical, K+ and pH gradients eventually causing irreversible damage to the midgut wall which gives rise to epithelial cell lysis [6, 14] paving way to gut paralysis, cessation of feeding and finally (typically after 1–3 days) death from starvation and/or septicaemia. These toxins are mainly targeted enzymes and lectins of digestive systems of Lepidopteron and coleopteran pests.

2.1.1.1.1 Codon optimization for higher expression

Specificity, efficiency and insecticidal activity of toxins are more in codon optimised proteins than native form of Bt toxin and hence the former was introduced into various plants for increased level of expression [14]. Plants harbouring modified codon with plant preference rather than bacterial preference in which G:C rich codons are preferred over A:T rich codons, and undesirable mRNA secondary structure and polyadenylation signals eliminated, produced 100X higher than plants were transformed with unmodified (native) Bt gens [15, 16]. Genes are frequently inserted with constitutive promoters such as maize ubiquitin, CaMV35S, or rice actin 1, promoted protein expression at all times and in all parts of the plant allowing broader-spectrum ICPs to be targeted at different components of the pest complex. The use of tissue specific (e.g., RsS1 promoter for phloem-specific expression or PHA-L promoter for seed-specific expression) or inducible (e.g., potato pin2 wound-induced promoter) promoters is recommended to conserve the space and time of expression of toxic proteins and thus avoid unfavourable interactions with the beneficial insect's ecosystem [17].

The highest risk of resistance development would most likely arise from prolonged exposure to ineffective levels of the transgene, a situation that farmers would not tolerate and would almost certainly necessitate additional (different) control measures (deployment of which would in fact reduce the risk of resistance development). CaMV35S has a notoriously low/no expression in pollen. Temporal and spatial promoters are used to target the pest at its most vulnerable stage and time of infestation.

2.1.1.1.2 Cry protein classification

Cry protein classification is based on the degree of homology of Cry proteins. Primary Cry protein group: Cry proteins with less than 45 percent amino acid similarity fall into this category. Cry 1 to Cry 78, for example (in 2018). Secondary Cry protein group: Cry proteins with less than 78 percent amino acid similarity fall into this category. Cry 1A and Cry 1B are two examples. Tertiary Cry protein group: Cry proteins with less than 95% amino acid similarity fall into this category. Cry 1Aa, Cry 1Ac, etc. Cry proteins that share more than 95% of their amino acid sequences are

Classification	Percentage of homology of amino acids (%)	examples
Primary group	< 45	Cry 1 to Cry 78
Secondary group	< 78	Cry 1A, Cry 1B
Tertiary group	< 95	Cry 1Aa, Cry 1Ac
Quandary group	> 95	Cry 1Aa1, Cry 1Aa2

Table 1.
Classification of cry proteins based on their amino acids classification.

classified as part of the Quandary Cry protein group. Cry 1Aa1, Cry 1Aa2 (Cry 1Aa25 was recently discovered in 2019). Each Cry protein of the Bt bacterium has a distinct host range cry protein (**Table 1**) [18].

Even within the Cry protein subfamily, the toxic spectrum varies depending on the host. Cry1A and Cry1C proteins, for example, are toxic to larvae of lepidopteran pests such as the codling moth (*Cydia pomonella*), the European corn borer (*Ostrinia nubilalis*) [19], and heliothine bollworms, respectively. The Cry3A protein, on the other hand, is toxic to coleopteran pests such as the Colorado potato beetle (*Leptinotarsa decemlineata*) [20]. So far, 26 plant species have been genetically engineered and expressed for the Bt toxin [18].

2.1.1.2 Ipt gene from Agrobacterium tumefaciens

The introduction of *Isopentenyl transferase gene* (*ipt*) isolated from *Agrobacterium tumefacien* encoding a key rate limiting enzyme in the cytokinin biosynthetic pathway into the tobacco with wound inducible promoter recorded reduced consumption of leaves by the tobacco hornworm (*Manduca sexta*) and reduced survival of the peach potato aphid (*Myzus persicae*) leaving negative effects on plant development, such as an underdeveloped root system and a reduced total chlorophyll content [21].

2.1.1.3 Cholesterol-oxidase gene from streptomycete fungus

A *cholesterol-oxidase* gene from a streptomycete fungus has also been engineered into tobacco that was highly toxic to larvae of the boll weevil (*Anthonomus grandis*) and retarded the growth of the tobacco budworm (*Heliothis virescens*) by damaging the membranes of the insect-midgut epithelium [21–23].

2.1.1.4 Chitinase gene from Serratia marcesens

A bacterial *endochitinase* (from *Serratia marcesens*) has been shown to work in tandem with Bt toxin against *S. littoralis* larvae [24], but not (yet) in transgenic plants.

2.1.2 Resistance gene from higher plants

Plants have co-evolved with insects for millions of years, and have developed many adaptations in terms of antifeeding and anti-shelter, among which digestive enzyme inhibitors (proteinase and amylase inhibitors) and lectins have shown significant effect on insects, which have been transferred into crop plants without major alteration, and expression has been at a similar level to codon-optimised Bt toxins [25].

2.1.2.1 Proteinase inhibitors (PI)

Serine-like proteinases (trypsin-, chymotrypsin- and elastase-like endoproteases) dominant in lepidopteran larvae [26], a wider range of dominant gut proteinases in coleopteran species [27] and thiol proteases observed in corn rootworms (*Diabrotica spp.*) are some of the proteinases in insect which catalyse the release of amino acids from dietary protein, thereby providing nutrients essential for normal growth and development and thus the proteinase inhibitors plays an important part of the plant's natural defence system against herbivory by inhibiting the protein metabolism. Although the mode of action of PIs are not fully understood, it may be claimed that hypersecretion of digestive enzymes caused by the presence of the inhibitors, would result in depletion of essential amino acids [28, 29].

Serine and cysteine-proteinase inhibitors have been shown to inhibit the growth and development of a variety of insects, primarily lepidopteran and coleopteran species [29, 30]. The expression of a gene encoding a sweet potato trypsin protease inhibitor (TPI) in transgenic tobacco (at a relatively low level for a plant-derived ICP – 0.2%) results in severe growth retardation of *Spodoptera litura* caterpillars fed on it besides the presence of high levels of the same naturally in it [31].

The first instance of a plant-derived ICP gene being used in transgenic plants was the constitutive expression (through the CaMV35S gene promoter) of a trypsin inhibitor gene taken from cowpea (*Vigna unguiculata*) and in tobacco [32]. Proteinase inhibitors do not just alter gut digestive enzymes; they can also affect insect water balance, moulting, and enzyme regulation [33]. The majority of research has focused on serine-proteinase inhibitors derived from the plant families Fabaceae, Solanaceae, and Poaceae, which are mostly aimed against not only lepidopteran pests but also some coleopteran and orthopteran pests. The cowpea trypsin inhibitor (CpTI), which has been introduced into at least 10 different plant species, is the most active inhibitor discovered to date the protection provided by CpTI was less pronounced and consistent than that of tobacco containing a truncated Bt-toxin gene [34].

2.1.2.2 α-Amylase inhibitors

To block carbohydrate metabolism, a-amylase inhibitors from wheat (WAAI) and common bean (*Phaseolus vulgaris*) (BAAI) are utilised. When introduced into transgenic tobacco, the former increased mortality of lepidopteran larvae fed on it by 30–40% [35], while the latter, when expressed in transgenic pea seeds and driven by the pha1 gene promoter, elevated resistance to bean weevils [36, 37].

2.1.2.3 Lectins

Lectins are a diverse group of carbohydrate-binding proteins, that are toxic to insects of the orders Homoptera, Coleoptera, Lepidoptera, and Diptera. The very first demonstration of enhanced resistance of transgenic plants expressing a foreign lectin used is the gene encoding the glucose/mannose-binding lectin from pea (*Pisum sativum*) [38]. The mode of action of lectins against insects is unknown, but it has been shown that some bind to midgut-epithelial cells [39], and some insecticidal lectins also show significant mammalian toxicity, including lectins from *P. vulgaris* (phaeton haemagluttinin, PHA), and the greater insecticidal activity is shown by chitin-binding lectins from wheatgerm (WGA) expression in transgenic maize [40].

Recent interest has primarily focused on the mannose specific lectin from snowdrop (GNA), which has shown activity against peach potato aphids, potato tuber moths [41] and the rice brown planthopper (*Nilaparvata lugens*) [42]. GN expressed in potato and tomato significantly reduced fecundity and enhanced resistance, respectively, in laboratory experiments [42]. When overexpressed in different species of tobacco, tomato, and sweet gum, the tobacco anionic peroxidase, which is involved in crosslinking and polymerisation, inhibition of digestive enzymes, and the generation of highly reactive, toxic species, led to significant levels of resistance to several lepidopteran, coleopteran, and peach potato aphid [43]. The expression of tryptamine and tryptamine-derived alkaloids in plants may serve as anti-oviposition and anti-feedant agents, or as inhibitors of larval and pupal development, and when expressed in tobacco, inhibition of reproduction of the whitefly *Bemisia tabaci* by observed up to 97 percent [44].

2.1.3 Resistance genes from animals

Based on in vitro testing of proteolysis inhibition by several lepidopteran larvae midgut extracts, as potential insect-resistance proteins like bovine pancreatic trypsin inhibitor (BPTI), a1-antitrypsin (a1AT), and spleen inhibitor (SI) were identified and introduced into a variety of crop. Proteinase inhibitors derived from *M. sexta* and expressed in cotton and tobacco were reported to impede *B. tabaci* reproduction [45, 46]. Despite the introduction of chitinase (from the tobacco hornworm) into tobacco, these plants only exhibited a limited level of resistance to lepidopteran larvae and peach potato aphids [47].

2.1.4 Microbial proteins

The bulk screening of microbial culture supernatants against specific pests has been one strategy to the discovery of novel insecticidal proteins. Two proteins, Vip1 and Vip2, were isolated from vegetative *Bacillus cereus* culture supernatants and have been shown to be acutely poisonous to maize rootworms [47]. Some *B. thuringiensis* vegetative culture supernatants include a protein (Vip3A) that is acutely poisonous to Agrotis and Spodoptera caterpillars [48]. These proteins activity is extremely similar to that of Bt-endotoxins, yet they are distinctly separate from them.

2.1.5 Predator toxins

Genes producing neurotoxins from predatory mites [49] and scorpions [50] have been inserted into recombinant baculoviruses, where they effectively boost the rate of killing.

3. Disease resistance

Since the identification of the chemicals and genes involved in disease resistance in plants, attempts have been made to develop permanent disease resistance in commercially significant crop plants. Unfortunately, many of these efforts have failed because to the complexities of disease-resistance signalling and the wide range of infection routes employed by various pathogens. Although disease-resistant transgenic plants or seeds are not currently commercially accessible, future product development looks to be feasible as our understanding of pathophysiology and plant defence deepens.

In general, plants are protected by structural defence (plant cell wall, thick and waxy epidermis, trichomes, thorns) and chemical defence (production of secondary metabolites, proteins, and digestive enzymes) [51]. Plant defence response genes are classified into susceptibility genes (S), resistance genes (R), and non-host resistance genes (NHR). Pathogen-associated molecular pattern (PAMP)-triggered immunity (PTI) is broad-spectrum resistance, effective against non-host pathogens that fail to establish virulence [52, 53], and ETI is regarded to be the basis of R-gene resistance (host resistance). NHR is regarded as a quantitative trait encompassing numerous genes and varied pathways, whereas R gene resistance is regarded as a qualitative trait that adheres to Flor's gene-for-gene model and is dependent on a complex regulatory mechanism for pathogen detection and defence response [54, 55].

The successful transfer of R genes that express specific effector proteins in a gene-for-gene manner among plant species has resulted in long-term resistance and crop protection against a variety of pests and diseases. Polymorphisms in the coiled-coil (CC) and nucleotide-binding (NB) domains have been shown to be critical for recognition specificity, hence R genes could be edited [56]. In wheat, a two-amino acid mutation in the NB domain of the R gene (PM3F) protein was known to improve the resistance spectrum and HR response. S gene mutations can give both broad-spectrum and pathogen-specific resistance. Most S genes play functional roles in the plant, and mutations in them can cause serious pleiotropic defects. For example, CRISPR–Cas9-mediated disruption of the *OsSEC3A gene* in rice improved defence response against *Magnaporthe oryzae*, but the resistant plant also had dwarf stature and a lesion-mimic phenotype [57]. Similarly, leaf chlorosis was observed in wheat plants that had been MLO gene-edited for resistance to powdery mildew [58–60]. Rice *OsSWEET14* and 11 mutations resulted in shorter height and pollen abortion [61, 62].

Understanding the processes of NHR, which has a long lifespan and a broad spectrum, is essential for breeding disease-resistant cultivars. Type I NHR is asymptomatic [63, 64], whereas Type II NHR is similar to host resistance ("gene-for-gene") and is associated with the hypersensitive response triggered by pathogen penetration into plant tissue and activation of the resistance gene [65–68]. For conferring and developing disease resistance technologies are RNAi, TALENS, CRISPR/Cas, ZFN used transfer gene in various crops (**Table 2**) [76].

4. Herbicide resistance

Herbicides are essential components of today's integrated weed management strategy. To address the issue of repeated use of the same herbicide resulting in the evolution of resistant weeds, complicating their control, the development of diverse herbicide-tolerant crops is critical.

4.1 Photosystem based herbicides

Herbicides of various types disrupt various organelles and parts of plants. For example, now recognised by numerous names, including QB (Quinone), D1, encoded by the psbA gene,the 32-kDa weighted protein found in the thylakoid membrane acts

Crop	Disease	Gene and its function	GE Methods used	Ref.
Wheat	*Blumeria graminis*	*TaMLO* Repress immunity against powdery mildew	RNAi	[69]
Rice	*Xanthomonas oryzae pv. Oryzae*	EBS of *OsSWEET14* and *OsSWEET12*, Transcriptional induction	TALEN	[70–73]
Wheat	powdery mildew	Three homoallele *TaMLOof A1, B1, D1* homoeologs	TALEN	[59, 60, 74]
Rice	Rice tungro spherical virus (RTSV)	*eIF4G* Translation of viral RNA genomes	CRISPR/ Cas9	[75]
Rice	*Magnaphorthe oryzae*	*OsERF922* Negative regulator of blast resistance	RNAi CRISPR/Cas9	[59, 60]

Table 2.
RNAi, CRISPR/Cas 9 and TALEN mediated gene transfer for various diseases in different crops.

Amino acid position in Q_B protein	Wild type	Mutant
264	Serine	Alanine
264	Serine	Glycine
219	Valine	isoleucine
251	Alanine	Valine
255	Phenyl alanine	Tyrosine
275	Leucine	Phenyl alanine

Table 3.
Amino acid exchanges in wild and mutant plant proteins.

as a herbicide binding protein for a variety of proteins was identified using a photoaffinity marker (azidoatrazine) [73, 75]. This is the first gene found for herbicide tolerance research. Atrazine, the most commercially important PS II herbicide, is the most extensively used herbicide in maize and sorghum as these crops can extensively degrade atrazine by glutathione conjugation. Herbicide-tolerant weeds (*Senecio vulgaris*) discovered in maize for atrazine, simazine, and other s-traizine category herbicides, which were previously susceptible for these herbicides, demonstrated maternal inheritance and conferred 100 times less binding affinity for herbicide in thylakoid membrane [77–79]. There have been six recorded amino acid exchanges (**Table 3**) [80–83]. The 3D arrangement of the amino acids revealed that different herbicide classes had different binding sites on the QB protein.

4.1.1 Glyphosate

Glyphosate N-(phosphonomethyl) glycine is a post emergence herbicide and is a potent competitive inhibitor of the enzyme 5-enol-pyruvyl shikimic acid 3-phosphate (EPSP) synthase and it is rate limiting enzyme for aromatic amino acids synthesis pathway [84]. A gene (*aroA*) from gram negative bacteria *Salmonella typhinutrium* (now the same found in *Aerobacter aerogenes* [85], possessing mutation on 101th position of amino acid wherein proline is changed to serine a gene (*aroA*) encoding resistant EPSP enzyme against glyphosate was transformed to tobacco with octopine synthase promoter through agrobacterium transformation which led to insensitive response to herbicide [86]. The mutated gene was fused to an octopine synthetase promotor to enable expression in plants leading to regeneration of transgenic plants with glyphosate tolerance and impaired growth due to compartmentation of aromatic amino acid biosynthesis and disruption of EPSP synthase and other shikimic acid pathway enzymes located in chloroplasts. The precursor protein is directed into the chloroplast by amino-terminal regions known as transit peptides, which are eliminated during the absorption process. EPSP synthase is also a precursor protein that is enzymatically active and binds glyphosate. A fusion gene encoding the sequence of a plastidic transit peptide before the aroA sequence, which exhibits a 40-fold increase in EPSP-synthase activity in petunia plants, is likely to provide a greater level of resistance [87, 88].

4.1.2 Sulfonylurea and Imidazolinone herbicides

The herbicides imidazolinones and sulfonylureas operate by limiting acetolactate synthase (ALS), the first enzyme in the biosynthetic chain that results in the production of branched chain amino acids [89–91]. Thus, chlorsulfuron is readily detoxified by wheat, barley, or oats by phenyl ring hydroxylation and consequent conjugation with glucose. Sulfonylurea-resistant mutant strains have been isolated from species as varied as bacteria, fungi, and plant cell cultures. Using cell biology techniques, maize plants resistant to imidazolinone and tobacco plants resistant to sulfonylurea herbicides have been successfully established [90, 91].

4.1.3 Phosphinothricin

L-Phosphinothricinis (an analogue of glutamate), a component of the tripeptide "bialaphos" produced by several *Streptomyces viridochromogenes*, is a potent irreversible inhibitor of glutamine synthetase with herbicidal activity causing a rapid increase of ammonia concentration in plants which leads to the inhibition in photosynthesis, which derives the plants to death of the plant cell [92, 93]. Resistance to L-phosphinothricin in alfalfa was established via overproduction of a glutamine synthetase gene connected to the promoter of the 35S transcript of cauliflower mosaic virus, and the construct was incorporated into the genome of N. Tobaccum var. W38 by *A. tumefaciens* [94, 95]. Transgenic plants demonstrated superior resistance to high doses of commercial formulations of phosphinothricin and bialophos after integrating the resistance gene from *Streptomyces hygroscopicus* to tobacco and regulating it with the 35S promotor of the cauliflower mosaic virus.The treated plants showed no elevation in ammonia concentration, signifying the total shielding of the plant glutamine synthetase from the herbicide's activity [94]. Herbicide resistance is acquired via TALENs-mediated gene mutation of OsALS in rice and ALS (SurA and SurB) in tobacco [96, 97] (**Table 4**).

5. Stress

Abiotic stress is a highly complicated phenomena that involves biochemical and physiological changes in plant cells causing increased amounts of ROS (reactive oxygen species), that are extremely reactive, toxic and impact chlorophyll

S. No	Herbicide	Gene	Source of gene	Mutation
1.	Glyphosate	*aroA*	*Salmonella spp.*	P101 → S
2.	Sulphonylurea	*ilvGM*	*Escherichia coli*	A26 → V
		ILV2	*Saccharomyces cerevisiae*	P192 → S
		SURBHra	*Nicotiana tabacum*	P196 → A W573 → L
		SURBc3		P196 → Q
		Csr1	*A. thaliana*	P197 → S
3.	Phosphinothricin	*GS*	*M. sativa*	G245 → S,C,R R264 → K
4.	Triazines	*psbA*	*Amaranthus*	S264 → G
			S. nigrum	S264 → G
			Chlamydomonas	S264 → A F255 → Y V219 → I

Table 4.
Genes from various sources responsible for mutation in herbicide tolerance.

production, photosynthetic capability, and carbohydrate, protein, lipid, and antioxidant enzyme activities.

Genes that code for synthesis of osmolytes (proline, betain, etc.), water uptake and transport (ion transporters and aquaporin), transcriptional regulation and signal transduction mechanisms (MAPK) are identified from a variety of organisms and transformed into sensitive genotypes for generation of stress-protecting chemicals [98, 99]. The most efficient candidates for genetic transformation are transcription factors (TFs) that regulate the expression of a number of genes involved in conferring abiotic stress tolerance in plants [100].

5.1 Drought tolerance

Stress-related ring finger protein 1 (*OsSRFP1*), drought-induced SINA protein 1 (*OsDIS1*), and dry- and salt-tolerant protein 1 (*OsDST*) are negative regulators of drought tolerance, whose silencing increased antioxidant enzyme levels, reduced H_2O_2 concentrations, and enhanced drought tolerance in rice.

The CRISPR–Cas9 system was used to introduce novel alleles in Arabidopsis *OPENSTOMATA 2 (OST2)*-encoding gene—a key plasma membrane H^+ ATPase causing two significant mutations at the *ost2* locus that led to constitutive functioning of proton pumps, induction of necrotic lesions and exhibiting a substantially higher rate of stomatal closure coupled with a lower rate of transcriptional water loss which resulted in enhanced drought tolerance (**Table 5**).

6. Quality

Over two billion people worldwide are malnourished as a result of nutritional stress. Genetically modified (GM) crops have the potential to fulfil the worldwide

Crop	Gene	Stress	Reference
Rice	*AtDREB1A* from *Arabidopsis thaliana*	Resistance to drought	[101]
	BrCIPK1 gene from *Brassica rapa*	Enhanced abiotic stress tolerance	[102]
	A. thaliana transcriptional regulator DNA polymerase II subunit B3-1 *(DPB3-1)* gene	Induced heat tolerance	[103]
	Rice LSD1-like type ZFP gene *OsLOL5*	Tolerance against salt and oxidative	[104]
	JERF3 transcription factor (*Solanum lycopersicum*)	Drought tolerance	[105]
	OsDREB2A from *Oryza sativa*	Drought and salinity tolerance	[106]
	OsERF4a from *O. sativa*	Increased tolerance to drought stress	[107]
Wheat	*TaPIE1* from *Triticum aestivum*	Enhanced tolerance to Cold	[108]
A. thaliana	SPDS (Spermidine synthase) from *Cucurbita ficifolia*	Multiple environmental stress	[109]
	Tomato Monodehydroascorbate reductase (*LeMDAR*) gene	Increased tolerance to protoplast abiotic stresses	[110]
	Wheat *WRKY2* and *WRKY19* genes	Salt, drought and freezing stress tolerant	[111]
Maize	Rab28 LEA gene	Osmotic stress tolerance	[112]
Potato	Strawberry D-galacturonic acid reductase (*GalUR*) gene	Increased stress tolerance to methyl viologen, NaCl and mannitol	[113]
	Rat *GLOase*	Resistant to methyl viologen, NaCl and mannitol	[113]
	DHAR from *A. thaliana*	Increased tolerance to salt and drought	[114]
Soyabean	Salinity stress tolerance and fungal resistance	Overexpression of tobacco osmotin	[115]
	Soybean Osmotin-like protein isolated from *Solanum nigrum var. americanum* (*SnOLP*)	Increased drought tolerance	[116]
Chick pea	tolerance by expression of Vigna	Enhanced salinity	[117]
	Increased drought tolerance	overexpression of miR408	[118]
Red gram	Expressing *Vigna* pyrroline-5-carboxylate synthetase	Increased salt tolerance	[119]
Black gram	Overexpression of glyoxalase 1	Salt tolerance	[120]
	Expression of Arabidopsis alate transporter, *AtALMT1*	Increased aluminium tolerance	[121]
	ALDRXV4 gene	Tolerance to H_2O_2, drought, salt and methyl viologen induced stresses	[122]
Mung bean	Expression of *AtNHX1*	Increased salt tolerance	[123]
	Expressing *AtICE1*	Cold stress tolerance	[124]
Sugarcane	*EaDREB2* from Erianthusarundinaceus	Drought and salinity stress tolerance	[125]

Crop	Gene	Stress	Reference
Tobacco	SbpAPX from Salicornia brachiata	Resistance against salt, cold, drought, abscisic acid and salicylic acid stress	[126]
	GmERF7 transcription factor from *Glycine max*	Increased tolerance to salinity	[127]

Table 5.
Genes responsible for abiotic stress tolerance in various crops.

need for high-quality food through genetic engineering by doing more than merely boosting nutritional quality.

6.1 Golden rice

The carotenoid biosynthesis pathway in plants is a multistep process and is accomplished via four desaturation reactions to produce all-trans-lycopene from 15-cis-phytoene by phytoene desaturase (*PDS*), ζ-carotene isomerase (*Z-ISO*), ζ-carotene desaturase (*ZDS*), and carotenoid isomerase (*CRTISO*) whereas, a single enzyme encoded by the crtI gene accomplishes all the four reaction steps to produce lycopene in bacteria [128, 129]. In the carotenoid pathway, lycopene is an important branch point because it functions as the substrate for two competing enzymes, lycopene β-cyclase (*LYCB*), and lycopene ε-cyclase (*LYCE*). Lycopene is converted into γ-carotene, which is rapidly converted into β-carotene (has pro-vitamin A activity) by *LYCB* in a single pathway. The bacterial gene *crtY* encodes *LYCB* to produce β-carotene. Alternatively, with *LYCE* lycopene generates δ-carotene (which has no pro-vitamin A activity) [130–132].

About 60–80% by weight of total seed protein in rice is glutelin, and about 20–30% are prolamins which are the prime choice of promoter sequences responsible for expression of carotenoids in rice endosperm [133]. Six promoters of rice glutelin genes *(GluA-1, GluA-2, GluA-3, GluB-3, GluB-5, and GluC*) were isolated and examined in rice and listed in tables with their site of expression (**Table 6**) [134]. Newly reported promoters like *PROLAM26 RAL2 (LOC_Os07g11330), RAL4 (LOC_Os07g11380), and CAPIP (LOC_Os06g33640*) could be useful in the future [135].

A combination of Daffodils *phytoene synthase* (psy) gene, *lycopene-cyclase* (lcy) gene and *crtI* gene from bacteria *(Erwinia uredovora*) PDS (phytoene desaturase) are used to generate japonica rice with ß-carotene expression (mentioned here as GR1) [136, 137]. Gtu-1 promoter was used for psy, lcy and a constitutive CaMV35S promoter with a plastid-specific transit peptide (TP) used for the expression of *CrtI* gene [138, 139] in GR1, Under Gtu-1 promoter, an alliance of maize phytoene synthase (*Zmpsy*) gene with bacterial *crtI* has been expressed in rice to develop an improved golden rice variety (mentioned here as GR2) [140]. The synthetic gene constructions of two carotenoid biosynthetic genes, *psy* from Capsicum (*Capsicum annuum*) and *crtI* from Pantoea, were also reported for golden rice development. To create the PIC (Psy-IRES-CrtI) and PAC (Psy-2A-CrtI) constructions, coupling of two genes using either the synthetic codon-optimised 2A sequence (from foot-and-mouth disease virus) or the IRES sequence (the internal ribosome entry site) were utilised (**Figure 1**) [141, 142].

ZFN and TALENs mediated gene mutation of *IPK1* gene of the rice *OsBADH2* gene respectively encoded inositol1,3,4,5,6-penta-kisphosphate 2-kinase, resulted in both herbicide tolerance and reduction of phytate in developing seeds and production

S. No	Promoter	Expressing region
1.	*GluA1, GluA2 &GluA3*	Peripheral region of the endosperm
2.	*GluB5 and GluC*	The whole endosperm
3.	*GluB3*	Aleurone and subaleurone layers of rice grain

Table 6.
List of promoters and expressing region involved in Golden rice.

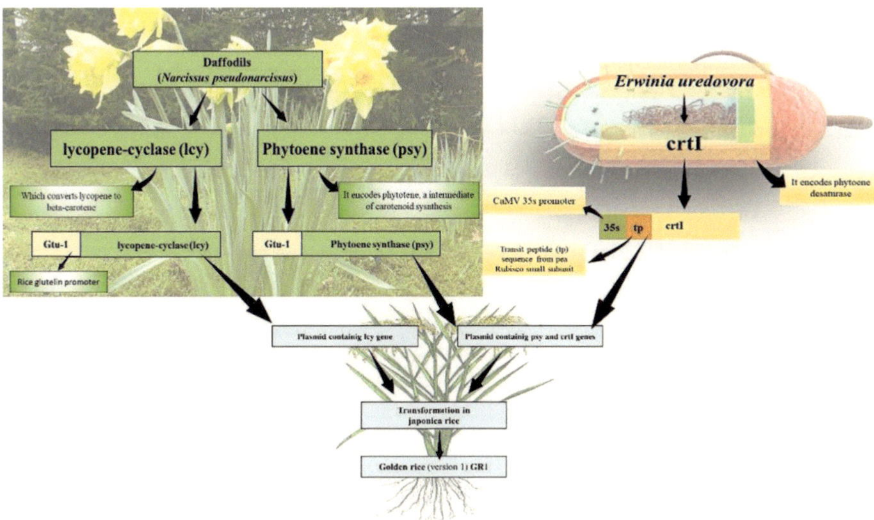

Figure 1.
A simplified version of ß-carotene expression in golden rice.

of fragrant rice by increased synthesis of fragrance compound 2-acetyl-1-pyrroline (2AP) [143]. In monocot plants, the *k1C* gene in sorghum plant was targeted using CRISPR for gene disruption at the N-terminal ER signal peptide region, resulting in greater Lysine content and improved protein digestibility [144]. Naim et al. deployed CRISPR-Cas to target the PDS, Phytoene desaturase expressing gene in Cavendish banana (*Musa acuminata*) for gene knockout to target the Albinism phenotype. CRISPR-Cas9 was used to disrupt rice genes critical for defining amylose concentration, fine structure of amylopectin, and physiochemical characteristics of starch, resulting in a larger proportion of long chains in amylopectin. The GBSS gene, which encodes Granule-bound starch synthase and is responsible for amylose production in the Potato plant, was targeted using CRISPR-Cas for Gene knockouts, resulting in a product with elevated amylopectin content [145].

In *Camelina sativa* the *FAD2* gene, which is important for fatty acid production, was targeted using CRISPR-Cas9 for gene deletion to enhance seed Oleic acid content. In tomato, CRISPR-Cas9 was used to target *SlAGL6*, a transcription factor that plays important roles in flower meristem and floral organ development, for gene deletion, resulting in a parthenocarpic phenotype [146]. The *CAO1* and *LAZY1* genes, which are responsible for synthesis of chlorophyll b from chlorophyll a and regulating shoot gravitropism, were disrupted using CRISPR-Cas to target the faulty synthesis of chlorophyll b and tiller spreading phenotypes, respectively (**Table** 7) [147].

Crop	Gene transformed	Quality character	Reference
Rice	Soybean ferritin gene	Increased iron content	[148]
	Phaseolus vulgaris ferritin gene	Enhancement in Fe content, cysteine and phytase level	[149]
	Lactoferrin gene from human	Increased iron content	[150]
	Ubi1-P-int (maize) & *GmFAD3 cDNA* (soyabean)	Enhanced α-linolenic acid content	[151]
	amino-deoxychorismate synthase (*ADCS 1*) & GTP cyclohydrolase I (*GTPCHI*) genes	100 times enhanced accumulation of vitamin B9	[152]
	BiP, lysine-rich binding protein	Increased lysine content	[153]
Maize	*SacB* gene (*Bacillus amyloliquefaciens*)	Stable accumulation of fructan	[154]
	Lysine feedback-insensitive *DHDPS*	Increased lysine content	[155]
	HGGT gene from barley	Tocotrienol content increase	[156]
	Wheat *DHAR* gene	Increased ascorbic acid	[157]
	Ferritin gene & phytase gene	Increased iron content	[158]
Red gram	Dihydrodipicolinate synthase overexpression	improve lysine content in seeds	[159]
soyabean	*DHDPS* and aspartokinase from *E. Coli*	Increased lysine content	[155]
	15-kDa Maize zein protein	Increased accumulation of C &M	[160]
Chick pea	Raffinose synthase 2 Silencing	Nutritional quality improvement	[161]

Table 7.
List of different genes transformed for quality character in different crops.

7. Male sterility

The use of genetically engineered male sterility has a variety of applications, ranging from hybrid seed production to transgenic bioconfinement in genetically modified crops. The influence of this technique has aided in dealing with global food security concerns. The production of transgenic male sterile plants through the expression of a ribonuclease gene under the direction of an anther- or pollen-specific promoter has shown to be an efficient method of producing pollen-free elite cultivars.

7.1 Male sterility due to mutation on nuclear genes

Mariani et al. [162] achieved the first success in developing genetically engineered male sterility in crop plants by transforming tobacco and rapeseed plants with a chimeric dominant gene *barnase* (bacterial RNase from *Bacillus amyloliquefaciense*) driven by a tapetum-specific promoter (TA29) from tobacco. The coding sequences of RNase T1 from *Aspergillus oryzae* and *barnase* from *B. amyloliquefaciens* used to manipulate the trait were fused with the tapetum-specific TA29 promoter which is responsible for the expression of the *barnase* gene specifically to anther tapetal cells, causing selective destruction of the tapetal cell layer that surrounds the pollen sac by hydrolysing the tapetal cells, causing abnormal pollen formation (**Figure 2**). Male sterile anther carries empty exine [162]. Mariani et al. [163] demonstrated fertility

DOI: http://dx.doi.org/10.5772/intechopen.105937

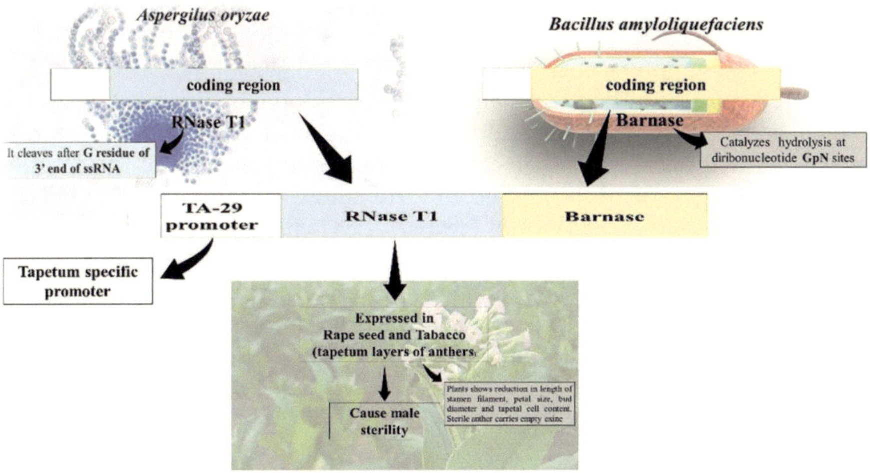

Figure 2.
Simplified version of expression of transgenic male sterile system.

restoration in TA29-barnase male sterile plants by gene encoding the barnase-specific RNase inhibitor called barstar which was isolated from same bacteria *B amyloliquefa-ciense*. When genetically engineered, male sterile plant is crossed with plant carrying TA29- barstar gene the F1 progeny shows co-expression of both genes in the anther of male fertile plants. In this system fertility restoration is due to the formation of tapetal cell-specific barnase/barstar protein complexes which completely inactivate the barnase enzyme [163]. This dominant nuclear genetic male sterility system faces same drawback as GMS system. During hybrid seed production the plants in female rows segregate in the ratio of 1:1 for male sterility and male fertility [164]. To counter the drawback of nuclear genetic male sterility system problem the *barnase* gene was linked to a dominant herbicide resistant gene (bar) under control of the constitutive promoter CaMV 35S which conferred resistance to broad-spectrum herbicide Basta (active ingredient is phosphinothricin or PPT). When seedlings are sprayed with Basta only the male sterile plants survive and the male fertile plants are killed as they lack bar gene.The use of bar gene allows elimination of male fertile segregants from female rows in the hybrid seed production plot thus assuring 100 per cent pure hybrid seed production [165].

7.2 Male sterility due to mutation on chloroplast genes

The genetic transformation of the plastid genome has various advantages, including high level transgenic expression, expression of multigene operons, transgene maternal inheritance, and expression of bacterial genes without codon optimization [166]. Ruiz and Daniell [167] elucidated that, with chloroplast transformation, hyper-expression of β-ketothiolase encoded by the phaA gene of Acinetobacter sp. in the leaves, flower, and anther of transgenic lines gets in the way of pollen development and results in male sterility. This was restored by exposing transgenic male sterile plants to continuous illumination, which allows acetyl CoA carboxylase (ACCase) to access acetyl CoA, restoring normal fatty acid synthesis and minimising PHB production through β-ketothiola.

7.3 Male sterility due to altering metabolic process

Callose is a plant polysaccharide comprised of β-1-3 glucan that is deposited around microspore tetrads during meiosis. Tight developmental regulation and the timing of callase activity are required for optimal microspore development. The expression of modified PR-b-1-3 glucanase in transgenic tobacco plants led in the premature disintegration of the microsporocyte callose wall, resulting in mild to total male sterility [168]. Chang et al. [169] created a rice hybrid breeding method employing the rice nuclear gene *Oryza sativa* No Pollen 1 (*OsNP1*), which encodes a putative glucose–methanol–choline–oxidoreductase with involvement in tapetum degeneration and pollen exine production. The ethyl methane sulfonate-induced rice mutant, *osnp1-1*, was completely male sterile.

7.4 Conditional male sterility

Conditional male sterility is a situation in which plants are typically fertile, but when a specific circumstance is applied, male sterility occurs. Hawkes et al. [170] revealed the use of inactive D-glufosinate as a male sterility inducer in transgenic plants expressing a modified (F58 K, M213S) version of *Rhodosporidium toruloides* Damino acid oxidase (DAAO) that converts oxidised D-glufosinate to its 2-oxo-derivative (2-oxo-4-methyl phosphiny to create transgenic plants, the modified DAAO encoding gene was coupled with the TAP1 promoter from *Antirrhinum majus* and transformed into tobacco plants. When D-glufosinate was sprayed on these transgenic plants, it caused full male sterility that lasted two or more weeks while having no effect on female fertility [170]. Guerineau et al. [171] expressed the temperature-sensitive diphtheria toxin A-chain polypeptide gene sequence under the tapetum-specific A9 promoter and generated transgenic Arabidopsis plants that were fully fertile at 26 C, but when the temperature was decreased to 18°C, male sterility was induced [171].

7.5 Male sterility through post transcriptional gene silencing

Jasmonic acid (JA), a plant hormone, is involved in several developmental signalling events in plants, including senescence, fruit ripening, anther dehiscence, and pollen maturation [172, 173]. Bae et al. [174] reported inducing male sterility by inhibiting OsAOS1 and OsAOS2 activity with the promoters of the anther-specific genes Osc4 and Osg6b, respectively. RNAi (pSK124) constructs were designed and converted into rice calli independently, concluding that the OsAOS2-RNAi vector driven by Osg6b promoter is potent enough for generating male sterility in rice.

7.6 Male sterility through modification of flavonoids

Any disruption in flavonoid production changes pigmentation and causes male sterility in plants. Fischer et al. [175] discovered the expression of a stilbene synthase (STS) gene from grape vine (*VstI*) driven by a 35S RNA promoter with duplicated enhancer region and a tapetum-specific promoter (Tap1) of *A. majus* produced male sterility strives for the substrates,4-coumaroyl CoA and malonyl CoA, which are required for sporopollenin and fatty acid biosynthesis, and hypothesised that there was a decrease in p-coumaroyl availability, resulting in impaired sporopollenin production and pollen wall formation, causing male sterility(**Figure 3**).

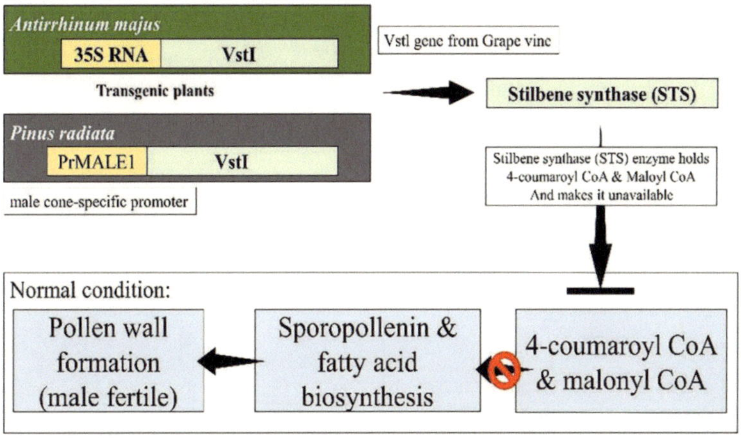

Figure 3.
Simplified version of male sterility through modification of flavonoids.

7.7 Male sterility through RNA editing

Nucleotide alterations or insertion of a nucleotide, leading to a change in the sequence of amino acid in polypeptide denotes RNA editing. Male sterility in CMS plants is connected to mitochondrial DNA rearrangement, causing the formation of novel chimeric open reading frames (ORFs), resulting in mitochondrial malfunction, such as the chimeric gene pcf-S of petunia, ORFB and ORF224 of polima in rapeseed [176]. The overexpression of unedited mitochondrial orfB gene in a transgenic strain of indica rice led to a decrease in activity of ATPase in F1F0-ATP synthase resulting in dose-dependent male sterility [177].

7.8 Heterologous male sterility

The association of CMS and new chimeric ORFs in mitochondrial DNA sequences, as well as mitochondrial dysfunction, is documented. Nizampatnam et al. [178] engineered transgenic tobacco plants to produce orfH522, a pet1-CMS-associated mitochondrial gene from sunflower that is driven by the TA29 promoter. Approximately 35% of the modified tobacco plants were completely sterile. Subsequently, by decreasing orfH522 transcripts using the RNAi approach, male fertility was restored [179].

8. GE for yield contributing characters

To address growing food demand as well as the challenges posed by climate change, major increases in yields of vital food crops employing transgenic technology are required. Using CRISPR-Cas9, researchers were able to increase grain number, dense erect panicles, and grain size in rice by disrupting the DEP1, Gn1a, IPA1, and GS3 genes, which are regulators of grain number, panicle architecture, grain size, and plant architecture [180]. CRISPR-Cas9 gene deletion targets the wheat genes TaGW2-B1, TaGW2-D1, and TaGW2-A1 that govern grain weight and protein content, leading to an increase in grain weight and protein content [181]. In maize, the gene ARGOS8 responded to water stress by increasing grain output [182].

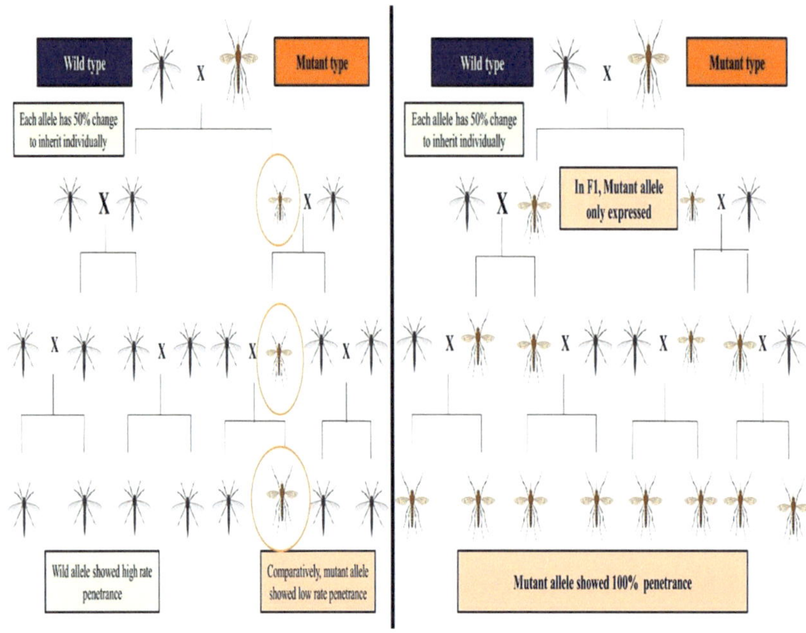

Figure 4.
Simplified version of comparison between normal mosquito population and genetic drive mosquitos' population. There is a huge discrepancy among the rapid adoption of GM crops for production, global markets, and consumer approval. However, the following is a list of transgenic crops that have been worldwide authorised and released for various characteristics (adopted from ISAAA database).

9. Gene drive

Gene drive is a genetic phenomenon of naturally occuring skewed inheritance mechanism that implies sexual reproduction and is essentially concerned with population suppression and population modification. Synthetic gene drives deploy genome editing technologies such as CRISPR to maximise the likelihood of a certain gene being inherited from 50% to almost 100% (**Figure 4**).

9.1 Working model of gene drive

When a gene drive is inserted into the genome of an animal, the progeny inherits the drive on one chromosome and a normal gene from the other parent. During early development, the CRISPR component of the drive shears the other copy. The cut is subsequently repaired using the drive as a template, resulting in two copies of the change being passed down to the progeny [183]. By suppressing the fertility gene termed doublesex on the usage of drive and thereby crashing a population of caged *Anopheles gambiae* mosquitos, Crisanti and his colleagues were able to prevent female mosquitos from biting or laying eggs while the drive was in place [184].

In *C. capitata* [185] and *Aedes aegypti*, a bisex RIDL system containing a tetracycline-repressible positive feedback transactivator (tTA) was successfully constructed, which does not require a specific promoter derived from the target species, but rather a minimal promoter used in conjunction with oligomerised tetO, the binding sequence. Under tight conditions, particularly in the absence of tetracycline, tTA (transactivator and lethal effector) accumulates to deadly levels in both sexes of the transgenic insect. Females must be removed from a 'male-only' release programme using an independent approach based on an underlying molecular mechanism that is female-specific. Functional Tra protein is only produced by females as it is encoded by a splice variant exclusively produced in females leading to functional tTA protein only produced in females rendering the system female-specific. Fu et al. [185] integrated the first intron of the sex determining gene Cctra into the DNA sequence coding for tTA to provide a sex selective component to the positive feedback system. If this Cctra fragment is spliced in the same as it is in its native context, tTA production is only allowed when the intron is spliced in the female-specific form, as the continuous coding frame of tTA is only restored in this variant. As this intron's full splicing in its native gene is exclusive to females, tTA expression was only expected in transgenic females. The analysis of tTA transcription in transgenic *C. capitata* revealed a sex-specific pattern similar to the natural Cctra gene. The inserted intron spliced to produce three distinct tTA transcripts: one female-specific (F1) and two nonsex-specific (M1 and M2). The female-specific transcript was the only one that encoded a complete tTA. As a result, when grown under harsh conditions, all of the transgenic female progeny died as larvae or pupae [186]. Thus, the utilisation of gene drivers in agriculture may be primarily harnessed by implementing sterile insect techniques for successful insect pest management (**Table 8**).

10. Discussion and conclusion

The advent of advanced targeted, 'customizable,' and precise new technologies for insect resistant plants, in their various embodiments and combinations, symbolises a sustainable option countering the emergence of resistant weeds, lessening agrochemical use, and curtailing adverse effects on nontarget organisms. The direct application of chimeric ODNs or siRNAs to plant cells enables for the generation of technically non-GM organisms. Inevitably, the refinement of promising techniques that are not currently destined to assign insect resistance, such as RNA manipulation with pentatricopeptide repeat proteins, the use of polygalacturonase-inhibiting proteins (PGIPs), ribozymes, and riboswitches, will likely expand the defence mechanism against pests available to researchers and farmers. However, thanks to recent advancements like RenSeq and directed molecular evolution, which enable the rapid identification of novel immune receptor genes, the pool of deployable genes for enhanced resistance to other microorganisms has grown significantly. In the near future, developments in molecular stacking and targeted gene insertion by genome editing are projected to predominate in establishing broad-spectrum resistance against both viral and nonviral diseases. Moreover, increasingly diverse, accurate, and economical genome-editing techniques like CRISPR-Cas allow for precise change of endogenous genes for disease resistance, such as susceptibility and decoy genes. On the flip side, the persistent cultivation of herbicide crops is estimated to have resulted in the resurgence of herbicide resistance in numerous weeds. Nevertheless, by benefiting from the shortcomings of the issue, the adverse consequences of herbicide-based technologies might be substantially minimised by introducing variety in weed control using alternative approaches, with an emphasis on crop rotation, herbicide rotation, and herbicide formulations. Stress-resistant plants may be generated with or without tissue culture by using simple knock-in, knock-out, replacement, fine-tuning of gene regulation, and point mutations at any gene locus. Although genome editing technology is in in its beginning phases, disruptions in specific genes can have unintended

Crops	IR	HR	DR	Abiotic stress	Quality	Pollination control	Altered growth / yield
Maize	119	128	—	6	8	6	2
Cotton	49	44	—	—	1	—	—
Cowpea	1	—	—	—	—	—	—
Bean	—	—	1	—	—	—	—
Brinjal	1	—	—	—	—	—	—
Popular	2	—	—	—	—	—	—
Potato	30	4	19	—	18	—	—
Rice	3	3	—	—	1	—	—
rose	—	—	—	—	2	—	—
Soyabean	6	32	—	2	9	—	1
Sugar cane	3	—	—	3	—	—	—
Tomato	1	—	1	—	9	—	—
Papaya	—	—	4	—	—	—	—
Plum	—	—	1	—	—	—	—
Petunia	—	—	—	—	2	—	—
Squash	—	—	2		—	—	—
Cucumis melo	—	—	—	—	2	—	—
Sweet pepper	—	—	1	—	—	—	—
Alfalfa	—	4	—	—	2	—	—
apple	—	—	—	—	3	—	—
Canola	—	33	—	—	10	20	—
Carnation	—	4	—	—	19	—	—
Chicory	—	3	—	—	—	3	—
Creeping Bentgrass	—	1	—	—	—	—	—
Flax	—	1	—	—	—	—	—
Brassica	—	4	—	—	—	—	—
Sugar Beet	—	3	—	—	—	—	—
Pine apple	—	—	—	—	1	—	—
Tobacco	—	1	—	—	1	—	—
Wheat	—	1	—	—	—	—	—
Safflower	—	—	—	—	2	—	—
Eucalyptus	—	—	—	—	—	—	1

Table 8.
List of globally approved and released transgenic crops for various characters.

negative consequences for plant growth and development. More research is needed to fully exploit the ability of the CRISPR-Cas System in regulating abiotic stress. One rationale could be that the genes governing beneficial qualities in crops are largely quantitative trait genes, which necessitates a deeper knowledge of gene regulatory networks. As a result, expanding our knowledge of gene regulatory networks is the foundation for greater crop development. Furthermore, the approach of knocking out target genes via NHEJ has attracted a strong interest for boosting agricultural output and quality. Gene targeted insertion or substitution enable the genes to integrate more easily. The HR repair pathway is ineffectual, constraining the use of site-specific insertion and substitution. It is also expected that eliminating the HR approach to editing would lead to more precise and effective crop improvement. If the seed is the most valuable portion of the plant and the crop is mostly self-fertilised, a good fertility restoration mechanism is necessary. It is likely that by coupling inducible male sterility with apomixis, fertile plants may be obtained once the trait was fixed. It is also necessary to guarantee that apomixis is inducible, encouraging apomixis to restore to sexual reproduction and enabling the breeder the opportunity of further enhancing the hybrids by establishing appropriate combiner lines.

The world has already experienced two technical revolutions and is currently undergoing a third revolution based on biotechnology and genomics, which is predicted to yield a plethora of transgenic crops for the benefit of humanity. Genetic engineering is a radical departure from traditional breeding since it allows scientists to transfer genetic material across organisms that could not be bred earlier. The degree of public knowledge regarding the benefits and drawbacks of transgenic plants should be strengthened, laying the groundwork for the effective dissemination of research findings to real time application. The central emphasis should be on the advancement of technologies competent of bridging the gaps in modern day technology. Nonetheless, developing countries are now required to assess genetically modified (GM) crops, and they will subsequently be expected to investigate the potential use of GM trees, cattle, and fish. These advancements may provide prospects for greater output, productivity, product quality, and adaptive fitness, but they will almost likely pose challenges to developing countries' research and regulatory capability.

Author details

Palanisamy Dhamotharan[1*], Sakthivel Surya Krishna[1], Sampathkumar Sruthi[2] and Subburaj Monica[3]

1 Tamilnadu Agricultural University, Coimbatore, Tamil Nadu, India

2 Annamalai University, Chidambaram, Tamil Nadu, India

3 Professor Jayashankar Telangana State Agricultural University, Hyderabad, Telangana, India

*Address all correspondence to: palanisamydhamotharan@gmail.com

References

[1] Schnepf HE, Whiteley HR. Cloning and expression of *bacillus thuringiensis* crystal protein gene in *Escherichia coli*. Proceedings of the National Academy of Sciences of the United States of America. 1981;**78**:2893-s2897

[2] Barton K, Whitely H, Yang NS. *Bacillus thuringiensis* d- endotoxin in transgenic *Nicotiana tabacum* provides resistance to lepidopteran insects. Plant Physiology. 1987;**85**:1103-1109

[3] Fischho! DA, Bowdish KS, Perlak FJ, Marrone PG, McCormick SM, Niedermeyer JG, et al. Insect tolerant transgenic tomato plants. Bio/ Technology. 1987;**5**:807-812

[4] Vaeck M, Reynaerts A, Hofte H, Jansens S, DeBeuckleer M, Dean C, et al. Transgenic plants protected from insect attack. Nature. 1987;**327**:33-37

[5] Peferoen M. Progress and prospects for field use of Bt genes in crops. Trends in Biotechnology. 1997;**15**(5):173-177

[6] Crook NE, Jarrett P. Viral and bacterial pathogens of insects. Journal of Applied Bacteriology. 1991;**70**:91S-96S

[7] Zhang Q, Hua G, Adang MJ. Effects and mechanisms of bacillus thuringiensis crystal toxins for mosquito larvae. Insect Science. 2017;**24**(5):714-729

[8] Hodgman TC, Ellar DJ. Models for the structure and function of the bacillus thuringiensis d-endotoxins determined by compilational analysis. Journal of DNA Sequencing Mapping. 1990;**1**:97-106

[9] Li J, Carroll J, Ellar DJ. Crystal structure of insecticidal d-endotoxin from bacillus thuringiensis at 2.5A_ resolution. Nature. 1991;**353**:815-821

[10] Knight PJ, Crickmore N, Ellar DJ. The receptor for bacillus thuringiensis Cry1A(c) d-endotoxin in the brush border membrane of the lepidopteran Manduca sexta is aminopeptidase-N. Molecular Microbiology. 1994;**11**:429-436

[11] Pease EA, English LH, Donovan WP. Characterization of the binding of CryIIIB2 to coleopteran brush border proteins. The FASEB Journal. 1995;**9**:A1313

[12] Luo K, Lu Y, Adang MJ. A 106Kd form of aminopeptidase is a receptor for bacillus thuringiensis cry1C d-endotoxin in the brush border membrane of Manduca sexta. Insect Biochemistry and Molecular Biology. 1996;**26**:783-791

[13] Yaoi K, Kadotani T, Kuwana H, Shinkawa A, Takahashi T, Iwahama H, et al. Aminopeptidase N of Bombyx mori as a candidate for the receptor of bacillus thuringiensis Cry1Aa toxin. European Journal of Biochemistry. 1997;**246**:652-657

[14] Gill SS, Cowles EA, Pietrantonio PV. The mode of action of bacillus thuringiensis endotoxins. Annual Review of Entomology. 1992;**37**(1):615-634

[15] Perlak FJ, Fuchs RL, Dean DA, McPherson SL, Fischhoff DA. Modification of the coding sequence enhances plant expression of insect control protein genes. Proceedings of the National Academy of Sciences of the United States of America. 1991;**88**:3324-3328

[16] Wong EY, Hironaka CM, Fischho! DA. Arabidopsis thaliana small subunit leader and transit peptide enhance

expression of bacillus thuringiensis proteins in transgenic plants. Plant Molecular Biology. 1992;**20**:81-93

[17] Hilder VA, Boulter D. Genetic engineering of crop plants for insect resistance—A critical review. Crop Protection. 1999;**18**(3):177-191

[18] Crickmore N, Baum J, Bravo A, Lereclus D, Narva K, Sampson K, Schnepf E, Sun M, Zeigler DR. *Bacillus thuringiensis* Toxin Nomenclature. 2018. Available from: http://www.btnomenclature.info/

[19] Armstrong CL, Parker GB, Pershing JC, Brown SM, Sanders PR, Duncan DR, et al. Field evaluation of European cornborer control in progeny of 173 transgenic corn events expressing an insecticidal protein from bacillus thuringiensis. Crop Science. 1995;**35**:550-557

[20] Perlak FJ, Stone TB, Muskopf YN, Petersen LJ, Parker GB, McPherson SA, et al. Genetically improved potatoes: Protection from damage by Colorado potato beetles. Plant Molecular Biology. 1993;**22**:313-321

[21] Smigocki A et al. . In: Carozzi N, Koziel M, editors. Advances in Insect Control: The Role of Transgenic Plants. Taylor & Francis; 1997. pp. 225-236

[22] Purcell JP, Greenplate JT, Jennings MG, Ryerse JS, Pershing JC, Sims SR, et al. Cholesterol oxidase: A potent insecticidal protein active against boll weevil larvae. Biochemical and Biophysical Research Communications. 1993;**196**(3):1406-1413

[23] Cho HJ, Choi KP, Yamashita M, Morikawa H, Murooka Y. Introduction and expression of the Streptomyces cholesterol oxidase gene (ChoA), a potent insecticidal protein active against boll weevil larvae, into tobacco cells. Applied Microbiology and Biotechnology. 1995;**44**(1):133-138

[24] Rigev A, Keller M, Strizhov M, Sneh B, Prudovsky E, Chet I, et al. Synergistic activity of a bacillus thuringinensis d-endotoxin and bacterial endochitinase against Spodoptera littoralis larvae. Applied Environmental Microbiology. 1996;**62**:3581-3586

[25] Gatehouse AM, Davison GM, Newell CA, Merryweather A, Hamilton W, Burgess EP, et al. Transgenic potato plants with enhanced resistance to the tomato moth, Lacanobia oleracea: Growth room trials. Molecular Breeding. 1997;**3**(1):49-63

[26] Christeller JT, Laing WA, Markwick NP, Burgess EPJ. Midgut protease activities in 12 phytophagous lepidopteran larvae: Dietary and protease inhibitor interactions. Insect Biochemistry and Molecular Biology. 1992;**22**(7):735-746

[27] Wolfson JL, Murdock LL. Diversity in digestive proteinase activity among insects. Journal of Chemical Ecology. 1990;**16**(4):1089-1102

[28] Gatehouse AMR, Hilder VA, Powell KS, Wang M, Davison GM, Gatehouse LN, et al. Insect-resistant transgenic plants: Choosing the gene to do the 'job'. Biochemical Society Transactions. 1992;**22**(4):944-949

[29] Broadway RM, Duffey SS. Plant proteinase inhibitors: Mechanism of action and effect on the growth and digestive physiology of larval Heliothis zea and Spodoptera exiqua. Journal of Insect Physiology. 1986;**32**(10):827-833

[30] Orr GL, Strickland JA, Walsh TA. Inhibition of Diabrotica larval growth by a multicystatin from potato

tubers. Journal of Insect Physiology. 1994;**40**(10):893-900

[31] Yeh KW, Lin MI, Tuan SJ, Chen YM, Lin CJ, Kao SS. Sweet potato (Ipomoea batatas) trypsin inhibitors expressed in transgenic tobacco plants confer resistance against Spodoptera litura. Plant Cell Reports. 1997;**16**(10):696-699

[32] Hilder VA, Gatehouse AMR, Sheerman SE, Barker RF, Boulter D. A novel mechanism of insect resistance engineered into tobacco. Nature. 1987;**330**:160-163

[33] Boulter D. Insect pest control by copying nature using genetically engineered crops. Phytochemistry. 1993;**34**(6):1453-1466

[34] Hoffmann MP, Zalom FG, Wilson LT, Smilanick JM, Malyj LD, Kiser J, et al. Field evaluation of transgenic tobacco containing genes encoding bacillus thuringiensis∂-endotoxin or cowpea trypsin inhibitor: Efficacy against Helicoverpa zea (Lepidoptera: Noctuidae). Journal of Economic Entomology. 1992;**85**(6):2516-2522

[35] Carbonero P, Royo J, Diaz I, Garcia-Maroto F, GonzalezHidalgo E, Gutierez C, et al. Cereal inhibitors of insect hydrolases (a-amylases and trypsin): Genetic control, transgenic expression and insect tests. In: Workshop on engineering plants against pests and pathogens (Vol. 1). Madrid: Instituto Juan March de Estudios Investigaciones; 1993;**10**:71

[36] Shade RE, Schroeder HE, Pueyo JJ, Tabe LM, Murdock LL, Higgins TJV, et al. Transgenic pea seeds expressing the a-amylase inhibitor of the common bean are resistant to bruchid beetles. Bio/Technology. 1994;**12**:793-796

[37] Schroeder HE, Gollasch S, Moore A, Tabe LM, Craig S, Hardie DC, et al. Bean alpha-amylase inhibitor confers resistance to pea weevil (Bruchus pisorum) in transgenic peas (Pisum sativum L.). Plant Physiology. 1995;**107**:1233-1239

[38] Boulter D, Edwards GA, Gatehouse AMR, Gatehouse JA, Hilder VA. Additive protective e!ects of incorporating two di!erent higher plant derived insect resistance genes in transgenic tobacco plants. Crop Protection. 1990;**9**:351-354

[39] Gatehouse AMR, Hilder VA. Genetic manipulation of crops for insect resistance. In: Molecular Biology in Crop Protection. Dordrecht: Springer; 1994. pp. 177-201

[40] Maddock SE. Expression in maize plants of wheatgerm agglutinin, a novel source of insect resistance. In: Third International Congress of Plant Molecular Biology. USA: Tucson, Arizona; 1991

[41] Down RE, Gatehouse AM, Hamilton WD, Gatehouse JA. Snowdrop lectin inhibits development and decreases fecundity of the glasshouse potato aphid (Aulacorthum solani) when administered in vitro and via transgenic plants both in laboratory and glasshouse trials. Journal of Insect Physiology. 1996;**42**(11-12):1035-1045

[42] Powell KS, Gatehouse AMR, Hilder VA, Van Damme EJM, Peumans WJ, Boonjawat J, et al. Different antimetabolic effects of related lectins towards nymphal stages of Nilaparvata lugens. Entomologia Experimentalis et Applicata. 1995;**75**(1):61-65

[43] Dowd PF, Lagrimini LM. In: Carozzi N, Koziel M, editors. Advances in Insect Control: The Role of Transgenic Plants. Taylor & Francis; 1997. pp. 195-223

[44] Thomas JC, Adams DG, Nessler CL, Brown JK, Bohnert HJ. Tryptophan

decarboxylase, tryptamine, and reproduction of the whitefly. Plant Physiology. 1995;**109**(2):717-720

[45] Thomas JC, Adams DG, Keppenne VD, Wasmann CC, Brown JK, Kanost MR, et al. Protease inhibitors of Manduca sexta expressed in transgenic cotton. Plant Cell Reports. 1995;**14**(12):758-762

[46] Thomas JC, Adams DG, Keppenne VD, Wasmann CC, Brown JK, Kanost MR, et al. Manduca sexta encoded protease inhibitors expressed in Nicotiana tabacum provide protection against insects. Plant Physiology and Biochemistry-Paris. 1995;**33**:611-611

[47] Estruch JJ, Carozzi NB, Desai N, Duck NB, Warren GW, Koziel MG. Transgenic plants: An emerging approach to pest control. Nature Biotechnology. 1997;**15**(2):137-141

[48] Estruch JJ, Warren GW, Mullins MA, Nye GJ, Craig JA, Koziel MG. Vip3A, a novel bacillus thuringiensis vegetative insecticidal protein with a wide spectrum of activities against lepidopteran insects. Proceedings of the National Academy of Sciences of the United States of America. 1996;**93**:5389-5394

[49] Tomalski MD, Miller LK. Insect paralysis by baculovirus mediated expression of a mite neurotoxin gene. Nature. 1991;**352**:82-85

[50] Stewart LMD, Hirst M, Ferber ML, Merryweather AT, Cayley PJ, Possee RD. Construction of an improved baculovirus insecticide containing an insect specifc toxin gene. Nature. 1991;**352**:85-88

[51] Freeman BC, Beattie GA. An Overview of Plant Defenses against Pathogens and Herbivores. The Plant Health Instructor. 2008. DOI: 10.1094. PHI-I-2008-0226-01

[52] Boller T, Felix G. A renaissance of elicitors: Perception of microbe-associated molecular patterns and danger signals by pattern-recognition receptors. Annual Review of Plant Biology. 2009;**60**:379-407

[53] Schulze-Lefert P, Panstruga R. molecular evolutionary concept connecting nonhost resistance, pathogen host range, and pathogen speciation. Trends in Plant Science. 2011;**16**(3):117-125

[54] Flor HH. Current status of the gene-for-gene concept. Annual Review of Phytopathology. 1971

[55] Jones J, Dangl JL. The plant immune system. Nature. 2006;**444**(7117):323-329

[56] Ashikawa I. Regions outside the leucine-rich repeat domain determine the distinct resistance specificities of the rice blast resistance genes Pik and Pik-m. Molecular Breeding. 2012;**30**(3):1531-1535

[57] Ma J, Chen J, Wang M, Ren Y, Wang S, Lei C, et al. Disruption of OsSEC3A increases the content of salicylic acid and induces plant defense responses in rice. Journal of Experimental Botany. 2018;**69**(5):1051-1064

[58] Acevedo-Garcia J, Spencer D, Thieron H, Reinstadler A, Hammond Kosack K, Phillips AL, et al. Mlo-based powdery mildew resistance in hexaploid bread wheat generated by a nontransgenic TILLING approach. Plant Biotechnology Journal. 2017;**15**(3):367-378

[59] Wang Q, Liu Y, He J, Zheng X, Hu J, Liu Y, et al. durable resistance to rice stripe virus. Nature Communications. 2014;**5**:1-8

[60] Wang Y, Cheng X, Shan Q, Zhang Y, Liu J, Gao C, et al. Simultaneous editing

of three homoeoalleles in hexaploid bread wheat confers heritable resistance to powdery mildew. Nature Biotechnology. 2014;**32**(9):947-951

[61] Antony G, Zhou J, Huang S, Li T, Liu B, White F, et al. Rice xa13 recessive resistance to bacterial blight is defeated by induction of the disease susceptibility gene Os-11N3. Plant Cell. 2010;**22**(11):3864-3876

[62] Chu Z, Yuan M, Yao J, Ge X, Yuan B, Xu C, et al. Promoter mutations of an essential gene for pollen development result in disease resistance in rice. Genes & Development. 2006;**20**(10):1250-1255

[63] Collins NC, Thordal-Christensen H, Lipka V, Bau S, Kombrink E, Qiu JL, et al. SNARE-protein-mediated disease resistance at the plant cell wall. Nature. 2003;**425**(6961):973-977

[64] Lu M, Tang X, Zhou JM. Arabidopsis NHO1 is required for general resistance against pseudomonas bacteria. Plant Cell. 2001;**13**(2):437-447

[65] Mysore KS, Ryu CM. Nonhost resistance: How much do we know? Trends in Plant Science. 2004;**9**(2):97-104

[66] Oh SK, Lee S, Chung E, Park JM, Yu SH, Ryu CM, et al. Insight into types I and II nonhost resistance using expression patterns of defense-related genes in tobacco. Planta. 2006

[67] Kamoun S, Van West P, Vleeshouwers VGAA, De Groot KE, Govers F. Resistance of Nicotiana benthamiana to Phytophthora infestans is mediated by the recognition of the elicitor protein INF1. Plant Cell. 1998;**10**(9):1413-1425

[68] Lindgren PB, Peet RC, Panopoulos NJ. Gene cluster of pseudomonas syringae pv. "Phaseolicola" controls pathogenicity of bean plants and hypersensitivity on nonhost plants. Journal of Bacteriology. 1986;**168**(2): 512-522

[69] Várallyay É, Giczey G, Burgyán J. Virus-induced gene silencing of Mlo genes induces powdery mildew resistance in Triticum aestivum. Archives of Virology. 2012;**157**(7):1345-1350

[70] Li T, Liu B, Spalding MH, Weeks DP, Yang B. High-efficiency TALEN-based gene editing produces disease-resistant rice. Nature Biotechnology. 2012;**30**(5):390-392

[71] Li T, Huang S, Zhou J, Yang B. Designer TAL effectors induce disease susceptibility and resistance to Xanthomonas oryzae pv. Oryzae in rice. Molecular Plant. 2013;**6**(3):781-789

[72] Macovei A, Sevilla NR, Cantos C, Jonson GB, Slamet-Loedin I, Čermák T, et al. Novel alleles of rice eIF4G generated by CRISPR/Cas9-targeted mutagenesis confer resistance to Rice tungro spherical virus. Plant Biotechnology Journal. 2018;**16**(11):1918-1927

[73] Trebst A, Draber W. Structure activity correlation of recent herbicides in photosynthetic reactions. In: Geissbühler H, editor. Advances in Pesticide Science. Vol. 2. New York: Pergamon Press; 1978. p. 223

[74] Shan Q, Wang Y, Li J, Zhang Y, Chen K, Liang Z, et al. Targeted genome modification of crop plants using a CRISPR-Cas system. Nature Biotechnology. 2013

[75] Pfister K, Steinback KE, Gardner G, Arntzen CJ. Photoaffinity labeling of an herbicide receptor protein in chloroplast

membranes. Proceedings. National Academy of Sciences. United States of America. 1981;**78**:981

[76] Li JF, Norville JE, Aach J, McCormack M, Zhang D, Bush J, et al. Multiplex and homologous recombinationmediated genome editing in Arabidopsis and Nicotiana benthamiana using guide RNA and Cas9. Nature Biotechnology. 2013;**31**(8):688-691

[77] Ryan GF. Resistance of common groundsel to simazine and atrazine. Weed Science. 1970;**18**:614

[78] Arntzen CJ, Ditto CL, Brewer PE. Chloroplast membrane alterations in triazine-resistant Amaranthus retroflexus biotypes. Proceedings of the National Academy of Sciences of the United States of America. 1979;**76**:278

[79] Radesorich SR, Steinback KE, Arntzen CJ. Effect of photosystem II inhibitors on thylacoid membranes of two common groundsel (Senico vulgaris) biotypes. Weed Science. 1979;**27**:216

[80] Hirschberg J, McIntosh L. Molecular basis of herbicide resistance in Amaranthus hybridus. Science. 1983;**222**:1346

[81] Golden S, Haselkorn R. Mutation to herbicide resistance maps within the psbA gene of Anacystis nidulans R2. Science. 1985;**229**:1104

[82] Erickson JM, Rahire M, Bennoun P, Delepelaire P, Diner B, Rochaix JD. Herbicide resistance in Chlamydomonas reinhardii results from a mutation in the chloroplast gene for the 32- Kilodalton protein of photosystem II. Proceedings. National Academy of Sciences. United States of America. 1984;**81**:3617

[83] Erickson JM, Rochaix JD. In: Galan GA, editor. Abstracts 1st International Congress of Plant Molecular Biology. Vol. 54. Athens, GA: University of Georgia Center for the Education for the International Society for Plant Molecular Biology; 1985

[84] Jaworski EG. Mode of action of N-phosphonomethylglycine: Inhibition of aromatic amino acid biosynthesis. Journal of Agricultural and Food Chemistry. 1972;**20**:1195

[85] Schulz A, Sost D, Amrhein N. Intensitivity of 5-enol-pyruvylshikimic acid 3-phosphate synthase to glyphosate confers resistance to this herbicide in a strain of Aerobacter aerogenes. Archives of Microbiology. 1984;**137**:121

[86] Steinrücken HC, Amrhein N. The herbicide glyphosate is a potent inhibitor of 5-enol-pyruvylshikimic acid 3-phosphate-synthase. Biochemical and Biophysical Research Communications. 1980;**94**:1207

[87] Shah DM, Horsch RB, Klee HJ, Kishore GM, Winter JA, Tuner NE, et al. Engineering herbicide tolerance in transgenic plants. Science. 1986;**233**:478

[88] Comai L, Facciotti D, Hiatt WR, Thompson G, Rose RE, Stalker PM. Expression in plants of a mutant aroA gene from salmonella typhimurium confers tolerance to glyphosate. Nature. 1985;**317**:741

[89] Mukitch MJ, Shaner DL, Stidham MA. Imidazolinones and acetohydroxyacid synthase from high plants. Plant Physiology. 1987;**83**:451

[90] La Rossa RA, Schloss IV. The sulfonylurea herbicide sulfometuron methyl is an extremely potent and selective inhibitor of acetolactate synthase in salmonella typhymurium.

The Journal of Biological Chemistry. 1984;**259**:8753

[91] Ray TB. Site of action of Chlorsulfuron-inhibition of valine and isoleucine biosynthesis in plants. Plant Physiology. 1981;**75**:827

[92] Leason M, Cunliffe D, Parkin D, Lea PJ, Miflin BJ. Inhibition of piston sativum leaf glutamine synthetase EC-6.3.1.2 by methionine sulfoximine phosphinotricin and other glutamate analogues. Phytochemistry. 1982;**21**:855

[93] Wild A, Manderscheid R. The effect of phosphinothricin on the assimilation of ammonica in plants. Zeitschrift für Naturforschung. 39c;**500**:1984

[94] Eckes P. Isolation of L-phosphinothricin resistant plants, Abstr. NATO advanced study institute. In: Plant Molecular Biology. Copenhagen, June: Carlsberg Laboratory; 1987

[95] Tischer E, Das Sarma S, Goodman HM. Nucleotide sequence of an alfalfa glutamine synthetase gene. Molecular & General Genetics. 1986;**203**:221

[96] Zhang Y, Zhang F, Li X, Baller JA, Qi Y, Starker CG, et al. Transcription activator-like effector nucleases enable efficient plant genome engineering. Plant Physiology. 2013;**161**(1):20-27

[97] Li T, Liu B, Chen CY, Yang B. TALEN-mediated homologous recombination produces site-directed DNA base change and herbicide-resistant rice. Journal of Genetics and Genomics. 2016;**43**(5):297-305

[98] Bhatnagar-Mathur P, Vadez V, Sharma KK. Transgenic approaches for abiotic stress tolerance in plants: Retrospect and prospects. Plant Cell Reports. 2008;**27**:411-424

[99] Tsai-Hung H, Jent-turn L, Yee-yung C, Ming-Tsair C. Heterology expression of the Arabidopsis C-repeat/dehydration response element binding factor 1 gene confers elevated tolerance to chilling and oxidative stresses in transgenic tomato. Plant Physiology. 2002;**130**:618-626

[100] Pasquali G, Biricolti S, Locatelli F, Baldoni E, Mattana M. Osmyb4 expression improves adaptive responses to drought and cold stress in transgenic apples. Plant Cell Reports. 2008;**27**:1677-1686

[101] Ravikumar G, Manimaran P, Voleti SR, Subrahmanyam D, Sundaram RM, Bansal KC, et al. Stress-inducible expression of AtDREB1A transcription factor greatly improves drought stress tolerance in transgenic indica rice. Transgenic Research. 2014;**23**:421-439

[102] Abdula SE, Lee HJ, Ryu H, Kang KK, Nou I, Sorrells ME, et al. Enhances abiotic stress tolerance by increasing proline biosynthesis in rice. Plant Molecular Biology Reporter. 2016;**34**:501-511

[103] Sato H, Todaka D, Kudo M, Mizoi J, Kidokoro S, Zhao Y, et al. The Arabidopsis transcriptional regulator DPB3-1 enhances heat stress tolerance without growth retardation in rice. Plant Biotechnology Journal. 2016;**14**(8):1756-1767

[104] Guan QJ, Ma HY, Wang ZJ, Wang ZY, Bu QY, Liu SK. A rice LSD1-like-type ZFP gene OsLOL5 enhances saline-alkaline tolerance in transgenic Arabidopsis thaliana, yeast and rice. BMC Genomics. 2016;**17**:142

[105] Zhang H, Liu W, Wan L, Li F, Dai L, Li D, et al. Functional analyses of ethylene response factor JERF3 with the aim of improving tolerance to drought

and osmotic stress in transgenic rice. Transgenic Research. 2010;**19**:809-818

[106] Mallikarjuna G, Mallikarjuna K, Reddy MK, Kaul T. Expression of OsDREB2A transcription factor confers enhanced dehydration and salt stress tolerance in rice (Oryza sativa L.). Biotechnology Letters. 2011;**33**:1689-1697

[107] Joo J, Choi HJ, Lee YH, Kim YK, Song SI. A transcriptional repressor of the ERF family confers drought tolerance to rice and regulates genes preferentially located on chromosome 11. Planta. 2013;**238**:155-170

[108] Zhu W, Yang L, Yang S, Gai J, Zhu Y. Overexpression of rice phosphate transporter gene OsPT2 enhances nitrogen fixation and ammonium assimilation in transgenic soybean under phosphorus deficiency. Journal of Plant Biology. 2016;**59**:172-181

[109] Kasukabe Y, He L, Nada K, Misawa S, Ihara I, Tachibana S. Overexpression of spermidine synthase enhances tolerance to multiple environmental stresses and upregulates the expression of various stress-regulated genes in transgenic Arabidopsis thaliana. Plant and Cell Physiology. 2004;**45**:712-722

[110] Li F, Wu QY, Sun YL, Wang LY, Yang XH, Meng QW. Overexpression of chloroplastic monodehydroascorbate reductase enhanced tolerance to temperature and methyl viologen-mediated oxidative stresses. Physiologia Plantarum. 2010;**139**:421-434

[111] Niu CF, Wei W, Zhou QY, Tian AG, Hao YJ, Zhang WK, et al. Wheat WRKY genes TaWRKY2 and TaWRKY19 regulate abiotic stress tolerance in transgenic Arabidopsis plants. Plant, Cell and Environment. 2012;**35**:1156-1170

[112] Amara I, Capellades M, Ludevid MD, Pages M, Goday A. Enhanced water stress tolerance of transgenic maize plants over-expressing LEA Rab28 gene. Journal of Plant Physiology. 2013;**170**:864-873

[113] Upadhyaya CP, Akula N, Young KE, Chun SC, Kim DH, Park SW. Enhanced ascorbic acid accumulation in transgenic potato confers tolerance to various abiotic stresses. Biotechnology Letters. 2010;**32**:321-330

[114] Eltayeb AE, Yamamoto S, Habora MEE, Yin L, Tsujimoto H, Tanaka K. Transgenic potato overexpressing Arabidopsis cytosolic AtDHAR1 showed higher tolerance to herbicide, drought and salt stresses. Breeding Science. 2011;**61**:3-10

[115] Seo JS, Sohn HB, Noh K, Jung C, An JH, Donovan CM, et al. Expression of the Arabidopsis AtMYB44 gene confers drought/salt-stress tolerance in transgenic soybean. Molecular Breeding. 2012;**29**:601-660

[116] Subramanyam K, Arun M, Mariashibu TS, Theboral J, Rajesh M, Singh NK, et al. Overexpression of tobacco osmotin (Tbosm) in soybean conferred resistance to salinity stress and fungal infections. Planta. 2012;**236**:1909-1925

[117] Ghanti SKK, Sujata KG, Kumar BV, Karba NN, Janardhan Reddy K, Rao MS, et al. Heterologous expression of P5CS gene in chickpea enhances salt tolerance without afecting yield. Biologia Plantarum. 2011;**55**:634

[118] Hajyzadeh M, Turktas M, Khawar KM, Unver T. miR408 overexpression causes increased drought tolerance in chickpea. Gene. 2015;**555**:186-193

[119] Surekha CH, Kumari KN, Aruna LV, Suneetha G, Arundhati A, Kishor PK. Expression of the Vigna aconitifolia P5CSF129A gene in transgenic pigeonpea enhances proline accumulation and salt tolerance. Plant Cell, Tissue and Organ Culture. 2014;**116**:27-36

[120] Bhomkar P, Upadhyay CP, Saxena M, Muthusamy A, Prakash NS, Pooggin M, et al. Salt stress alleviation in transgenic Vigna mungo L. Hepper (blackgram) by overexpression of the glyoxalase I gene using a novel cestrum yellow leaf curling virus (CmYLCV) promoter. Molecular Breeding. 2008;**22**:169-181

[121] Saha B, Swain D, Borgohain P, Rout GR, Koyama H, Panda SK. Enhanced exudation of malate in the rhizosphere due to AtALMT1 overexpression in blackgram (Vigna mungo L.) confers increased aluminium tolerance. Plant Biology. 2020;**22**:701-708

[122] Sidhu H, Schmidt ME, Cornelius JG, Thamilselvan S, Khan SR, Hesse A, et al. Direct correlation between hyperoxaluria/oxalate stone disease and the absence of the gastrointestinal tract-dwelling bacterium Oxalobacter formigenes: Possible prevention by gut recolonization or enzyme replacement therapy. Journal of the American Society of Nephrology. 1999;**10**:S334-S340

[123] Sahoo DP, Kumar S, Mishra S, Kobayashi Y, Panda SK, Sahoo L. Enhanced salinity tolerance in transgenic mungbean overexpressing Arabidopsis antiporter (NHX1) gene. Molecular Breeding. 2016;**36**:144

[124] Rout GR, Bansal A, Swain D, Jadhao KR, Shelke RG, Panda SK. Overexpression of ICE1 gene in mungbean (Vigna radiata L.) for cold tolerance. Plant Cell, Tissue and Organ Culture. 2020;**143**(3):593-608

[125] Augustine SM, Narayan AJ, Syamaladevi DP, Appunu C, Chakravarthi M, Ravichandran V, et al. Overexpression of EaDREB2 and pyramiding of EaDREB2 with the pea DNA helicase gene (PDH45) enhance drought and salinity tolerance in sugarcane (Saccharum spp. hybrid). Plant Cell Reports. 2015;**34**:247-263

[126] Singh N, Mishra A, Jha B. Over-expression of the peroxisomal ascorbate peroxidase (SBPAPX) gene cloned from halophyte Salicornia brachiata confers salt and drought stress tolerance in transgenic tobacco. Marine Biotechnology. 2014;**16**:321-332

[127] Zhai Y, Wang Y, Li Y, Lei T, Yan F, Su L, et al. Isolation and molecular characterization of GmERF7, a soybean ethylene-response factor that increases salt stress tolerance in tobacco. Gene. 2013;**513**:174-183

[128] Li F, Murillo C, Wurtzel ET. Maize Y9 encodes a product essential for 15-cis-zeta-carotene isomerization. Plant Physiology. 2007;**144**:1181-1189

[129] Chen Y, Li F, Wurtzel ET. Isolation and characterization of the Z-ISO gene encoding a missing component of carotenoid biosynthesis in plants. Plant Physiology. 2010;**153**:66-79

[130] Bai C, Twyman RM, Farre G, et al. A golden era-pro-vitamin A enhancement in diverse crops. In Vitro Cellular & Developmental Biology. Plant. 2011;**47**:205-202

[131] Tian L, Magallanes-Lundback M, Musetti V, DellaPenna D. Functional analysis of beta- and epsilon-ring carotenoid hydroxylases in Arabidopsis. Plant Cell. 2003;**5**:1320-1332

[132] Seo M, Koshiba T. Complex regulation of ABA biosynthesis in plants. Trends in Plant Science. 2002;**7**:41-48

[133] Kawakatsu T, Yamamoto MP, Hirose S, et al. Characterization of a new rice glutelin gene GluD-1 expressed in the starchy endosperm. Journal of Experimental Botany. 2008;**59**:4233-4245

[134] Qu LQ, Xing YP, Liu WX, et al. Expression pattern and activity of six glutelin gene promoters in transgenic rice. Journal of Experimental Botany. 2008;**59**(9):2417-2242

[135] Xu R, Li D, Li H, et al. Isolation of four rice seed-specific promoters and evaluation of endosperm activity. Plant Cell, Tissue and Organ Culture. 2017;**128**:125-132

[136] Burkhardt PK, Beyer P, Wunn J, et al. Transgenic rice (Oryza sativa) endosperm expressing daffodil (Narcissus pseudonarcissus) phytoene synthase accumulates phytoene, a key intermediate of provitamin a biosynthesis. The Plant Journal. 1997;**11**(5):1071-1078

[137] Ye X, Al-Babili S, Kloti A, et al. Engineering the provitamin A (beta-carotene) biosynthetic pathway into (carotenoid-free) rice endosperm. Science. 2000;**287**(5451):303-305

[138] Rai M, Datta K, Parkhi V, et al. Variable T-DNA linkage configuration affects inheritance of carotenogenic transgenes and carotenoid accumulation in transgenic indica rice. Plant Cell Reports. 2007;**26**:1221-1231

[139] Parkhi V, Rai M, Tan J, et al. Molecular characterization of marker free transgenic indica rice lines that accumulate carotenoids in seed endosperm. Molecular Genetics and Genomics. 2005;**274**:325-336

[140] Paine JA, Shipton CA, Chagger S, et al. Improving the nutritional value of Golden rice through increased pro-vitamin A content. Nature Biotechnology. 2005;**23**:482-487

[141] Ha S-H, Liang YS, Jung H, et al. Application of two bicistronic systems involving 2A and IRES sequences to the biosynthesis of carotenoids in rice endosperm. Plant Biotechnology Journal. 2010;**8**:928-938

[142] Ha S-H, Kim JK, Jeong YS, et al. Stepwise pathway engineering to the biosynthesis of zeaxanthin, astaxanthin and capsanthin in rice endosperm. Metabolic Engineering. 2019;**52**:178-189

[143] Chaudhary A, Teotia S, Singh D. Tools for engineering resistance against pathogens in plants. Journal of Plant Biochemistry and Biotechnology. 2022;**31**:459-488

[144] Li A, Jia S, Yobi A, Ge Z, Sato SJ, Zhang C, et al. Editing of an alpha-kafirin gene family increases, digestibility and protein quality in sorghum. Plant Physiology. 2018a;**177**: 1425-1438

[145] Zorrilla-Fontanesi Y, Pauwels L, Panis B, Signorelli S, Vanderschuren H, Swennen R. Strategies to revise agrosystems and breeding to control fusarium wilt of banana. Nature Food. 2020;**1**(10):599-604

[146] Klap C, Yeshayahou E, Bolger AM, Arazi T, Gupta SK, Shabtai S, et al. Tomato facultative parthenocarpy results from SlAGAMOUSLIKE 6 loss of function. Plant Biotechnology Journal. 2017;**15**:634-647

[147] Andersson M, Turesson H, Nicolia A, Falt AS, Samuelsson M, Hofvander P. Efficient targeted multiallelic mutagenesis in tetraploid potato (Solanum tuberosum) by transient CRISPR-Cas9 expression

in protoplasts. Plant Cell Reports. 2017;**36**:117-128

[148] Vasconcelos M, Datta K, Oliva N, Khalekuzzaman M, Torrizo L, Krishnan S, et al. Enhanced iron and zinc accumulation in transgenic rice with the ferritin gene. Plant Science. 2003;**164**:371-378

[149] Lucca P, Hurrell R, Potrykus I. Genetic engineering approaches to improve the bioavailability and the level of iron in rice grains. Theoretical and Applied Genetics. 2001;**102**(2):392-397

[150] Suzuki YA, Kelleher SL, Yalda D, Wu L, Huang J, Huang N, et al. Expression, characterization and biological activity of recombinant human lactoferrin in rice. Journal of Pediatric Gastroenterology and Nutrition. 2003;**36**:190-199

[151] Anai T, Koga M, Tanaka H, Kinoshita T, Rahman SM, Takagi Y. Improvement of rice (Oryza sativa L.) seed oil quality through introduction of a soybean microsomal omega-3 fatty acid desaturase gene. Plant Cell Reports. 2003;**21**:988-992

[152] Storozhenko S, De Brouwer V, Volckaert M, Navarrete O, Blancquaert D, Zhang GF, et al. Folate fortification of rice by metabolic engineering. Nature Biotechnology. 2007;**25**:1277-1279

[153] Kawakatsu T, Wang S, Wakasa Y, Takaiwa F. Increased lysine content in rice grains by over-accumulation of BiP in the endosperm. Bioscience, Biotechnology, and Biochemistry. 2010;**74**:2529-2531

[154] Caimi PG, McCole LM, Klein TM, Kerr PS. Fructan accumulation and sucrose metabolism in transgenic maize endosperm expressing a bacillus amyloliquefaciens SacB gene. Plant Physiology. 1996;**110**:355-363

[155] Falco SC, Guida T, Locke M, Mauvais J, Sandres C, Ward RT, et al. Transgenic canola and soybean seeds with increased lysine. Biotechnology. 1995;**13**:577-582

[156] Cahoon EB, Hall SE, Ripp KG, Ganzke TS, Hitz WD, Coughlan SJ. Metabolic redesign of vitamin E biosynthesis in plants for tocotrienol production and increased antioxidant content. Nature Biotechnology. 2003;**21**:1082-1087

[157] Chen Z, Young TE, Ling J, Chang SC, Gallie DR. Increasing vitamin C content of plants through enhanced ascorbate recycling. Proceedings of the National Academy of Sciences of the United States of America. 2003;**100**:3525-3530

[158] Drakakaki G, Marcel S, Glahn RP, Lund EK, Pariagh S, Fischer R, et al. Endosperm-specific co-expression of recombinant soybean ferritin and aspergillus phytase in maize results in significant increases in the levels of bioavailable iron. Plant Molecular Biology. 2005;**59**:869-880

[159] Thu TT, Dewaele E, Claeys M, Jacobs M, Angenon G. Increasing lysine levels in pigeonpea (Cajanus cajan (L.) Millsp) seeds through genetic engineering. Plant Cell, Tissue and Organ Culture. 2007;**91**:135-143

[160] Dinkins RD, Reddy MSS, Meurer CA, Yan B, Trick H, Thibaud-Nissen F, et al. Increased sulfur amino acids in soybean plants overexpressing the maize 15 kD azein protein. In Vitro Cellular & Developmental Biology. 2001;**37**:742-747

[161] Valentine MF, De Tar JR, Mookkan M, Firman JD, Zhang ZJ. Silencing of soybean rafnose synthase gene reduced rafnose family oligosaccharides and increased true metabolizable energy of poultry feed. Frontiers in Plant Science. 2017;**8**:692

[162] Mariani C, De Beuckeleer M, Truettner J, Leemans J, Goldberg RB. Induction of male sterility in plants by a chimeric ribonuclease gene. Nature. 1990;**347**:737-741

[163] Mariani C, Gossele V, Beuckeleer MD, Block MD, Goldberg RB, Greef WD, et al. A chimaeric ribonuclease-inhibitor gene restores fertility to male sterile plants. Nature. 1992;**357**:384-387

[164] Bedinger P. The remarkable biology of pollen. Plant Cell. 1992;**4**:879-887

[165] Denis M, Delourme R, Gourret JP, Mariani C, Renard M. Expression of engineered nuclear male sterility in Brassica napus. Plant Physiology. 1993;**101**(4):1295-1304

[166] De Cosa B, Moar W, Lee S-B, Miller M, Daniell H. Overexpression of the Bt cry2Aa2 operon in chloroplasts leads to formation of insecticidal crystals. Nature Biotechnology. 2001;**19**:71-74

[167] Ruiz ON, Daniell H. Engineering cytoplasmic male sterility via the chloroplast genome by expression of b-ketothiolase. Plant Physiology. 2005;**138**:1232-1246

[168] Worrall D, Hird DL, Hodge R, Paul W, et al. Premature dissolution of the microsporocyte callose wall causes male sterility in transgenic tobacco. The Plant Cell. 1992;**4**:759-771

[169] Chang Z, Chen Z, Wang N, Xie G, Lu J, Yan W, et al. Construction of a male sterility system for hybrid rice breeding and seed production using a nuclear male sterility gene. Proceedings of the National Academy of Sciences. 2016;**113**:14145-14150

[170] Hawkes T, Pline-Srnic W, Dale R, Friend E, et al. Dglufosinate as a male sterility agent for hybrid seed production. Plant Biotechnology Journal. 2011;**9**:301-314

[171] Guerineau F, Sorensen AM, Fenby N, Scott R. Temperature sensitive diphtheria toxin confers conditional malesterility in Arabidopsis thaliana. Plant Biotechnology Journal. 2003;**1**: 33-42

[172] Ishiguro S, Kawai-Oda A, Ueda J, Nishida I, Okada K. The DEFECTIVE IN ANTHER DEHISCENCE1 gene encodes a novel phospholipase A1 catalyzing the initial step of jasmonic acid biosynthesis, which synchronizes pollen maturation, anther DEHISCENCE, and flower opening in Arabidopsis. The Plant Cell. 2001;**13**:2191-2209

[173] Sanders PM, Lee PY, Biesgen C, Boone JD, et al. The Arabidopsis DELAYED DEHISCENCE1 gene encodes an enzyme in the jasmonic acid synthesis pathway. The Plant Cell. 2000;**12**:1041-1061

[174] Bae HK, Kang HG, Kim GJ, Eu HJ, et al. Transgenic rice plants carrying RNA interference constructs of AOS (allene oxide synthase) genes show severe male sterility. Plant Breeding. 2010;**129**:647-651

[175] Fischer R, Budde I, Hain R. Stilbene synthase gene expression causes changes in flower colour and male sterility in tobacco. The Plant Journal. 1997;**11**:489-498

[176] Handa H, Gualberto JM, Grienenberger J-M. Characterization

of the mitochondrial orfB gene and its derivative, orf224, a chimeric open reading frame specific to one mitochondrial genome of the "Polima" male-sterile cytoplasm in rapeseed (Brassica napus L.). Current Genetics. 1995;**28**:546-552

[177] Chakraborty A, Mitra J, Bhattacharyya J, Pradhan S, et al. Transgenic expression of an unedited mitochondrial orfB gene product from wild abortive (WA) cytoplasm of rice (Oryza sativa L.) generates male sterility in fertile rice lines. Planta. 2015;**241**:1463-1479

[178] Nizampatnam NR, Doodhi H, Narasimhan YK, Mulpuri S, Viswanathaswamy DK. Expression of sunflower cytoplasmic male sterility-associated open reading frame, orfH522 induces male sterility in transgenic tobacco plants. Planta. 2009;**229**:987-1001

[179] Nizampatnam NR, Kumar VD. Intron hairpin and transitive RNAi mediated silencing of orfH522 transcripts restores male fertility in transgenic male sterile tobacco plants expressing orfH522. Plant Molecular Biology. 2011;**76**:557-573

[180] Wang F, Wang C, Liu P, Lei C, Hao W, Gao Y, et al. Enhanced rice blast resistance by CRISPR/Cas9-targeted mutagenesis of the ERF transcription factor gene OsERF922. PLoS One. 2016;**11**:e0154027

[181] Zhang Y, Li D, Zhang D, Zhao X, Cao X, Dong L, et al. Analysis of the functions of TaGW2 homoeologs in wheat grain weight and protein content traits. The Plant Journal. 2018;**94**(5):857-866

[182] Shi J, Gao H, Wang H, Lafitte HR, Archibald RL, Yang M, et al. ARGOS8 variants generated by CRISPR-Cas9 improve maize grain yield under field drought stress conditions. Plant Biotechnology Journal. 2017;**15**:207-216. DOI: 10.1111/pbi.12603

[183] Scudellari MEGAN. Hijacking evolution. Nature. 2019;**571**(7764):160-162

[184] Kyrou K, Hammond AM, Galizi R, Kranjc N, Burt A, Beaghton AK, et al. A CRISPR–Cas9 gene drive targeting doublesex causes complete population suppression in caged Anopheles gambiae mosquitoes. Nature Biotechnology. 2018;**36**(11):1062-1066

[185] Fu G, Condon KC, Epton MJ, Gong P, Jin L, Condon GC, et al. Female-specific insect lethality engineered using alternative splicing. Nature Biotechnology. 2007;**25**(3):353-357

[186] Phuc HK, Andreasen MH, Burton RS, Vass C, Epton MJ, Pape G, et al. Late-acting dominant lethal genetic systems and mosquito control. BMC Biology. 2007;**5**(1):1-11